Cisgenics and Transgenics

Bhupendra Koul

Cisgenics and Transgenics

Strategies for Sustainable Crop Development and Food Security

Springer

Bhupendra Koul
School of Bioengineering & Biosciences
Lovely Professional University
Jalandhar, Punjab, India

ISBN 978-981-19-2118-6 ISBN 978-981-19-2119-3 (eBook)
https://doi.org/10.1007/978-981-19-2119-3

This Springer imprint is published by the registered company Springer Nature Singapore Pte Ltd.
The registered company address is: 152 Beach Road, #21-01/04 Gateway East, Singapore 189721, Singapore

Preface

Biotechnology, a multidisciplinary science, has the potential to find solutions for major socioeconomic problems, like nutritious food, fiber, and fuel, and other requirements of life through plant improvement. Since the conventional techniques deployed in crop improvement may not keep pace with the growing demands of expanding population, decreasing land resources, and environmental stresses, the in vitro technologies and genetic modification technologies in crop improvement have great relevance. Biotechnologies have proved to be a boon and can cater to the needs of a layman. In the twenty-first century, it has become easier to transfer any gene regardless of its source into desired plant species as a routine procedure by exploiting the novel capabilities of *Agrobacterium tumefaciens*, the natural "genetic engineer," and alternative strategies based on transfection of plant protoplast or by biolistic devices.

The remarkable advances in gene technology have not only helped in understanding the gene structure and functions, but also offered access to unlimited gene pool together with incorporation and expression of gene from taxonomically unrelated species, in "tailoring" crop plants with useful genetically engineered characteristics to combat stress, drought, and disease including several qualitative and quantitative traits. Nowadays, genetic transformation of plants has become crucial for plant biotechnologists, biochemists, and physiologists, for expressing a foreign gene from a heterologous system into a suitable organism by using the techniques and tools of genetic engineering, for either qualitative or quantitative improvement of organism(s) or their products. But the success of biotechnology depends on the expression level, purity, economical aspect, and sustained recovery of the desired products. Gene cloning, genetic engineering, and biotechnological tools will always guide us for further research in plant improvement. There are three major advantages of developing transgenic plants: (i) adding a gene to a crop plant can improve its agricultural, horticultural, or nutrient value; (ii) GM plants can act as living bioreactors for the feasible and cost-effective production of industrially/pharmaceutically important proteins or metabolites; and (iii) genetic transformation of plants acts as a model for studying the regulation and expression of genes. On the other hand, cisgenics is an eco-friendly crop improvement technique (alternative to transgenics), which involves the genetic modification of a plant transferring one or more genes from any related source or from a sexually compatible plant. Cisgenic crops (i) are accepted by the general public; (ii) do not involve hectic and costly time-consuming procedure of approvals; (iii) are not a threat to the biodiversity and are not a potential risk to the health and the ecosystem; (iv) are safer than those raised by conventional breeding; (v) and preclude linkage drag.

The objectives of this book are:

- To acquaint young plant biotechnologists with the available genetic transformation techniques for introducing value-added traits in crop plants.
- To provide explicit information on transgenic technology as an indispensable tool for biotechnologists.
- To promote cisgenics as an eco-friendly technology for crop improvement.
- To be informed regarding the application of transgenic plants as natural and living bioreactors for the feasible production of pharmaceuticals.
- To be informed about the current scenario and future prospects of GM crops.

This book entitled "**Crop Improvement Strategies for Food and Health Security: Cisgenics and Transgenics**" summarizes the current scenario, major challenges, and future prospects of crop improvement strategies (for combating biotic and abiotic stresses and enhancing crop yield, nutritional quality, and quantity), in the light of cisgenics and transgenics, in six major chapters as described below.

Chapter 1—**Plant Transformation Techniques**: Genetic transformation refers to the introduction of one or more genes into the genome of an organism using specialized precautionary procedure(s) (as opposed to conventional procedures), and the changed organism is referred to as a transgenic organism. These remarkable advances in gene technology have not only helped in understanding the gene structure and functions, but also offered access to unlimited gene pool together with incorporation and expression of gene(s) from taxonomically unrelated species, in "tailoring" crop plants with useful genetically engineered characteristics (qualitative and quantitative) to combat various environmental stresses (biotic and abiotic) and human nutritional deficiencies. Adding a gene to a crop plant can increase its agricultural, horticultural, or ornamental value, and it can also be used as a living bioreactor for the cost-effective synthesis of pharmaceutically useful metabolites and proteins. Genetically modified phytoremediator/hyperaccumulator plants can remediate contaminated soils and waters. Plant genetic transformation may also be used to investigate how genes are regulated and expressed throughout development and other biological processes. In general, there are two methods for artificial DNA transfer: vector-less or direct DNA transfer (physical or chemical gene transfer) and (ii) vector-mediated or indirect DNA transfer method (*Agrobacterium*-mediated transformation, *Agrobacterium*-mediated virus infection). Among these methods, *Agrobacterium*-mediated transformation is thought to induce less rearrangement of the transgene and introduces a lower copy number than the other direct DNA delivery methods. This chapter focuses on explicit information on various vector-less/direct (physical, chemical) and vector-mediated/indirect (*Agrobacterium*-mediated) plant transformation techniques.

Chapter 2—**Strategies to Enhance the Expression of the Transgene**: It is difficult to transfer a foreign gene from an AT-rich to a GC-rich species. To enhance the expression of heterologous genes in plants, several molecular and genetic techniques have been devised and deployed, including avoiding of sequence motifs/codons that cause mRNA degradation/low expression (putative polyA/mRNA instability sequences/RNA polymerase II termination signals/cryptic splicing sites/secondary structures, and so on), subcellular targeting of proteins to compartments (ER/vacuole/apoplast, and so on) suitable for their accumulation and stability, and incorporating elements that ensures high-level expression (strong promoters: either modified natural or synthetic/5′ untranslated leader sequence/TIC, etc.). This chapter summarizes various strategies that aim at preparing/improving a naturally occurring gene from lower organism to be expressed in higher plants and also synthetic gene designing for higher expression. It also focuses on the use of strong promoters to drive the expression of transgene(s) and improve the stability of gene product through subcellular targeting.

Chapter 3—**Cisgenics and Crop Improvement**: The word "cis" in cisgenics means "on the same side/same kingdom" and "genics" means "pertaining to genes." Cisgenesis is the process of genetically modifying a plant species with a natural gene produced from a sexually crossable plant. Transgenesis involves the genetic alteration of a plant with one or more genes from any non-plant organism or from a plant that is sexually incompatible. Cisgenics and intragenics are eco-friendly alternatives to transgenics. In cisgenesis, the inserted gene remains unchanged and has its own introns and regulatory elements. The concept of cisgenesis was brought by Dutch researchers Schouten, Krens, and Jacobsen in the year 2006. Cisgenic crops are accepted by the general public; they do not involve hectic and costly time-consuming procedure of approvals; are not a threat to the biodiversity and are not a potential risk to the health and the ecosystem; are safer than those raised by conventional breeding; and preclude linkage drag. This chapter encompasses the advantages and drawbacks of cisgenesis, their implications towards sustainable crop improvement, and their future prospects.

Chapter 4—**Transgenics and Crop Improvement**: In developing nations, where arable land per capita is declining but human and animal populations are constantly expanding, the key constraint for food and nutritional security for the human population in the next years will be sustained plant productivity and crop yield(s). Apart from the genetic potential of plant species, agricultural plant output is quite variable and is impacted by a variety of physical, abiotic, and biotic factors. Transgenic technology has the potential to cope with these situations and to feed the teeming millions through crop improvement strategies. The basic objective of any transformation technique is to get the desired gene into the cell's nucleus without compromising the cell's capacity to live. The plant is considered to be transformed if the inserted gene is functional and the gene product is produced. The plant is called transgenic after the gene introduced is stable, inherited, and expressed in following generations. Therefore, transgenic plants are plants that have been genetically engineered with novel traits and are identified as a class of genetically modified organism (GMO). Several GM crops such as corn, cassava, soybean, canola, squash, tobacco, mustard, tomato, rice, papaya, cotton, alfalfa, sugar beet, and brinjal have been commercialized worldwide, and some are under pipeline. These crops and their products have now gained acceptance in several countries across the world. Previously, the focus of development of transgenic plants was to develop water, salinity, temperature, insect, and disease tolerance, but now the focus of this technology is enhancement of nutritional components in edible crops so as to improve people's health. Transgenic technology is an indispensable tool for the biotechnologists and has a bright future.

Chapter 5—**Molecular Pharming**: There are different expression platforms (transgenic plants, plant cell culture, bacteria, yeast, microalgae, animals, and animal cell culture) for the production of pharmaceuticals. The usage of plants as bioreactors for the manufacture of therapeutically and industrially important molecules like proteins, carbohydrates, and lipids in plants is termed as "molecular pharming." Plant molecular pharming has become a lucrative biotechnology industry as it is a safe, cost-effective, and eco-friendly technique with reduced risk of contamination with human and animal pathogens; permits comprehensive posttranslational modification; needs minimal growth requirements with infinite scalability in the field; and allows recombinant proteins to be targeted to various subcellular compartments for accumulation. Over the past few years, several reports on the plant-based expression of pharmaceutically important proteins have been published. To amplify the yield and quality of biopharmaceuticals, several factors including the type of host plant, gene construct, subcellular localization, posttranslational modifications, protein extraction, and downstream processing are crucial factors. Several plant-expressed molecules have been commercialized, and the rest are in the pipeline under clinical trials. This chapter deals with the importance, limitations, challenges, recent developments, and future prospects of molecular pharming.

Chapter 6—**Future Prospects of GM Plants**: The global population is increasing by 83 million annually and is expected to cross 8.6 billion by the year 2030. To cope with this alarming situation and to feed the teeming millions, the food supplies must increase from 25 to 70% by the year 2050. To attain food and health security, political commitments must be implemented in letter and spirit and quantitative objectives for boosting food production must be set. The GM crop productivity achieved in the last 21 years (1996–2016) is noteworthy and indicates that the conventional crop technology alone cannot meet the demands of the growing population. The continued acceptance, adoption, and cultivation of GM crops are indicators of farmer's and consumer's satisfaction towards mitigating the challenges of biotic and abiotic stresses, conservation of biodiversity, and economic, health, and social benefits. Since the commercialization (1996 to the present) of biotech crops (major: soybean, corn, cotton; minor: alfalfa, sugar beet, papaya, squash, and eggplant), the global acreage (28 countries) has reached 189.8 million hectares. The biotech crops will continue to contribute immensely towards sustainable, bountiful, and profitable agricultural yield and shall deliver significant environmental, economic, health, and social benefits to the farmers. This chapter outlines GM crops' key contributions to global food security and the economy, as well as the obstacles and risks connected with their release and the future of GM technology.

Punjab, India Bhupendra Koul

Contents

1 Plant Transformation Techniques 1
1.1 Introduction 2
1.2 Direct DNA Transfer 5
1.2.1 Physical Gene Transfer Methods 5
1.2.2 Chemical Methods of Gene Transfer 13
1.2.3 DNA Imbibition by Seeds, Embryos, Cells, and Tissues 14
1.3 Indirect DNA Transfer 14
1.3.1 *Agrobacterium*-Mediated Genetic Transformation 14
1.3.2 Ti Plasmid of *Agrobacterium* 16
1.3.3 Organization of T-DNA 16
1.3.4 Organization of *vir* Region 17
1.3.5 T-DNA Transfer Process 18
1.3.6 Advantages of *Agrobacterium*-Mediated Genetic Transformation 19
1.3.7 Disadvantages of *Agrobacterium*-Mediated Genetic Transformation 19
1.3.8 Selectable Markers 20
1.4 Transformation Protocols 20
1.4.1 Genetic Transformation of *Arabidopsis* 20
1.4.2 Genetic Transformation of Rice 20
1.4.3 Genetic Transformation of Chickpea 21
1.4.4 Genetic Transformation of Tomato 27
1.4.5 Genetic Transformation of Potato 34
1.4.6 Genetic Transformation of Cotton 34
1.4.7 Genetic Transformation of Stevia 34
1.4.8 Genetic Transformation of Sugar Beet 34
1.4.9 Genetic Transformation of Maize 37
1.4.10 Genetic Transformation of Melon 37
1.4.11 Genetic Transformation of Poplar 41
1.4.12 Genetic Transformation of Sugarcane 41
1.4.13 Genetic Transformation of Apple 42
1.4.14 Genetic Transformation of Flax 42

1.4.15 Genetic Transformation of Sweet Pepper 42
1.4.16 Genetic Transformation of Soybean 45
1.4.17 Genetic Transformation of Canola 45
1.4.18 Genetic Transformation of Alfalfa 48
1.4.19 Genetic Transformation of Squash 48
1.4.20 Genetic Transformation of Eggplant 48
1.5 Conclusions 52
References 67

2 Strategies for Enhancement of Transgene Expression 75
2.1 Introduction 75
2.2 Designing of Coding Sequence of the Gene 76
2.2.1 Removal of Destabilizing Elements for Optimal Expression 77
2.2.2 Cryptic Splicing Sites 77
2.2.3 Codon Bias 78
2.3 Incorporation of Elements for High Expression 79
2.3.1 Role of Promoters in Eukaryotic Gene Expression 79
2.3.2 Untranslated Regions (UTR) and Sequences 83
2.3.3 Translation Initiation Context (TIC) 83
2.4 Improvement in Foreign Protein Accumulation and Stability 84
2.4.1 Subcellular Targeting of Recombinant Protein 84
2.5 Conclusions 92
References 92

3 Cisgenics and Crop Improvement 107
3.1 Introduction 107
3.2 Difference Between Cisgenics and Transgenics 108
3.3 The Limitations of Cisgenesis 109
3.4 Cisgenesis and Sustainable Crop Improvement 111
3.4.1 Techniques Involved in the Development of Cisgenic/Intragenic Crop 111
3.4.2 Sources of Genes for Cisgenic/Intragenic Technology 112
3.4.3 Cisgenesis/Intragenesis as a Novel Biotechnology in Plant Breeding 115
3.5 Merits of Cisgenesis Over Conventional Breeding Methods 115
3.5.1 Time-Saving Technique 115
3.5.2 Maintenance of Plant Genetic Constitution 116
3.5.3 Overcomes the Problem of Linkage Drag 116
3.5.4 Traits with Limited Allelic Variability Are Improved . . . 116
3.5.5 The Decreased Use of Pesticides 117
3.6 The Potential Roles of Cisgenesis in Other Breeding Techniques 117
3.7 Rules and Regulations on the Use of Cis/Intragenic Plants 118

3.7.1 Rejections of Cisgenics Exemption from GMOs 118
3.8 Issues Associated with Genetic Modification 125
3.9 Conclusions 125
References 126

4 Transgenics and Crop Improvement 131
4.1 Introduction 132
4.2 Crop Improvement Through Transgenic Technology 132
4.3 Advantages of Transgenic Techniques in Crop Improvement 134
4.3.1 Improved Crop Yields 134
4.3.2 Enhancement in Crop Protection 135
4.3.3 Improvement in Food Processing 136
4.3.4 Improved Nutritive Value 136
4.3.5 Improved Shelf Life 137
4.3.6 Environmental Benefits 138
4.3.7 Benefits for Developing Countries 138
4.4 Disadvantages of Transgenic Crops 139
4.4.1 Biosafety-Related Issues 139
4.4.2 Antibiotic Resistance 140
4.4.3 Environmental Effects 140
4.4.4 Impacts on Nontarget Organism 141
4.4.5 Cost for Commercialization 141
4.5 Safety and Regulations 142
4.6 Transgenics for Herbicide Resistance 142
4.6.1 Herbicide-Tolerant Crops 142
4.6.2 Yields of Herbicide-Resistant (HR) Crops 143
4.6.3 Methods for Developing Herbicide-Resistant (HR) Crops 143
4.6.4 Advantages of Herbicide-Resistant (HR) Crops 144
4.6.5 Disadvantages of Herbicide-Resistant (HR) Crops 144
4.6.6 Herbicide-Resistant Crops (HRCs) and Crop Disease 145
4.6.7 Glyphosate-Resistant Crops 145
4.6.8 The Future of HRCs 145
4.7 The Story of Transgenic Mustard 146
4.7.1 Genetic Modification of Mustard 146
4.7.2 Aphid/Insect Pest-Resistant Mustard 147
4.7.3 Disease-Resistant Mustard 148
4.7.4 Herbicide-Tolerant Mustard 148
4.7.5 Transgenic Mustard for Improved Nutrient-Use Efficiency 149
4.7.6 In Planta Modification in *B. juncea* 149
4.8 Transgenics for Pest Resistance 150
4.9 Bt Technology 155
4.9.1 History of Bt 155

4.9.2 The Structure, Variety, and Toxicity of Bt Proteins 156
4.9.3 Types of Receptors 157
4.9.4 Mechanism of Bt Toxin Action 161
4.9.5 Mode of Action of Vegetative Insecticidal Proteins (VIPs) 163
4.10 Transgenic Plants with Bt Crystal Protein Genes 167
4.10.1 Effectiveness of Bt-*cry1Ab* and *Ac* Genes in Genetically Modified Crops 168
4.11 The Story of Bt Cotton 169
4.11.1 Development of Bt Cotton 170
4.11.2 Bollgard I Cotton (First-Generation Bt Cotton) 172
4.11.3 Bollgard II (Second-Generation Bt Cotton) 173
4.11.4 Bollgard III 174
4.11.5 Impact/Benefits of Bt Cotton 174
4.12 The Story of Bt Brinjal 175
4.12.1 Development of Bt Brinjal 176
4.12.2 Benefits of Bt Brinjal 176
4.13 Safety Issues Related to Bt Crops 177
4.13.1 Disadvantages of Bt Crops 178
4.14 Transgenics for Disease Resistance 182
4.14.1 Transgenic Crops Developed for Disease Resistance . . . 184
4.15 Pathogenesis-Related Proteins (PR Proteins) 186
4.15.1 Discovery and Categorization of PR Proteins 188
4.15.2 PR Proteins and Pathogenic Resistance 189
4.16 Antimicrobial Peptides and Disease Tolerance 190
4.17 Ribosomal Inactivating Proteins 192
4.17.1 Role of RIPs in Plant Pathogen Resistance 193
4.18 Use of Antimicrobial Proteins 195
4.18.1 Classification and Functions of AMPs 196
4.19 Pathogen-Derived Resistance for Viral Diseases 199
4.19.1 Strategies of PDR 199
4.19.2 Protection Conferred by Nucleic Acids 199
4.19.3 Protection Through Movement Proteins (MP) 200
4.19.4 RNA (or DNA)-Mediated Resistance 200
4.19.5 Tolerance Conferred by a Coat Protein 201
4.19.6 Resistance Modulated by a Replicase 201
4.20 Non-pathogen-Derived Resistance for Viral Diseases 202
4.20.1 Protection Against Plant Viruses Through RNA Silencing 202
4.20.2 CRISPR-Cas-Based Plant Viral Disease Resistance . . . 203
4.20.3 TALEN/ZFN-Based Resistance Against Viral Diseases 204
4.21 Transgenic for Stress Tolerance 206

4.21.1 Transgenic Crops Developed for Resistance to Abiotic Stresses 207
4.21.2 Production of Osmoprotectants in Plants 238
4.21.3 Na^+/H^+ Antiporters for Improved Salt Tolerance 239
4.21.4 COR and Heat-Shock Regulons 240
4.21.5 CBF Route/Pathway Regulation 240
4.21.6 Acclimatization to Cold Temperatures Without Activation of CBF Transcripts 240
4.21.7 Thermotolerance via HSF and HSP 241
4.21.8 Thermotolerance Mediated by Temperature-Sensitive Transcription Factors 242
4.21.9 Expression of Enzymes Involved in Scavenging of ROS 242
4.21.10 The SOD Enzyme Family 242
4.21.11 Production of Antioxidants 243
4.21.12 Nonenzymatic Antioxidant Components 245
4.21.13 Enzymatic Components 246
4.22 Transgenics for Nutrient Biofortification and Yield 248
4.22.1 Crop Plants Biofortified for Nutrients 248
4.23 Engineering Plant Protein Composition 248
4.23.1 Plant Proteins Improved Through Genetic Engineering Techniques 253
4.24 Engineering Plants for Vit A Composition 255
4.24.1 Biosynthesis of Vitamin A (Retinol) 256
4.24.2 Biofortification of Vitamin A in Crops 257
4.24.3 Functions of Vitamin A 259
4.25 Biofortified Rice 259
4.25.1 Development of "Golden Rice" (GR) 261
4.25.2 Storing of "Golden Rice" 262
4.26 Biofortified Maize and Cassava 262
4.26.1 Biofortified Cassava 264
4.27 Engineering Plant Mineral Composition 265
4.27.1 Genetically Engineered Crops with Improved Mineral Contents 266
4.27.2 Biofortified Wheat 268
4.28 Enhancement of Photosynthesis for Improved Yield 270
4.29 Conclusions 273
References 275

5 Molecular Pharming 349
5.1 Introduction 350
5.2 Benefits of Molecular Pharming 352
5.3 Platforms for Molecular Pharming 352
5.3.1 Plants with Transgenes 353
5.3.2 Suspension Cell Cultures 353

5.3.3 Temporal Expression Systems/Platforms 354
5.4 Molecular Pharming in Mammalian Organisms 356
5.5 Plant-Based Molecular Pharming 356
5.5.1 Molecular Pharming of Proteins 357
5.5.2 Molecular Pharming of Carbohydrates 360
5.5.3 Molecular Pharming of Lipids 372
5.6 Conclusions 376
References 376

6 Future Prospects of GM Plants 387
6.1 Introduction 388
6.2 The Current Status of Transgenic Crops 390
6.2.1 Production of Insect-Resistant Crops 393
6.2.2 Herbicide-Resistant Transgenic Crops 394
6.2.3 Virus-Resistant Transgenic Crops 394
6.2.4 Transgenic Crops Tolerant to Abiotic Stress 402
6.2.5 Transgenic Crops and Nutrition 403
6.2.6 Transgenic Crops with Therapeutic Functions 404
6.3 Ethical Issues and Risks Associated with Transgenic Crops 405
6.3.1 Transgenic Crops and Biosafety Issues 405
6.3.2 Destruction of Resistance 407
6.3.3 Harmful Effects on the Nontarget Organisms 407
6.4 Transgenic Crops and Patent Rights 408
6.5 Commercialization-Related Issues 408
6.6 Frontiers in Plant Genetic Manipulations 408
6.7 Advances in Genome Editing for Crop Improvement 409
6.8 Synthetic and Artificial Chromosomes 411
6.9 Targeted Epigenetic Modification 413
6.10 The Future of GM Crops 413
6.11 Conclusions 414
References 414

About the Author

Bhupendra Koul is an Associate Professor at the Department of Biotechnology, School of Bioengineering and Biosciences, Lovely Professional University (LPU), Punjab, India. During his Ph.D. at the CSIR-NBRI, Lucknow, he worked on the optimization, introduction, and expression of modified full-length and truncated versions of Bt-*cry1Ab* and *1Ac* genes in tomato for developing non-chimeric and stable transgenic lines resistant to two lepidopteran insects (*Helicoverpa armigera* and *Spodoptera litura*) and evaluated the performance of both the versions of *cry1Ab* and *1Ac* genes for the stability and efficacy of insecticidal toxin in transgenic plants. He also evaluated the performance and role of various cis-motifs of synthetic promoters in the overexpression of genes in tomato and performed comparative in silico analyses of several *cry1A* genes for toxicity to target insects. He has also optimized the regeneration and *Agrobacterium*-mediated transformation of Stevia (*Stevia rebaudiana* Bertoni) and has developed herbicide-resistant transgenic Stevia for effective weed management in Stevia cultivation. He has 5 years of research experience and was awarded the "CSIR-Senior Research Fellowship (SRF)" in the year 2013. He also has 8 years of teaching experience and received the "Teacher Appreciation Award 2016" from LPU in the Discipline of Biotechnology through the MHRD Minister, Government of India. He has designed the full-length synthetic *cry1Ac* gene (GenBank: KP195020.1) and has published 45 research papers in national and international journals as well as 21 book chapters and 2 authored books with Springer Nature.

Abbreviations

2,4-D	2,4-Dichlorophenoxyacetic acid
2D-PAGE	Two-dimensional polyacrylamide gel electrophoresis
2iP	N6-(2-isopentyl) adenine
4-MU	Methylumbelliferone
4-MUG	Methylumbelliferyl-β-D-glucuronide
ABA	Abscisic acid
ABRE	Abscisic acid response element
ACC	1-Amino-cyclopropane-1-carboxilic acid
ACCase	Acetyl-CoA carboxylase
ACMV	African cassava mosaic virus
ACP	Acyl carrier protein
AdoMetDC	S-adenosyl-L-methionine decarboxylase
AFLP	Amplified fragment length polymorphism
ALS	Acetolactate synthase
AMPA	Aminomethylphosphonic acid
AMV	Alfalfa mosaic virus
APHIS	Animal and plant health inspection service
ArMV	Arabis mosaic virus
AuxRE	Auxin response element
BADH	Betaine aldehyde dehydrogenase
BAP	6-Bezylaminopurine
BNYVV	Beet necrotic yellow vein virus
BP	Base pair
BSE	Bovine spongiform encephalitis
BT	*Bacillus thuringiensis*
BYDV	Barley yellow dwarf virus
bZIP	Basic leucine zipper
cab	Chlorophyll a/b-binding promoter
CaMV	Cauliflower mosaic virus
CaMV35S	Cauliflower mosaic virus 35S promoter
CAT	Chloramphenicol acetyltransferase
CDK	Cyclin-dependent kinase
cDNA	Complementary DNA

CE	Coupling element
Cm	Centimorgan
CMO	Choline monooxygenase
CMS	Cytoplasmic male sterility
CMV	Cucumber mosaic virus
CoA	Coenzyme A
COR	Cold-responsive
CP	Capsid or coat protein
CPMR	Coat protein-mediated resistance
CPMV	Cowpea mosaic virus
CpTl	Cowpea trypsin inhibitor
DAHP	3-Deoxy-D-arabino-heptulosonate 7-phosphate
DCL	Dicer-like protein
DEAE	Diethylaminoethyl
DEFRA	Department for Environment, Food and Rural Affairs
DFR	Dihydroflavonol-4-reductase
DHFR	Dihydrofolate reductase
DMSO	Dimethyl sulfoxide
DNA	Deoxyribonucleic acid
DP	Degree of polymerization
DPE	Downstream promoter element
DRE	Dehydration-responsive element
Ds	Double stranded
DST	Downstream element
DTT	Dithiothreitol
EDTA	Ethylenediaminetetraacetic acid
EFSA	European Food Safety Authority
ELISA	Enzyme-linked immunosorbent assay
EPA	Environmental protection agency
EPSPS	5-Enolpyruvylshikimate-3-phosphate synthase
ER	Endoplasmic reticulum
ERA	Environmental risk assessment
EST	Expressed sequence tag
EU	European Union
FDA	Food and Drug Administration
GBSS	Granule-bound starch synthase
GEAC	Genetic Engineering Appraisal Committee
GFP	Green fluorescent protein
GGDP	Geranylgeranyl diphosphate
GlcNAc	*N*-acetylglucosamine
GM	Genetically modified
GMO	Genetically modified organism
GNA	*Galanthus nivalis* (snowdrop plant) agglutinin
GOX	Glyphosate oxidase
GSH	Glutathione

GSSG	Oxidized glutathione
GST	Glutathione S-transferase
GUS	β-Glucuronidase
HAS	Human serum albumin
HEAR	High-erucic-acid rapeseed
HIV	Human immunodeficiency virus
HPP	Hydroxyphenylpyruvate
HPPD	Hydroxyphenylpyruvate dioxygenase
hpRNA	Hairpin RNA
HR	Hypersensitive response
HSE	Heat-shock element
HSF	Heat-shock factor
HSP	Heat-shock protein
IAA	Indole-3-acetic acid
ICP	Insecticidal crystal protein
Ig	Immunoglobulin
Inr	Initiator region
IPR	Intellectual Property Rights
IRGSP	International Rice Genome Sequencing Project
ISAAA	International Service for the Acquisition of Agri-biotech Applications
JIP	Jasmonate-induced protein
kb	Kilobase(s)
kDa	Kilodalton
Km	Michaelis-Menten constant
LEAR	Low-erucic-acid rapeseed
LRE	Light-responsive element
LRR	Leucine-rich repeat
LTRE	Low-temperature-response element
MAPK	Mitogen-activated protein kinase
MAR	Matrix attachment region
mas	Mannopine synthase promoter
MCS	Multiple cloning site
MES	2-(N-morpholino) ethanesulfonic acid
miRNA	MicroRNA
MP	Movement protein
mRNA	Messenger RNA
MS	Murashige and Skoog (culture medium)
MUG	4-Methylumbelliferyl glucuronide
NAA	1-Naphthylacetic acid
NADPH	Reduced nicotinamide adenine dinucleotide phosphate
NB	Nucleotide binding
NLS	Nuclear localization signal
nos	Nopaline synthase
NPC	Nuclear pore complex
NSP	Nuclear shuttle protein

NVCP	Norwalk virus capsid protein
ocs	Octopine synthase promoter
ORFs	Open reading frames
PAC	Plant artificial chromosomes
PAMP	Pathogen-associated molecular pattern
PCR	Polymerase chain reaction
PDR	Pathogen-derived resistance
PEG	Polyethylene glycol
PEP	Phosphoenolpyruvate
PEPC	Phosphoenolpyruvate carboxylase
PG	Polygalacturonase
PHA	Polyhydroxyalkanoate
PHB	Polyhydroxybutyrate
Phy	Phytochrome
PLRV	Potato leafroll virus
PME	Pectin methylesterase
PPT	Phosphinothricin (also known as glufosinate)
PR	Pathogenesis related
PRSV	Papaya ringspot virus
PTGS	Posttranscriptional gene silencing
pv	Pathovar
PVX	Potato virus X
PVY	Potato virus Y
QTL	Quantitative trait locus
RAPID	Random-amplified polymorphic DNA
rbc	Ribulose-1,5-bisphosphate carboxylase promoter
rbcL	Ribulose-1,5-bisphosphate carboxylase/oxygenase large subunit
rbcS	Ribulose-1,5-bisphosphate carboxylase/oxygenase small subunit
RdRp	RNA-dependent RNA polymerase
RFLP	Restriction fragment length polymorphism
RIP	Ribosome-inactivating protein
RISC	RNA-induced silencing complex
RNA	Ribonucleic acid
RNAi	RNA interference
ROS	Reactive oxygen species
rRNA	Ribosomal RNA
Rubisco	Ribulose-1,5-bisphosphate carboxylase/oxygenase
S	Svedberg unit
SA	Streptococcal antigen
SAG	Senescence-activated gene
SAGE	Serial analysis of gene expression
SAR	Systemic acquired resistance
SAUR	Small auxin-up RNA
SBE	Starch-branching enzyme
scFv	Single-chain variable fragment antibody

sg	Subgenomic
siRNA	Small interfering RNA (or guide RNA)
slgA	Secretory IgA
SNP	Single nucleotide polymorphism
SOD	Superoxide dismutase
ss	Single stranded
SSR	Simple sequence repeat
TAG	Triacylglycerol
TBP	TATA-binding protein
T-DNA	Transfer DNA
TetC	Tetanus toxin fragment C
TetR	Tetracycline repressor
TEV	Tobacco etch virus
Ti	Tumor inducing
TIC	Translation initiation context
TILLING	Targeted induced local lesions in genomes
TLCV	Tomato leaf curl virus
TMV	Tobacco mosaic virus
TPS	Technology protection system
tRNA	Transfer DNA
TSWV	Tomato spotted wilt virus
tTA	Tetracycline transactivator
TTSS	Type III protein secretion system
UAS	Upstream activation sequences
ubi-1	Ubiquitin promoter
*uid*A	β-Glucuronidase gene
uORF	Upstream ORF
URS	Upstream repression sequences
USDA	United States Department of Agriculture
UTR	Untranslated region
VIP 1	Vire2-interacting protein 1
VLP	Viruslike particle
VPg	Viral genome-linked protein
X-gluc	5-Bromo-4-chloro-3-indolyl glucuronide
ZYMV	Zucchini yellow mosaic virus

Plant Transformation Techniques

1

Abstract

Plant genetic transformation is the intentional alteration and modification of its genome by introducing one or more foreign gene(s) through a variety of methods (other than traditional procedures), and the modified plant is referred to as transformed or transgenic. These remarkable advances in gene technology have not only helped in understanding the gene structure and functions, but also offered access to unlimited gene pool together with incorporation and expression of gene(s) from taxonomically unrelated species, in "tailoring" crop plants with useful genetically engineered characteristics (qualitative and quantitative) to combat various environmental stresses (biotic and abiotic) and human nutritional deficiencies. Adding a gene to a crop plant can increase its agricultural, horticultural, or ornamental value, and it can also be used as a living bioreactor for the cost-effective synthesis of pharmaceutically useful metabolites and proteins. Genetically modified phytoremediator/hyperaccumulator plants can remediate contaminated soils and waters. Plant genetic transformation may also be used to investigate how genes are regulated and expressed throughout development and other biological processes. In general, there are two methods for artificial DNA transfer: vector-less or direct DNA transfer (physical or chemical gene transfer) and (ii) vector-mediated or indirect DNA transfer method (*Agrobacterium*-mediated transformation; *Agrobacterium*-mediated virus infection). Among these methods, *Agrobacterium*-mediated transformation is thought to induce less rearrangement of the transgene and introduces a lower copy number than the other direct DNA delivery methods. This chapter focuses on explicit information on various vector-less/direct (physical, chemical) and vector-mediated/indirect (*Agrobacterium*-mediated) plant transformation techniques.

Keywords

Agrobacterium · Genetic transformation · Biolistics · Macroinjection · Electroporation

B. Koul, *Cisgenics and Transgenics*, https://doi.org/10.1007/978-981-19-2119-3_1

1.1 Introduction

The deliberate altering of the genome of an organism (bacteria, plant, or animal) is referred to as genetic transformation, and the genetically altered organism is referred to as a GMO. These days, transgenic plants and GM crops are the new buzzwords, which are attracting the attention of scientists, farmers, plant breeders, and the general public worldwide. It is indeed a new revolution in plant biotechnology and agriculture on a global scale. Crop yield will be a major concern for food and nutritional security for the world's burgeoning billions by 2050. Crop production and productivity are extremely variable and subject to a variety of physical, abiotic, and biotic restrictions. To cope with these predictable challenges, the plant biotechnologists have geared up and focused their attention towards sustainable agriculture in terms of crop improvement through transgenic approach in a myriad of traits that are not able to be achieved by plant breeding techniques. Some of the key research areas or the purpose for the crop improvement through plant transformation technique(s) are shown in Fig. 1.1. These are (i) improvement in plant tolerance to biotic and abiotic stresses (maintenance of plant productivity); (ii) enhancement of nutritional value (protein, amino acids, lipids, and vitamins); (iii) enhancement of agronomic traits (seed dormancy, disease resistance: bacterial, fungal, viral, and nematode resistance); (iv) introduction of novel traits (fruit ripening, fruit and flower color, etc.); (v) enhancement of phytoremediation potential, nitrogen-use efficiency (NUE), salt tolerance, and heavy metal stress tolerance (biomining, etc.); (vi) cost-effective production of pharmaceutical compounds (molecular pharming: production of antibodies, vaccines, therapeutic proteins); and (vii) metabolic pathway regulation (nutrient capture, carbohydrate production, essential oil production, biopolymer production).

Thus, in the twenty-first century, biotechnologists have been able to develop transgenic plants through *biofortification* [high β-carotene in *Zea mays* (Yan et al. 2010), *Oryza sativa* (Singh et al. 2017), *Manihot esculenta* (La Frano et al. 2013), and *Solanum tuberosum* (Ducreux et al. 2005; Song et al. 2016); lysine in *Oryza sativa* (Yang et al. 2016), *Brassica napus* (Falco et al. 1995), and *Glycine max* (Kim et al. 2012); folic acid in *Oryza sativa* (Storozhenko et al. 2007); iron in *Triticum aestivum* (Borg et al. 2012); oleic acid in *Glycine max* (Kinney and Knowlton 1998)]; *improving nutritional quality* [improving protein in staple vegetables, *Manihot esculenta* and *Solanum tuberosum*; removing allergens and antinutrients; removing cyanide from the roots of *Manihot esculenta* (Siritunga and Sayre 2003; Siritunga et al. 2004); removing glycoalkaloid toxin from *Solanum tuberosum* (Arnqvist et al. 2003) and allergenic proteins from *Oryza sativa* (Tada et al. 1996; Ogo et al. 2014) and *Triticum aestivum* (Kalunke et al. 2020); increasing antioxidant content (Verhoeyen et al. 2002) and lycopene content (Römer et al. 2000) in *Solanum lycopersicum*]; *biotic* and *abiotic stress tolerance* [pest-resistant *Solanum lycopersicum* (Koul et al. 2012, 2014, 2015; Koul 2020), *Gossypium* sp. (Tabashnik et al. 2002; Torres et al. 2009), *Oryza sativa* (Xu et al. 1996; Shu et al. 2000), and *Brassica oleracea* (Jin et al. 2000); aphid-resistant *Brassica juncea* (Kanrar et al. 2002; Dutta et al. 2005); pod borer-resistant *Cicer arietinum* (Sanyal et al. 2005);

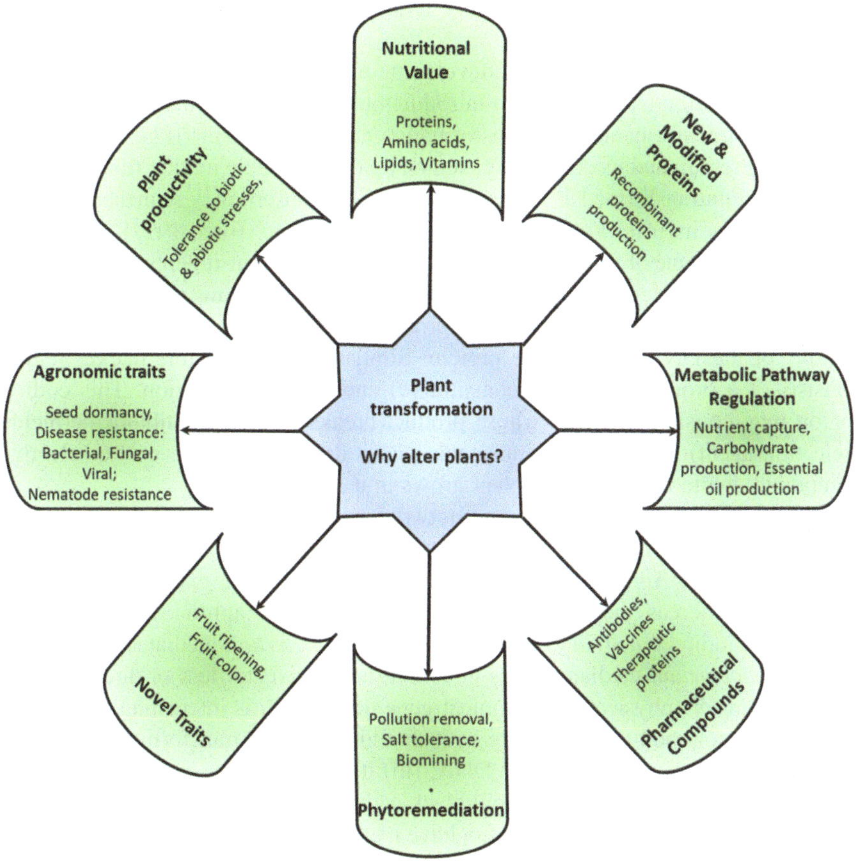

Fig. 1.1 Purpose of plant transformation

brown plant hopper (BPH)-resistant *Oryza sativa* (Tang et al. 2001; Xiaofen et al. 2001; Boddupally et al. 2018); drought-tolerant *Oryza sativa* (Gao et al. 2011; Zhang et al. 2011) and *Triticum aestivum* (Gao et al. 2009); drought-tolerant *Zea mays* (Zhang et al. 2011); salt-tolerant *Oryza sativa* (Nagamiya et al. 2007; He et al. 2011, Tang et al. 2019); herbicide-tolerant *Stevia rebaudiana* (Taak et al. 2020, 2021); enhanced nutrient-use efficiency, heterosis, and apomixis in crops, etc.]; *biopharming and production of edible vaccines* [expression of hepatitis B surface antigen in *Musa* sp. (Kumar et al. 2005; Elkholy et al. 2009), expression of cholera toxin subunits in potato (Arakawa et al. 1997), etc.]; *floriculture* [manipulation of flower color (Hanumappa et al. 2007; He et al. 2013), shelf life of cut flowers (Noman et al. 2017), novel pigmentation-violet *Dianthus caryophyllus* (Nakamura et al. 2020), blue rose (Katsumoto et al. 2007), etc.]; and *phytoremediation* [e.g., removal of heavy metals (Grichko et al. 2000; Mosa et al. 2016; Koul and Taak

2018), *Populus* sp. expressing mercury reductase (Rugh et al. 1998; Choi et al. 2007), etc.].

The two major requirements for developing transgenic plants are (a) preparation of a transformation cassette and (b) introduction of the transformation cassette and selection of transformants. The transformation cassette contains (i) gene of interest (the coding region and its controlling elements), (ii) selectable marker (distinguishes transformed/untransformed plants), and (iii) insertion sequences (aids *Agrobacterium* insertion)/robust transformation procedure (Birch 1997). As already mentioned, the gene of interest contains promoter region, transit peptide, and coding region. The promoter region controls when, where, and how much the gene is expressed; the transit peptide targets the protein to correct organelle; and the "coding region" of the DNA encodes the protein. Similarly, the selectable marker region consists of a promoter (usually constitutive) and the coding region. This coding region may consist of a gene whose product breaks down an antibiotic/herbicide (selection agent), etc. The non-transgenic plants die in the presence of a selective compound, while the transgenic plant grows in its presence.

Gene transfer can be accomplished in two ways: (a) naturally and (b) artificially. Bacterial conjugation, transformation, transposition, phage transduction, retroviral transduction, and *Agrobacterium*-mediated gene transfer are examples of natural techniques, whereas the artificial method involves two approaches: (i) vector-mediated or indirect method of DNA transfer (*Agrobacterium*-mediated transformation; *Agrobacterium*-mediated virus infection) and (ii) vector-less or direct method of DNA transfer (physical and chemical gene transfer). It is interesting to note the properties of a good host, which are (i) it should be easy to transform, (ii) it should support replication of recombinant DNA, (iii) it should be free from elements that interfere with replication, (iv) it should lack active restriction enzymes (e.g.: *E. coli* k12 substrain HB101), (v) it should not have methylase activity, and (vi) it should be deficient in normal recombination function.

The following elements are crucial to a successful gene transfer procedure: (1) a cost-effective, simple, reproducible, genotype-independent regeneration process; (2) target tissues that are capable of both transformation and regeneration; (3) a method for delivering DNA that is effective; (4) the ability to recover viable plants while avoiding somaclonal variation in transgenic plants; and (5) the technique for selection of transgenic tissues (Velcheva et al. 2005; Thi Van et al. 2010).

A high-frequency, genotype-independent in vitro regeneration system that is receptive to *Agrobacterium*-mediated transformation is required for the development of transgenic lines (Birch 1997). Genotype, type of explant, explant orientation, wounding process, cocultivation length, role of phenolic chemicals, *Agrobacterium* strain, bacterial cell density, and other parameters influencing genetic transformation all play a part in determining total transformation efficiency. Improving transformation efficiency and, more importantly, generating non-chimeric transgenic plants need optimization, selection, and screening processes. The various methods of artificial gene transfer are mentioned in Fig. 1.2.

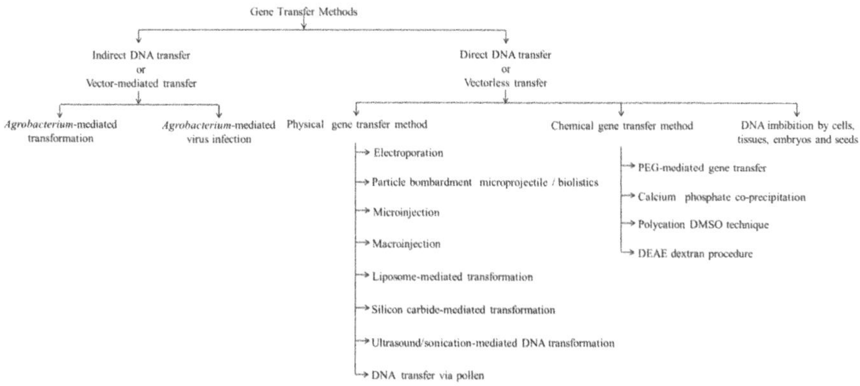

Fig. 1.2 Flowchart showing gene transfer techniques

1.2 Direct DNA Transfer

Direct DNA transfer or vector-less transfer regimes are simple and effective DNA transfer techniques. It has three categories: (i) physical method, (ii) chemical method, and (iii) DNA imbibition by cells, tissues, embryos, and seeds.

1.2.1 Physical Gene Transfer Methods

These transfer methods are species and genotype independent and do not utilize a vector for the transfer of DNA into the plant cells/tissues/embryos/cotyledons, etc. These methods involve (a) electroporation, (b) particle bombardment/ microprojectile, (c) macroinjection, (d) microinjection, (e) lipofection, (f) silicon carbide (SiC)-mediated transformation, (g) ultrasound/sonication-mediated transformation, and (h) pollen-mediated DNA transfer.

1.2.1.1 Electroporation

The biological membranes/cells when exposed to high electric fields lead to great electric conductivity and permeability. This method may be either reversible or irreversible (Kotnik et al. 2015). Thus, electroporation is the technique of using high-field-strength electrical impulses to reversibly permeabilize cell membranes so that big molecules such as DNA can be feasibly taken up. Electroporation is used to transform protoplasts in a temporary or integrative manner. Linear DNA, a field strength of 1.25 kV/cm, and PEG (aids in DNA interaction with the cell membrane) must be employed to improve protoplast transformation efficiency. Transformation of intact plant cells (rice, sugar beet, etc.) has also been standardized. However, the efficiency of transformation is dependent on wounding or macerozyme treatment of the cells or tissues to be transformed. The process has the advantages of being quick,

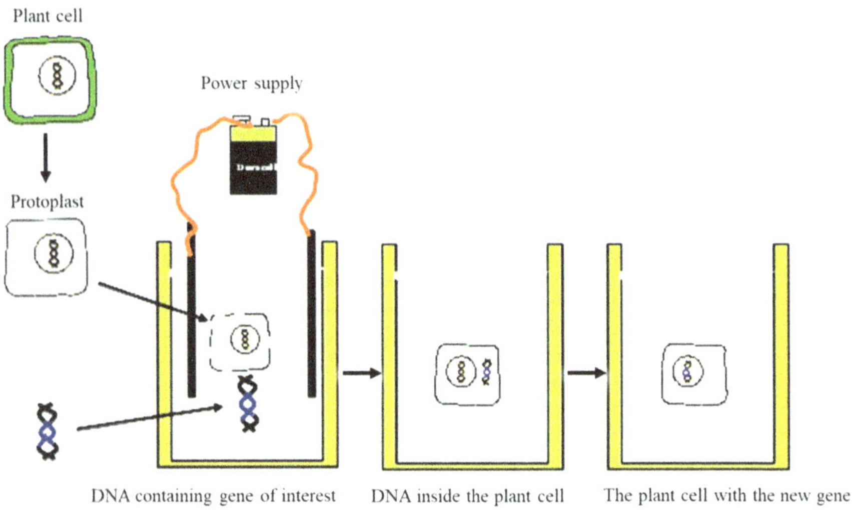

Fig. 1.3 Basic plan of electroporation in plant protoplast

nontoxic, and uncomplicated. It can also be applied to a large number of cells at once, resulting in a high percentage of stable transformants (Fig. 1.3). The sole disadvantage is the difficulty in regeneration of plantlets from protoplasts.

1.2.1.2 Particle Bombardment/Microprojectile

A technique known as particle bombardment, particle gun method, gene gun, particle acceleration method, or biolistics method can be used to make transgenic organisms such as microbes, mammalian cells, or plant materials. This is made possible by introducing nucleic acid into the aforementioned cells/tissues by applying helium pressure (through an evacuated chamber), which drives the entry of microprojectiles (DNA-bearing particles: 1–3 μm in diameter) into the cells. Klein et al. (1987) and Sanford et al. (1987) were the first to describe this approach. PDS1000 (gunpowder-driven device) or PDS1000/He (helium driven particle gun) is used in the basic system (Fig. 1.4). The plant tissue is transformed using gold or tungsten particles that have been coated with DNA. As previously stated, the particles are driven at high speeds into the target plant material, releasing DNA that can subsequently be incorporated into the genome. The sample is placed in a bombardment chamber that has been evacuated to subatmospheric pressure. When the instrument is triggered (in the case of the PDS1000/He), helium is injected into the gas acceleration tube and held there until the rupture disc's specific pressure is attained. The disc bursts, and the helium shock waves that follow propel the coated microparticle-carrying microcarrier disc a short distance towards the halting screen. The microcarrier is retained by the stopping screen, while the microparticles pass through the screen and into the bombardment chamber, penetrating the target cells.

It is worth noting that the microcarriers'/microprojectiles' launch velocity is influenced by a number of variables: (i) helium pressure (rupture disc selection,

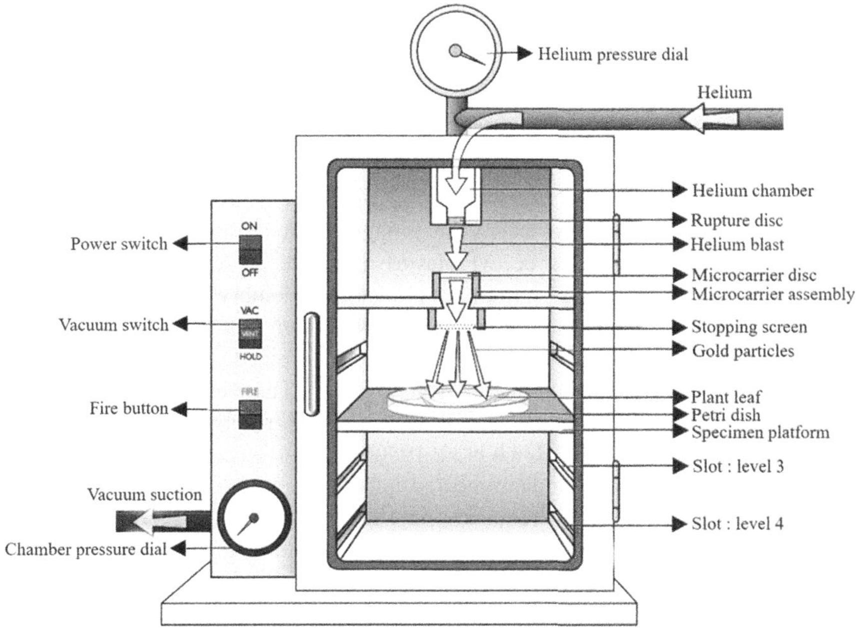

Fig. 1.4 Biolistics-mediated DNA delivery

450–2200 psi), (ii) vacuum, (iii) rupture disc distance from the microcarrier, (iv) distance between the "microcarrier launch assembly" and the stopping screen, and (v) distance between the target cells and the stopping screen. By adjusting these factors, one can generate a variety of velocities that will ideally transform a variety of cell kinds.

When it was discovered that tungsten particles were phytotoxic, they were substituted with gold particles (Russell et al. 1992). The efficiency of biolistics-based transformation is determined by physical characteristics such as particle velocity, size, and number, as well as the amount of DNA present (Sanford 1988). The type of explants, their osmotic condition, and the length of preculture prior to bombardment are all factors that influence success (Klein and Jones 1999).

Microinjection of nucleic acids into plant cells or embryos is far more difficult than biolistics, which enables the transformation of animal cells with specific growth requirements that are incompatible with other methods of gene transfer. It uses less DNA and cells than previous approaches and can be used for both transient and stable transformation. It is a pioneering method for successful cereal transformation that has been in use for many years. The tissues commonly used for transformation are immature zygotic embryos, isolated scutella, unmerged inflorescences, and embryogenic callus tissues. It is also used for transforming the microspores of barley to raise transgenic plants. This technology has the following advantages: (i) it is clean and safe; (ii) it can alter organized tissue; (iii) it can operate as a universal delivery system; (iv) it can transform refractory species; and (v) it may be used to examine basic plant processes. The downsides include nonhomologous integration into the chromosome, which is characterized by many copies and certain degree of rearrangement in plant gene transfer. It could result in chimerical plants. Furthermore, uncontrolled bombardment velocity causes significant damage to the target cells.

1.2.1.3 Microinjection

Microinjection is a technique where the foreign DNA is mostly transferred into (specifically into the nucleus or cytoplasm) animal cells (e.g., egg, oocyte, or embryo) or plant cells (e.g., protoplasts, ovary, or embryo) using fine-tipped (0.5–1.0 μm diameter) micropipette or glass needle (Zhang and Yu 2008). It involves micromanipulation, such as suction-mediated holding pipette, an injecting needle, and specially designed inverted microscope (Fig. 1.5). There have been reports that protoplasts can survive microinjection and are an ideal material for transformation due to absence of a cell wall (Reich et al. 1986).

Agrobacterium DNA was microinjected into tobacco protoplasts. When nuclei were injected, transformation frequency percentage was 14%, whereas it was 6% when cytoplasm was injected. Pollen grains, in addition to protoplasts, can be microinjected and successfully used for in vitro fertilization (Crossway 1989; Kranz and Lörz 1990). The *npt* marker gene was microinjected into microspore-derived canola embryos at the 12-cell stage. Up to 57% of the plants obtained were transgenic. Most transgenics, on the other hand, were chimeric, and pure transformants could only be obtained by secondary embryogenesis (Yeung 2002). In genetic engineering and transgenics, the method is widely utilized as a vector to insert genetic material into a single cell. The efficiency of this approach has been improved owing to the computerized control of the holding pipette, microscopy, and visualization.

Microinjection's applications include the following: (i) the process is applicable to both plant and animal cells, but the technique is ideally useful for producing transgenic animals in a cost-effective manner; (ii) the embryonic cells can be used

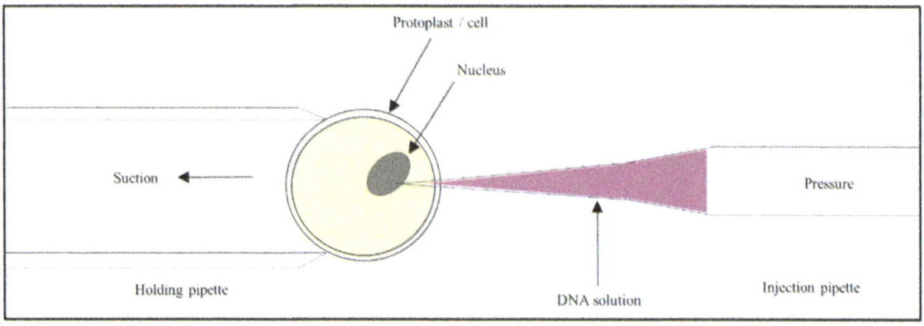

Fig. 1.5 Microinjection technique

for gene transfer; (iii) protoplasts of *Nicotiana tabacum*, alfalfa, etc. have been successfully transformed with this method; and (iv) it might be a beneficial way for introducing several bioactive molecules into the same target single cells at the same time, such as antibodies, peptides, RNAs, plasmids, diffusion markers, elicitors, Ca^{2+}, as well as nuclei and artificial micro- or nanoparticles holding those chemicals.

Thus, microinjection has the following advantages: (i) other approaches have a far lower incidence of stable DNA integration; (ii) the approach may be used to transform both primary cells and cells from established cultures; (iii) in this method, the DNA inserted is exposed to fewer significant changes; and (iv) only precise integration of DNA and that too in limited copy number can be obtained. The main disadvantages of the microinjection technique are that it is a costly procedure, it requires skilled personnel, it is more useful for animal cells, it is more useful for embryonic cells, it requires knowledge of mating timing and oocyte recovery, and it is only useful for protoplasts rather than walled cells (Rivera et al. 2012).

1.2.1.4 Macroinjection

The plant materials like immature embryos, immature pollens, and germinating pollens can be injected with DNA using a syringe needle, having greater diameter than the needle used for microinjection (Rivera et al. 2012). Macroinjection is an experimental approach for transferring artificial DNA to cereal plants that are unable to regenerate and grow into complete plants from cultivated cells. Nearly 0.3–0.5 ml of DNA solution is injected immediately above the tiller node. The timing of injecting the DNA solution should be optimized. It should be done 14 days before meiosis. Transgenic rye, tobacco, and other cereals are produced by this method (Pena et al. 1987). It is advantageous as it does not require the use of protoplasts and the technique is simple and cost effective. The main disadvantage is that the procedure is less precise and less effective and has a ten times lower frequency of transformation than biolistics, which might result in the production of chimeric plants (Rivera et al. 2012).

1.2.1.5 Liposome-Mediated Transformation

Liposomes are lipid-based spheres that can carry chemicals into cells (Fig. 1.6). These are synthetic vesicles that can be used to transfer exogenous molecules like transgenes. They are a sphere of lipid bilayers that surround the molecule to be transported and enable it to get there after fusing with the cell membrane (Rakoczy-Trojanowska 2002). The liposomes take up plasmids which are artificially prepared, and the desired DNA is packed with the liposomes; thereafter, they are mixed with protoplasts. After that, plasmids are released in the protoplast. The ionizable lipids that undergo phase change in response to the pH of the cytoplasm, e.g., dioleoyl phosphatidylethanolamine (DOPE) to construct liposomes [DOPE and phosphatidyl choline (PC)], release DNA into the cytoplasm once they are phagocytosed (Felgner et al. 1987). Cationic lipids are positive-charged lipids that are used for nucleic acid transfer. As compared to the uncharged liposomes, these positive ones may interact more readily with the negatively charged cell membrane, which allows DNA to be

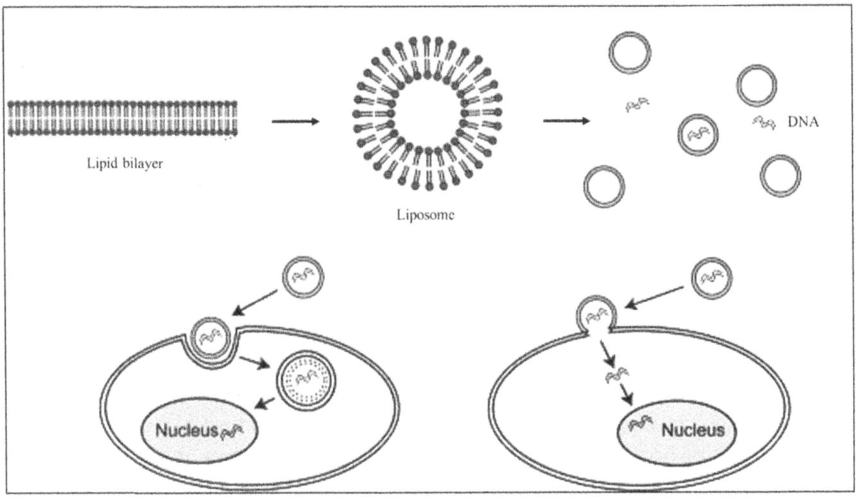

Fig. 1.6 Lipofection technique

transported directly across the plasma membrane by cationic liposome-cell surface fusion.

Lipofectin, an in vitro transfecting agent, is one example of cationic lipids that may be utilized to generate cationic liposomes. Recently, lipofection was employed to convert mycelial filamentous fungus (Chai et al. 2013). This technique is straightforward to use, has a long shelf life, is low in toxicity, and inhibits degradation of nucleic acids (Mannino and Gould-Fogerit 1988).

1.2.1.6 Silicon Carbide (SiC)-Mediated Transformation

No specialized equipment is required for a simple procedure employing silicon carbide fibers. Plant material (such as suspension of the cultured cells, embryos, and calluses derived from embryos) is placed in a buffer containing DNA and SiC fibers and vortexed (Rakoczy-Trojanowska 2002). The fibers (0.3–0.65 μm in diameter and 10–100 μm in length) penetrate the cell wall and the membrane, allowing DNA to enter the cell (Rivera et al. 2012). Intercellular collisions that occur violently in the vortex mill are likely to improve cellular penetration.

The DNA linked to the fiber surface is transported into the penetrated cell and incorporated into the nuclear genome, transforming plant cells permanently (Fig. 1.7). The disadvantages of this method include a lack of suitable plant material and the inherent risks of the fibers, which necessitate careful handling (Rakoczy-Trojanowska 2002). Although the process was used with maize and oat genotypes, many cereals, including rice, wheat, and barley, have recently been transformed using this technology without the use of cell suspension cultures (Thompson et al. 1995).

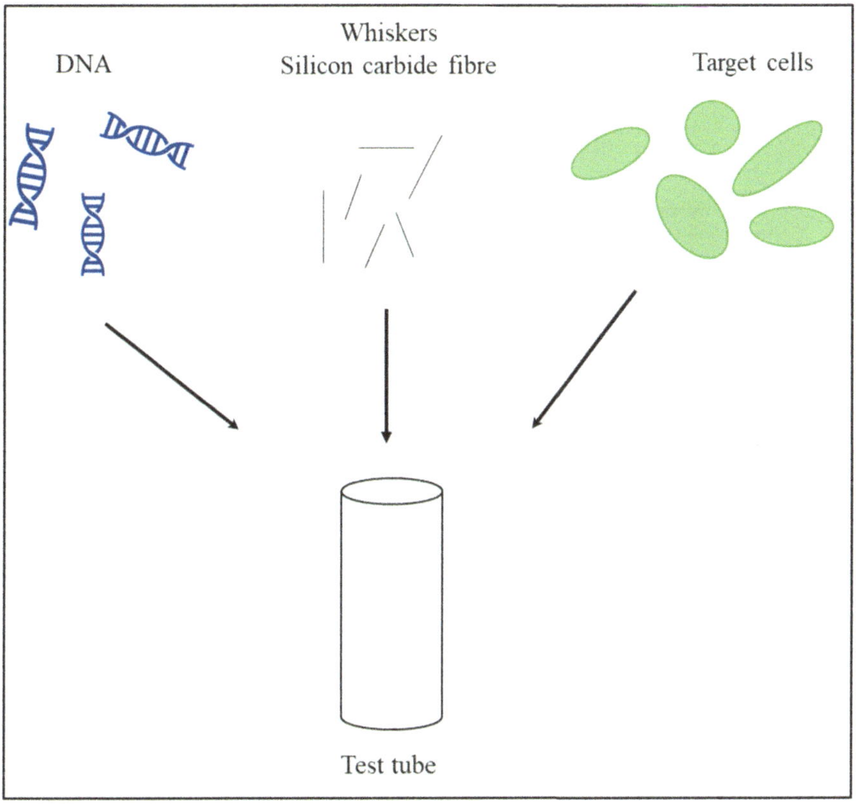

Fig. 1.7 Whisker-mediated plant cell transformation

1.2.1.7 Ultrasound/Sonication-Mediated Transformation

Ultrasound-mediated membrane perforation/sonoporation can be used to transfer DNA into the excised/injured tissue (leaf discs, cotyledonary nodes, somatic embryos, bacterial cell, etc.). This method of gene transfer is used by the scientists for monocots, dicots, gymnosperms, etc. The cultured explants are sonicated with plasmid DNA containing marker genes for transformation. The transformed explants when transferred to selection medium undergo shoot induction. In sugar beet and tobacco, mild sonication of 20 KHz ultrasound was used to enable the entry of chloramphenicol acetyltransferase (CAT) gene in protoplasts (Rivera et al. 2012). This method is less labor intensive and cost effective for transforming the plant tissue (Finer and Trick 1997); however, the duration of sonication requires optimization. Figure 1.8 shows the parts of an ultrasonic bath.

1.2.1.8 DNA Transfer by Pollen

After pollination in which styles are cut, they are inserted with the foreign DNA. The DNA will reach the ovary and then to ovule by the formation of the pollen tube. This

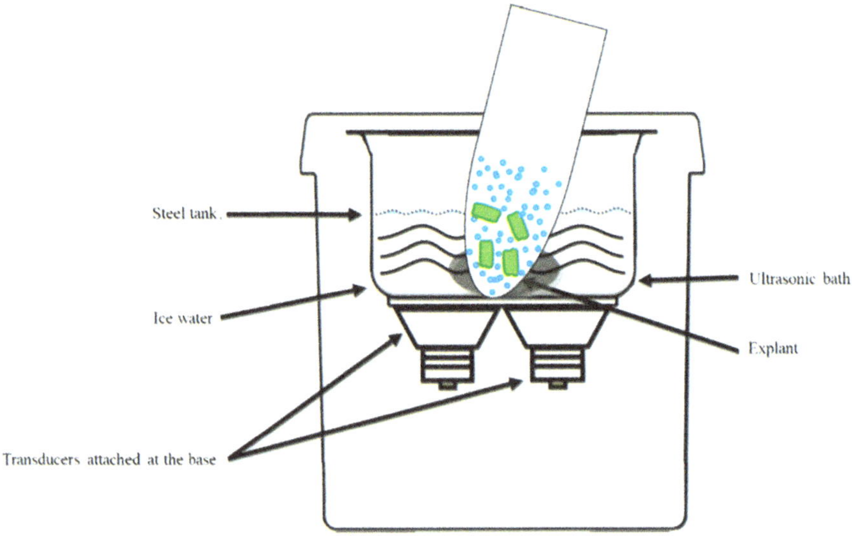

Fig. 1.8 General characteristics of the ultrasonic bath

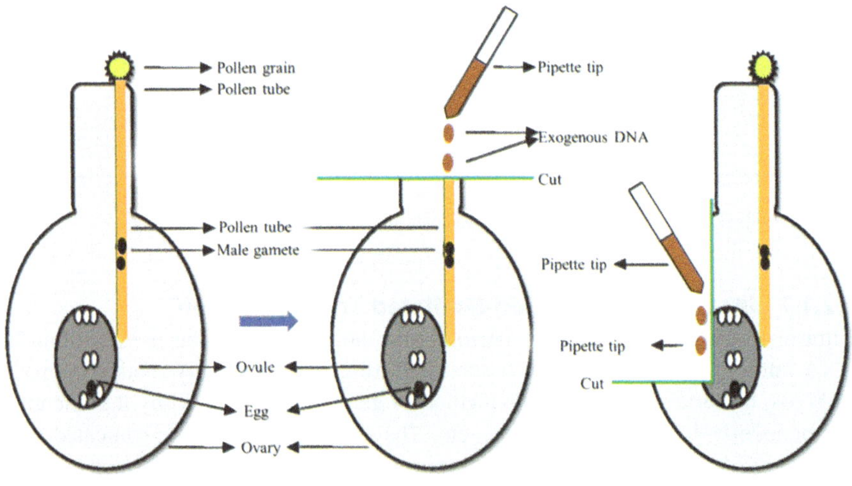

Fig. 1.9 Pollen tube-mediated gene transfer (Modified after Ali et al. 2015). (**a**) A pollen grain germinates, and a pollen tube develops through the style, finally piercing the ovule; (**b**) application of exogenous DNA on decapitated style to promote PTT; and (**c**) direct delivery of DNA into the ovule (ovary drip technique)

method is known as pollen tube pathway or DNA transfer by pollen (Fig. 1.9). This method was firstly used for transforming the rice. After that, many transgenics were produced like rye and watermelon (Chen et al. 1998). The plasmid DNA or inoculum of bacteria can be transferred into the pollen mother cells without removing the stigmas. It is done at the premeiotic stage. This type of transformation was done in the rye plant (Pena et al. 1987). The transformation through this technique is ten times better than microprojectile technique (Rakoczy-Trojanowska 2002).

1.2.2 Chemical Methods of Gene Transfer

1.2.2.1 PEG-Mediated Gene Transfer

It is a chemical method for transferring the exogenous DNA into the protoplasts using polyethylene glycol (PEG). Protoplast, when exposed to exogenous DNA in the presence of PEG and other chemicals like calcium chloride, facilitates the entry of DNA into protoplasts. The cells which are formed as a result of the formation of walls around them are cultured. This method is simple, and transformed cell will have high survival and division rate (Mathur and Koncz 1998). This method has also been reported for genetic transformation of filamentous fungi like *Stagonospora nodorum* (Liu and Friesen 2012).

1.2.2.2 Calcium Phosphate Coprecipitation

In this technique, DNA preparation used in transfection is first dissolved in calcium chloride solution, which is added slowly to the phosphate buffer, which leads to the formation of insoluble calcium phosphate which is coprecipitated with DNA. It is then kept undisturbed for 30 min so that proper precipitation can take place. It is then added to the cells to be transfected. The precipitated particles are taken in by the cell by phagocytosis. In the small portion of transfected cells, DNA becomes integrated into the cell's genome, which in response leads to a stable and permanent transfection (Roy et al. 2003).

1.2.2.3 The Polycation DMSO Technique

This technique is simple and produces a high number of stable transfectants. These stable transfectants are formed in the monolayer cultures, which have low quantity of DNA to be transferred. This involves basically two stages that are adsorption and internalization. The adsorption technique is done by polycation polymer, which will lead to the uniform and proper coating of targeted cells. After that, cells are permeabilized by exposing the dimethyl sulfoxide (DMSO) to take up DNA complexes. The main advantage is that it works completely and independently for each cell line (Aubin et al. 1994).

1.2.2.4 DEAE Dextran Procedure

It stands for diethylaminomethyl (DEAE) dextran. It is a polycationic derivative of carbohydrate dextran, which is a polymer. It is used to transfer DNA into cells of plants. The DEAE is a cationic molecule, which binds with the negatively charged backbone of the DNA results in the net positive charge of DNA and DEAE dextran molecule complex. This positive charge will let the molecule to attach to the cell membrane and thus by the process of endocytosis by the use of glycerol or DMSO (Vaheri and Pagano 1965). The main advantages are low cost and high reproducibility. The main disadvantage is that it cannot be able to produce stable cell lines (Gulick 2003). DEAE dextran treatment has been used for the isolation of vacuoles from mesophyll cells of *Bryophyllum daigremontianum*. The isolated vacuoles contained malic acid (Buser and Matile 1977).

1.2.3 DNA Imbibition by Seeds, Embryos, Cells, and Tissues

It has been reported that DNA can be inserted into the dry cells of plants, embryos, seeds, and tissues by the process of imbibition. The somatic embryos (vegetative analogues of the seeds) have apical and basal meristematic regions (bipolar structures). These two regions are similar to the zygotic embryo except the fact that they emerged from the vegetative cells of plant not from fertilized cells (Mckersie et al. 1989). In a study, the somatic embryos of alfalfa (*Medicago sativa*) were maintained at less than 15% moisture and were further dipped in medium containing plasmid DNA harboring the GUS gene (Senaratna et al. 1990; Senaratna et al. 1991). Transient GUS expression confirmed the integration of the transgene. However, DNA imbibition by intact plant cells has not been successful because the cell wall acts not only as a barrier but also as an efficient trap for the DNA.

1.3 Indirect DNA Transfer

1.3.1 *Agrobacterium*-Mediated Genetic Transformation

Agrobacterium tumefaciens (Gram-negative) is a soil phytopathogen belonging to the Rhizobiaceae family that produces crown gall in a wide range of dicotyledonous plants (Fig. 1.10a, b). Crown gall is a kind of plant tumor that appears as a mass of undifferentiated tissue at the junction of the root and the stem of afflicted plants (Smith and Townsend 1907). The bacteria transfer a little portion of its own plasmid

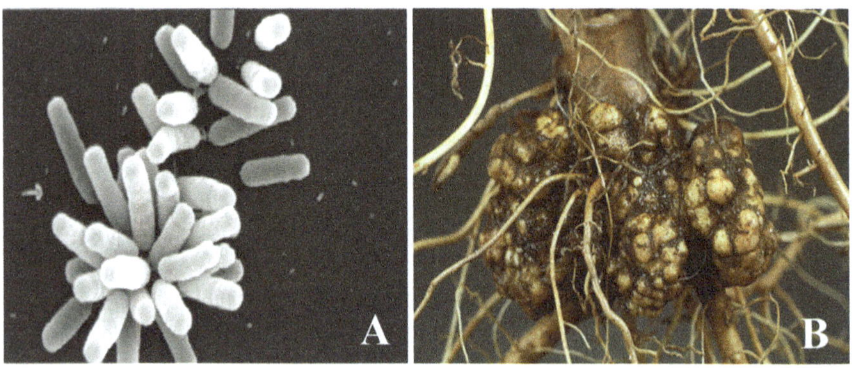

Fig. 1.10 *Agrobacterium tumefaciens*. (**a**) Electron micrograph of *A. tumefaciens*. (**b**) Plant root with crown galls

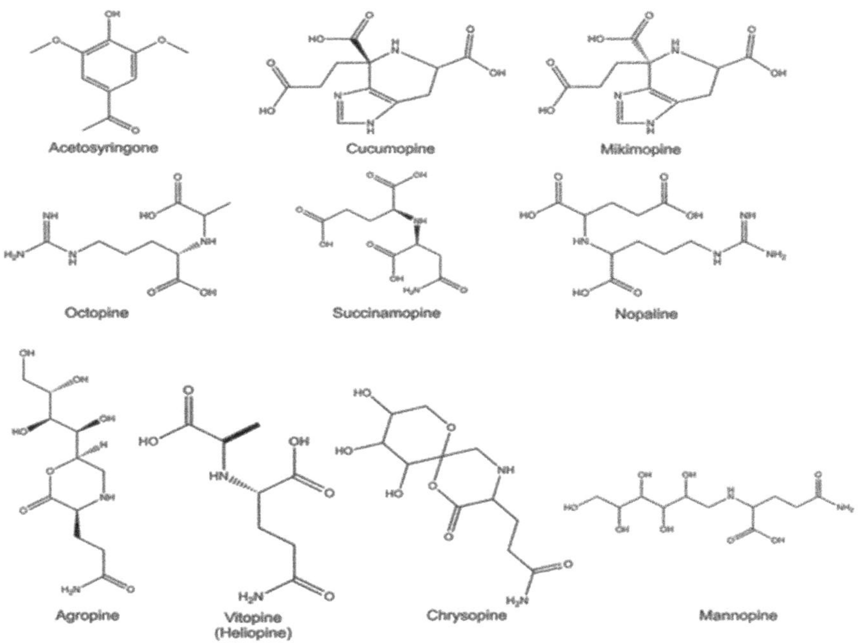

Fig. 1.11 Opines released by different strains of *Agrobacterium*

DNA termed T-DNA (transfer DNA) into the plant cell after infection at the wound site, resulting in two important processes:

(i) The plant cell starts to multiply and develop tumors, and it gains the capacity to grow in cultures with no growth regulator.
(ii) They start making opines (octopine, nopaline, and other arginine derivatives) that are not seen in normal tissues.

Bacteria are divided into octopine, nopaline, agropine, succinamopine, and chrysopine strains (octopine is the condensation result of arginine and pyruvic acid) (Fig. 1.11). Crown gall disease is characterized by the metabolism of opines. The bacterial strain, not the host plant, determines the kind of opine generated. In general, the bacteria cause the production of an opine, which it may catabolize and use as its only carbon and nitrogen energy source. Clearly, an intriguing interplay has developed, in which *A. tumefaciens* subverts the plant's metabolism to produce amino acids that can only be used by the bacterium as food and energy.

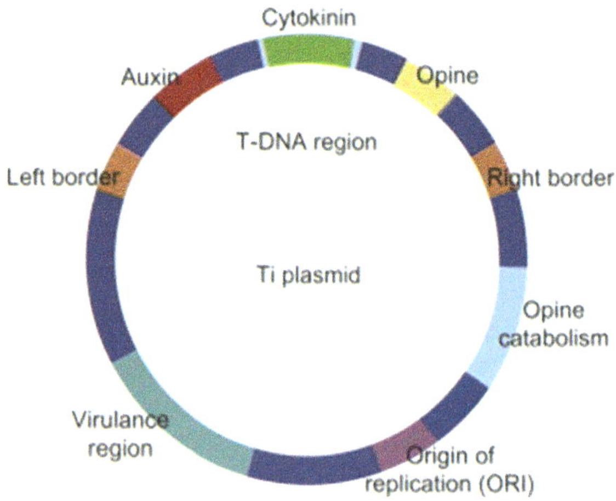

Fig. 1.12 Ti plasmid harboring the T-DNA. (Image source: www.science direct.com)

1.3.2 Ti Plasmid of *Agrobacterium*

Agrobacterium tumefaciens' ability to cause crown gall disease in plants is controlled by genetic material carried on a large conjugative plasmid (approximately 200 kb in size) known as Ti plasmid for its tumor-inducing capabilities (Fig. 1.12). When the bacteria are cured for the plasmid, it loses its virulence, and cured strains lose their ability to use octopine or nopaline. Ti plasmids have temperature-sensitive replication, which means that high temperatures (more than 30 °C) cause plasmids to cure. Replication (origin of replication), conjugal transfer, pathogenicity, and T-DNA are all present in Ti plasmids.

T-DNA transmission to plants requires three bacterial genetic components:

(i) T-DNA bordering sequences (25 bp direct repeats).
(ii) Ti plasmid encoding virulence (*vir*) genes in a location outside of the T-DNA.
(iii) Number of chromosomal genes, some of which are required for bacterium-plant cell adhesion.

1.3.3 Organization of T-DNA

During *Agrobacterium* infection, T-DNA (transfer DNA), which is a 23 kb section of Ti plasmid, is transported into the plant genome. T-DNA is characterized on both sides by 25 bp repeats termed border sequences, which are essential for T-DNA excision and transfer, and contains the gene for constitutive synthesis of auxins, cytokinins, and opines. The deletion of either border sequence prevents T-DNA from entering the plant cell. Mutational study, on the other hand, reveals that the right repeat is necessary for T-DNA transfer and that it functions in both a cis and polar manner. T-DNA has two borders: TL (left of T-DNA) and TR (right of T-DNA).

In nopaline plasmids, both TL and TR are always transferred together and incorporated into the plant genome as a single segment. The TL and TR are transported separately in octopine plasmids; thus, a single cell can carry one or both of these segments. T-DNA contains three genes that are involved in the production of crown galls. Two of these genes, *iaaM* and *iaaH*, encode tryptophan 2-monooxygenase and indoleacetamide hydrolase, respectively, which convert tryptophan to indole 3-acetic acid (IAA). The locus was previously known as the "shooty" locus, whose genes were termed as *tms1* (tumor with shoots) and *tms2*. The third gene, *ipt*, encodes isopentenyl transferase, a zeatin-type cytokinin; the locus was formerly known as the "rooty" locus and was renamed *tmr* (tumor having roots). Near the right border of T-DNA are genes involved in opine production. Eukaryotic regulatory sequences are found in every gene in T-DNA. As a result, these genes are only expressed in plant cells, and neither *Agrobacterium* nor *E. coli* express them.

1.3.4 Organization of *vir* Region

The Ti plasmid's *vir* region has eight operons that cover around 40 kb of DNA and contain 25 genes. This area is required for virulence and T-DNA transfer because it facilitates both cis and trans T-DNA transfer into the plant genome (Hooykaas and Beijersbergen 1994). Four of the eight *vir* operons, *vir*A, *vir*B, *vir*D, and *vir*G, are required for virulence, whereas the remaining four serve a supporting function in T-DNA transfer. The constitutively expressed *vir*A and *vir*G govern the expression of other *vir* loci. The activation of *vir*G by *vir*A, in response to the activation of *vir*A by plant phenolics such as acetosyringone and -hydroxy acetosyringone, initiates signal transduction. *vir*G dimerizes after activation and stimulates the transcription of additional *vir* genes (Zambryski et al. 1989). Table 1.1 lists the roles of several *vir* genes.

Table 1.1 Functions of different *vir* genes

Vir region	No. of genes	Function
*vir*A	1	Encodes a sensor protein; receptor for acetosyringone and functions as an autokinase; also phosphorylates *vir*G protein; constitutive expression
*vir*B	11	Membrane proteins; role in conjugal tube formation
*vir*C	2	Helicase activity
*vir*D	4	*vir*D1 has topoisomerase activity and *vir*D2 is an endonuclease
*vir*E	2	Single-strand binding protein (SSBP)
*vir*F	1	Not well understood
*vir*G	2	DNA-binding protein, induces the expression of all *vir* operon; constitutive expression
*vir*H	2	Not well known

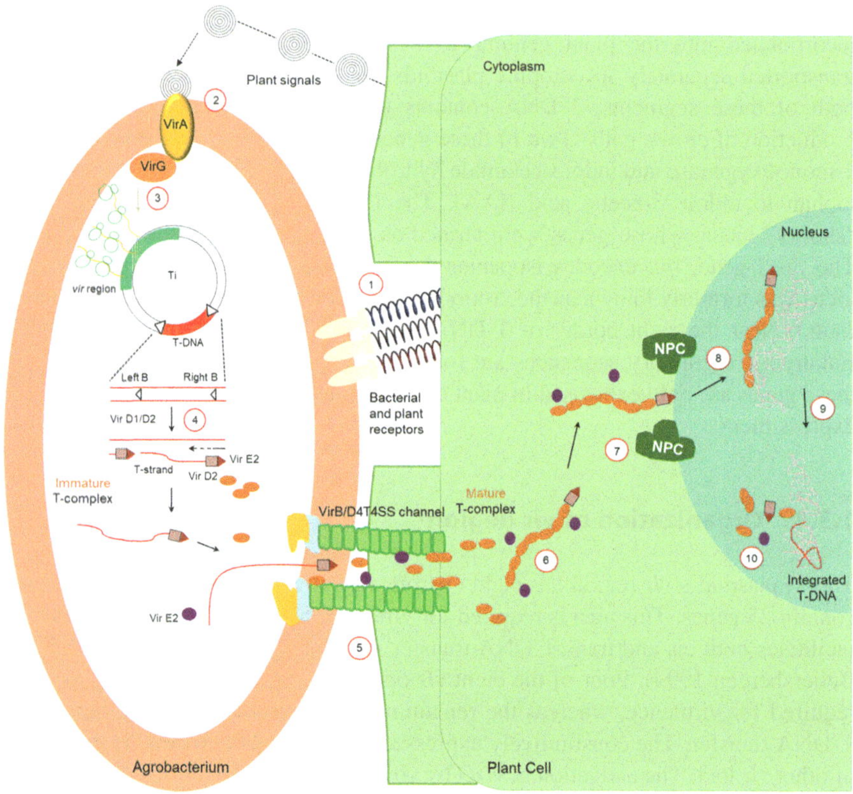

Fig. 1.13 Plant genetic transformation mediated by *Agrobacterium*. The identification and attachment of the *Agrobacterium* to the host cells are the first stage in the transformation process, which consists of ten key processes: (1) the virA/virG detects specific plant signals; (2) activation of *vir* gene region; (3) the virD1/D2 protein complex facilitates the generation of a mobile copy of the T-DNA; (4) transport of virD2–DNA complex (immature T-complex) into the host cell cytoplasm along with several other vir proteins; (5) virE2 attaches to the T-strand (mature T-complex), travels through the cytoplasm of the host cell (6), and is actively imported into the nucleus of the host cell; (7) once within the nucleus, T-DNA is attracted to the integration site; (8) escorting proteins are stripped off (9) and integrated into the host genome (10). (Redrawn from Tzfira and Citovsky 2006)

1.3.5 T-DNA Transfer Process

T-DNA transfer begins when bacteria are introduced into a plant wound (Fig. 1.13). Wounding is an essential step in the process, and it may be required in part for the plant to manufacture particular chemicals that activate the expression of the *vir* genes. Acetosyringone and -hydroxy acetosyringone are two of the most active compounds discovered. The T-DNA transfer process begins with the attachment of the virD1 gene product to the right border (RB) sequence; virD1 contains topoisomerase activity, which aids virD2, an endonuclease, in nicking at the right border and covalently attaching to the 5′ end. The 3′ end formed at the nick location acts as a primer for replacement DNA synthesis in the 5′-3′ direction, causing the T-strand to be displaced from the DNA duplex.

The *vir*E2 protein is a single-strand DNA-binding protein that attaches to single-stranded T-DNA in around 600 copies, shielding it from nuclease activity. The *vir*B operon produces membrane-bound proteins that aid in the construction of conjugal tubes between bacterial and plant cells, providing a conduit for T-DNA transfer, whereas *vir*B11 has ATPase activity, which provides the energy required for T-DNA transport into plant cells (Zambryski et al. 1989). T-DNA is driven towards the nucleus of the plant cell by nuclear localization signals on the *vir*D2 and *vir*E2 proteins. This process explains the relevance of the right border repeat in T-DNA transfer, as well as the polarity and cis-acting character of the border repeat sequences.

Apart from Ti plasmid, chromosomal virulence genes (*chv*) are also involved in T-DNA transfer from *Agrobacterium* to plants. The *chv* genes are required for the synthesis of cyclic glucans, which are involved in plant cell-binding *chv* A, *chv* B, and *psc* A that are involved in the synthesis and export of cyclic β-1,2-glucan. A more direct role in attachment has been demonstrated for rhicadhesin, a calcium-binding protein located on bacterial cell surface. The induction of *Agrobacterium vir* genes in response to plant wound-specific compounds implies that a bacterial recognition system must detect the plant signal and transmit the information inside the bacterial cells. This process is mediated by products of *vir*A and *vir*G.

1.3.6 Advantages of *Agrobacterium*-Mediated Genetic Transformation

Agrobacterium-facilitated transformation is the much-deployed method for gene transfer in explants such as cotyledonary leaves, vegetative leaves, hypocotyl, and stem explants (McCormick et al. 1986). It is a more natural manner of transferring DNA and is more acceptable to those who feel that natural is superior. It can infect plant cells, tissues, and organs that are otherwise healthy. As a result, tissue culture restrictions are much less of a worry. Plants that have been changed have a high rate of regrowth. It has the potential to transfer large amounts of DNA without creating major rearrangements. T-DNA integration is a relatively accurate approach for insertional mutagenesis, as it introduces one or more copies of the transgene into the genome at one or more sites. T-DNA integration is a precise method of introducing one or more copies of the transgene into the genome at one or a few loci, making it an excellent vehicle for insertional mutagenesis.

1.3.7 Disadvantages of *Agrobacterium*-Mediated Genetic Transformation

It has limitation of host range. Although much research has been done (development of highly virulent strains), some important food crops are still not amenable to *Agrobacterium*-mediated transformation.

Table 1.2 Selectable marker genes used in plant transformation regimes

Substrates used for selection	Selectable markers	Gene
Bromoxynil	Bromoxynil nitrilase	*bxn*
G418, kanamycin, neomycin, paromomycin	Neomycin phosphotransferase	*npt*II
Gentamycin	Gentamycin acetyl transferase	*aacC1*
Glyphosate	5-Enolpyruvyl shikimate 3 phosphate (EPSP) synthase	*aroA*
Hygromycin B	Hygromycin phosphotransferase	*hpt*
L-phosphinothricin (PPT)	Phosphinothricin acetyl transferase	*pat*
Methotrexate	Dihydrofolate reductase	*dhfr*
Streptomycin	Streptomycin phosphotransferase	*aph*E
Sulfonyl urea, imidazolinones	Acetolactate synthase mutant form	*als*

1.3.8 Selectable Markers

A critical aspect for genetic transformation is the selection of transformed cells. This is accomplished by the use of selectable marker genes contained in the vector with the gene of interest. Plant transformation schemes must include selectable markers. Table 1.2 lists the most commonly used selectable marker genes in plants. Each selectable marker has both favorable and unfavorable characteristics. As a result, the marker should be chosen depending on the plant species and other factors in the study. Resistance to the aminoglycoside antibiotics kanamycin, neomycin, and G418 is conferred by the *npt*II gene from transposon Tn5. The *npt*II gene product, neomycin phosphotransferase, phosphorylates these antibiotics and renders them inactive (Bevan et al. 1983). Because no endogenous level has been documented in green plants, this marker is the most extensively used approach for plant selection and screening.

1.4 Transformation Protocols

1.4.1 Genetic Transformation of *Arabidopsis*

The protocol for *Agrobacterium*-mediated genetic transformation of *Arabidopsis* (Fig. 1.14) has been redrawn from the reports of Zhang et al. (2006) and Clough and Bent (1998).

1.4.2 Genetic Transformation of Rice

The protocol for *Agrobacterium*-mediated transformation of rice (Fig. 1.15; Table 1.3) has been redrawn from the report of Nishimura et al. (2006) and Sahoo et al. (2011).

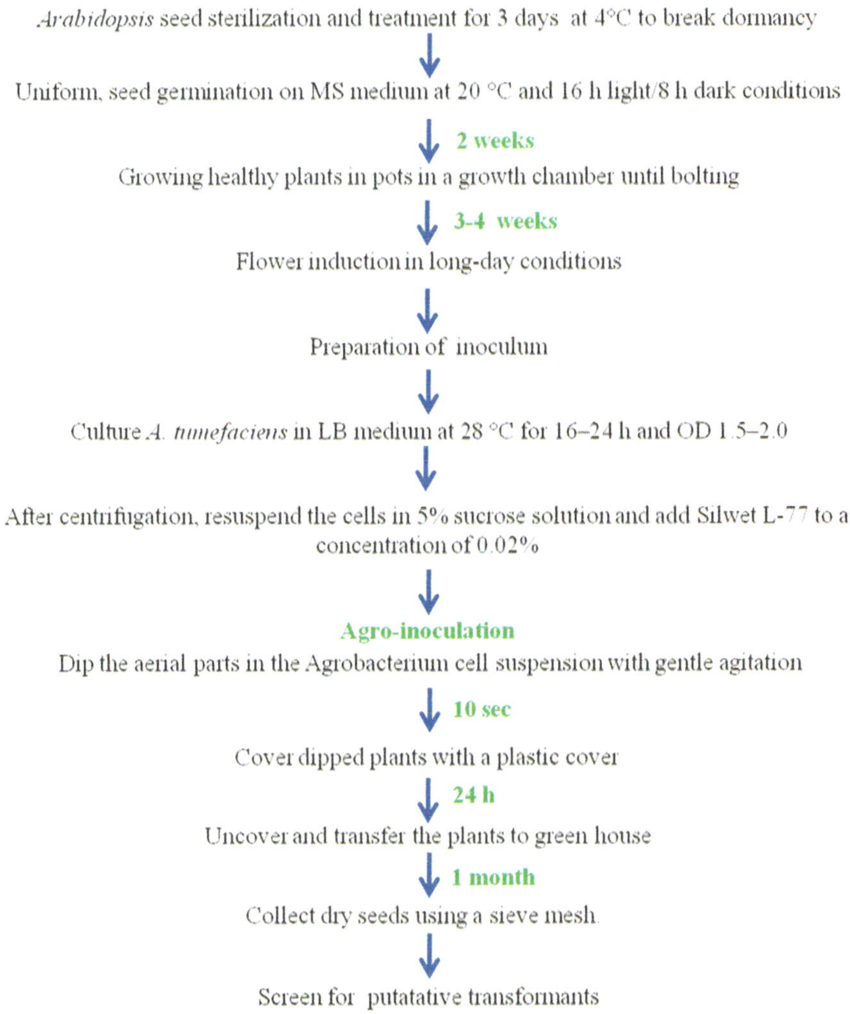

Fig. 1.14 *Agrobacterium*-mediated genetic transformation of *Arabidopsis*

1.4.3 Genetic Transformation of Chickpea

The protocol for chickpea seed sterilization and *Agrobacterium*-mediated genetic transformation of chickpea (Fig. 1.16a, b) has been redrawn from the reports of Yadav et al. (2015) and Bhowmik et al. (2019).

The procedure for *Agrobacterium*-mediated genetic transformation of chickpea is mentioned below:

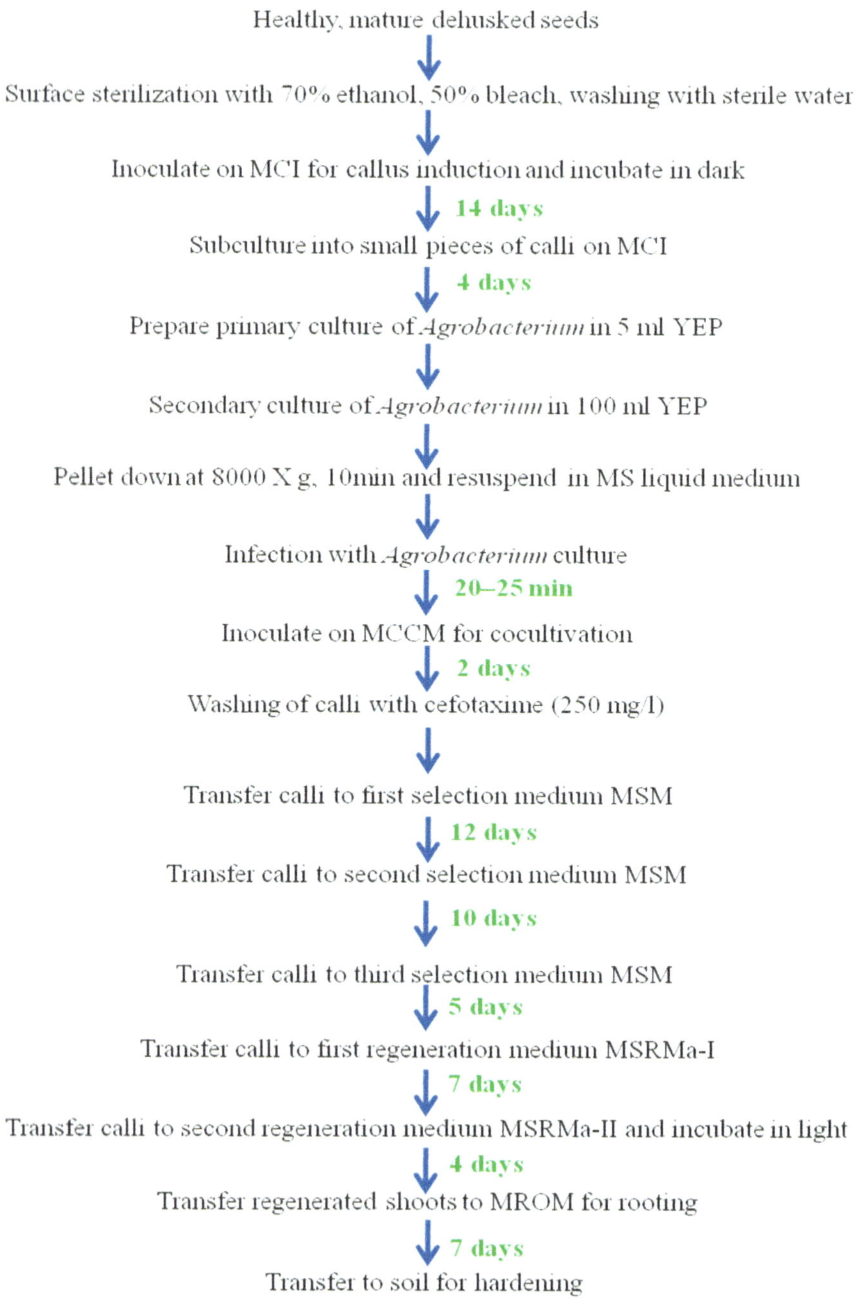

Fig. 1.15 *Agrobacterium*-mediated genetic transformation of rice

Table 1.3 Different media and their composition for rice transformation

Media	Composition
YEP medium	Bacto peptone (10 g/L) + yeast extract (10 g/L) + sodium chloride (5 g/L) + pH 7.0
MS resuspension medium	MS salts + sucrose (68 g/L) + glucose (36 g/L) + KCl (3 g/L) + $MgCl_2$ (4 g/L) + pH 5.2 + acetosyringone (150 μM, freshly prepared at a concentration of 1 M in 100% DMSO)
MCCM	MS salts containing maltose (30 g/L) + casein hydrolysate (0.3 g/L) + L-proline (0.6 g/L) + glucose (10 g/L) + 2,4-D (3 mg/L) + BAP (0.25 mg/L) + pH 5.2 + phytagel (3 g/L) + acetosyringone (150 μM)
MSM	MS salts containing maltose (30 g/L) + casein hydrolysate (0.3 g/L) + L-proline (0.6 g/L) + 2,4-D (3 mg/L) + BAP (0.25 mg/L) + pH 5.8 + phytagel (3 g/L) + cefotaxime (250 mg/L) + hygromycin (50 mg/L)
MSRMa-I	MS salts containing maltose (30 g/L) + kinetin (2 mg/L) + NAA (0.2 mg/L) + pH 5.8 + agarose (10 g/L) + cefotaxime (250 mg/L) + hygromycin (30 mg/L)
MSRMa-II	MS salts containing maltose (30 g/L) + kinetin (2 mg/L) + NAA (0.2 mg/L) + pH 5.8 + agarose (8 g/L) + cefotaxime (250 mg/L) + hygromycin (30 mg/L)
MSRMb	MS salts containing maltose (30 g/L) + BAP (2.7 mg/L) + kinetin (1.2 mg/L) + NAA (0.5 mg/L) + pH 5.8 + agarose (4 or8 or 10 g/L) + cefotaxime (250 mg/L) + hygromycin (30 mg/L)
MROM	1/2 MS salts + sucrose (30 g/L) + pH 5.8 + phytagel (3 g/L) + cefotaxime (250 mg/L) + hygromycin (30 mg/L)

1.4.3.1 Sterilization of Seeds

Healthy seeds were handpicked, fully washed with running tap water, and then transferred to sterilized conical flasks containing a 1% Tween-80 solution for 10 min, followed by 4–5 rinses with sterile Milli-Q water, and finally 0.1% $HgCl_2$ (w/v) solution in a laminar flow hood. For 5 min, the contents of the flask were briskly stirred. The $HgCl_2$ solution was decanted, and the seeds were carefully washed for 4–5 times with sterile Milli-Q water. The seeds were soaked in sterile Milli-Q water overnight.

1.4.3.2 Preparation of Explants

Cotyledonary Nodes (CN)

Surface-sterilized mature seeds were soaked in sterile distilled water supplemented with BAP (2 mg/L) overnight, and the seeds were allowed to germinate under culture room conditions. Following an incubation of 21 days, germinated seeds were collected and root, cotyledons, developed multiple shoots, and adjoining axillary and hypocotyl regions were precisely excised through transverse plane with removal of 1.5–2.0 mm thickness of layers containing epidermal and subepidermal cells and preexisting meristematic tissues. Twenty-one-day-old explants were processed, devoid of preexisting meristems, and were inoculated on PGR-supplemented medium for regeneration.

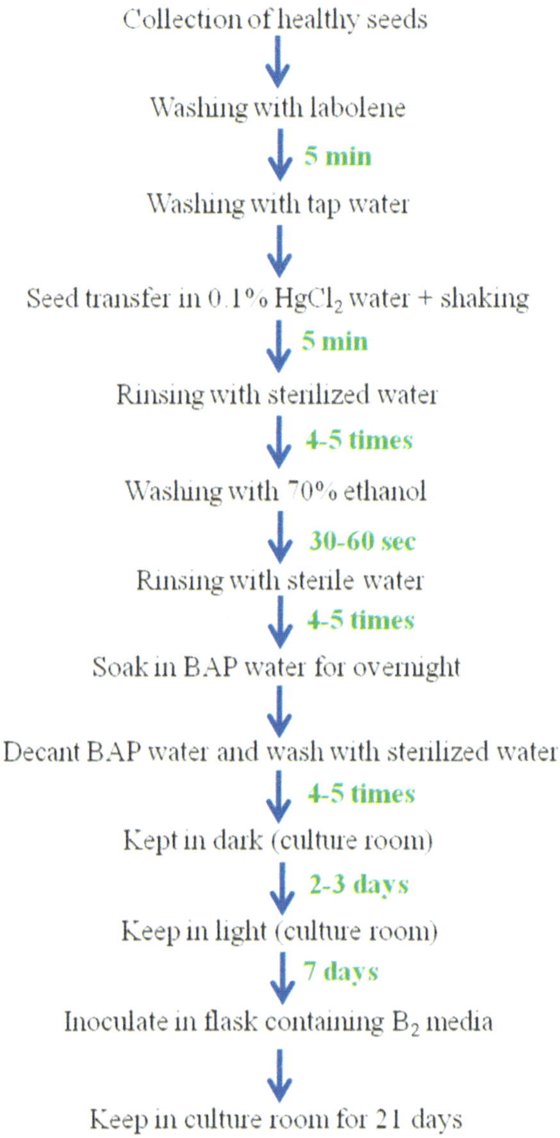

Fig. 1.16 (**a**) Procedure for chickpea seed sterilization. (**b**) *Agrobacterium*-mediated genetic transformation of chickpea using cotyledonary node explants

In YEB flask (25 ml) add appropriate antibiotics

↓

Inoculate single clone of *Agrobacterium* (appropriate construct)

↓

Kept overnight on shaker 28°C, 200 rpm

↓

Cut epicotyl and hypocotyl (~left 2-3 min) of explant

↓

Prepare cotyledonary node explant

↓

Pelletize *Agrobacterium* culture 5 min, 5000 rpm, 4°C

↓

Suspend the pellet with co-cultivation medium having acetosyringone O.D. = 0.7 - 0.8

↓

Remove primary existing meristem and cotyledonary shoots from explant

↓

Sonicate the cotyledonary node explant with co-cultivation medium

↓ **30–60 sec**

Resuspend the explant in fresh co-cultivation medium

↓

Vacuum treatment

↓ **5 min**

Mild agitation, 60 rpm, 24°C

↓ **20 min**

Decant co-cultivation suspension culture

↓

Dry the explant on sterilized blotting paper

↓

Transfer the dried cotyledonary node on B_1As plate

↓

Plate kept in dark at 18°C

↓ **2 days**

Transfer on B_1Cf plate in light

↓ **8 days**

Explants with buds sub-cultured in flasks having B_1CfK medium

↓ **10-12 days**

Shoots sub-cultured on fresh MK_{100} selection medium

↓ **10-12 days**

Sub-culture on ZAgC selection medium

↓ **10-12 days**

Shoots sub-cultured on basal medium

↓ **10-12 days**

Shoots ready for grafting

Fig. 1.16 (continued)

In Vitro Root Induction

Individually developed shoots with a height of 2.5–3.0 cm were transferred to root induction medium (RIM), which included 1/2 MS salts, and the rest of the ingredients were same. Medium was supplemented with B5 vitamins, sucrose 1.5% (w/v), IBA (0.5 mg/L), and agar 0.6% (w/v). Well-developed roots appeared after 2 weeks of culture under normal conditions.

Hardening and Acclimatization of In Vitro Plantlets

From the culture tubes, the rooted plantlets were removed, and to eliminate any residues of agar, the roots were rinsed under running water. Plantlets were transplanted into sterilized soilrite-filled paper pots and irrigated with half-strength liquid MS media devoid of sucrose. The pots were stored in a hardening chamber with a plexiglass canopy and a bottom plastic tray (BASCO, India). The hood features four sliding panes that may be adjusted to control relative humidity. The hardening chamber was kept in a culture room with a temperature of 24 ± 1 °C and a photoperiod of 16 h. Inside the chamber, an atomizer was used to generate mist. The plexiglass hood was removed after 2–3 weeks of hardening, and the plants were transplanted in the earthen pots filled with soil, sand, and farmyard manure in a 3:1:1 ratio and shifted to a glass house maintained at 24 ± 1 °C under natural light for flowering and seed setting.

Cocultivation of Explants with *Agrobacterium*

For genetic transformation of chickpea, several explants of chickpea were cocultivated with *Agrobacterium* strains containing chimeric plasmid constructions, and the precise processes are as follows:

1. *Agrobacterium* culture grown in 25 ml of YEB medium at 24–28 °C was harvested. The bacterial pellet was resuspended in liquid MS medium to obtain on OD_{600} nm between 0.8 and 1.0.
2. Various explants were excised, sonicated for 30–60s, and infected in bacterial suspension with constant shaking at 75 rpm at 24 °C for 20 min.
3. The explants were blotted-dried on sterile blotting sheets and incubated on semisolid MS medium supplemented with required PGRs. The Petri dishes were kept in culture room conditions and incubated in the dark for 2 days.
4. After cocultivation period (2–4 days), the explants were transferred to Petri dishes having MS medium supplemented with appropriate PGRs and cefotaxime (500 mg/L) and cultured 1 week under normal growth conditions.
5. The explants were thereafter subcultured on MS supplemented with 100 mg/L kanamycin, as well as sufficient doses of required PGRs and cefotaxime (250 mg/L) for 2 weeks.

6. The emerging shoots from the explants cultured on kanamycin-supplemented medium were excised and subcultured on MS medium supplemented with required PGRs and kanamycin (100 mg/L) for 2 weeks. The shoots screened on kanamycin selection medium were subcultured onto fresh antibiotic-supplemented medium for screening of the transformants. The shoots which survived three cycles of kanamycin selection were transferred to shoot elongation medium, prior to rooting. Stem nodes were dissected and incubated on slants, for clonal propagation, and the remaining portion of shoot was either transferred to root-inducing medium or grafted on seedling stocks for further development. The rooted plant stocks were subjected to hardening and acclimatization, before their transfer to glasshouse for growth and their molecular characterization.

1.4.4 Genetic Transformation of Tomato

The protocol for tomato seed sterilization and *Agrobacterium*-mediated genetic transformation of tomato (Fig. 1.17a, b) has been redrawn from the report of Koul et al. (2014, b).

The media components have been summarized in Table 1.4.

1.4.4.1 Procedure for *Agrobacterium*-Mediated Genetic Transformation of Tomato

1. **Tomato seed sterilization** ● TIMING **40 min for steps 1–3**
   In a sterile conical flask, place 10 g of tomato seeds and sterilize them under a laminar flow. Rinse the seeds 4–5 times with sterile distilled water after washing with a labolene neutral detergent solution.
   ▲ CRITICAL STEP Do not use excess detergent as it will increase the number of washing steps. For 3 g seeds, 1 ml 5% labolene is sufficient.
2. After that, soak the seeds in a 70% (v/v) ethanol solution for 90 s while shaking them constantly, and then rinse them 4–5 times with water.

   ▲ CRITICAL STEP Ethanol washing should not be longer than 90 s because it will reduce seed viability during the germination stage. Avoid bringing the flask's mouth too close to the flame when doing this step.
3. Apply a 4% (v/v) sodium hypochlorite solution to the seeds for 10–15 min. Place the conical flask on a rotatory shaker set to 150 revolutions per minute (3 g). Rinse the seeds in water 4–5 times.

   ▲ CRITICAL STEP If the seeds do not turn white within 10–15 min (maximum 20 min), discard the seeds.
4. **Seed germination and seedling inoculation** ● TIMING **8–9 days**
   Dry the seed on sterile blotting sheets after sterilization. In a culture room, incubate the seeds on semisolid MS medium (Murashige and Skoog 1962) containing B5 vitamins, 3% sucrose, and 0.8% agar at 24 ± 2 °C in the dark. Place the Petri plates at 20 ± 2 °C for 3 days with a 16-h photoperiod, a light

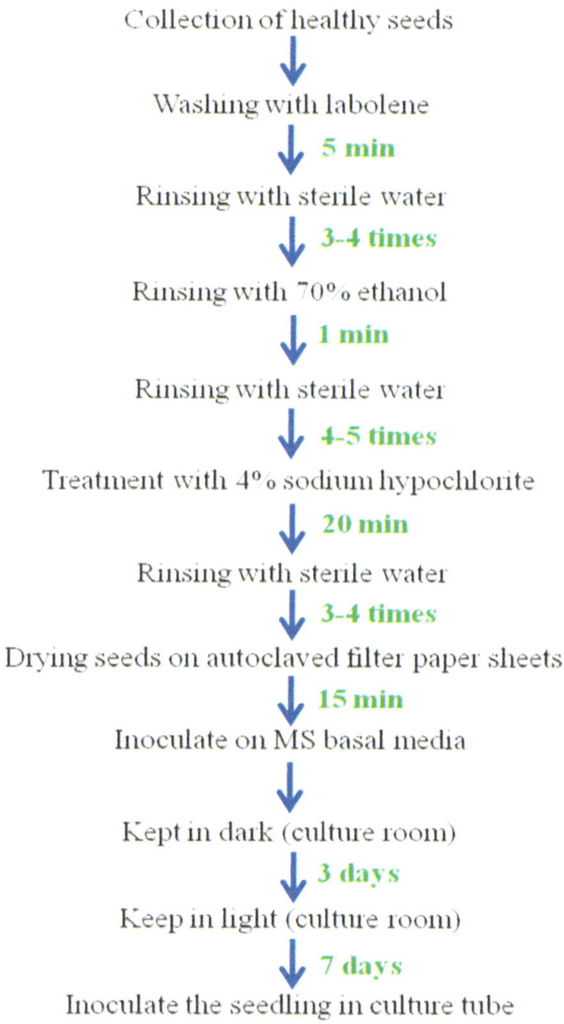

Fig. 1.17 (**a**) Procedure for tomato seed sterilization. (**b**) *Agrobacterium*-mediated genetic transformation of tomato using vegetative leaf disc explant

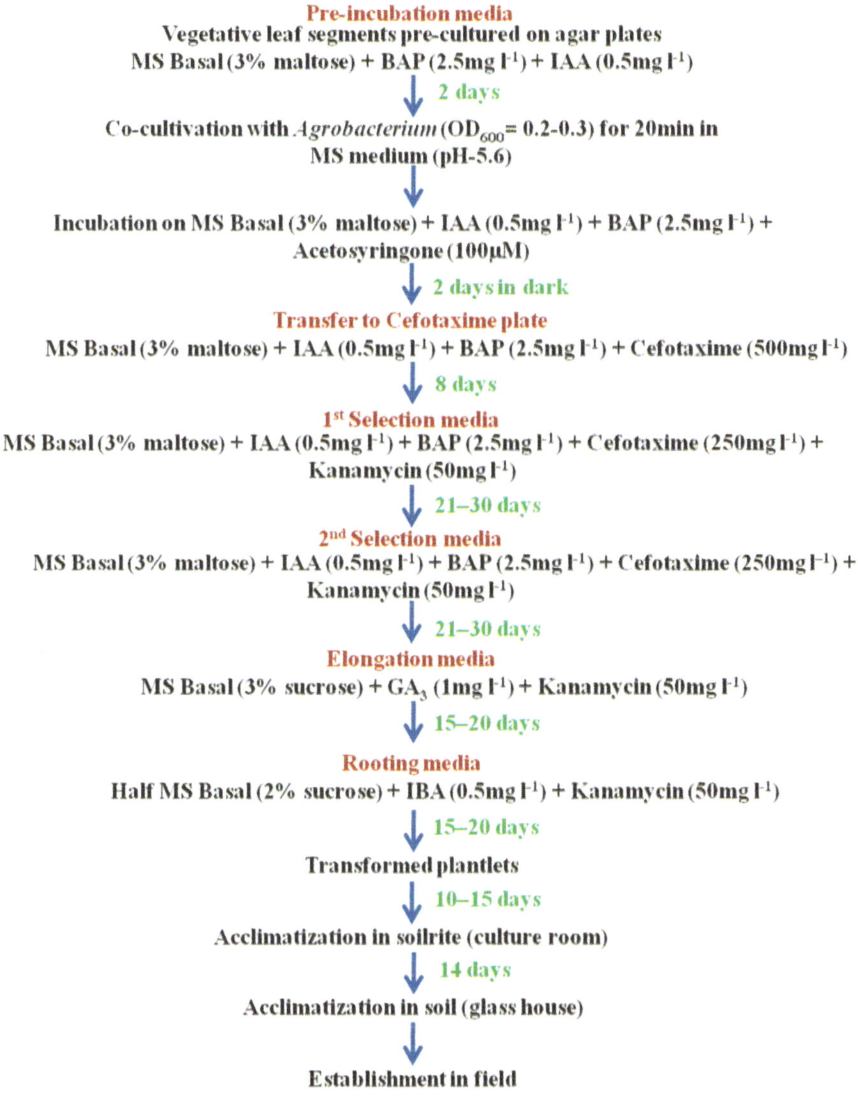

Fig. 1.17 (continued)

intensity of 100 molm^{-2} s^{-1}, and a relative humidity of 78 ± 4%. After 7–8 days, tomato seeds germinate. Inoculate the seedlings in MS basal media in culture tubes (pH 5.8).

▲ CRITICAL STEP If the blotting sheets are not properly autoclaved, fungal contamination will appear within 4 days, and it is advisable to discard the whole plate. If somehow localized contamination is removed from the plate, it is seen that it again reappears during culture tube stage.

Table 1.4 Different media and their composition for tomato transformation

Media	Composition
Preincubation media (IB)	MS + maltose (3%) + agar (0.8%) + BAP (2.5 mg/L) + IAA (0.5 mg/L)
Cocultivation medium (IBAs)	MS + maltose (3%) + agar (0.8%) + acetosyringone (100 μM) + BAP (2.5 mg/L) + IAA (0.5 mg/L)
Cefotaxime media (IBCf)	MS + maltose (3%) + cefotaxime (500 mg/L) + BAP (2.5 mg/L) + IAA (0.2 mg/L)
Shoot induction media (IBCfK) (SIM–1 and SIM–2)	MS + maltose (3%) + agar (0.8%) + Cf (250 mg/L) + BAP (2.5 mg/L) + IAA (0.5 mg/L) + kanamycin (50 mg/L)
Shoot elongation media (SEM)	MS + sucrose (3%) + agar (0.8%) + GA_3 (1.0 mg/L) + kanamycin (50 mg/L)
Root induction media (RIM)	½ MS + sucrose (2%) + IBA (0.5 mg/L) + agar (0.6%) + kanamycin (50 mg/L)

5. **Preincubation of vegetative leaves** • TIMING **2–3 days**
   After 16–18 days of seed inoculation, vegetative leaves are ready for preincubation. Use second and third leaves for the purpose. Preculture the excised tomato leaves on preincubation medium for 2–3 days.

   ▲ CRITICAL STEP Try not to damage the leaf lamina while placing it onto the medium. Do not press the leaves against the medium. Do not proceed to the next step if the leaves turn yellow during preincubation (this may happen if the handling of the explants or the culture room conditions is not appropriate).
6. **Preparation of bacterial inoculum** • TIMING **5 min for handling, 2 days for growth**
   Streak a loop of glycerol on YEB agar plates with suitable antibiotics (rifampicin (50 mg/L), kanamycin (50 mg/L), and streptomycin (50 mg/L)) and incubate for 2 days in the dark at 28 °C.
7. **Miniculture** • TIMING **5 min for handling, 16 h for growth**
   Pick a single colony with a sterilized toothpick, inoculate in 5 ml YEB in a screwcap tube (25 ml capacity), and incubate for 12–18 h in the dark at 28 °C on a rotatory shaker set to 200 rpm (overnight).
8. **Culture for transformation** • TIMING **5 min for handling, 16 h for growth**
   Inoculate 100–200 L of the overnight culture into a 25 ml YEB flask (150 ml capacity) and incubate for 12–18 h in the dark at 28 °C (overnight).

   ▲ CRITICAL STEP If bacterial flocks are seen in the medium, do not proceed further with the culture.
9. **Tomato transformation** • TIMING **5 min for handling**
   In a laminar flow, pour the bacterial culture into two sterile glass centrifuge tubes with a 15 ml capacity each (Corex, USA). Seal the mouths of the tubes with parafilm and centrifuge for 5 min at 4 °C at 5000 rpm (3024 g). Remove the supernatant and add 2 ml cocultivation media for agro-inoculation to the pellet.

To resuspend the bacterial pellet, lightly tap the tube. Set the O.D.600 of the cocultivation media to 0.20–0.22 ($\geq 2 \times 10^9$ cells ml^{-1}) using this culture.

10. **Explant preparation and inoculation/Cocultivation ● TIMING 15 min for preparation of explant and 15 min for inoculation**
For this, use a scalpel blade and forceps. Sterilize the blade and forceps by soaking them in 70% ethanol and then flame sterilizing them. Cut the leaves longitudinally and perpendicular to the midrib portion of the leaf disc explants and trim the periphery regions. As a result, leaf segments ranging in size from 2 to 3 mm are produced (Fig. 1.18a). With the help of forceps, suspend the leaf discs in cocultivation media for 15 min with mild agitation.
▲ CRITICAL STEP Dip the leaves in little cocultivation media while cutting, in a Petri plate, and then suspend leaf discs in cocultivation medium, in another plate, for 15 min, with (gentle) intermittent agitation. Make sure not to damage the explant while preparing it. Middle basal portion of the leaf is significantly better in producing normal elongated shoots and in terms of percent responding explants. For every experiment, a maximum of four plates of preincubated leaves are required. Three such experiments are enough to get 20 healthy transgenic plants.
11. **Transfer the explants to IBA media ● TIMING 2 days**
Place the explants on autoclaved blotting paper after inoculation and transfer to IBA plates (25 per plate) in a semidry state. Cocultivate for 2 days in the dark under culture room conditions. The media should be in touch with the abaxial surface.
▲ CRITICAL STEP One must be cautious while handling the explants at this stage as they are more prone to damage after agro-inoculation. Abaxial orientation of explant on culture medium induces better shoot regeneration than its adaxial counterpart.
12. **Transfer the explants to IBCf media ● TIMING 5–7 days**

▲ CRITICAL STEP Excess *Agrobacterium* attaches to the abaxial surface of the explant and contaminates the media; hence, cefotaxime (bacteriostatic drug) supplementation is used to get rid of it.
13. **First selection ● TIMING 21 days**
Subculture the explants onto SIM1.
14. Examine the explants for color changes after they have been incubated for some days. A resistant transgenic explant remains green, but a sensitive (non-transgenic) explant goes yellow.
▲ CRITICAL STEP Explants that are still green produce leafy and true shoots. To continue with the experiment, we should use real shoots.
15. **Second selection ● TIMING 21 days**
Cut all the shoots from each explant into little pieces for selection of transformants and regeneration. Culture on SIM2 media (same as SIM1, supplemented with zeatin 2.5 mg/L, preferable to BAP) (21 days).
Cutting small independent regenerated shoots with a pair of vegetative leaves and placing them on SIM2 for the same amount of time are another option.

Fig. 1.18 *Agrobacterium*-mediated genetic transformation of tomato. (**a**) Explant preparation from vegetative leaf. (**b**) Explants kept on cocultivation medium. (**c**) Non-transformed explants subjected to antibiotic selection. (**d–e**) Stages during selection and screening. (**g** and **h**) Occurrence of leafy shoots. (**i**) Regenerating leaf disc explants on SIM1. (**j**) Regeneration of shoots from non-chimeric explants on second selection medium (SIM2). (**k**) Explant bearing true shoot after second selection on SIM2. (**l**) Shoot elongation on SEM. (**m**) Rooting on RIM. (**n**) Hardening of in vitro-raised plantlets. (**o**) A glasshouse-grown acclimatized transgenic tomato plant bearing fruits (Koul et al. 2014, b)

16. **Transfer to cefotaxime media** ● TIMING **7 days**
Remove any additional callus tissue from the base of the shoots and keep them on Cf-supplemented medium to avoid contamination of the SEM (same as SIM2).
▲ CRITICAL STEP Cefotaxime supplementation is used in this case to remove contamination from the cut shoot's base.
17. **Transfer to shoot elongation media** ● TIMING **14 days**
Shoot elongation media (SEM) was used to subculture the independent shoots that regenerated after the second selection cycle in culture tubes (Fig. 1.18l).
▲ CRITICAL STEP The tolerance and sensitivity of tomato explants to kanamycin antibiotic are 50 mg/L, but at this stage it may vary from 50 to 70 mg/L.
18. **Transfer to root induction media** ● TIMING **14 days**
Individually developed shoots (2.5–3.0 cm height) should be transferred to RIM containing kanamycin (25–50 mg/L) in culture tubes (Fig. 1.18m).
19. **Hardening of in vitro-developed plantlets** ● TIMING **2–3 weeks**
The rooted plantlets should be removed from the culture tubes. To remove residues of agar, thoroughly wash the roots under running water. In pots containing sterilized soilrite, irrigate the plantlets with half-strength liquid MS media (without sucrose and vitamins). In the hardening chamber, keep the cups in a tray with a plexiglass hood (hardening chamber) at high humidity levels (90–95%) (Fig. 1.18n).
▲ CRITICAL STEP It is recommended that the roots be rinsed in cold water and that the plantlets be moved to 150 ml culture containers at room temperature. These safeguards are critical for the survival of delicate plantlets. The sliding windows on the plexiglass cowl may be changed to control the relative humidity. The hardening chamber is kept in the culture room at a temperature of 22 ± 1 °C and a photoperiod of 16 h. Inside the chamber, an atomizer is utilized to create mist.
20. **Transfer of plants to pots**
After 2–3 weeks of hardening, remove the plexiglass hood; transplant the plants to clay pots filled with a 3:1:1 ratio of soil, sand, and farmyard manure; and then transfer them to a glasshouse maintained at 24 ± 1 °C and optimum light for flowering, fruiting, and seed setting (Fig. 1.18o).
21. **Collection and storage of transgenic seeds**
Collect the seeds from T_0 transgenic plants. Dry the collected seeds and keep in separate bags in a desiccator.

1.4.4.2 Critical Steps During the Process of Transformation

Step 5

Try not to damage the leaf lamina while placing it onto the medium. Do not press the leaves against the medium. Do not proceed to the next step if the leaves turn yellow during preincubation (this may happen if the handling of the explants or the culture room conditions is not appropriate).

Step 10
Dip the leaves in little liquid cocultivation media while cutting, in a Petri plate, and then suspend leaf discs in cocultivation medium, in another plate, for 15–20 min, with (gentle) intermittent agitation. Make sure not to damage the explant while preparing it. Middle basal portion of the leaf is significantly better in producing normal elongated shoots and in terms of percent of responding explants. For each experiment, a minimum of 100 explants were used for agro-inoculation.

Step 11
One must be cautious while handling the explants at this stage as they are more prone to damage after agro-inoculation.

Table 1.5 shows a step-by-step solution to the basic issues related to *Agrobacterium*-mediated genetic transformation of tomato.

1.4.5 Genetic Transformation of Potato

The protocol for *Agrobacterium*-mediated genetic transformation of potato (Fig. 1.19; Table 1.6) has been redrawn from the reports of Duan et al. (2012) and Craze et al. (2018).

1.4.6 Genetic Transformation of Cotton

The protocol for seed sterilization and *Agrobacterium*-mediated genetic transformation of cotton (Fig. 1.20a, b) has been redrawn from the report of Kumar et al. (2009).

The different media used in cotton transformation are tabulated in Table 1.7.

1.4.7 Genetic Transformation of Stevia

The protocol for *Agrobacterium*-mediated transformation of stevia (Figs. 1.21 and 1.22) has been redrawn from the report of Taak et al. (2020, 2021).

1.4.8 Genetic Transformation of Sugar Beet

The protocol for *Agrobacterium*-mediated transformation of sugar beet (Fig. 1.23; Table 1.8) has been redrawn from the report of D'Halluin et al. (1992) and Norouzi et al. (2005).

Table 1.5 Troubleshooting guide to *Agrobacterium*-mediated genetic transformation of tomato

Procedure step	Problem	Possible reason	Solution
4	Bacterial/fungal contamination in plates kept for seed germination	Media was not properly sterilized Seeds were not properly sterilized Blotting sheets were not properly autoclaved Forceps not flame sterilized Laminar flow bench not clean	Presterilize glassware before media sterilization Autoclave the blotting sheets twice or thrice Give time to properly sterilize the forceps Laminar flow bench needs to be fumigated
16	Leaf discs turn yellow during IBCf stage	Incubation in the dark was prolonged Cefotaxime concentration must be very high Leaves from which leaf disc was prepared were too old	Strictly follow the protocol
13	All the explants stay green All the explants turn yellow	Kanamycin in the media must have degraded (selection not operating) Cocultivation event was not proper Kanamycin concentration was above the prescribed limit	Subject the explants to fresh selection media Compare the results with positive and negative controls Strictly follow the protocol
13–15	Excess callusing and leafy shoots but no true shoots	IAA and BAP imbalance Excess BAP used	Compare the results with the controls Use the hormone concentration as described in the protocol
10	Glassy shoots rather than normal shoots (hyperhydricity), in petri plates	High cytokinin concentration used Excess BAP used High relative humidity in the culture vessel (petri plates)	Strictly follow the cytokinin concentration as described in the protocol The use of less tightly fitting closure may sometimes solve the problem
9–10	Bacterial contamination persists at the base of each explant during SIM1 stage	Specified IBCf treatment was not followed O.D. of inoculation media must be inappropriate	Strictly follow the protocol

(continued)

Table 1.5 (continued)

Procedure step	Problem	Possible reason	Solution
17	Shoots do not respond to SEM	Forgot to add gibberellic acid in the media Bacterial contamination at the base of the shoot	Trim the base of the shoots and inoculate on fresh SEM If there is contamination at the base of the explant, trim the base and subject the explant to cefotaxime-supplemented media for at least 5 days and proceed again to SEM stage
18	Elongated shoots do not root properly	Forgot to add indole-3-butyric acid in the media Bacterial contamination at the base of the shoot	Prepare fresh IBA media Trim the base of the shoots and subculture them onto fresh RIM

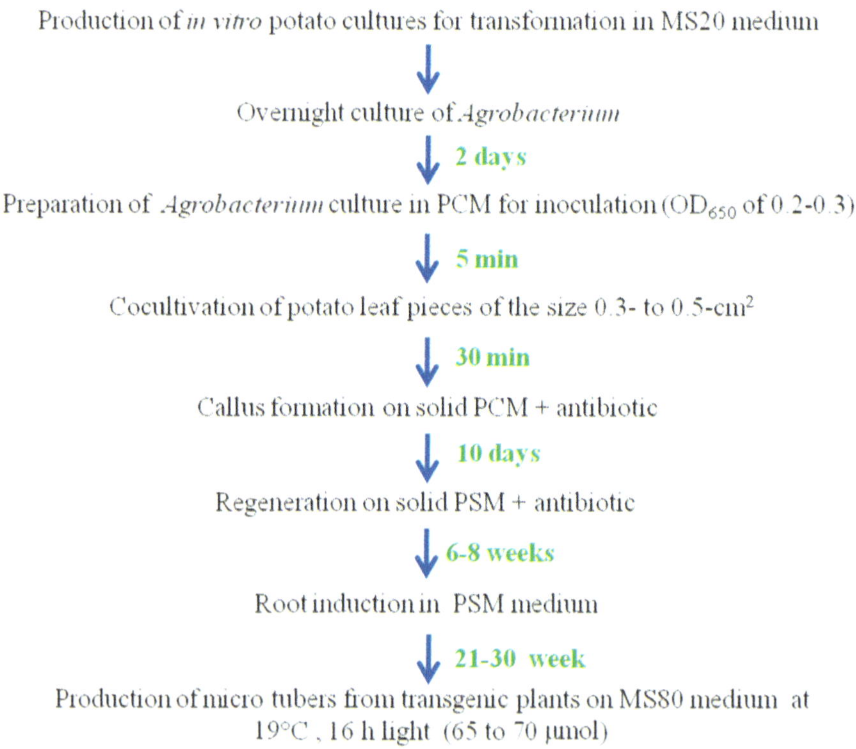

Fig. 1.19 *Agrobacterium*-mediated genetic transformation of potato using leaf explants

Table 1.6 Different media and their composition for potato transformation

Media	Composition
MS20	MS medium (4.71 g L^{-1}) + sucrose (20 g L^{-1}) + agar (7 g L^{-1}) + adjust to pH to 5.8 + autoclave (at 121 °C for 15 min) and dispense into Beatson jars (50 ml/jar)
MS80	MS medium (4.71 g L^{-1}) + sucrose (80 g/L) + Gelrite (2.5 g/L) + adjust to pH 5.7 + autoclave (at 121 °C for 15 min) and dispense into Beatson jars (50 ml/jar)
PCM (coculture medium)	MS medium (4.71 g/L) + sucrose (20 g/L) + MES (500 mg/L) + 2,4-D (2 mg/L) + sodium salt (from stock) + agar (5 g/L) + adjust to pH 5.7 + autoclave (at 121 °C for 15 min) + allow to cool to 60 °C and add zeatin (0.5 g/L) from stock solution + dispense into sterile petri dishes
PCM (liquid)	MS medium (4.71 g/L) + sucrose (20 g/L) + 2,4-D (2 mg/L) + zeatin (0.5 mg/L) + adjust to pH 5.8 + filter sterilize using a 0.22 μm filter
PCM ± 100 K	MS medium (4.71 g/L) + sucrose (20 g/L) + MES (500 mg/L) + 2,4-D (2 mg/L) + agar (5 g/L) + adjust to pH 5.7 + autoclave (at 121 °C for 15 min) + allow to cool to approx. 60 °C + add zeatin (0.5 mg/L) + timentin (150 mg/L) + dispense into sterile petri dishes
PSM 100 K	MS medium (4.71 g/L) + sucrose (20 g/L) + MES (500 mg/L) + agar (5 g/L) + adjust to pH 5.7 + autoclave (at 121 °C for 15 min) + allow to cool to approx. 60 °C + add zeatin (0.5 mg/L) + gibberellic acid (2 mg/L) + timentin (150 mg/L) + kanamycin (100 mg/L) + dispense into sterile petri dishes
YEP medium for *Agrobacterium* growth	Yeast extract (5 g/L) + bacto peptone (10 g/L) + sodium chloride (5 g/L) + Bacto agar (15 g/L) + adjust pH to 6.8 + autoclave at 121 °C for 15 min + store up to 6 months at room temperature

1.4.9 Genetic Transformation of Maize

The protocol for *Agrobacterium*-mediated transformation of maize (Fig. 1.24; Table 1.9) has been redrawn from the report of Ishida et al. (2007).

1.4.10 Genetic Transformation of Melon

The protocol for *Agrobacterium*-mediated transformation of melon (Fig. 1.25; Table 1.10) has been redrawn from the report of Choi et al. (2012).

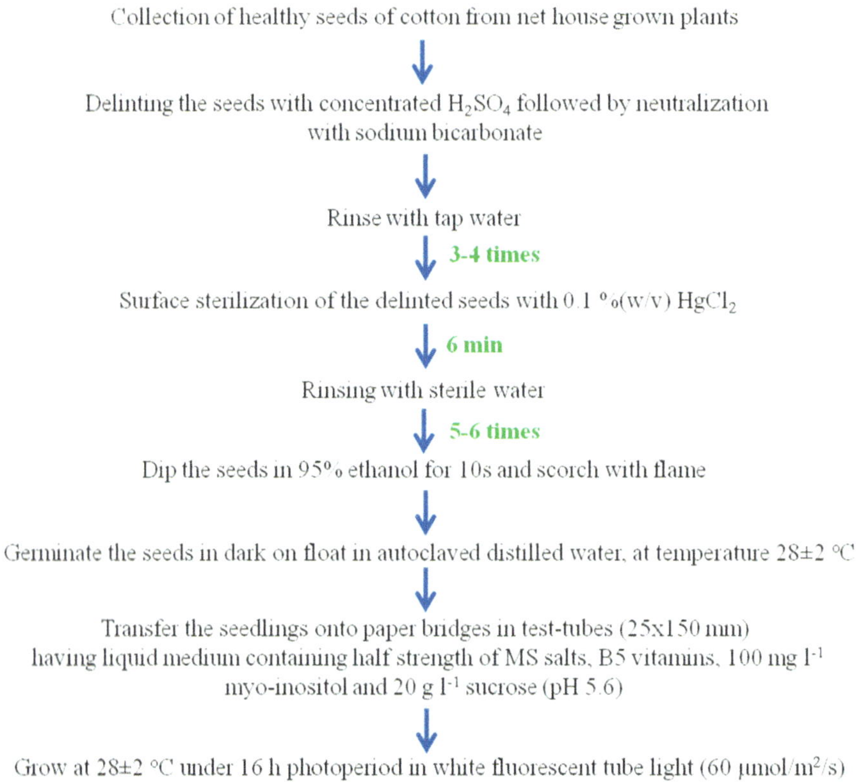

Fig. 1.20 (**a**) Procedure for cotton seed sterilization. (**b**) *Agrobacterium*-mediated genetic transformation of cotton using cotyledonary node explants

Streak *Agrobacterium tumefaciens* harbouring binary vector on YEB medium

↓

Pick a single isolated colony for primary culture in 5 ml YEP medium containing antibiotics (rifampcin 50 mgL^{-1}, streptomycin 200 mgL^{-1} and kanamycin 50 mgL^{-1})

↓

Inoculated the secondary culture in 50 ml of YEP medium at 28 °C, 200 rpm, in dark.

↓

Harvest the secondary culture at an O.D. of 1.4 at 6,000 rpm for 5 min

↓

Resuspended the pellet in 100 ml IM medium (induction; pH 6.0) for 4 h at 175 rpm and 26 °C

↓

Pellet and resuspended the *Agrobacterium* pellet in 100 ml MSO (pH 5.65) for 2 h at 150 rpm and 25°C

↓

Use 8 mm long hypocotyl segment and 1 cm^2 cotyledon pieces as explants (taken from 5-7 days old seedlings) for *Agrobacterium*-mediated transformation

↓

Agro-inoculation of the explants

↓ **15–20 min**

Blot dry the explants on sterile Whatman filter paper and transfer onto co-cultivation medium at 28°C in diffuse light

↓ **3 days**

Wash the explants with cefotaxime (250 mgL^{-1}) for 5 min followed by four rinses with autoclaved distilled water

↓ **15 days**

Blot-dry the explants and transfer to callus induction medium (pH 5.8)

↓ **3 weeks**

Sub-culture the explants in callus induction medium

↓

After 2-3 cycles on growth regulator medium transfer the actively growing callus onto hormone free medium (MS salts, B5 vitamins, myoinositol: 100 mgL^{-1}, glucose: 30 gL^{-1} and agar: 0.8%) containing hygromycin B:50 mgL^{-1} and augmentin: 250 mgL^{-1} for further callus formation

↓ **3 weeks**

Sub-culture the embryogenic calli on hormone free medium with hygromycin B: 50 mgL^{-1} and augmentin: 250 mgL^{-1}

↓

Maintain the globular, heart-shaped and torpedo-shaped embryos on hormone free medium

↓

Fig. 1.20 (continued)

Transfer the embryo longer than 5mm onto germination medium (½ strength MS salts, ½ strength B5 vitamins, myo-inositol 100 mgL^{-1}, sucrose 20 gL^{-1}, $MgCl_2$ 750 mgL^{-1} and phytagel 2.2 gL^{-1})

↓

Incubate the cultures in petri-plates sealed with Parafilm under 60 $\mu molm^{-2}s^{-1}$ light intensity and 16 h photoperiod

↓

Transfer the rooted embryos which produced shoot-apex to 7.5 x 7.5 x 10 cm (l x b x h) Planton boxes containing 100 ml of MSG medium

↓

Transfer the plantlets to pots for hardening

↓

Acclimatize the regenerated plantlets in plastic pots containing sterile potting mix [garden (sandy loam) soil, sand, vermiculite and peat moss in a 2:1:1:1 ratio]

↓

During the 14-18 days hardening period (at 28±2°C, 60 $\mu mol/m^2/s$ light intensity and 16 hours photoperiod) cover the plants with polythene bags to maintain high humidity and irrigated with sterile water

↓

For the next 2 days, raise the polythene bags partly, before removing them fully.

↓ 20 days

Transfer the plants to a glasshouse for further growth in earthen pots

↓ 4-6 weeks

Transfer the glasshouse grown plants to net-house to be planted in soil

Fig. 1.20 (continued)

Table 1.7 Different media and their composition for cotton transformation

Media	Composition
YEB medium	Beef extract (5.0 g/L) + yeast extract (1.0 g/L) + peptone (5.0 g/L) + $MgSO_4.7H_2O$ (0.05 g/L) + sucrose (5.0 g/L) + agar (15.0 g/L), pH 7.2
YEP medium	Yeast extract (10.0 g/L) + peptone (10.0 g/L) + NaCl (5.0 g/L) + pH 7.2
Induction medium	NH_4Cl (1.0 g/L) + $MgSO_4.7H_2O$ (0.33 g/L) + KCl (0.15 g/L) + $CaCl_2.2H_2O$ (0.01 g/L) + $FeSO_4.7H_2O$ (0.0025 g/L) + NaH_2PO_4 (0.24 g/L) + MES (0.39 g/L) + D-glucose (5.0 g/L) + pH 6.0
MSO medium	MS salts + B5 vitamins + MES (1.95 mg/L) + glucose (20 g/L) + pH 5.65
Cocultivation medium	MS salts + B5 vitamins + myoinositol (100 mg/L) + glucose (30 g/L) + agar (0.8%).
Callus induction medium	MS salts + B5 vitamins + myoinositol (100 mg/L) + 2,4 D (0.5 mg/L) + BAP (0.2 mg/L) + $MgCl_2$ (0.75 g/L) + phytagel (2.2 g/L)
Germination medium (MSG)	½ Strength MS + ½ strength B5 vitamins + myoinositol (100 mg/L) + sucrose (20 g/L) + $MgCl_2$ (750 mg/L) + phytagel (2.2 g/L)

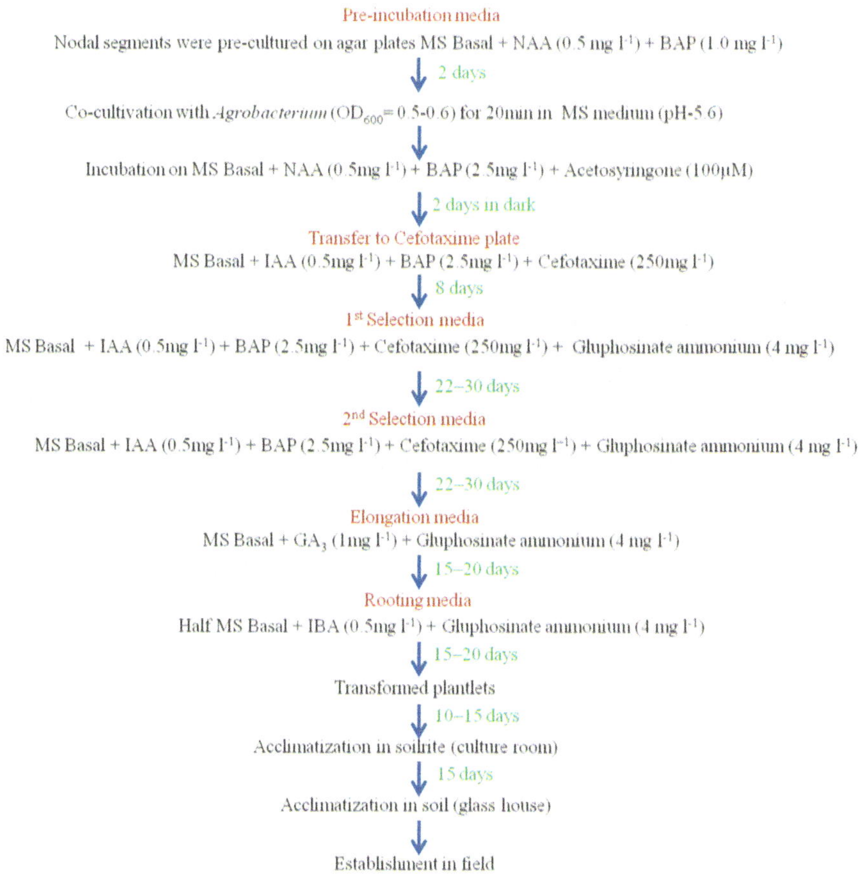

Fig. 1.21 *Agrobacterium*-mediated genetic transformation of stevia using nodal explants

1.4.11 Genetic Transformation of Poplar

The protocol for *Agrobacterium*-mediated transformation of poplar (Fig. 1.26; Table 1.11) has been redrawn from the report of Kumar et al. (2019).

1.4.12 Genetic Transformation of Sugarcane

The protocol for *Agrobacterium*-mediated transformation of sugarcane (Fig. 1.27; Table 1.12) has been redrawn from the report of Basso et al. (2017).

Fig. 1.22 Transformation and in vitro regeneration of stevia nodal explants (Taak et al. 2020). (**a**) Explant preparation for agro-inoculation. (**b**) Explants incubated on cocultivation media. (**c**) Selection. (**d**) Shoot elongation in SEM. (**e**) Rooting in RIM. (**f**) Hardening of in vitro-raised plantlets. (**g**) Greenhouse-grown acclimatized transgenic stevia plants

1.4.13 Genetic Transformation of Apple

The protocol for *Agrobacterium*-mediated genetic transformation of apple (Fig. 1.28; Table 1.13) has been redrawn from the report of De Bondt et al. (1996) and Zhang et al. (2006)).

1.4.14 Genetic Transformation of Flax

The protocol for *Agrobacterium*-mediated genetic transformation of flax (Fig. 1.29) has been redrawn from the report of Bastaki and Cullis (2014) and Mandal et al. (2018).

1.4.15 Genetic Transformation of Sweet Pepper

The protocol for *Agrobacterium*-mediated genetic transformation of sweet pepper (Fig. 1.30; Table 1.14) has been redrawn from the report of Zhu et al. (1996) and Kumar et al. (2012).

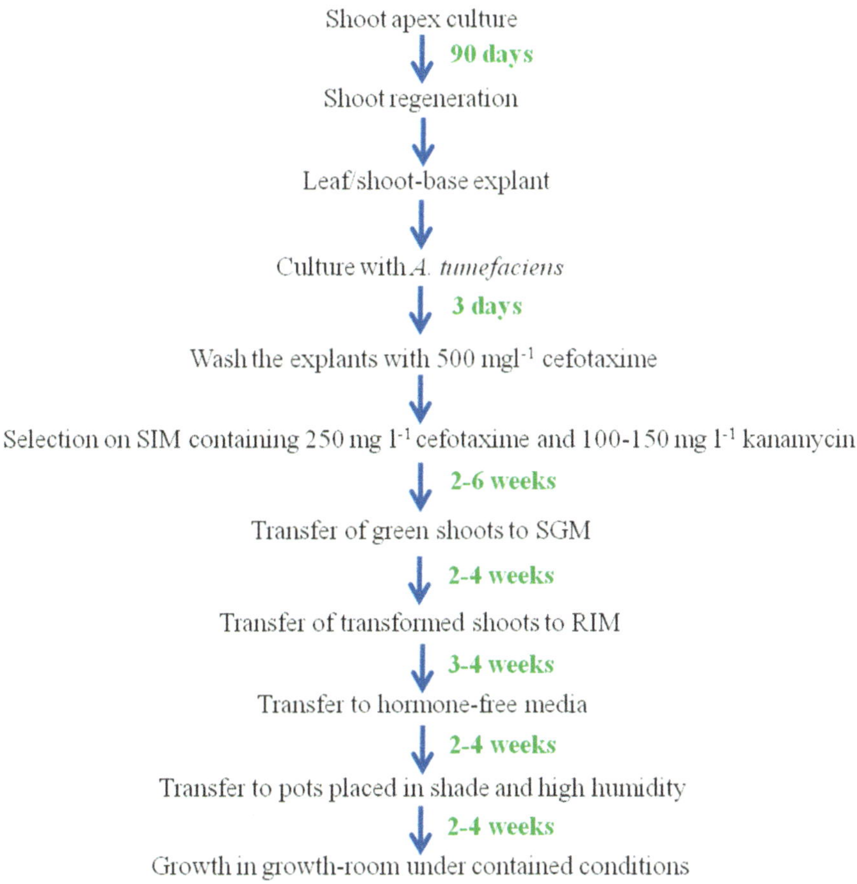

Fig. 1.23 *Agrobacterium*-mediated genetic transformation of sugar beet

Table 1.8 Different media and their composition for sugar beet transformation

Media	Composition
Half MSB medium	MS salts + B5 vitamins + sucrose (10 g/L) + agar (8 g/L)
PGoB medium	BA (8.9 μM) + NAA (2.7 μM) + TIBA (1 μM) + sucrose (30 g/L) + agar (8 g L^{-1})
SIM	PGoB medium + BA (4.4 μM) + NAA (5.4 μM) + sucrose (30 g/L), agar (8 g L^{-1}) + covered by 1 ml of tobacco nursing cells
SGM	MSB medium + sucrose (30 g/L), agar (8 g L^{-1}), and cefotaxime (250 mg L^{-1}), with or without kanamycin (50 mg L^{-1})
RIM	Half-strength MSB + sucrose (20 g/L) + NAA (16.1 μM) + agar (8 g L^{-1})

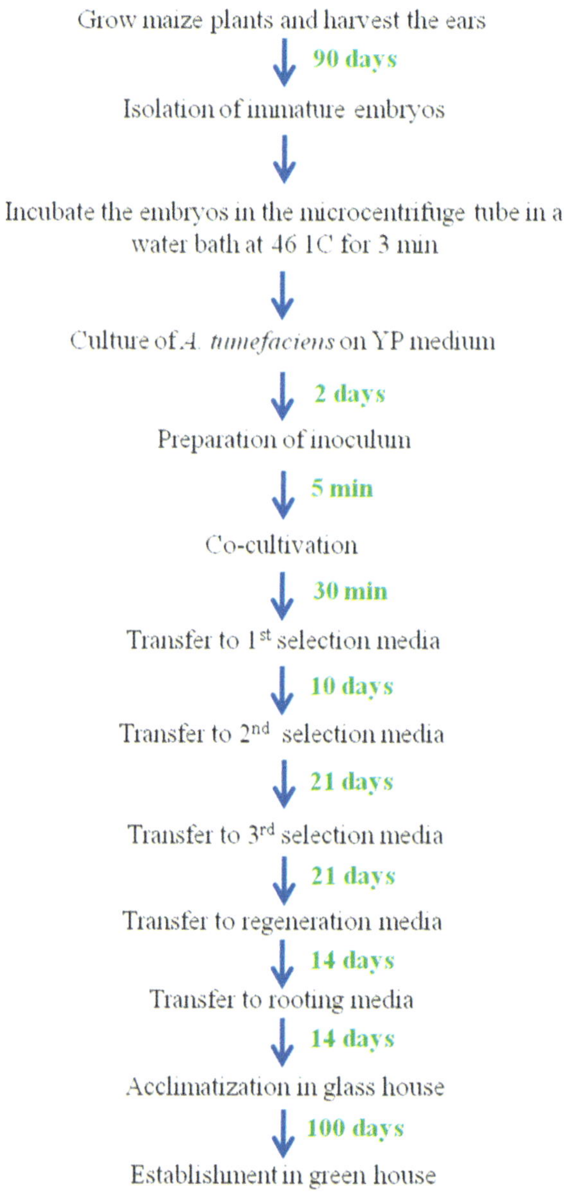

Fig. 1.24 *Agrobacterium*-mediated genetic transformation of maize

Table 1.9 Different media and their composition for maize transformation

Media	Composition
YP medium	Yeast extract (5 g) + peptone (10 g) + sodium chloride (5 g) + pH to 6.8 + agar (15 g)
LS-inf medium (for preparation of immature embryo)	Add 100 ml of 10 LS major salts + FeEDTA (100×) + LS minor salts + modified LS vitamins +2,4-D (1.5 mg L^{-1}) + sucrose (68.46 g) + glucose (36.04 g) + casamino acids (1.0 g) + pH to 5.2
LS-inf-AS medium (for infection)	Acetosyringone (100 mM) + LS-inf medium
LS-AS medium (for cocultivation)	LS major salts + FeEDTA (100×) + LS minor salts, modified LS vitamins +2,4-D (1.5 1.5 mg L^{-1}) + $CuSO_4$ (100 mM) + sucrose (20 g), glucose (10 g), proline (0.7 g) + MES (0.5 g) + pH to 5.8
LSD1.5A medium (for first selection of transformed cells)	LS major salts + FeEDTA + LS minor salts, modified LS vitamins +2,4-D (1.5 mg L^{-1}) + sucrose (20 g) + proline (0.7 g) + MES (0.5 g) + pH to 5.8 agar (8 g) + carbenicillin (250 mg L^{-1}) + cefotaxime (100 mg L^{-1}) + $AgNO_3$ (100 mM) + phosphinothricin (5 mg L^{-1}) + hygromycin (15 mg L^{-1})
LSD1.5B medium (for second and third selection of transformed cells)	LSD1.5A medium + phosphinothricin (10 mg L^{-1}) + hygromycin (30 mg L^{-1})
LSZ medium (for regeneration of shoots	LS major salts + FeEDTA + LS minor salts + modified LS vitamins + zeatin (5 mg L^{-1}) + CuSO4 (100 mM) + sucrose (20 g) + MES (0.5 g) + pH 5.8 + agar (8 g) + carbenicillin (250 g L^{-1}) + cefotaxime (100 mg L^{-1}) + phosphinothricin (5 mg L^{-1}) + hygromycin (30 mg L^{-1})
LSF medium (for rooting of transformed plants)	LS major salts + FeEDTA + LS minor salts + modified LS vitamins + IBA (0.2 mg L^{-1}) + sucrose (15 g) + MES (0.5 g) + pH to 5.8 + gellan gum (3 g) + autoclave

1.4.16 Genetic Transformation of Soybean

The protocol for *Agrobacterium*-mediated genetic transformation of soybean (Fig. 1.31; Table 1.15) has been redrawn from the report of Li et al. (2017) and Chen et al. (2018).

1.4.17 Genetic Transformation of Canola

The protocol for *Agrobacterium*-mediated genetic transformation of canola (Fig. 1.32; Table 1.16) has been redrawn from the report of Cardoza and Stewart (2004) and Zhang et al. (2005).

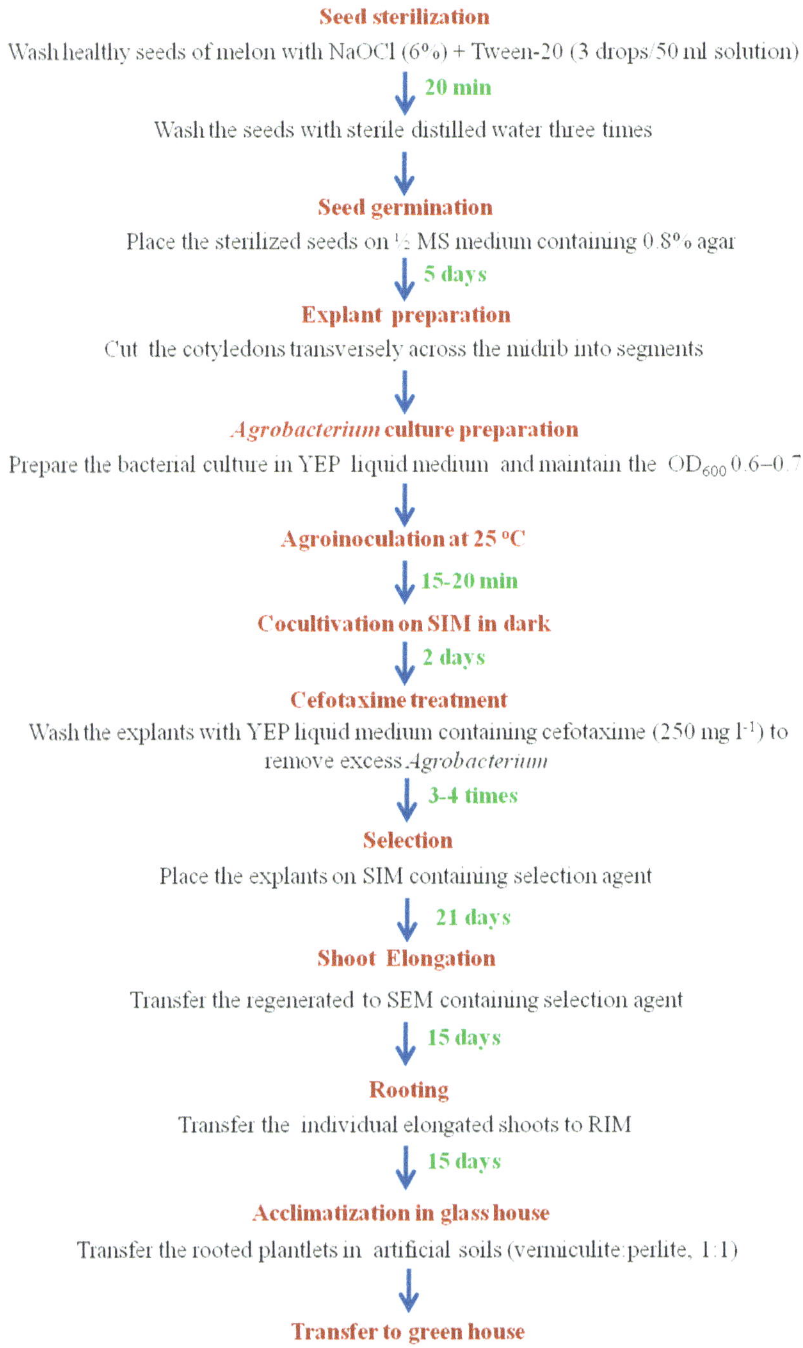

Fig. 1.25 *Agrobacterium*-mediated genetic transformation of melon

Table 1.10 Different media and their composition for melon transformation

Media	Composition
Shoot induction medium (SIM)	MS + sucrose (30 g L^{-1}) + agar (7 g L^{-1}) + BAP (1.0 mg L^{-1}) + IAA (0.01 mg L^{-1})
Shoot elongation medium (SEM)	MS + sucrose (30 g L^{-1}) + agar (7 g L^{-1}) + BAP (0.5 mg L^{-1})
Root induction medium (RIM)	MS + sucrose (30 g L^{-1}) + agar (7 g L^{-1})

Preparation of *Agrobacterium*-culture (OD: 0.3–0.8) one day before, from glycerol stock

↓

Harvest the leaves from 21-28 days old invitro grown sterile plants

↓

Dip the leaves in water before dividing into two halves

↓

Blot-dry on sterile filter paper

↓

Add 25 μL of 0.1 M acetosyringone to *Agrobacterium* suspension 30 min before co-cultivation

↓

Agro-inoculate the leaf-discs for 10 sec and shake at 100–200 rpm for 30 min in dark at 28°C

↓

Place the blot-dried leaf-disc on solidified WPM medium in sterile petri dishes.

↓

Co-cultivate in dark in culture-room at 24–25°C

↓ **2–3 days**

Wash the explants with sterile aqueous cefotaxime solution (500 mgl^{-1})

↓ **5 min**

Microcalli with shoots will appear

↓ **4-8 weeks**

Transfer the leaf-discs on solid regeneration medium and incubate in culture room at 23–25°C under appropriate light conditions

↓

Excise shoots and transfer to WPM medium devoid of PGRs but containing antibiotics and cefotaxime for rooting

↓

Growth in growth-room under contained conditions

Fig. 1.26 *Agrobacterium*-mediated leaf disc transformation of poplar

Table 1.11 Media used for poplar transformation

Media	Composition
Woody plant medium (WPM)	MS salts + sucrose (20 g/L) + agar (7 g/L)
Callus induction medium	Half-strength MS + sucrose (20 g/L) + NAA (16.1 μM) + agar (8 g L^{-1}) After autoclaving, add NAA (1.5 mg L^{-1}), BAP (0.25 mg L^{-1}), and TDZ (2.2 μg L^{-1})
Yeast extract broth (YEB) medium	Dissolve 5 g of meat extract, yeast extract (1 g), peptone (5 g), and sucrose (5 g) in 700 ml distilled water; stir gently until the medium gets completely dissolved; and fill up with distilled water to 1 L. Sterilize the medium by autoclaving at 121 °C for 20 min. Cool the autoclaved YEB medium to about 45 °C. Add 1 ml of 2 M $MgSO_4$. To prepare YEB plates, add bacterial agar to 1.5% before autoclaving

1.4.18 Genetic Transformation of Alfalfa

The protocol for *Agrobacterium*-mediated genetic transformation of alfalfa (Fig. 1.33; Table 1.17) has been redrawn from the report of Tohidfar et al. (2013) and Jiang et al. (2019).

1.4.19 Genetic Transformation of Squash

The protocol for *Agrobacterium*-mediated genetic transformation of squash (Fig. 1.34; Table 1.18) has been redrawn from the report of Shah et al. (2008).

1.4.20 Genetic Transformation of Eggplant

The protocol for seed sterilization and *Agrobacterium*-mediated genetic transformation of eggplant (Fig. 1.35; Table 1.19) has been redrawn from the report of Filippone and Lurquin (1989), Franklin and Sita (2003), Pratap et al. (2011), and Jadhav et al. (2015).

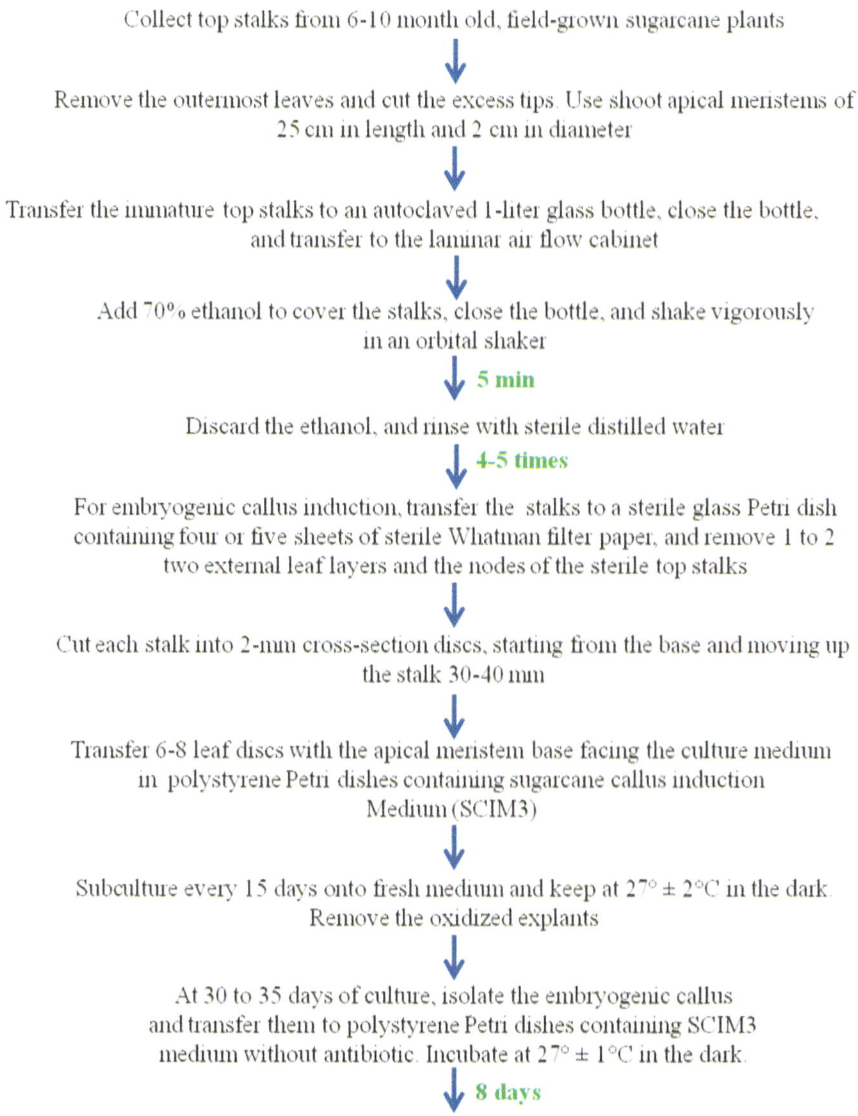

Fig. 1.27 *Agrobacterium*-mediated genetic transformation of sugarcane

At 2 to 3 h prior to transformation, select embryogenic callus, wash away excess culture medium, and remove oxidized tissues

↓

Fragment embryogenic callus into pieces of 3 mm in size, transfer to sterile polystyrene Petri dishes without cover, and leave in the laminar air flow cabinet for approximately 20 to 30 min to allow partial dehydration of the callus

↓

Streak the inoculum of *A. tumefaciens* transformed with the desired binary vector from a glycerol stock onto YEB plates containing the appropriate antibiotic

↓

Transfer a single colony to a sterile 50-ml conical tube containing 15 ml of YEB medium containing 100 µg/ml rifampicin and the appropriate concentration of antibiotic to select the binary vector (see recipe). Grow the culture to OD_{600} 0.5 at 28°C on an orbital shaker at 150 rpm

↓

Transfer 1 ml of the pre-inoculum to a sterile Erlenmeyer flask containing 100 ml YEB medium supplemented with selective antibiotics, and incubate with shaking at 28°C in the dark to OD_{600} 0.5

↓

Transfer the bacterial culture to sterile 50-ml conical tubes, centrifuge for 15 min at 3500 × g, 20°C, and discard the supernatant

↓

Suspend the pellet in *Agrobacterium* minimal medium containing the selective antibiotics and 100 µM acetosyringone to an OD_{600} of 0.4

↓

Transfer the culture to a sterile 200-ml Erlenmeyer flask and incubate the culture for 12 to 16 h at 22°C, with shaking at 150 rpm, in the dark

↓

Transfer the bacterial culture to a sterile 50-ml conical tube, centrifuge for 10 min at 3500 × g, 20°C

↓

Working in a laminar air flow cabinet, discard the supernatant and suspend the pellet in 25 ml of half-strength Murashige and Skoog basal liquid medium

↓

Pellet the bacteria by centrifugation for 10 min at 3500 × g, 20°C. Discard the supernatant and suspend the cell pellet to OD_{600} 0.4 in half-strength Murashige and Skoog basal liquid medium containing 100 µM acetosyringone

↓

Fig. 1.27 (continued)

Transfer the embryogenic callus to a sterile 50-ml conical tube, until the tube is a quarter full

↓

Add the Agrobacterium culture from step 21 to the embryonic callus

↓

For each 1 ml of bacterial culture, add 10 μl of 10% Poloxamer 188 solution and 1 μl of 100 mM acetosyringone stock solution

↓

Homogenize gently, and incubate in the dark for 5 min without shaking

↓

Discard the *Agrobacterium* culture, dry the excess plant tissue on sterile Whatman filter paper (in laminar air flow cabinet, blot dry the callus on sterile Whatman filter paper for 20 min), and transfer the agro-inoculated callus to polystyrene Petri dishes containing solid co-cultivation medium (SCCM)

↓

Incubate the Petri dishes for three days at 22°C in the dark

↓

Transfer the agro-inoculated callus to Petri dishes containing callus pre-selection medium (CPSM) with 300 mg/l ticarcillin, but without selective agent

↓

Incubate for seven days at 25°C in the dark

↓

Transfer agro-inoculated callus to Petri dishes containing callus regeneration medium containing 6-benzylaminopurine (CRM-BAP)

↓

Seal the Petri dishes with PVC film and transfer to a growth room set at 26° ± 1°C, under a 16/8-h light/dark photoperiod at 50 μmol $m^{-2} s^{-1}$ cool white light for three days, followed by 100 μmol $m^{-2} s^{-1}$ for a three days, and then 150 μmol $m^{-2} s^{-1}$ to the end

↓

Subculture after every 12 to 15 days in CRM-BAP, and keep the Petri dishes in a growth chamber at 26 ± 1°C under a 16/8-h light/dark photoperiod at 150 $m^{-2} s^{-1}$ light illumination

↓

Fig. 1.27 (continued)

Transfer the regenerated putative transgenic plants to Magenta vessel GA-7 containing sugarcane rooting medium (SRM)

↓

Keep the Magenta vessel in a growth chamber at 26°±1°C under a 16/8-h light/dark photoperiod at 150 m^{-2} s^{-1} light until the desired rooting for acclimatization

↓

Transfer the regenerated plants to soil/Plantmax substrate/vermiculite mixture (3:1:0.5 proportions), and acclimate in the growth chamber at 26° ± 2°C, under a 16/8-h light/dark photoperiod at 500 μmol m^{-2} s^{-1} and 65% relative humidity

↓ 8-12 days

Transfer the regenerated plants to the greenhouse

Fig. 1.27 (continued)

1.5 Conclusions

Transgenic plants were commonly used to study about various processes like pathogenesis, development processes, genomic organization, reception of light, and signal transduction (Alves et al. 1999). These days, transgenic crops exhibit 10% of the crops worldwide. It is the source of income for various countries. Genome manipulation technology is one of the promising areas, which is the source of real revolution in the field of biotechnology. The genetic engineering tools include the restriction enzymes that have the function to permit the cleavage in the DNA, while the ones which permit the joining of DNA are the ligases. The genome modification can be done by inserting, deleting, and replacing targeted DNA sequences (Shah et al. 2018). Over the past 10–15 years, plant transformation techniques have progressed tremendously (De Block 1993; Koul et al. 2014, b). Transgenic technologies have made crop improvement feasible and fast. Thus, it can produce plants with useful phenotypes and genotypes that are not able to be achieved alone by plant breeding techniques. Among the vector-mediated plant transformation techniques, the use of *Agrobacterium tumefaciens* as a DNA delivery vehicle has revolutionized the plant transformation regimes, as it ensures stable transformation/transgenesis.

It is also important to note that there are certain factors which regulate the success of explant transformation and regeneration (organogenesis) after transformation. These include the explant genotype; plant part used as an explant; age status of the mother plant; health status (biotic/abiotic stress or healthy plant) of the mother plant; size of the explant; endogenous hormone level of the explant; explant part/side (dorsal or ventral) in contact with the media; time of subculturing; media

Table 1.12 Different media and their composition for sugarcane transformation

Media	Composition
Agrobacterium (AB) minimal medium (1 L), for preconditioning bacteria	100 ml 10× phosphate stock solution +100 ml 10× saline stock solution +10 ml 20% (w/v) glucose, filter-sterilized +1 ml acetosyringone stock solution (100 mM) + prepare fresh
Callus preselection medium (CPSM) with or without selective agent (herbicide)	To prepare 1 liter, combine the following in distilled or milli-Q water: MS basal salts +1 ml each 1000× sugarcane vitamin stock solution +50 ml commercial coconut water (5% final) + hydrolyzed casein (500 mg) + L-cysteine (54.4 mg) + citric acid (150 mg) + myoinositol (100 mg) + 1 ml kinetin stock solution (1 mg/ml) + 1 ml 1000× copper sulfate stock solution +3 ml 2,4-dichlorophenoxyacetic acid stock solution (1 mg/ml) + adjust pH to 5.8 using 1 M KOH and then add phytagel (2.5 g) + autoclave at 121 °C for 20 min + cool to 60 °C, and then add 1 ml 1000× silver nitrate stock solution +100 ml maltose (2%)/ glucose stock solution (1%) + 1 ml ticarcillin stock solution (300 mg/ml) + 3 ml glufosinate-ammonium stock solution (1 mg/ml)
Callus regeneration medium containing 6-benzylaminopurine (CRM BAP)	MS basal salts +1 ml 1000× sugarcane vitamin stock solution +50 ml commercial coconut water (5% final) + L-cysteine (54.4 mg) + citric acid (150 mg) + myoinositol (100 mg) + 1 ml 6-benzylaminopurine (BAP) stock solution (1 mg/ ml) + adjust pH to 5.8 using 1 M KOH and add phytagel (2.5 g) + autoclave at 121 °C for 20 min + cool to 60 °C and then add 1 ml 1000× silver nitrate stock solution +100 ml maltose (2%)/ glucose stock solution (1%) + 1 ml ticarcillin stock solution (300 mg/ml) + 3 ml glufosinate-ammonium stock solution (1 mg/ml)
Solid cocultivation medium (SCCM)	MS basal salts +1 ml 1000× sugarcane vitamin stock solution +50 ml commercial coconut water (5% final) + hydrolyzed casein (500 mg) + L-cysteine (54.4 mg) + citric acid (150 mg) + myoinositol (100 mg) + 1 ml kinetin stock solution (1 mg/ml) + 3 ml 2,4-dichlorophenoxyacetic acid stock solution (1 mg/ml) + adjust pH to 5.8 using 1 M KOH + phytagel (2.5 g) + autoclave at 121 °C for 20 min + cool to 60 °C, and then add 1 ml 1000× silver nitrate stock solution +100 ml maltose (2%)/ glucose stock solution (1%) + 1 ml acetosyringone stock solution (100 mM)
Sugarcane callus induction medium (SCIM3)	MS basal salts +1 ml 1000× sugarcane vitamin stock solution + sucrose (27.25 g, 2.72% final) + commercial coconut water (50 ml, 5% final) + hydrolyzed casein (500 mg) + L-cysteine

(continued)

Table 1.12 (continued)

Media	Composition
	(54.4 mg) + citric acid (150 mg) + myoinositol (100 mg) + 1 ml 1000× copper sulfate stock solution +3 ml 2,4-dichlorophenoxyacetic acid stock solution (1 mg/ml) + adjust pH to 5.8 using 1 M KOH solution, and then add phytagel (2.5 g) + autoclave at 121 °C for 20 min + cool to approximately 60 °C, and then add 1 ml 1000× silver nitrate stock solution +1 ml cefotaxime stock solution (250 mg/ml)
Sugarcane rooting medium (SRM)	MS basal salt powder mixture +1 ml 1000× sugarcane vitamin stock solution + sucrose (30 g, 3% final) + myoinositol (100 mg) + isoleucine (132 mg) + 1 ml indole-3-butyric acid stock solution (1 mg /ml) + adjust pH to 5.8 using 1 M KOH, and then add phytagel (2.5 g) + autoclave at 121 °C for 20 min + cool to 60 °C, and then add 1 ml ticarcillin stock solution (300 mg/ml) + 3 ml glufosinate-ammonium stock solution + pour 50 ml per Magenta vessel GA-7
Sugarcane vitamin stock solution (1000×)	Thiamine (1 g/L) + pyridoxine (500 mg/L) + nicotinic acid (500 mg/L) + glycine (2 g/L) + L-arginine (50 g/L) + citric acid (150 mg/L)
Half-strength Murashige and Skoog basal liquid medium	MS basal salts +1 ml 1000× sugarcane vitamin stock solution + myoinositol (100 mg) + adjust pH to 5.8 using 1 M KOH, autoclave at 121 °C for 20 min and then add 100 ml maltose (2%)/glucose stock solution (1%) + add 1 ml acetosyringone stock solution (100 mM) immediately before use

composition; aseptic conditions during culture of explant; light, humidity, and temperature during cocultivation/culture; and type of plant transformation strategy (if *Agrobacterium*-mediated transformation, then *Agrobacterium* cell density, agro-inoculation duration, cocultivation duration, acetosyringone concentration, etc.) (Koul et al. 2014, b; Taak et al. 2020).

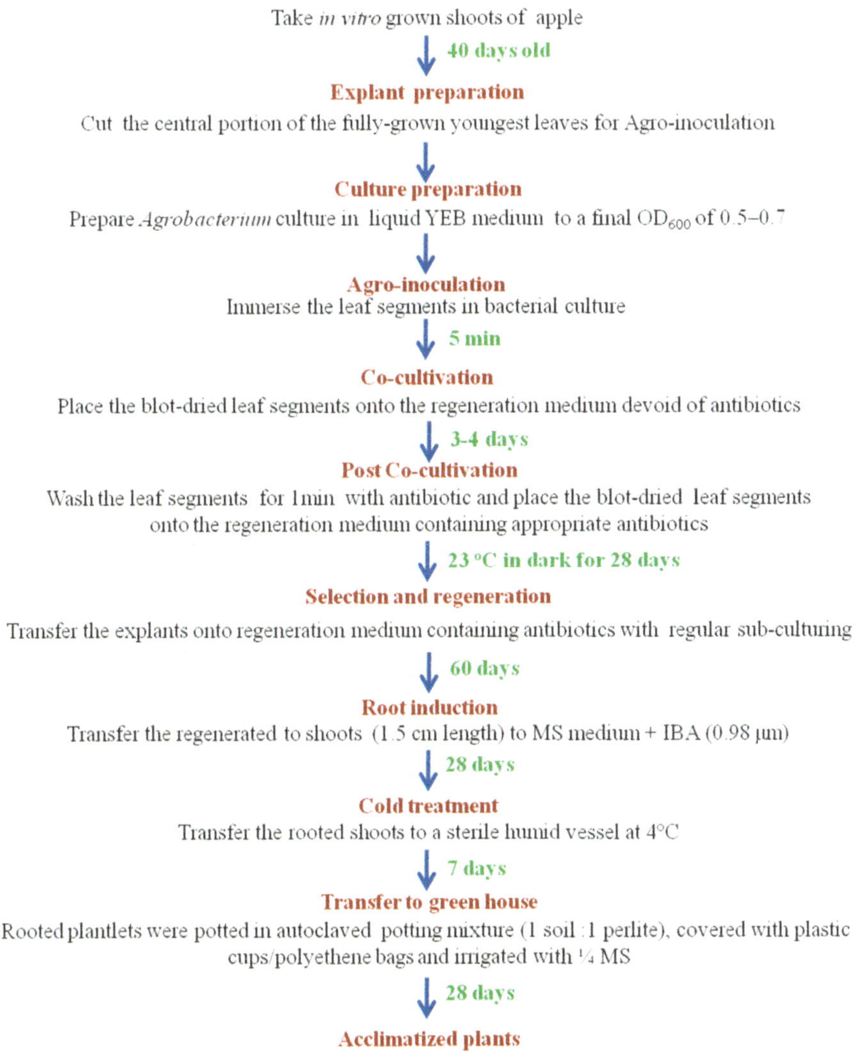

Fig. 1.28 *Agrobacterium*-mediated leaf disc transformation of apple

Table 1.13 Different media and their composition for apple transformation

Media	Composition
Shoot regeneration medium	MS + sucrose (30 g/L) + agar (7 g/L) + BAP (17.8 µM) + NAA (1.1 µM)
Root induction medium (RIM)	MS + sucrose (30 g/L) + agar (7 g/L) + IBA (0.98 µM)

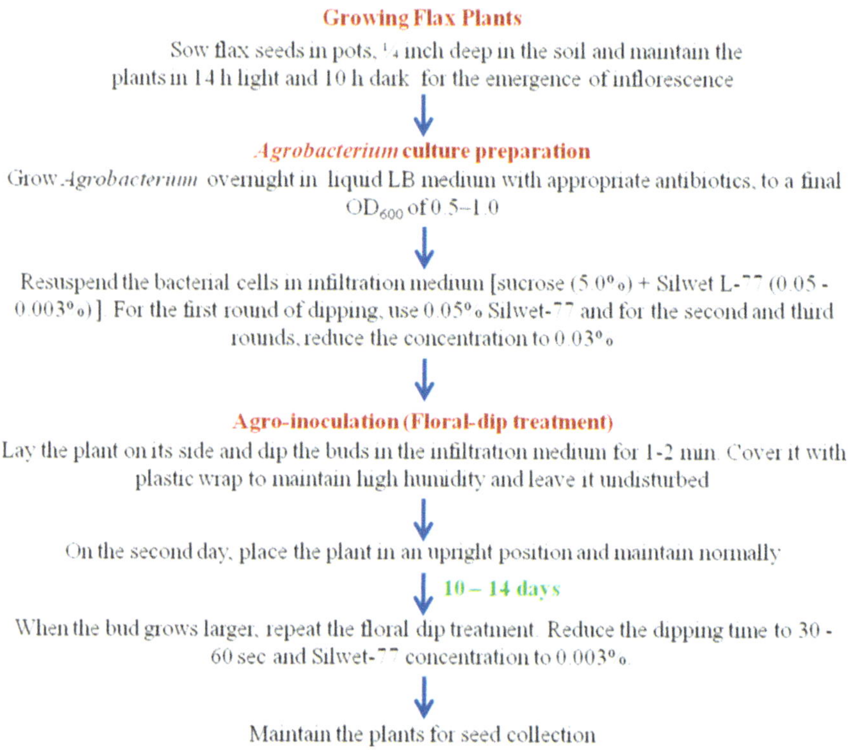

Fig. 1.29 *Agrobacterium*-mediated floral dip transformation of flax

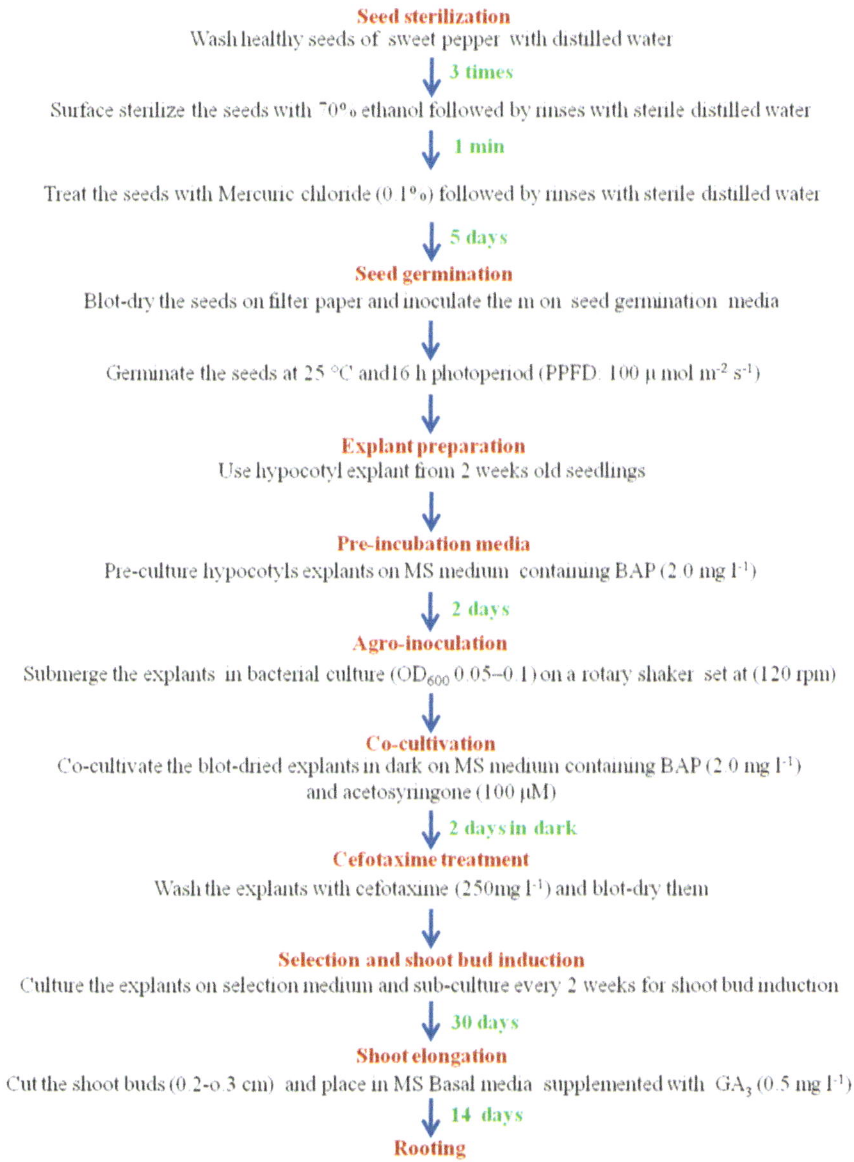

Fig. 1.30 *Agrobacterium*-mediated floral dip transformation of sweet pepper

Table 1.14 Different media and their composition for sweet pepper transformation

Media	Composition
Seed germination medium	MS + sucrose (30 g/L) + phytagel (2.5 g/L)
Preculture medium	MS + sucrose (30 g/L) + BAP (2 mg/L) + phytagel (2.5 g/L)
Cocultivation medium	MS + sucrose (30 g/L) + BAP (2 mg L^{-1}) + acetosyringone (100 μM) + phytagel (2.5 g/L)
Shoot regeneration medium	MS + sucrose (30 g/L) + BAP (4 mg/L) + phytagel (2.5 g/L)
Shoot elongation medium	MS + sucrose (30 g/L) + GA3 (0.5 mg/L) + phytagel (2.5 g/L)
Root induction medium (RIM)	MS + sucrose (30 g/L) + NAA (0.5 mg/L) + phytagel (2.5 g/L)

Seed sterilization

Wipe the healthy seeds of soybean with cloth dipped in ethanol (75%)

↓

Surface sterilize the seeds with NaClO (100 ml) + HCl (10 ml) followed by rinses with sterile distilled water

↓ **6 hours**

Explant preparation

Soak the sterilized seeds in sterile water to get the half-seed cotyledonary explants

↓ **16–18 hours; 23 °C in dark**

Cut the imbibed seeds longitudinally along the hilum using a scalpel blade and remove the seed coats

↓

***Agrobacterium* culture preparation**

Add 200ul of the starter culture to 200 ml YEB culture and grown at 28°C (250 rpm) in an incubator shaker until 0.6 OD_{650}

↓

Centrifuge the *Agrobacterium* culture at 5000 rpm (22°C) for 10 min, and resuspend the pellet in liquid CCM until OD_{650} reaches 0.6. Shake the resuspended-infection medium for 0.5 h (70 rpm, 22°C) before use

↓

Agro-inoculation

Trim the hypocotyls to 3mm size. Put the half-seed cotyledonary explants from 1 day old seedlings in *Agrobacterium* infection medium (OD_{650} 0.6)

↓

Co-cultivation

Co-cultivate the blot-dried explants in co-cultivation medium at 23°C under a photoperiod of 16 h/8 h (light/dark)

↓ **5 days**

Selection and shoot bud induction

Insert the explants (45 degrees) in SIM with the adaxial side facing upward, and maintain in a walk-in chamber under the photoperiod of 16 h/8 h (light/dark) and temperature of 23°C

↓ **28 days**

Shoot elongation

Transfer the explants to SEM. Subculture the explants on SEM every 14 days until the shoot attaints a height of 3cm

↓ **14 days**

Rooting

Place the elongated shoots in B_5 basal media supplemented with IBA (1 mg l^{-1}) + MES (0.59 g l^{-1}) until the roots attain a length of 2-3 cm

↓ **14 days**

Acclimatization

Transfer the plants into pots (soil: vermiculite = 1:1) and maintain in greenhouse at 28/24°C with a photoperiod of 16 h/8h (light/dark)

Fig. 1.31 *Agrobacterium*-mediated genetic transformation of soybean

Table 1.15 Different media and their composition for soybean transformation

Media	Composition
Cocultivation medium (CCM)	1/10× B5 basal medium + MES (3.9 g/L) + sucrose (3%) + pH = 5.4 + BAP (1.67 mg/L) + GA3 (0.25 mg/L), acetosyringone (40 mg/L) + DTT (0 or 154.2 mg/L)
Shoot induction medium (SIM)	B5 basal medium + sucrose (3%) + phytagel 0.35%, MES (0.58 g/L) + pH 5.6 + 6-BA (1.67 mg/L), cefotaxime (50 mg/L)
Shoot elongation medium (SEM)	MS basal medium + sucrose (3%) + phytagel (0.35%) + MES (0.58 g/L) + pH 5.6 + GA3 (0.5 mg/L) + IAA (0.1 mg/L) + zeatin (1 mg/L) + asparagine (50 mg/L) + L-pyroglutamic acid (100 mg/L) + cefotaxime (75 mg/L)
Rooting medium	B5 basal medium + sucrose (1.5%), agar powder (0.8%), MES (0.59 g/L), pH 5.7 + IBA (1 mg/L)

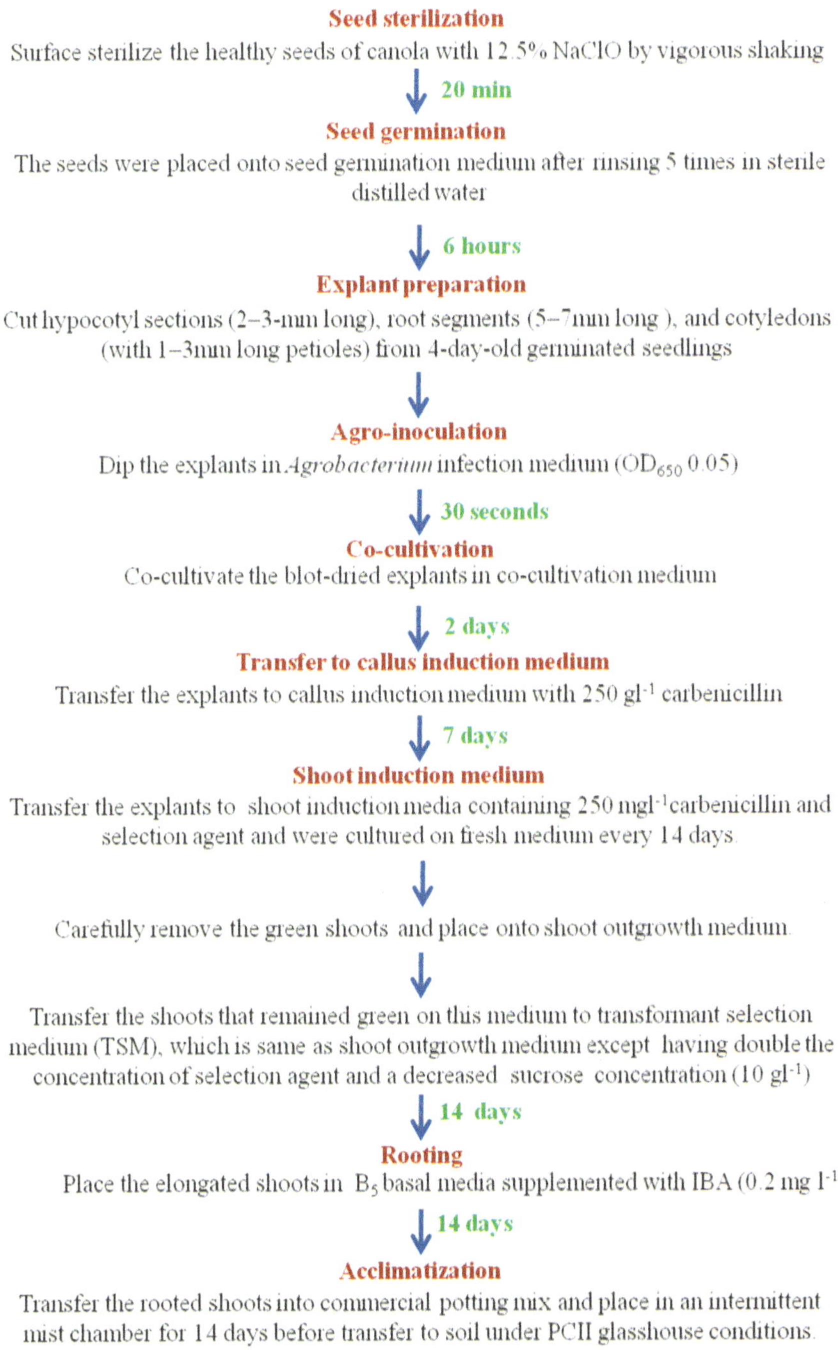

Fig. 1.32 *Agrobacterium*-mediated genetic transformation of canola

Table 1.16 Different media and their composition for canola transformation

Media	Composition
Seed germination medium	MS medium + sucrose (20 mg/L) + phytagel (7 g/L)
Callus induction and shoot initiation medium	MS medium + vitamins + nicotinic acid (1 mg/L) + thiamine-HCl (10 mg/L), pyridoxine-HCl (1 mg/L), myoinositol (100 mg/L), BAP (3 mg/L), silver nitrate (0.5 mg/L), NAA (0.2 mg/L), GA3 (0.01 mg/L), + sucrose (20 g/L) + phytagel (7 g/L)
Shoot outgrowth medium	MS salts and vitamin + adenine hemisulfate (40 mg/L), polyvinyl pyrrolidone (500 mg/L), BAP (1 mg/L), carbenicillin (250 mg/L), and selection agent
Rooting medium	Root induction medium was germination medium containing IBA (0.2 mg/L)

Seed sterilization

Take healthy seeds of alfalfa and surface sterilize by soaking in ethanol (70%) followed by rinsing five times in sterile distilled water

↓ 30 sec

Treat the seeds with sodium hypochlorite (1.5 %) and rinse three times in sterile distilled water before use.

↓ 12 min

Seed germination/ plantlet formation

Grow alfalfa plants on MS basal medium in a growth chamber at 25°C with 16/8h light/dark photoperiod

↓ 28-42 days

Explant preparation

Cut petioles and leaves from 28-42 days old plants into pieces

↓

Agro-inoculation

Dip the explants in *Agrobacterium* infection medium (OD_{650} 0.6-0.7) at 26°C

↓ 3 min

Co-cultivation

Co-cultivate the blot-dried explants in co-cultivation medium at 25 ± 2°C in dark

↓ 4-8 days

Wash the explants in sterile distilled MS, blot on filter paper, and transfer to co-culture medium supplemented with selection agent and cefotaxime (200 mg l^{-1}), under 40-50 μE $m^{-2}s^{-1}$ at 25 ± 2°C

↓ 21-28 days

Selection /callus induction medium

Transfer the explants on selection media to induce callus under fluorescent light at 26 °C and 16/8 h light/dark photoperiod with sub-culturing on fresh medium every 10 days

↓ 28 days

Transfer calli and somatic embryos to Boi2Y medium containing selection agent and cefotaxime (200 mg l^{-1})

↓

Rooting/plantlet formation

Pick the germinated embryos and transfer to MMS medium

↓

Acclimatization

Transfer the individual rooted plantlets to pots containing a (1:1 v/v) mixture of sterile soil and sand in plastic cups covered with polyethylene bags.

Fig. 1.33 *Agrobacterium*-mediated genetic transformation of alfalfa

Table 1.17 Different media and their composition for alfalfa transformation

Media	Composition
Seed germination medium	MS medium + sucrose (30 mg/L) + agar (8 g/L)
Coculture medium	Solid MS medium + kinetin (1 mg/L) + 2,4-D (10 mg/L) + selection agent + cefotaxime (200 mg/L)
Callus induction medium	MS medium containing selection agent + cefotaxime (200 mg/L)
Boi2Y medium	Blaydes medium + selection agent + cefotaxime (200 mg/L)
MMS medium	MS medium + Nitsch's vitamin + myoinositol (100 mg/L) + sucrose (30 mg/L) + cefotaxime (200 mg/L) + pH 5.9 + agar 0.8% (w/v)

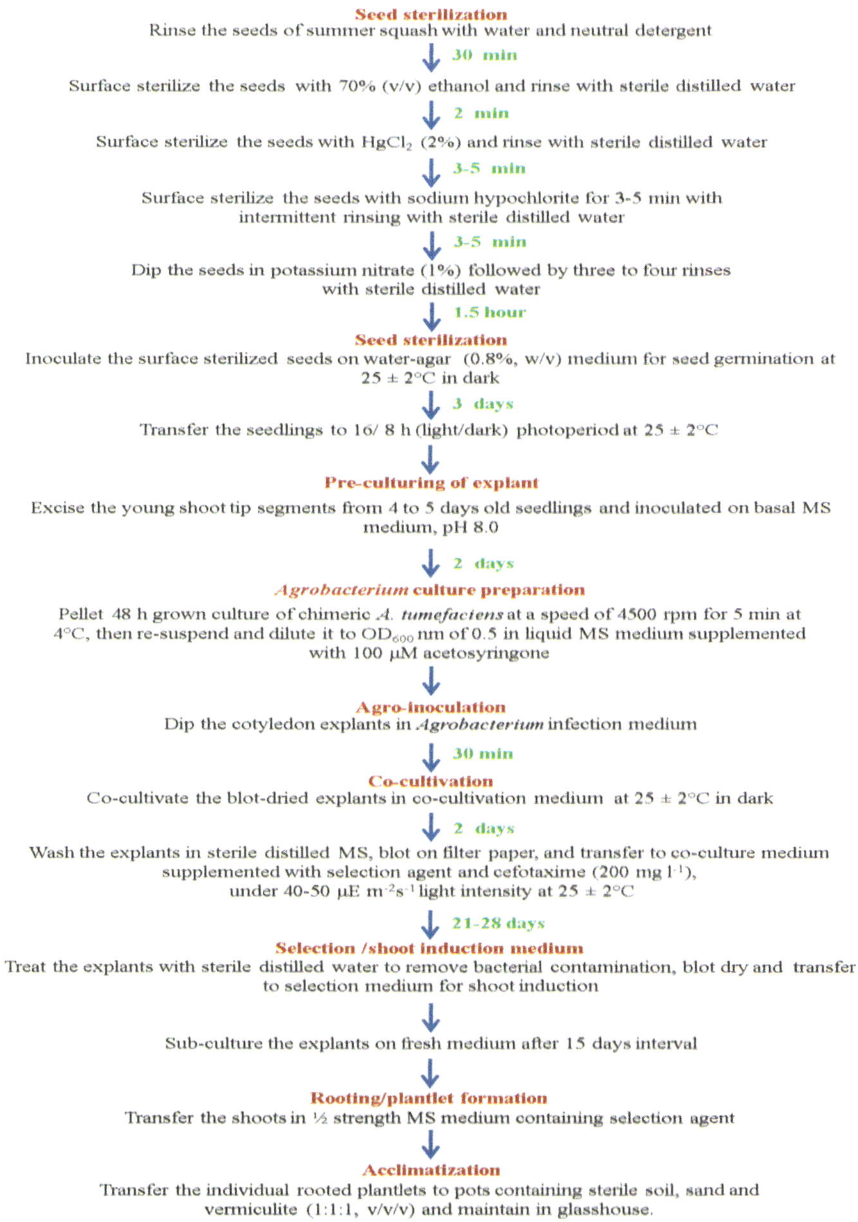

Fig. 1.34 *Agrobacterium*-mediated genetic transformation of squash

Table 1.18 Different media and their composition for squash transformation

Media	Composition
Regeneration medium	MS basal + TDZ (0.1 mg L^{-1}) + BAP (2.5 mg L^{-1}) + IAA (0.5 mg L^{-1}) + sucrose (3%), agar (0.8%), kanamycin (50 mg L^{-1}) + cefotaxime (250 mg L^{-1})
Rooting medium	1/2MS + kanamycin (50 L^{-1}) + selection agent + cefotaxime (200 mg L^{-1})

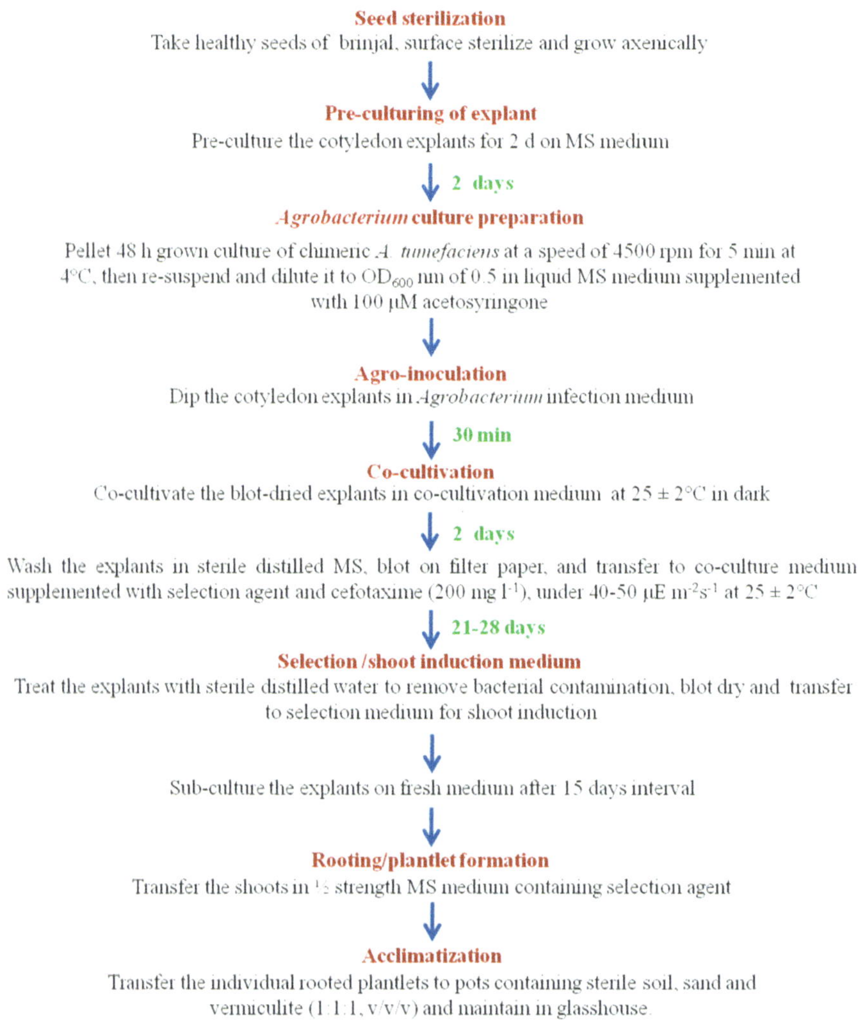

Fig. 1.35 *Agrobacterium*-mediated genetic transformation of eggplant

Table 1.19 Media used for brinjal transformation

Media	Composition
Regeneration medium	MS basal + BAP (2.5 mg/L) + TDZ (0.1 mg/L) + IAA (0.5 mg/L) + sucrose (3%), agar (0.8%), kanamycin (50 mg/L) + cefotaxime (250 mg/L)
Rooting medium	1/2MS + kanamycin (50 mg/L) + selection agent + cefotaxime (200 mg/L)

References

Ali A, Bang SW, Chung SM et al (2015) Plant transformation via pollen tube-mediated gene transfer. Plant Mol Biol Rep 33:742–747

Alves A, Quecini V, Vieria M (1999) Plant transformation: advances and perspectives. SciELO Analytics 56:1–8

Arakawa T, Chong DK, Lawrence Merritt J, Langridge WH (1997) Expression of cholera toxin B subunit oligomers in transgenic potato plants. Transgenic Res 6(6):403–413

Arnqvist L, Dutta PC, Jonsson L, Sitbon F (2003) Reduction of cholesterol and glycoalkaloid levels in transgenic potato plants by overexpression of a type 1 sterol methyltransferase cDNA. Plant Physiol 131(4):1792–1799

Aubin RA, Weinfeld M, Mirzayans R, Paterson MC (1994) Polybrene/DMSO-assisted gene transfer. Mol Biotechnol 1(1):29–48

Basso MF, da Cunha BADB, Ribeiro AP, Martins PK, de Souza WR, de Oliveira NG, Nakayama TJ, Augusto das Chagas Noqueli Casari R, Santiago TR, Vinecky F, Cancado LJ, de Sousa CAF, de Oliveira PA, de Souza SACD, de Almeida Cancado GM, Kobayashi AK, Molinari HBC (2017) Improved genetic transformation of sugarcane (*Saccharum* spp.) embryogenic callus mediated by *Agrobacterium tumefaciens*. Curr Protoc Plant Biol 2:221–239

Bastaki NK, Cullis CA (2014) Floral-dip transformation of flax (*Linum usitatissimum*) to generate transgenic progenies with a high transformation rate. J Vis Exp 19(94):e52189

Bevan MW, Flavell RB, Chilton MD (1983) A chimaeric antibiotic resistance gene as a selectable marker for plant cell transformation. Nature 304(5922):184–187

Bhowmik D, Shekhar S, Cheng AY, Long H, Tan GZ, Hoang TM, Karbaschi MR, Williams B, Higgins TJ, Mundree SG (2019) Robust genetic transformation system to obtain non-chimeric transgenic chickpea. Front Plant Sci 10:524

Birch RG (1997) Plant transformation: problems and strategies for practical application. Annu Rev Plant Biol 48(1):297–326

Boddupally D, Tamirisa S, Gundra SR, Vudem DR, Khareedu VR (2018) Expression of hybrid fusion protein (Cry1Ac:: ASAL) in transgenic rice plants imparts resistance against multiple insect pests. Sci Rep 8(1):1

Borg S, Brinch-Pedersen H, Tauris B, Madsen LH, Darbani B, Noeparvar S, Holm PB (2012) Wheat ferritins: improving the iron content of the wheat grain. J Cereal Sci 56(2):204–213

Buser C, Matile P (1977) Malic acid in vacuoles isolated from Bryophyllum leaf cells. Z Pflanzenphysiol 82(5):462–466

Cardoza V, Stewart CN (2004) *Agrobacterium*-mediated transformation of canola. In: Transgenic crops of the world. Springer, Dordrecht, pp 379–387

Chai R, Zhang G, Sun Q, Zhang M, Zhao S, Qiu L (2013) Liposome mediated mycelia transformation of filamentous fungi. Fungal Biol:577–583

Chen WS, Chiu CC, Liu HY, Lee TL, Cheng JT, Lin CC, Wu YJ, Chang HY (1998) Gene transfer via pollen-tube pathway for anti-fusarium wilt in watermelon. IUBMB Life 46(6):1201–1209

Chen L, Cai Y, Liu X, Yao W, Guo C, Sun S, Wu C, Jiang B, Han T, Hou W (2018) Improvement of soybean *Agrobacterium*-mediated transformation efficiency by adding glutamine and asparagine into the culture media. Int J Mol Sci 19(10):3039

Choi YI, Noh EW, Lee HS, Han MS, Lee JS, Choi KS (2007) Mercury-tolerant transgenic poplars expressing two bacterial mercury-metabolizing genes. J Plant Biol 50(6):658–662

Choi JY, Shin JS, Chung YS, Hyung NI (2012) An efficient selection and regeneration protocol for *Agrobacterium*-mediated transformation of oriental melon (*Cucumis melo* L. var. makuwa). Plant Cell Tissue Organ Cult 110(1):133–140

Clough SJ, Bent AF (1998) Floral dip: a simplified method for *Agrobacterium*-mediated transformation of *Arabidopsis thaliana*. Plant J 16(6):735–743

Craze M, Bates R, Bowden S, Wallington EJ (2018) Highly efficient *Agrobacterium*-mediated transformation of potato (*Solanum tuberosum*) and production of transgenic microtubers. Curr Protoc Plant Biol 3:33–41

Crossway A (1989) Microinjection of cells and protoplasts: integration of foreign DNA. In: Plant protoplasts and genetic engineering II. Springer, Berlin, Heidelberg, pp 228–240

De Block M (1993) The cell biology of plant transformation, current state, problems, prospects and implications for the plant breeding. Euphytica 71(1):1–14

De Bondt A, Eggermont K, Penninckx I, Goderis I, Broekaert WF (1996) *Agrobacterium*-mediated transformation of apple (malus × domestica Borkh.): an assessment of factors affecting regeneration of transgenic plants. Plant Cell Rep 15(7):549–554

D'Halluin K, Bossut M, Bonne E, Mazur B, Leemans J, Botterman J (1992) Transformation of sugarbeet (*Beta vulgaris* L.) and evaluation of herbicide resistance in transgenic plants. Bio Technol 10(3):309

Duan H, Richael C, Rommens CM (2012) Overexpression of the wild potato eIF4E-1 variant Eva1 elicits potato virus Y resistance in plants silenced for native eIF4E-1. Transgenic Res 21(5): 929–938

Ducreux LJ, Morris WL, Hedley PE, Shepherd T, Davies HV, Millam S, Taylor MA (2005) Metabolic engineering of high carotenoid potato tubers containing enhanced levels of β-carotene and lutein. J Exp Bot 56(409):81–89

Dutta I, Majumder P, Saha P, Ray K, Das S (2005) Constitutive and phloem specific expression of *Allium sativum* leaf agglutinin (ASAL) to engineer aphid (*Lipaphis erysimi*) resistance in transgenic Indian mustard (*Brassica juncea*). Plant Sci 169(6):996–1007

Elkholy SF, Ismail RM, Bahieldin A, Sadik AS, Madkour MA (2009) Expression of hepatitis B surface antigen (HBsAg) gene in transgenic banana (Musa Sp.). Arab J Biotechnol 12:291–302

Falco SC, Guida T, Locke M, Mauvais J, Sanders C, Ward RT, Webber P (1995) Transgenic canola and soybean seeds with increased lysine. Bio/technology 13(6):577–582

Felgner PL, Gadek TR, Holm M, Roman R, Chan HW, Wenz M, Northrop JP, Ringold GM, Danielsen M (1987) Lipofection: a highly efficient, lipid-mediated DNA-transfection procedure. Proc Natl Acad Sci 84(21):7413–7417

Filippone E, Lurquin PF (1989) Stable transformation of eggplant (*Solanum melongena* L.) by cocultivation of tissues with *Agrobacterium tumefaciens* carrying a binary plasmid vector. Plant Cell Rep 8(6):370–373

Finer JJ, Trick HN (1997) U.S. Patent No. 5,693,512. Washington, DC: U.S. Patent and Trademark Office

Franklin G, Sita GL (2003) *Agrobacterium tumefaciens*-mediated transformation of eggplant (*Solanum melongena* L.) using root explants. Plant Cell Rep 21(6):549–554

Gao SQ, Chen M, Xia LQ, Xiu HJ, Xu ZS, Li LC, Zhao CP, Cheng XG, Ma YZ (2009) A cotton (*Gossypium hirsutum*) DRE-binding transcription factor gene, GhDREB, confers enhanced tolerance to drought, high salt, and freezing stresses in transgenic wheat. Plant Cell Rep 28(2):301–311

Gao T, Wu Y, Zhang Y, Liu L, Ning Y, Wang D, Tong H, Chen S, Chu C, Xie Q (2011) OsSDIR1 overexpression greatly improves drought tolerance in transgenic rice. Plant Mol Biol 76(1): 145–156

Grichko VP, Filby B, Glick BR (2000) Increased ability of transgenic plants expressing the bacterial enzyme ACC deaminase to accumulate Cd, Co, Cu, Ni, Pb, and Zn. J Biotechnol 81(1):45–53

Gulick T (2003) Transfection using DEAE-Dextran. Curr Protoc Cell Biol 19(1):20–24

Hanumappa M, Choi G, Ryu S, Choi G (2007) Modulation of flower colour by rationally designed dominant-negative chalcone synthase. J Exp Bot 58(10):2471–2478

He X, Hou X, Shen Y, Huang Z (2011) TaSRG, a wheat transcription factor, significantly affects salt tolerance in transgenic rice and Arabidopsis. FEBS Lett 585(8):1231–1237

He H, Ke H, Keting H, Qiaoyan X, Silan D (2013) Flower colour modification of chrysanthemum by suppression of F3′H and overexpression of the exogenous *Senecio cruentus* F3′5′H gene. PLoS One 8(11):e74395

Hooykaas PJ, Beijersbergen AM (1994) The virulence system of Agrobacterium tumefaciens. Annu Rev Phytopathol 32:157–179

Ishida Y, Hiei Y, Komari T (2007) *Agrobacterium*-mediated transformation of maize. Nat Protoc 2(7):1614

Jadhav M, Jadhav A, Pawar B, Kale A, Kute N (2015) *Agrobacterium*-mediated genetic transformation of brinjal with *cry1F* gene for resistance against shoot and fruit borer. J Crop Improv 29(5):518–527

Jiang Q, Fu C, Wang ZY (2019) A unified *Agrobacterium*-mediated transformation protocol for alfalfa (*Medicago sativa* L.) and *Medicago truncatula*. In: Transgenic plants. Humana Press, New York, NY, pp 153–163

Jin RG, Liu YB, Tabashnik BE, Borthakur D (2000) Development of transgenic cabbage (*Brassica oleracea* var. capitata) for insect resistance by *Agrobacterium tumefaciens*-mediated transformation. In Vitro Cell Dev Biol P 36(4):231–237

Kalunke RM, Tundo S, Sestili F, Camerlengo F, Lafiandra D, Lupi R, Larré C, Denery-Papini S, Islam S, Ma W, D'amico S (2020) Reduction of allergenic potential in bread wheat RNAi transgenic lines silenced for CM3, CM16 and 0.28 ATI genes. Int J Mol Sci 21(16):5817

Kanrar S, Venkateswari J, Kirti P, Chopra V (2002) Transgenic Indian mustard (*Brassica juncea*) with resistance to the mustard aphid (*Lipaphis erysimi* Kalt.). Plant Cell Rep 20(10):976:– 981

Katsumoto Y, Fukuchi-Mizutani M, Fukui Y, Brugliera F, Holton TA, Karan M, Nakamura N, Yonekura-Sakakibara K, Togami J, Pigeaire A, Tao GQ (2007) Engineering of the rose flavonoid biosynthetic pathway successfully generated blue-hued flowers accumulating delphinidin. Plant Cell Physiol 48(11):1589–1600

Kim MJ, Kim JK, Kim HJ, Pak JH, Lee JH, Kim DH, Choi HK, Jung HW, Lee JD, Chung YS, Ha SH (2012) Genetic modification of the soybean to enhance the β-carotene content through seed-specific expression. PLoS One 7(10):e48287

Kinney AJ, Knowlton S (1998) Designer oils: the high oleic acid soybean. In: Genetic modification in the food industry. Springer, Boston, MA, pp 193–213

Klein TM, Jones TJ (1999) Methods of genetic transformation: the gene gun. In: Molecular improvement of cereal crops. Springer, Dordrecht, pp 21–42

Klein TM, Wolf ED, Wu R, Sanford JC (1987) High-velocity microprojectiles for delivering nucleic acids into living cells. Nature 327(6117):70–73

Kotnik T, Frey W, Sack M, Meglič SH, Peterka M, Miklavčič D (2015) Electroporation-based applications in biotechnology. Trends Biotechnol 33(8):480–488

Koul B (2020) Genetically modified (GM) crops harbouring *Bacillus thuringiensis* (BT) gene(s) to combat biotic stress caused by insect pests. In: Kumar M, Kumar V, Prasad R (eds) Phyto-microbiome in stress regulation. Environmental Microbial Biotechnology. Springer, Singapore. https://doi.org/10.1007/978-981-15-2576-6_2

Koul B, Taak P (2018) Biotechnological strategies for effective remediation of polluted soils. Springer

Koul B, Srivastava S, Amla DV, Sanyal I (2014) Establishment and optimization of *Agrobacterium*-mediated transformation and regeneration of tomato (*Solanum lycopersicum* L.). Int J Biosci 4:51–69

Koul B, Srivastava S, Sanyal I, Tripathi B, Sharma V, Amla DV (2014) Transgenic tomato line expressing modified *Bacillus thuringiensis* cry1Ab gene showing complete resistance to two lepidopteran pests. Springer Plus 3(1):1–3

Koul B, Yadav R, Sanyal I, Amla DV (2015) Comparative performance of modified full-length and truncated *Bacillus thuringiensis-cry1Ac* genes in transgenic tomato. Springerplus 4(1):1–4

Koul B, Yadav R, Sanyal I, Sawant S, Sharma V, Amla DV (2012) Cis-acting motifs in artificially synthesized expression cassette leads to enhanced transgene expression in tomato (*Solanum lycopersicum* L.). Plant Physiol Biochem 61:131–141

Kranz E, Lörz H (1990) Micromanipulation and in vitro fertilization with single pollen grains of maize. Sex Plant Reprod 3(3):160–169

Kumar GB, Ganapathi TR, Revathi CJ, Srinivas L, Bapat VA (2005) Expression of hepatitis B surface antigen in transgenic banana plants. Planta 222(3):484–493

Kumar RV, Sharma VK, Chattopadhyay B, Chakraborty S (2012) An improved plant regeneration and *Agrobacterium*-mediated transformation of red pepper (*Capsicum annuum* L.). Physiol Mol Biol Plants 18(4):357–364

Kumar M, Shukla AK, Singh H, Tuli R (2009) Development of insect resistant transgenic cotton lines expressing *cry1EC* gene from an insect bite and wound inducible promoter. J Biotechnol 140:143–148

Kumar S et al (eds) (2019) Transgenic plants: methods and protocols, methods in molecular biology, vol 1864. Springer Nature, Singapore

La Frano MR, Woodhouse LR, Burnett DJ, Burri BJ (2013) Biofortified cassava increases β-carotene and vitamin A concentrations in the TAG-rich plasma layer of American women. Br J Nutr 110(2):310–320

Li S, Cong Y, Liu Y, Wang T, Shuai Q, Chen N, Gai J, Li Y (2017) Optimization of *Agrobacterium*-mediated transformation in soybean. Front Plant Sci 8:246

Liu Z, Friesen TL (2012) Polyethylene glycol (PEG)-mediated transformation in filamentous fungal pathogens. In: Plant fungal pathogens. Humana Press, Totowa, NJ, pp 365–375

Mandal C, Dutta S, Meena K, Mandal A (2018) Efficient plantlet regeneration using hypocotyl explants and *Agrobacterium*-mediated genetic transformation in flax (*Linum usitatissimum* L.). Plant Cell Biotechnol Mol Biol 19:358–371

Mannino RJ, Gould-Fogerite S (1988) Liposome mediated gene transfer. Biotechniques 6(7):682–690

Mathur J, Koncz C (1998) PEG-mediated protoplast transformation with naked DNA. In: Arabidopsis protocols. Humana Press, New York, NY, pp 267–276

McCormick S, Niedermeyer J, Fry J, Barnason A, Horsch R, Fraley R (1986) Leaf disc transformation of cultivated tomato (L. esculentum) using Agrobacterium tumefaciens. Plant Cell Rep 5 (2):81–84

McKersie BD, Senaratna T, Bowley SR, Brown DCW, Krochko JE, Bewley JD (1989) Application of artificial seed technology in the production of hybrid alfalfa (*Medicago sativa* L.). In Vitro Cell Dev Biol 25(12):1183–1188

Mosa KA, Saadoun I, Kumar K, Helmy M, Dhankher OP (2016) Potential biotechnological strategies for the cleanup of heavy metals and metalloids. Front Plant Sci 7:303

Murashige T, Skoog F (1962) A revised medium for rapid growth and bioassays with tobacco tissue cultures. Physiol Plant 15:473–497

Nagamiya K, Motohashi T, Nakao K, Prodhan SH, Hattori E, Hirose S, Ozawa K, Ohkawa Y, Takabe T, Takabe T, Komamine A (2007) Enhancement of salt tolerance in transgenic rice expressing an *Escherichia coli* catalase gene, katE. Plant Biotechnol Rep 1(1):49–55

Nakamura N, Suzuki T, Shinbo Y, Chandler S, Tanaka Y (2020) Development of violet transgenic carnations and analysis of inserted transgenes. In: The carnation genome. Springer, Singapore, pp 135–146

Nishimura A, Aichi I, Matsuoka M (2006) A protocol for Agrobacterium-mediated transformation in rice. Nat Protoc 1(6):2796–2802

Noman A, Aqeel M, Deng J, Khalid N, Sanaullah T, Shuilin H (2017) Biotechnological advancements for improving floral attributes in ornamental plants. Front Plant Sci 8:530

Norouzi P, Malboobi MA, Zamani K, Yazdi-Samadi H (2005) Using a competent tissue for efficient transformation of sugarbeet (*Beta vulgaris* L.). In Vitro Cell Dev Biol Plant 41(1):11–16

Ogo Y, Wakasa Y, Hirano K, Urisu A, Matsuda T, Takaiwa F (2014) Generation of transgenic rice with reduced content of major and novel high molecular weight allergens. Rice 7(1):1–9

Pena A, Lorz H, Schell J (1987) Transgenic rye plants obtained by injecting DNA into young floral tillers. Nature 325(6101):274–276

Pratap D, Kumar S, Raj SK, Sharma AK (2011) *Agrobacterium*-mediated transformation of eggplant (*Solanum melongena* L.) using cotyledon explants and coat protein gene of cucumber mosaic virus. Indian J Biotechnol 10:19–24

Rakoczy-Trojanowska M (2002) Alternative methods of plant transformation-a short review. Cell Mol Biol Lett 7:849–858

Reich JJ, Lyer VN, Miki BL (1986) Efficient transformation of alfafa protoplasts by the internuclear microinjection of Ti plamids. Bio/Technology 4(11):1001–1004

Rivera AL, Gómez-Lim M, Fernández F, Loske AM (2012) Physical methods for genetic plant transformation. Phys Life Rev 9(3):308–345

Roy I, Mitra S, Maitra A, Mozumdar S (2003) Calcium phosphate nanoparticles as novel non-viral vectors for targeted gene delivery. Int J Pharm 250(1):25–33

Römer S, Fraser PD, Kiano JW, Shipton CA, Misawa N, Schuch W, Bramley PM (2000) Elevation of the provitamin a content of transgenic tomato plants. Nat Biotechnol 18(6):666–669

Rugh CL, Senecoff JF, Meagher RB, Merkle SA (1998) Development of transgenic yellow poplar for mercury phytoremediation. Nature Biotechnol 16(10):925–928

Russell JA, Roy MK, Sanford JC (1992) Major improvements in biolistic transformation of suspension-cultured tobacco cells. In Vitro Plant 28(2):97–105

Sahoo KK, Tripathi AK, Pareek A, Sopory SK, Singla-Pareek SL (2011) An improved protocol for efficient transformation and regeneration of diverse indica rice cultivars. Plant Methods 7:49

Sanford JC (1988) The biolistic process. Trends Biotechnol 6(12):299–302

Sanford JC, Klein TM, Wolf ED, Allen N (1987) Delivery of substances into cells and tissues using a particle bombardment process. Part Sci Technol 5(1):27–37

Sanyal I, Singh AK, Kaushik M, Amla DV (2005) *Agrobacterium*-mediated transformation of chickpea (*Cicer arietinum* L.) with *Bacillus thuringiensis cry1Ac* gene for resistance against pod borer insect *Helicoverpa armigera*. Plant Sci 168(4):1135–1146

Senaratna T, BD MK, Kasha KJ, Procunier JD (1991) Direct DNA uptake during the imbibtion of dry cells. Plant Sci 79:223–228

Senaratna T, McKersie B, Bowley S (1990) Artificial seeds of alfalfa (*Medicago sativa* L.) induction of desiccation tolerance in somatic embryos. In Vitro Cell Dev Biol 1(26):85–90

Shah T, Andleeb T, Lateef S, Noor M (2018) Genome editing in plants: advancing crop transformation and overview of tools. Plant Physiol Biochem 131:12–21

Shah P, Singh NK, Khare N, Rathore M, Anandhan S, Arif M, Singh RK, Das SC, Ahmed Z, Kumar N (2008) *Agrobacterium*-mediated genetic transformation of summer squash (*Cucurbita pepo* L. cv. Australian green) with cbf-1 using a two-vector system. Plant Cell Tissue Organ Cult 95(3):363–371

Shu Q, Ye G, Cui H et al (2000) Transgenic rice plants with a synthetic cry1Ab gene from *Bacillus thuringiensis* were highly resistant to eight lepidopteran rice pest species. Mol Breed 6:433–439

Singh SP, Gruissem W, Bhullar NK (2017) Single genetic locus improvement of iron, zinc and β-carotene content in rice grains. Sci Rep 7(1):1–1

Siritunga D, Arias-Garzon D, White W, Sayre RT (2004) Over-expression of hydroxynitrile lyase in transgenic cassava roots accelerates cyanogenesis and food detoxification. Plant Biotechnol J 2(1):37–43

Siritunga D, Sayre RT (2003) Generation of cyanogen-free transgenic cassava. Planta 217(3): 367–373

Smith EF, Townsend CO (1907) A plant-tumor of bacterial origin. Science 25(643):671–673

Song XY, Zhu WJ, Tang RM, Cai JH, Chen M, Yang Q (2016) Over-expression of StLCYb increases β-carotene accumulation in potato tubers. Plant Biotechnol Rep 10(2):95–104

Storozhenko S, De Brouwer V, Volckaert M, Navarrete O, Blancquaert D, Zhang GF, Lambert W, Van Der Straeten D (2007) Folate fortification of rice by metabolic engineering. Nat Biotechnol 25(11):1277–1279

Taak P, Koul B, Tiwari S (2021) Development of herbicide-resistant transgenic stevia (*Stevia rebaudiana* Bertoni) as an effective weed-management strategy in stevia cultivation. Sugar Tech 23(6):1340–1350

Taak P, Tiwari S, Koul B (2020) Optimization of regeneration and *Agrobacterium*-mediated transformation of Stevia (*Stevia rebaudiana* Bertoni): a commercially important natural sweetener plant. Sci Rep 10(1):1–12

Tabashnik BE, Dennehy TJ, Sims MA, Larkin K, Head GP, Moar WJ, Carriere Y (2002) Control of resistant pink bollworm (*Pectinophora gossypiella*) by transgenic cotton that produces *Bacillus thuringiensis* toxin Cry2Ab. Appl Environ Microbiol 68(8):3790–3794

Tada Y, Nakase M, Adachi T, Nakamura R, Shimada H, Takahashi M, Fujimura T, Matsuda T (1996) Reduction of 14–16 kDa allergenic proteins in transgenic rice plants by antisense gene. FEBS Lett 391(3):341–345

Tang Y, Bao X, Zhi Y, Wu Q, Guo Y, Yin X, Zeng L, Li J, Zhang J, He W, Liu W (2019) Overexpression of a MYB family gene, OsMYB6, increases drought and salinity stress tolerance in transgenic rice. Front Plant Sci 10:168

Tang K, Hu Q, Sun X, Wan B, Qi H, Lu X (2001) Development of transgenic rice pure lines with enhanced resistance to rice brown planthopper. In Vitro Cell Dev Biol Plant 37(3):334–340

Thi Van D, Ferro N, Jacobsen HJ (2010) Development of a simple and effective protocol for Agrobacterium tumefaciens mediated leaf disc transformation of commercial tomato cultivars. GM Crops 1–5:312–321

Thompson J, Drayton PR, Frame BR, Wang K, Dunwell J (1995) Maize transformation utilizing silicon carbide whiskers: a review. Euphytica 85:75–80

Tohidfar M, Zare N, Jouzani GS, Eftekhari SM (2013) *Agrobacterium*-mediated transformation of alfalfa (*Medicago sativa*) using a synthetic *cry3a* gene to enhance resistance against alfalfa weevil. Plant Cell Tissue Organ Cult 113(2):227–235

Torres JB, Ruberson JR, Whitehouse M (2009) Transgenic cotton for sustainable pest management: a review. In: Lichtfouse E (ed) Organic farming, pest control and remediation of soil pollutants. Sustainable agriculture reviews, vol 1. Springer, Dordrecht. https://doi.org/10.1007/978-1-4020-9654-9_4

Tzfira T, Citovsky V (2006) Agrobacterium-mediated genetic transformation of plants: biology and biotechnology. Curr Opin Biotechnol 17(2):147–154

Vaheri A, Pagano JS (1965) Infectious poliovirus RNA: a sensitive method of assay. Virology 27 (3):434–436

Velcheva M, Faltin Z, Flaishman M, Eshdat Y, Perl A (2005) A liquid culture system for Agrobacterium-mediated transformation of tomato (Lycopersicon esculentum L. Mill.). Plant Sci 168:121–130

Verhoeyen ME, Bovy A, Collins G, Muir S, Robinson S, De Vos CH, Colliver S (2002) Increasing antioxidant levels in tomatoes through modification of the flavonoid biosynthetic pathway. J Exp Bot 53(377):2099–2106

Xiaofen S, Kexuan T, Bingliang W, Huaxiong Q, Xinggui L (2001) Transgenic rice homozygous lines expressing GNA showed enhanced resistance to rice brown planthopper. Chi Sci Bull 46(20):1698–16703

Xu D, Duan X, Wang B, Hong B, Ho TH, Wu R (1996) Expression of a late embryogenesis abundant protein gene, HVA1, from barley confers tolerance to water deficit and salt stress in transgenic rice. Plant Physiol 110(1):249–257

Yadav R, Mehrotra M, Singh AK, Niranjan A, Kunti R, Singh R, Sanyal I, Lehri A, Pande V, Amla DV (2015) Improvement in *agrobacterium*-mediated transformation of chickpea (*Cicer arietinum* L.) by the inhibition of polyphenolics released during wounding of cotyledonary node explants. Protoplasma 254:253–269

Yan J, Kandianis CB, Harjes CE, Bai L, Kim EH, Yang X, Skinner DJ, Fu Z, Mitchell S, Li Q, Fernandez MG (2010) Rare genetic variation at *Zea mays* crtRB1 increases β-carotene in maize grain. Nat Genet 42(4):322–327

Yang QQ, Zhang CQ, Chan ML, Zhao DS, Chen JZ, Wang Q, Li QF, Yu HX, Gu MH, Sun SS, Liu QQ (2016) Biofortification of rice with the essential amino acid lysine: molecular characterization, nutritional evaluation, and field performance. J Exp Bot 67(14):4285–4296

Yeung EC (2002) The canola microspore derived embryo as a model system to study developmental processes in plants. J Plant Biol 3(45):119–133

Zambryski P, Tempe J, Schell J (1989) Transfer and function of T-DNA genes from Agrobacterium Ti and Ri plasmids in plants. Cell 56(2):193–201

Zhang X, Henriques R, Lin SS, Niu QW, Chua NH (2006) Agrobacterium-mediated transformation of *Arabidopsis thaliana* using the floral dip method. Nat Protoc 1(2):641

Zhang S, Li N, Gao F, Yang A, Zhang J (2011) Over-expression of TsCBF gene confers improved drought tolerance in transgenic maize. Mol Breed 26:455–465

Zhang Y, Singh MB, Swoboda I, Bhalla PL (2005) Agrobacterium-mediated transformation and generation of male sterile lines of Australian canola. Aust J Agric Res 56(4):353–361

Zhang Z, Sun A, Cong Y, Sheng B, Yao Q, Cheng ZM (2006) *Agrobacterium*-mediated transformation of the apple rootstock Malus micromalus Makino with the RolC gene. In Vitro Cell Dev Biol Plant 42(6):491–497

Zhang L, Xiao S, Li W, Feng W, Li J, Wu Z, Gao X, Liu F, Shao M (2011) Overexpression of a harpin-encoding gene hrf1 in rice enhances drought tolerance. J Exp Bot 62(12):4229–4238

Zhang Y, Yu LC (2008) Microinjection as a tool of mechanical delivery. Curr Opin Biotechnol 19(5):506–510

Zhu YX, Ou-Yang WJ, Zhang YF, Chen ZL (1996) Transgenic sweet pepper plants from *Agrobacterium* mediated transformation. Plant Cell Rep 16(1-2):71–75

Strategies for Enhancement of Transgene Expression

2

Abstract

Expressing a foreign gene from an AT-rich organism in a GC-rich organism is difficult. To enhance the expression of heterologous genes in plants, several molecular techniques have been discovered and implemented. These include deletion of sequence motifs and codons that cause mRNA degradation and, as a result, low expression (mRNA instability/putative polyadenylation sequences, cryptic splicing sites, RNA polymerase II termination signals, secondary structures, and so on), protein targeting to cellular compartments for optimal accumulation and stability (apoplast, ER, vacuole, and so on), and incorporation of elements that direct high-level expression (strong promoters, 5′UTR sequence, TIC, etc.). This chapter summarizes various strategies that aim at preparing/improving a naturally occurring gene from lower organism to be expressed in higher plants and also synthetic gene designing for higher expression. It also focuses on the use of strong promoters (modified natural or synthetic) to drive the expression of transgene(s) and improve the stability of gene product through subcellular targeting.

Keywords

Heterologous gene · Subcellular targeting · Gene designing · mRNA instability sequences · Translation initiation context

2.1 Introduction

Biotechnology, a multidisciplinary science, has the potential in resolving solutions and finding alternatives for major socioeconomic problems, like food, fiber, and fuel and other necessities of life through plant improvement. In times to come, the conventional techniques or strategies for crop improvement may not be able to fulfill the demands of teeming millions, shrinking land resources, and environmental

B. Koul, *Cisgenics and Transgenics*, https://doi.org/10.1007/978-981-19-2119-3_2

stresses; hence, we need to rely on the genetic manipulation of plants for sustainable crop improvement. Biotechnologies have proved to be a boon and can cater to the needs of a layman.

In the twenty-first century, it is now possible to transfer any gene irrespective of its origin into desired plant species as a routine procedure by exploiting the novel capabilities of *Agrobacterium tumefaciens*, the natural "genetic engineer," and alternative strategies based on transfection of plant protoplast or by biolistic devices. Nowadays, people are trying to express a foreign gene from a heterologous system into a suitable organism by using tools and techniques of genetic engineering for either qualitative or quantitative improvement of organism or their products. But the success of biotechnology depends on the expression level, purity, economical aspect, and sustained recovery of the desired products.

Transcription, translation, posttranslational changes, and accumulation of recombinant protein in the heterologous environment of the plant cell govern the expression of a transgene at varying levels. To increase the expression of heterologous genes in plants, a variety of molecular and genetic techniques can be used. These techniques do involve the use of computational biology, bioinformatics, and structural genomics. These approaches are described below.

2.2 Designing of Coding Sequence of the Gene

The success of the crop improvement strategies through transgenics depends on the stability, functionality, and expression level of the transgene. It is interesting to note that poor expression of heterologous foreign genes (interkingdom transfer) can occur due to improper codon utilization/mRNA instability/premature polyadenylation/ abnormal splicing and instability of expressed recombinant proteins coded by native genes (Murray et al. 1991; Perlak et al. 1991; Haseloff et al. 1997; Jarvis et al. 1997; Rouwendal et al. 1997; Gustafsson et al. 2004).

Avoiding sequence motifs that drive mRNA for degradation and low-level expression are two ways for increasing the expression of heterologous genes in plants. For higher expression in plants, extensive coding region modifications and several factors such as GC content (increasing the GC content of the gene from bacterial origin), codon bias, elimination of cryptic splicing sites, putative transcription termination signals (AAUAAA and variants), mRNA instability sequences (AUUUA and variants), and integration of 5′ or 3′ controlling sequences can be done (Perlak et al. 1991; Haseloff et al. 1997; Rouwendal et al. 1997; Streatfield 2007; Koul et al. 2012). To avoid secondary structures that can block the transcription process and destabilize the mRNA, long hairpin loops and A/T strings should be avoided. Codon usage is heavily biased towards T/A ending codons in dicotyledonous plants, whereas G/C ending codons in monocotyledonous plants. Plants avoid the CG and TA ending codons in particular, probably to decrease methylation and less energetic stability (Beutler et al. 1989; Fennoy and Bailey-Serres 1993).

2.2.1 Removal of Destabilizing Elements for Optimal Expression

Many variables influence the expression of a gene in a heterologous system, including sequence patterns that destabilize mRNA. Despite the use of strong or tissue-specific promoters to control their transcription, the genes for *Bacillus thuringiensis* crystal protein, T4 lysozyme, *Klebsiella pneumoniae* cyclodextrin glycosyl transferase, and bacterial mercuric ion reductases were expressed at very low levels in plants. As a result, in order to achieve high-level expression of foreign genes in plants, the coding region should be redesigned (Perlak et al. 1991; Koul et al. 2012), with unfavorable sequence motifs and rare codons either eliminated or changed, as illustrated in Table 2.1.

2.2.2 Cryptic Splicing Sites

The mRNA splicing machinery in plants is mainly conserved, although it can read a heterologous coding sequence incorrectly, resulting in mRNA instability and degradation. UA-rich regions are thought to be bound by unique protein factor(s) that interact with or recruit splicing factors to intron splicing signals early in the spliceosome assembly process (Perlak et al. 1991; Brown and Simpson 1998; Fiume et al. 2004; Ashraf et al. 2005).

Table 2.1 Factors affecting the expression of heterologous genes in plants

Sequence motifs	Factors
According to tRNA abundance-based codon table, avoid TA and CG ending codons	Codon bias
A + T stretches	Cryptic splicing sites
 54.20% 44.65%	G + C content Monocotyledons Dicotyledons
AATAAA, ATTTA, TTCTT	mRNA instability signals
AAAATA, AACCAA, AAGCAT, AAGGTGGAATAAA, AATAAT, AATACA, AATACA, AATCAA, AATTAA, AGTAAA, ATAAAA, ATACAT, ATACTA, ATATAA, ATTAAA, ATTAAT, ATTCTC, CATAAA, TTTTTT	Putative polyadenylation signals
Promoter, 5′ UTR, signal peptide, translation initiation context (TIC), 3′ UTR	Regulatory sequences
ATTCTC, TATTTTTT, TTCTT, CAN$_{7-9}$AGTNNA	RNA polymerase II termination signals
Hairpin loops especially at 3′ end	Secondary structure
 GC(A/C)(A/G)(A/C)(C/G)ATGGCG TAAA(A/C) AATGGC(G/T)	Translation initiation context Monocotyledons Dicotyledons

2.2.3 Codon Bias

Because of the genetic code's degeneracy, organisms can employ distinct codons at varying frequency (Murray et al. 1989). Almost every organism utilized for heterologous protein expression has some degree of variable codon usage or bias, and it is becoming increasingly obvious that codon bias has a significant impact on heterologous protein expression (Kane 1995; Gustafsson et al. 2004; Streatfield 2007). Between genes that are expressed at high and low levels, plants exhibit a different codon bias. Monocots have a high preference for G/C ending codons, while dicots prefer T/A ending codons (Fennoy and Bailey-Serres 1993).

A prominent strategy called as "codon optimization" is used to modify the unusual codons in the target gene to reflect the host's codon usage pattern without changing the amino acid sequence of the produced protein. Use of "plant-optimized" synthetic genes dramatically boosted transgenic expression, according to several studies (Kang et al. 2004; Ashraf et al. 2005; Peng et al. 2006; Suo et al. 2006; Agarwal et al. 2008).

The codon composition of genes within an organism tends to be the same for all genes and matches the concentration of isoacceptor tRNAs present in that organism (Zhou et al. 1999). Another strategy to improve recombinant protein expression is to optimize the coding sequence according to the host plant without changing the amino acid sequence of the generated protein (Daniell et al. 2009). Several studies have described the importance and the use of codon-optimized synthetic genes to improve recombinant protein expression (Perlak et al. 1991; Mason et al. 1998; Ashraf et al. 2005; Mishra et al. 2006; Peng et al. 2006; Suo et al. 2006; Maclean et al. 2007; Agarwal et al. 2008; Kim et al. 2008; Tsuboi et al. 2008).

As we are aware of that, intron sequences and intron splice sites should be avoided in transgenes; however, intron sequences have been shown to contribute to overexpression in specific situations (Rethmeier et al. 1997). It has been discovered that introns near the promoter can boost mRNA synthesis and translation (Rose 2004). Because of the location effect, transgene copy number, or gene silencing, the T0 transgenic population created under comparable conditions displayed variable transgene expression (Wilde et al. 2000; Bhat and Srinivasan 2002).

Several tactics have been used to sustain transgenic expression in plants, including the transfer of certain components together with the transgene and the identification of a low-copy-number transformant; particularly, those with single-copy insertions have used nuclear matrix attachment regions (MAR) to improve transgene transcriptional activity by reducing position effect, and plant artificial chromosomes (PAC) have been synthesized based on MAR sequence elements to improve transgene expression (Allen et al. 1993, 1996; Liu and Tabe 1998; Somerville and Somerville 1999; Nowak et al. 2001; Abranches et al. 2005; Butaye et al. 2005; Yunus et al. 2008; De Paepe et al. 2009). The identification and cloning of functioning Arabidopsis plant centromeres enabling the creation of stably inherited PACs have been described, which is a step forward in this domain (Copenhaver and Preuss 2000).

2.3 Incorporation of Elements for High Expression

By increasing the rates of transcription and translation, expression vector design can aid to achieve larger quantities of heterologous proteins in transgenic plants. Some essential features that control high-level expression are (Twyman et al. 2003) mentioned below:

2.3.1 Role of Promoters in Eukaryotic Gene Expression

The promoter region is a critical cis-acting regulatory element that regulates the transcription of surrounding coding sequences into mRNA, which is then translated into proteins. Gene expression in eukaryotes is controlled by a number of complicated systems that allow them to respond to a variety of biological demands/ stimulus. The promoter regulatory region, which governs the quantitative and qualitative expression of genes downstream, is the most significant component that regulates gene expression. This region is a collection of cis-regulatory sequences that interact with transcription factors (TFs) in a coordinated manner to control the recruitment of RNA polymerase (Hampsey 1998; Koul et al. 2012).

Core elements and regulatory elements are the two fundamental components of eukaryotic promoters. A TATA sequence positioned upstream of the transcription start site and an initiator sequence (Inr) enclosing the start site are core promoter elements that specify the site for assembly of the preinitiation complex (PIC). A TATA box, an Inr sequence, or both of these regulatory elements can be found in promoters. The TATA box (also known as the Goldberg-Hogness box) was the first core promoter element discovered in eukaryotic protein-coding genes. The downstream promoter element (DPE), a third core element, is roughly 30 bp downstream of the start site. The DPE appears to behave as a transcription factor (TFIID)-binding site for TATA-deficient promoters when combined with the Inr element (Hampsey 1998).

Upstream activation sequences (UAS) and upstream repression sequences (URS) are gene-specific regions that are positioned upstream of the core promoter and influence the rate of transcription initiation. They serve as binding sites for enhancers and repressors, respectively, for transcription. In plants, main regulatory sequences are often found within 1000 bp upstream (5′) of the transcription start site, while regulatory sequences can also be found further upstream (3′) or downstream (3′) of the coding region (Hampsey 1998; Koul et al. 2012).

To achieve high-level expression of transgenes in plants, naturally occurring strong promoters such as the *CaMV35S* promoter, octopine synthase (*ocs*) promoter, manopine synthase (*mas*) promoter, maize ubiquitin (*Ubi-1*) promoter, histone (H2B, H3 H4) promoter, and ribulose 1,5-bisphosphate carboxylase (shuffling their activator domains to generate chimeric promoters) have been used to develop new series of promoters to achieve greater and regulated expression of transgenes (Odell et al. 1985). In the *CaMV35S*, *ocs*, and *mas* promoters, upstream sequence subdomains that function as transcriptional activators have been used to develop synergistic combinations for multiple transcriptional activator elements for biotechnological applications (Di Rita and Gelvin 1987; Kay et al. 1987; Leisner and Gelvin 1988; Benfey and Chua 1990; Ni et al. 1995; Mitsuhara et al. 1996).

2.3.1.1 Significance of Strong Promoter(S)

The most difficult aspect of plant genetic engineering is achieving significant levels of foreign gene expression(s). As a result, the transformation cassettes are carefully constructed to ensure that the transgene is expressed in a controlled manner in genetically modified plants. The core element that governs basal transcription and upstream gene-specific cis-regulatory elements responsible for increased gene expression make up the promoter(s), which are a series of transcription control modules. The importance of core promoter elements in developmentally and environmentally controlled gene expression is now well established (Fang et al. 1989; Yoshida and Shinmyo 2000; Smale and Kadonaga 2003; Halpin 2005).

Constitutive promoters induce gene expression in all tissues of a plant in the same way, regardless of the environment or developmental stage. The native constitutive cauliflower mosaic virus promoter (*CaMV35S*) and its modified double-enhancer version (*DECaMV35S*) are the most widely utilized promoters for constitutive transgene expression in both dicot and monocot plants. In diverse dicot plant species, large levels of expression and accumulation of protein products have been reported using this promoter (Odell et al. 1985; Benfey et al. 1989; Battraw and Hall 1990; Gutiérrez-Ortega et al. 2005).

The maize polyubiquitin-1 (*ubi-1*) promoter is chosen for monocotyledonous species and has been shown to boost expression (Christensen et al. 1992; Stoger et al. 2000). Hernandez-Garcia et al. (2009) identified the soybean polyubiquitin promoter as a strong constitutive promoter. Mannopine synthase (*mas*), *mac* promoter (hybrid of mas promoter and *CaMV35S* enhancer region), actin promoter, C1 promoter, cassava vein mosaic virus promoter, and nopaline synthase (*nos*) promoter are among the other constitutive promoters (Sharma and Sharma 2009; Koul et al. 2012).

Plant growth, development, and regeneration may be hampered by constitutive expression of a foreign biomolecule. To limit any potential negative effect on plant growth, tissue-specific, developmental stage-specific, or inducible promoters are used to limit transgene expression to specific organs, tissues, or even cell types and accumulate the product in certain organs such as seeds, fruits, roots, and tubers to improve harvesting efficiency (Zuo and Chua 2000; Sharma and Sharma 2009). Patatin (tuber-specific), tomato E8 (fruit-specific), rbcS (leaf-specific), arcelin, legumin, maize globulin, zein, rice glutelin, soybean P-conglycinin a'-subunit promoter (seed-specific), and Amy 3 (inducible) promoter are some examples of these promoters.

Combinatorial engineering of cis elements, such as enhancers, activators, or repressors, upstream of a core promoter including a set of minimal elements like the TATA box, a transcription start site, and the CCAAT consensus sequence, has been used to create artificial promoters. Multiple copies of the CaMV35S promoter's enhancer element have been used to boost the expression level of several recombinant biomolecules in plants, and the promoter with duplicated enhancer sequence has been used to boost the expression level of several recombinant biomolecules in plants (Kay et al. 1987; Comai et al. 1990; Guerineau et al. 1992; Ni et al. 1995; Ruggiero et al. 2000; Sawant et al. 2001; Ashraf et al. 2005; Dong et al. 2005; Mishra et al. 2006; Agarwal et al. 2008; Koul et al. 2012). Bidirectional promoters have also been developed, which can simultaneously transcribe two genes in both directions, upstream and downstream. Plants with trans-acting factors co-expressed have higher levels of recombinant protein expression (Yang et al. 2001; Hull et al. 2005; Chaturvedi et al. 2006; Zhang et al. 2008; Lv et al. 2009).

Box 2.1: Case Study

The creation and combining of several cis-motifs can boost gene expression dramatically. On the other hand, using a promoter to stimulate the expression of numerous transgenes (in the same plant) repeatedly can reduce a transgene's inducibility. This is due to gene silencing based on homology. Because the CaMV35S promoter is the most frequent promoter for transgenic constitutive expression, it would be beneficial to evaluate additional promoters or construct synthetic expression modules for increased transgene constitutive expression (Curtis et al. 2000; Fagard and Vaucheret 2000; Sawant et al. 2005; Venter 2007).

By shuffling the activator domains, transcription activators, or duplication of enhancer sequences, as in *CaMV35S*, *ocs*, *mas* elements, and multiple transcriptional activator elements or domain swapping, several efforts have been made to develop novel promoters for high and tightly regulated expression of transgenes. Interactions between activator activation domains and general transcription machinery components aid in the recruitment and stabilization of the transcription initiation environment (Kay et al. 1987; Leisner and Gelvin 1988; Benfey and Chua 1990; Comai et al. 1990; Mitsuhara et al. 1996; Bhullar et al. 2003). As a result, with a better understanding of the general transcription components and promoter architecture, a plant promoter for increased transgenic expression can be designed (Liu et al. 1999; Bhullar et al. 2003).

Synthetic promoters with defined regulatory sequences are a promising field in transgenic research for achieving regulated increased transgene expression under both constitutive and inducible conditions. Various cis-regulatory elements and their fusion to optimize the expression features of a particular gene could be used to create artificial expression cassettes.

Several synthetic promoters have been developed previously by combining the cis-elements of different promoters of highly expressing genes, placing cis-elements in association with heterologous promoters, and domain swapping of inactive domains with functionally active domains from heterologous

(continued)

Box 2.1 (continued)

promoters (Comai et al. 1990; Mitsuhara et al. 1996; Sawant et al. 2001; Bhullar et al. 2003; Venter 2007).

A completely synthetic 450 bp DNA fragment designated as complete expression cassette (*Pcec*) was constructed from regulatory sequences compiled from a database of highly expressed plant genes showing constitutive expression in order to investigate the significance of specific sequences in the 5′region of genes (Sawant et al. 2001). For the development of the entire expression cassette, a 138 bp minimum expression cassette (*Pmec*) with a proximal TATA box region and a 312 bp transcription activation module (TAM) including several cis-elements were used in a progressive synthesis of *Pcec*. Because of the importance of TAM for improved expression, *Pdec* and *Ptec* promoters were developed in double and triple copies for the predicted response.

The expression profile of the *uidA* gene in transgenic tomato plants was examined using four synthetic promoters, and the results were compared to the naturally occurring *CaMV35S* constitutive promoter and its derivatives, such as the minimal expression cassette and double-enhancer cassettes. The best performing synthetic promoter (*Pcec*) was used to express the modified insecticidal *cry1Ac* gene of *Bacillus thuringiensis* to develop transgenic tomato resistant to lepidopteran insects, particularly *Helicoverpa armigera*, with optimal toxin expression and no observable detrimental effect on new shoot regeneration, which was otherwise showing low expression with *DECaMV35S* promoter in tobacco and cotton (Rawat et al. 2011; Koul et al. 2012).

2.3.1.2 Designing of Synthetic Promoters

To accomplish regulated gene expression, designing synthetic promoters with defined regulatory sequences is an appealing area (Sawant et al. 2001; Bhullar et al. 2003; Rushton et al. 2002). Multimerization and fusing of large activator pieces are required for the creation of a chimeric promoter. Sawant et al. (2001) established a unique gene expression module based on the consensus of nucleotide sequence analysis of a database of genes selected for the ability to express at high levels in plants in order to examine the significance of certain motifs, sequence in the 5′ region (b).

They created a number of chimeric gene expression modules by combining the TATA box nucleotide sequence, transcription start site, untranslated leader, and translation region in highly expressed genes in plants with a range of sequence properties discovered upstream of the TATA box (Sawant et al. 2001). A synthetic promoter can be separated into two parts: a minimal core promoter (−46 to −1) and an activation module (−350 to −38). The importance of differential expression of uidA with a combination of transcriptional activation modules has been demonstrated (Koul et al. 2012).

The *E. coli* glucuronidase gene *(uidA)* is a better reporter gene for examining promoter performance in transgenic plants because it is straightforward to quantify, highly sensitive, and detectable using histochemical tests to localize gene activity in specific cell types (Jefferson 1987). The *E. coli uidA (gusA)* gene encodes the enzyme glucuronidase (GUS, E.C. 3.2.1.31), which catalyzes the hydrolysis of β-D-glucuronides. External substrates 4-methylumbelliferyl glucuronide (MUG) for specific activity assessments and 5-bromo-4-chloro-3-indolyl glucuronide (X-gluc) for histological localization are used by the enzyme (Jefferson 1987). Microorganisms, vertebrates, and invertebrates all have GUS activity, whereas plants have relatively little background activity (Gilissen et al. 1998). The GUS enzyme is very stable within plants and is nontoxic when expressed at high levels and has been extensively used for functional evaluation of promoter or other regulatory sequences with transgenic approach.

2.3.2 Untranslated Regions (UTR) and Sequences

Transgene expression is regulated by the 5′ and 3′ UTR leader sequences, which regulate transcription and translation initiation of the foreign gene. The accumulation of recombinant proteins has been found to be considerably increased by viral leaders at the 5′ UTR. Tobacco mosaic virus (TMV), tobacco etch virus (TEV), and alfalfa mosaic virus (AMV) leader sequences are the most popular and are used to optimize the expression of a variety of foreign proteins in plants. The 3′ UTR contains a message for transcript polyadenylation, which has a direct impact on the stability of mRNA. To stabilize the transcript formation, heterologous 3′ UTRs from plants or plant viruses were employed (Tyc et al. 1984; (Dowson Day et al. 1993; Datla et al. 1993; Gallie et al. 1995; Haq et al. 1995; Hood et al. 1997; Chan and Yu 1998; Staub et al. 2000; Wang et al. 2001; Ko et al. 2003; Agarwal et al. 2008; Lu et al. 2008; Wang et al. 2008).

2.3.3 Translation Initiation Context (TIC)

At the translational level, proper protein biosynthesis start is a critical determinant of gene expression. A growing number of open reading frames (ORFs) with important translational regulatory features have been discovered in recent investigations. Translation initiation context (TIC) nucleotide sequences flanking the initiator codon ATG play a crucial role in the commencement of translation in all species and are determined as TAAACAATGG for highly expressed dicot plant genes. However, various additional parameters, such as the distance from the 5′ end of the transcript and the existence of secondary structures, influence the designation of an ATG codon as a translation initiation point (Kozak 1987; Boyd and Thummel 1993; Joshi et al. 1997; Kozak 2002; Miyasaka et al. 2002).

The current paradigm for eukaryotic translation initiation postulates that the 40S ribosomal initiation complex recognizes the first ATG codon in a favorable environment while scanning the mRNA sequence from the 5′ end. The addition of a plant-preferred TIC sequence motif to heterologous genes greatly improved transgenic plant expression levels (Kozak 2002; Ashraf et al. 2005; Mishra et al. 2006; Agarwal et al. 2008, Sharma et al. 2008). In plant mRNAs, the locations −3 and +4 are extensively conserved.

The most conserved nucleotides in plant genes are purines at both sites (ideally A at −3 and G at +4), and point mutations at these positions resulted in a dramatic reduction, if not complete removal, of foreign gene expression. In most highly expressed plant genes, alanine comes after N-terminal methionine, and insertion of this subsequent codon increased expression by severalfold (Kozak 1986, 1999; Agarwal et al. 2009; Sawant et al. 1999, Sawant et al. 2001). Tables 2.2 and 2.3 show information on natural and synthetic promoters, as well as their distinguishing characteristics. The promoters utilized in CRISPR/Cas9 genome editing are listed in Table 2.4.

2.4 Improvement in Foreign Protein Accumulation and Stability

Degradation of recombinant proteins in plants has a direct impact on the final yield, uniformity, and overall quality of the protein. Several ways for reducing protein hydrolysis and increasing the stability of foreign proteins in plant protein factories have been presented. Avoiding protein degradation-directing amino acid sequences, organ-specific expression, organelle-specific targeting, protein fusions, and co-expression of companion protease inhibitors are only a few examples (Doran 2006; Boehm 2007; Streatfield 2007; Benchabane et al. 2008).

2.4.1 Subcellular Targeting of Recombinant Protein

Subcellular targeting is critical for molecular farming since it influences not only product output but also structural characteristics and functional performance. The importance of protein targeting can be summed up in three words: (i) the storage compartment for recombinant proteins, which has a strong influence on the interconnected processes of folding, assembly, and posttranslational modification, all of which affect the protein's stability and yield (Schillberg et al. 2002; Twyman et al. 2003; Fischer et al. 2004; Streatfield 2007); (ii) if a protein needs to be modified, its destination is vital, for example, glycoproteins must be targeted to the secretory pathway because glycans are added to proteins in the ER and Golgi apparatus of plant cells (Horvath et al. 2000); and (iii) protein targeting could be crucial for plant survival: proteins accumulating in the cytosol can be hazardous, whereas proteins getting collected in the chloroplast or vacuole can typically be

Table 2.2 Natural promoters and their characteristics

Promoter	Characteristics	References
CaMV35S	Constitutive and high level of transgene expression in both monocots and dicots	Odell et al. (1985)
DECaMV35S	Constitutive expression of transgenes in dicots and monocots	Odell et al. (1985) Benfey et al. (1989)
β-Conglycinin	Directs embryo-specific expression	Chen et al. (1986)
CaMV 19S	Weak promoter than *CaMV35S*	Lawton et al. 1987
Adh1	Highly active in root tissue than in leaf	Zhang and Wu (1988)
Rol A/B/C	*Rol b* and *c* genes represent a bidirectional promoter	Schmülling et al. (1989)
OsACT1	Weak plant promoter	McElroy et al. (1991)
Ubi1 and *Ubi2*	Utilized in monocots specifically in cereals	Christensen et al. (1992)
ZmUbi1	Strong plant promoter	Cornejo et al. (1993)
a-amylase	Functions in all cell types of the mature leaves, stems, sheaths, and roots	Chan et al. (1993)
Act2/ACT8	Strong constitutive expression in vegetative tissues	An et al. (1996)
OsAct	Expression of *hva1* gene in the leaf and root tissues resulted in salt and water stress tolerance	Xu et al. (1996, b)
FMV FLt	Strong constitutive promoter, with strength comparable to or greater than that of the *CaMV35S* promoter	Maiti et al. (1997)
b-Phaseolin	Only used for the developing seed	Van der Geest and Hall (1997)
FLt	Constitutive expression	Maiti and Shepherd (1998)
SCBV	Reliable for expression in monocots	Tzafrir et al. (1998)
MMV FLt	Strong constitutive promoter and its strength is greater than that of the *CaMV35S* promoter	Dey and Maiti (1999)
FMV Sgt	Twice stronger than the *CaMV35S* promoter	Bhattacharyya et al. (2002)
MtHP	Expression level is similar to that of CaMV35S promoter	Xiao et al. (2005)
GmUbi	High level of constitutive expression in soybean tissues reported	Hernandez-Garcia et al. (2009)
APX, PGD1, and *R1G1B*	Promoters are highly active at all stages of plant growth with distinct level of activity	Park et al. (2010)
WM403	Useful in driving nucellus-specific gene expression in plants	Dwivedi et al. (2010)

(continued)

Table 2.2 (continued)

Promoter	Characteristics	References
HaFAD2–1	Strong seed-specific promoter useful for biotechnology applications	Zavallo et al. (2010)
EF1a/UBI10/ GLU1	Endosperm-specific and an excellent resource for promoters for transgenic research in cereal species	Coussens et al. (2012)
SUI1 and *L36*	Drives *uidA* expression in all tissues of *Arabidopsis* at levels comparable to that of the *CaMV35S* promoter	Koia et al. (2013)
GSP	Used for modification of seed phenotypes in agronomically important crops	Sunkara et al. (2014)
AV3	More active in plants	Wang et al. (2014)
EF1α	Better alternative for obtaining strong and ubiquitous transgene expression compared to the *CaMV35S* promoter	Althoff et al. (2014)
Rep and *CP*	Strong transient expression in tobacco and cotton leaves whereas *cp* promoter showed lower expression	Khan et al. (2015)
AtTCTP	Alternative to the *CaMV35S* promoter for developing GM crops	Han et al. (2015)
JcUEP	Constitutive transgene expression in *jatropha* and other plants has been reported	Tao et al. (2015)
AtSCPL30	Reduces transgene silencing and reliable for multigene transformation	Jiang et al. (2018)
CsGAPC2, CsEF1	In the vegetative tissues of transgenic sweet orange plants, constitutive expression was observed	Erpen-Dalla Corte et al. (2020)

tolerated since they do not interfere with the host's natural metabolism (Mavituna 2005).

Unexpected targeting and incomplete or incorrect processing of heterologous protein in transgenic plants with native signal sequence have been documented, as reported for heat-labile toxin of *E. coli* in maize granules, delta endotoxin of *Bacillus thuringiensis* in tobacco, and cholera toxin B subunit in tobacco. Using known signal sequences that can be fused to the target gene, targeting to all intracellular compartments is possible with the current state of the art and understanding (Wong et al. 1992; Arakawa et al. 1997; Kusnadi et al. 1997; Chikwamba et al. 2003; Jani et al. 2004; Teli and Timko 2004). The cytosol, endoplasmic reticulum (ER), apoplastic space, vacuole, and chloroplast have all been investigated and compared for foreign protein accumulation (Table 2.5). The biochemical environment of the compartment, such as pH, and the available space for protein storage are two factors that may influence the extent of accumulation in distinct subcellular sites (Smith et al. 2002; Streatfield 2007). Several studies comparing the efficacy of protein production via cytoplasmic, ER, and apoplastic accumulation in different plants have been conducted.

In contrast to cytosolic expression, which gives poor protein yields, ER targeting produces the highest yield of physiologically active protein (Owen et al. 1992; Tavladorak et al. 1993; Fecker et al. 1996; Schouten et al. 1996; Zimmermann

Table 2.3 Synthetic promoters and their characteristics

Promoter	Characteristics	References
2 X W2/2 X S/2 X D, 4 X W2/4 X S	Inducible expression	Rushton et al. (2002)
pOCSn-OCS, pLOCSn-OCS, pΔOCS, pLOCSΔOCS	Constitutive expression	Liu and Bao (2009)
A27znGlb1	Tissue-specific expressions of the chimeric promoter	Shepherd and Scott (2009)
VR-ACS1	Constitutive expression	Wever et al. (2010)
(SP), SP-EE, SPFF, and *SP-*	Inducible expression	Shokouhifar et al. (2011)
AtMYB60 promoter*: GUS* reporters	Tissue-specific expression (guard cells)	Cominelli et al. (2011)
MSgt-FSgt	Constitutive expression	Kumar et al. (2011)
pporRFP	Inducible expression	Liu et al. (2011)
FSuasFcp, FuasFScp	Constitutive expression	Ranjan et al. (2012)
EFCFS-HS-1, EFCFS-HS-2, EFCFSHS-3	Tissue-specific expression of EFCFSHS-3 in vascular tissues	Ranjan and Dey (2012)
Pmec, Pcec, Pdec, and *Ptec*	Enhanced transgene expression in tomato	Koul et al. (2012)
pCL	Expresses specifically and regulates the activity of acid vacuolar invertase in potato tubers at low temperature	Li et al. (2013)
4 × *CCTC*	Fungal colonization under low pi condition inducible	Lota et al. (2013)
4 × *GCC*	Jasmonic acid inducible	Van der Does et al. (2013)
Sab/sba	Cold inducible	Li et al. (2013)
4 × *ROSE1* ~ *7*	ROS inducible	Wang et al. (2013)
FfC (FUAS35SCP), FsFfCBD (FS5FUAS35SCP BD)	Bidirectional constitutive expression	Patro et al. (2013)
4x*RE/B4REA*	Hormonal and bacterial pathogen inducible	Liu et al. (2014)
FSgt-PFlt, MSgt, PFlt, PFlt-UAS-2X	Constitutive	Acharya et al. (2014)
4 × RSRE	Stress inducible	Benn et al. (2014)
p35S-PCHS -Ω, p35S-LCHS -Ω, pOCSPCHS-Ω, pOCS-LCHS-Ω	Transgenic enhancement of floral traits (flower specific)	Du et al. (2014)
MUASMSCP	Constitutive expression	Acharya et al. (2014)
MSgt-PFlt, FSgt-PFlt, PFlt-UAS-2X	Constitutive expression	Acharya et al. (2014)
CL	Tuber specific and cold inducible	Li et al. (2015)

(continued)

Table 2.3 (continued)

Promoter	Characteristics	References
GSSP1, GSSP3, GSSP5, GSSP6, GSSP7	Green tissue specific	Wang et al. (2015)
SynS1, SynS2	Constitutive expression	Chakravarthi et al. (2015)
MAMP-responsive synthetic promoter	Microbial pathogen attack	Kanofsky et al. (2016)
Saps (Sap11)	Constitutive expression	Scranton et al. (2016)
P_ DRE::35S	Inducible expression	Gerasymenko and Sheludko (2017)
SynR2 SynR1	Root specific	Mohan et al. (2017)
Ap, Dp, ANDp	Inducible expression	Ge et al. (2018)
PRSGA, P2RSGA, P2RSPA, PRSGPA, P2RSGPA, PR5SGPA, P2R5SGPA	Seed-specific bidirectional promoters	Liu et al. (2018)
BiGSSP2, BiGSSP3, BiGSSP6, and *BiGSSP7*	Bidirectional expression efficiencies specifically in green tissues	Bai et al. (2020)
SynP16	Soybean	Kummari et al. (2020)

et al. 1998; Conrad and Fiedler 1998; Schillberg et al. 1999). The high yield in the ER is due to the presence of several key posttranslational protein maturation factors in an oxidizing environment, such as chaperones, protein disulfide isomerase, and glycosylation enzymes, as well as very few proteases (Vitale et al. 1993; Shani et al. 1994; Bruyns et al. 1996). These are the most significant influences on protein folding, assembly, and stability. The cytosol lacks or has only a few of these characteristics.

The addition of a H/KDEL carboxy-terminal tetrapeptide tag to the ER results in yields that are 10–100 times higher. Vacuolar targeting and secretion into the apoplast (intercellular space beneath the cell wall) are similarly linked to ER/Golgi transit and the aforementioned maturation and stability features, as well as downstream processing advantages. However, when compared to ER accumulation, protein output is frequently found to be lower, owing to the presence of proteases and proteolytic destruction in these compartments (Pelham 1989; Herman et al. 1990; Sijmons et al. 1990; Van Engelen et al. 1994; Pueyo et al. 1995; Schouten et al. 1996; Conrad and Fiedler 1998; Gomord et al. 1999; Peeters et al. 2001; Sardana et al. 2002; Schillberg et al. 2003; Sojikul et al. 2003; Panahi et al. 2004; Ashraf et al. 2005; Ma et al. 2005; Yang et al. 2005; Mishra et al. 2006; Peng et al. 2006; Zheng et al. 2006; Agarwal et al. 2008).

Proteins directed to the ER are not changed in the Golgi apparatus; therefore, they have a lot of mannose glycans instead of the complicated form of glycosylation that some recombinant proteins need for stability and high biological activity.

Table 2.4 Promoters that have been used in CRISPR/Cas9 genome editing

Source	Cas9 promoters	Targeted gene	Cas9 (codon optimization)	Reference(s)
O. sativa	2x*CaMV35S*	*OsPDS*, *OsBADH2*	Rice	Shan et al. (2013)
A. thaliana	2x*CaMV35S*	*YFFP*	Human	Feng et al. (2013)
Fragaria vesca	*AtUBQ10*, *CaMV35S*	*TAA1*, *ARF8*	*Arabidopsis* and maize	Zhou et al. (2013)
A. thaliana	*PcUBI4–2*	*GUUS*, *UGUS*	*Arabidopsis*	Fauser et al. (2014)
A. thaliana	*PcUBI4–2*	*GUUS*, *UGUS*	*Arabidopsis*	Fauser et al. (2014)
A. thaliana	*PcUBI4–2*	*ADH1*	*Arabidopsis*	Schiml et al. (2014)
A. thaliana	*ICU2*	*AtFT*, *AtSPL4*	*Arabidopsis*	Hyun et al. (2015)
N. benthamiana	*CaMV35*	*NbPDS*, *NbPCNA*	Human	Ali et al. (2015)
O. sativa	*ZmUbi*	*GUUS*	Rice	Mao et al. (2016)
A. thaliana	*AtUBQ1*	*GUUS*	Human	Mao et al. (2016)
Z. mays	*ZmDMC1*	*Zmzb7*	Human	Feng et al. (2018)
G. hirsutum	2x*P35s*	*GUS*		Long et al. (2018)
Solanum lycopersicum, *N. benthamiana*	*CaMV35S*	*CP*, *rep*	Human	Tashkandi et al. (2018)
S. lycopersicum	*SlEF1α*, *Slp16*, *Pcubi4*, 2x*CaMV35S*	*SlNADK2A*	*Arabidopsis*	Hashimoto et al. (2018)
A. thaliana	*DD45*, *RPS5a*	*Lhcb1*	*Arabidopsis*	Ordon et al. (2020)

Recombinant proteins that require posttranslational modifications, such as glycosylation, have another essential compartment in the plant vesicular system. Glycosylation can alter a glycoprotein's physical properties, immunological interactions, biological activity, and pharmacokinetics (Bailey et al. 1998; Twyman et al. 2003).

The polypeptide backbone, the glycosylation site on the polypeptide, and the environment and type of the host cell are all elements that might influence the distribution of glycoforms formed on a protein. Co-translational transfer of an oligosaccharide precursor on the nascent polypeptide backbone at certain asparagine (Asn) residues of the consensus sequence Asn-X-Ser/Thr initiates N-glycosylation in the ER (where X is any amino acid, except proline or aspartic acid). The N-glycan is further processed by ER and Golgi processing enzymes as the protein migrates

Table 2.5 Subcellular targeting of recombinant proteins in transgenic plant cells

Plant	Recombinant protein	Expression level in subcellular compartment	Reference
Nicotiana tabacum L.	Secretory IgG	Leaf apoplast: 1.3% of TSP	Hiatt et al. (1989)
N. tabacum L.	Anti-phytochrome	Leaf cytosol: 0.1% of TSP	Owen et al. (1992)
N. tabacum L.	Vicilin	Leaf apoplast: 250 ng/mg percent of TSP Leaf ER: 25300 ng/mg percent of TSP	Wandelt et al. (1992)
Medicago sativa	Vicilin	Leaf apoplast: 50 ng/mg percent of TSP Leaf ER: 1020 ng/mg percent of TSP	Wandelt et al. (1992)
N. tabacum L.	Anti-phytochrome	Leaf apoplast: 0.5% of TSP	Firek et al. (1993)
N. tabacum L.	IgG	Root apoplast: 0.35% of TSP	Van Engelen et al. (1994)
N. tabacum L.	IgG-A hybrid	Leaf apoplast: 500 μg/g percent of FW	Ma et al. (1995)
N. tabacum L.	Anti-human creatine kinase	Leaf cytosol: 0.01% of TSP	Bruyns et al. (1996)
N. tabacum L.	Anti-beet necrotic yellow vein virus	Leaf apoplast:0.1% of TSP	Feckeret al. (1996)
N. tabacum L.	ScFv anti-cutinase	Leaf apoplast: 0.01% of TSP Leaf ER: 1% of TSP Leaf cytosol: 0% of TSP	Schouten et al. (1996)
N. tabacum L.	ScFv anti-oxazolone	Leaf apoplast: 0.1% of TSP Leaf ER: 1.1% of TSP	Fiedler et al. (1997)
N. tabacum L.	ScFv anti-oxazolone	Seed apoplast: 0.04% of TSP Seed ER: 0.9% of TSP	Fiedler et al. (1997)
Zea mays	Avidin	Seed apoplast: 2.3% of TSP	Hood et al. (1997)
S. tuberosum	Anti-β 1,4-endoglucanase	Root cytosol: 0.3% of TSP	Schouten et al. (1997)
S. tuberosum	ScFv anti-oxazolone	Tuber apoplast: 2.0% of TSP Tuber cytosol: 2.0% of TSP	Artsaenko et al. (1998)
N. tabacum L.	BiscFv 2429	Leaf apoplast: 0.014% of TSP Leaf ER: 0.43% of TSP Leaf cytosol: 0% of TSP	Fischer et al. (1999)
N. tabacum L.	Tomato mosaic virus ab-rAb29	Leaf apoplast: 8.5 μg/g FW Leaf cytosol: 0 μg/g FW	Schillberg et al. (1999)
Hordeum vulgare	β-Glucanase	Grain vacuole (protein storage vacuole): 54 μg/mg TSP	Horvath et al. (2000)
N. tabacum L.	Human growth hormone	Seed apoplast: 0.16% of TSP	Leite et al. (2000)
Arabidopsis thaliana	FAb MAK33	Leaf apoplast: 6.53% of TSP Leaf ER: 5.9% of TSP	Peeters et al. (2001)

(continued)

Table 2.5 (continued)

Plant	Recombinant protein	Expression level in subcellular compartment	Reference
N. tabacum L.	E1 endo-1,4 β-glucanase/cellulase	Leaf apoplast: 0.11% of TSP Leaf cytosol: 0.0003% of TSP	Ziegelhoffer et al. (2001)
N. tabacum L.	Avidin, streptavidin	Leaf vacuole: 1.5% of TSP	Murray et al. (2002)
N. tabacum L.	Hepatitis B surface antigen	Leaf apoplast: 0.031% of TSP Leaf ER: 0.22% of TSP Leaf cytosol: 0% of TSP Leaf vacuole: 0.032% of TSP	Ramírez et al. (2002)
N. tabacum L.	scFv anti-carcinoembryonic	Leaf apoplast: 1 μg/g Leaf ER: 25 μg/g Leaf vacuole:1 μg/g	Stoger et al. (2002)
N. tabacum L.	Ab anti-carcinoembryonic	Leaf apoplast: 0.9 mg/kg FW Leaf ER: 9 mg/kg FW	Vaquero et al. (2002)
S. tuberosum	Equistatin	Leaf ER: 1.9% of TSP Leaf cytosol: 0.36% of TSP	Outchkourov et al. (2003)
BY-2 cells	Hepatitis-B surface antigen	BY-2 cell apoplast: 226 ng/mg TSP BY-2 cell ER: 177 ng/mg TSP BY-2 cell cytosol: 128 ng/mg TSP	Sojikul et al. (2003)
N. tabacum L.	Human epidermal growth factor	Leaf apoplast: 0.11% of TSP Leaf cytosol: ~0% of TSP	Wirth et al. (2004)
N. tabacum L.	Endoglucanase (E1)	Leaf apoplast: 0.25% of TSP	Dai et al. (2005)
N. tabacum L.	Human growth hormone	Leaf apoplast: 10% of TSP Leaf cytosol: 0.01% of TSP	Gils et al. (2005)
Arabidopsis thaliana	Spider silk-like protein-DP1B	Leaf apoplast: 8.5% of TSP Leaf ER: 6.7% of TSP Leaf vacuole: 0% of TSP	Yang et al. (2005)
Arabidopsis thaliana	Spider silk-like protein-DP1B	Seed apoplast: 0% of TSP Seed ER: 18% of TSP Seed vacuole: 8.2% of TSP	Yang et al. (2005)
N. tabacum L.	Ab 14D9-κ and γ chain	Leaf apoplast: 2.9% of TSP Leaf ER: 5.2% of TSP	Petruccelli et al. (2006)
N. tabacum L.	HIV-1 Nef	Protoplast cytosol: 0.7% of TSP Protoplast apoplast: Unstable	Marusic et al. (2007)
N. tabacum L.	Human insulin-like growth factor	Seed vacuole (protein storage vacuole): 800 μg/g dry weight	Cheung et al. (2009)

down the secretory pathway, with the addition or removal of sugar residues (Dwek 1996).

While the early steps in N-glycosylation and the glycosylation machinery of the ER are common in most species and add comparable oligomannose glycans, the subsequent processing and modification of the glycans in the Golgi apparatus vary significantly among hosts. Yeast N-glycans are hyper-mannosylated, fungal N-glycans are high-mannose, and insect N-glycans are paucimannosidic, with a

Man3GlcNAc2 core replaced with α-(1,3)- and α-(1,6)-linked fucose at the proxy (GlcNAc) (Jarvis et al. 2003; Gerngross et al. 2004; Chen et al. 2005; Brooks 2006).

Plants can add complex N-linked glycans by substituting two GlcNAc residues in the core, similar to the glycosylation pattern seen in humans. Plants, on the other hand, do not add galactose and terminal sialic acid residues, but instead add α-(1,3)-fucose and α-(1,6)-xylose residues, which may elicit an unwanted immune response in humans. In order to humanize the glycan pattern formed by transgenic plants, many techniques have been devised (Lerouge et al. 1998; Wilson 2002; Bakker et al. 2006; Bardor et al. 2006; Cox et al. 2006, Loos et al. 2011; Frey et al. 2009; Karg et al. 2009; Matsuo and Matsumura 2010; Castilho et al. 2010).

2.5 Conclusions

The success of transgenic technologies depends solely on the expression, stability, and integrity of the transgene in the offsprings of the transgenic plants. In order to ensure it, the transgene must be codon optimized and directed to a safe subcellular target through a robust transformation technique. In a nutshell, the transgene must be deprived of the sequence motifs and codons that lead to mRNA degradation and low-level expression (putative polyadenylation and mRNA instability sequences and RNA polymerase II termination signals, cryptic splicing sites, secondary structures, etc.). Elements directing high-level expression (strong promoters: either modified natural or synthetic, 5′ untranslated leader sequence, translation initiation context, etc.) must be deployed, and the transgene product must be targeted to cellular compartments (ER, vacuole, apoplast, etc.) suitable for accumulation and stability of the protein. After following these precautions while gene designing, varying degree of success (expression level) has been reported in various transgenic plants.

References

Abranches R, Marcel S, Arcalis E, Altmann F, Fevereiro P, Stoger E (2005) Plants as bioreactors: a comparative study suggests that *Medicago truncatula* is a promising production system. J Biotechnol 120:121–134

Acharya S, Ranjan R, Pattanaik S, Maiti IB, Dey N (2014) Efficient chimeric plant promoters derived from plant infecting viral promoter sequences. Planta 239(2):381–396

Acharya S, Sengupta S, Patro S, Purohit S, Samal SK, Maiti IB, Dey N (2014) Development of an intra-molecularly shuffled efficient chimeric plant promoter from plant infecting *Mirabilis mosaic virus* promoter sequence. J Biotechnol 169:103–111

Agarwal S, Jha S, Sanyal I, Amla DV (2009) Effect of point mutations in translation initiation context on the expression of recombinant human α1-proteinase inhibitor in transgenic tomato plants. Plant Cell Rep 28:1791–1798

Agarwal S, Singh R, Sanyal I, Amla DV (2008) Expression of modified gene encoding functional human a-1-antitrypsin protein in transgenic tomato plants. Transgenic Res 17:881–896

Ali Z, Abulfaraj A, Idris A, Ali S, Tashkandi M, Mahfouz MM (2015) CRISPR/Cas9-mediated viral interference in plants. Genome Biol 16(1):1–1

Allen GC, Hall GE Jr, Childs LC, Weissinger AK, Spiker S, Thompson WF (1993) Scaffold attachment regions increase reporter gene expression in stably transformed plant cells. Plant Cell 5:603–613

Allen GC, Hall GE Jr, Michalowski S, Newman W, Spiker S, Weissinger AK (1996) High-level transgene expression in plant cells: effects of a strong scaffold attachment region from tobacco. Plant Cell 8:899–913

Althoff F, Kopischke S, Zobell O, Ide K, Ishizaki K, Kohchi T, Zachgo S (2014) Comparison of the *MpEF1α* and *CaMV35* promoters for application in Marchantia polymorpha overexpression studies. Transgenic Res 23(2):235–244

An YQ, McDowell JM, Huang S, McKinney EC, Chambliss S, Meagher RB (1996) Strong, constitutive expression of the *Arabidopsis ACT2/ACT8* actin subclass in vegetative tissues. Plant J 10(1):107–121

Arakawa T, Chong DK, Merritt JL, Langridge WH (1997) Expression of cholera toxin B subunit oligomers in transgenic potato plants. Transgenic Res 6(6):403–413

Artsaenko O, Kettig B, Fiedler U, Conrad U, During K (1998) Potato tubers as a biofactory for recombinant antibodies. Mol Breed 4:313–319

Ashraf S, Singh PK, Yadav DK, Shahnawaz M, Mishra S, Sawant SV, Tuli R (2005) High-level expression of surface glycoprotein of rabies virus in tobacco leaves and its immunoprotective activity in mice. J Biotechnol 119:1–14

Bai J, Wang X, Wu H, Ling F, Zhao Y, Lin Y, Wang R (2020) Comprehensive construction strategy of bidirectional green tissue-specific synthetic promoters. Plant Biotechnol J 18(3):668–678

Bailey JE, Prati EGP, Jean-Mairet J, Sburlati AR, Umana P (1998) Engineering glycosylation in animal cells. In: Merten OW (ed) New developments and new applications in animal cell technology. Kluwer Academic Publishers, Boston, MA

Bakker H, Rouwendal GJ, Karnoup AS, Florack DEA, Stoopen GM, Helsper J (2006) An antibody produced in tobacco expressing a hybrid beta-(1,4)-galactosyltransferase is essentially devoid of plant carbohydrate epitopes. Proc Natl Acad Sci U S A 103:7577–7582

Bardor M, Cabrera G, Rudd PM, Dwek RA, Cremata JA, Lerouge P (2006) Analytical strategies to investigate plant N-glycan profiles in the context of plant-made pharmaceuticals. Curr Opin Struck Biol 16:576–583

Battraw MJ, Hall TC (1990) Histochemical analysis of CaMV35S promoter β-glucuronidase gene expression in transgenic rice plants. Plant Mol Biol 15:527–538

Benchabane M, Goulet C, Rivard D, Faye L, Gomord V, Michaud D (2008) Preventing unintended proteolysis in plant protein biofactories. Plant Biotechnol J 6:633–648

Benfey PN, Chua NH (1990) The cauliflower mosaic virus 35S promoter: combinatorial regulation of transcription in plants. Science 250:959–966

Benfey PN, Ren L, Chua NH (1989) The *CaMV35S* enhancer contains at least two domains which can confer different developmental and tissue-specific expression patterns. EMBO J 8(8): 2195–2202

Benn G, Wang CQ, Hicks DR, Stein J, Guthrie C, Dehesh K (2014) A key general stress response motif is regulated non-uniformly by CAMTA transcription factors. Plant J 80(1):82–92

Beutler E, Gelbart T, Han J, Koziol JA, Beutler B (1989) Evolution of the genome and the genetic code: selection at the dinucleotide level by methylation and polyribonucleotide cleavage. Proc Natl Acad Sci U S A 86:192–196

Bhat SR, Srinivasan S (2002) Molecular and genetic analyses of transgenic plants: considerations and approaches. Plant Sci 163:673–681

Bhattacharyya S, Dey N, Maiti IB (2002) Analysis of cis-sequence of subgenomic transcript promoter from the *Figwort mosaic virus* and comparison of promoter activity with the *cauliflower mosaic virus* promoters in monocot and dicot cells. Virus Res 90(1–2):47–62

Bhullar S, Chakravarthy S, Advani S, Datta S, Pental D, Burma PK (2003) Strategies for development of functionally equivalent promoters with minimum sequence homology for transgene expression in plants: cis-elements in a novel DNA context versus domain swapping. Plant Physiol 132:988–998

Boddupally D, Tamirisa S, Gundra SR, Vudem DR, Khareedu VR (2018) Expression of hybrid fusion protein (Cry1Ac::ASAL) in transgenic rice plants imparts resistance against multiple insect pests. Sci Rep 8(1):1–10
Boehm R (2007) Bioproduction of therapeutic proteins in the 21st century and the role of plants and plant cells as production platforms. Ann N Y Acad Sci 1102:121–134
Borg S, Brinch-Pedersen H, Tauris B, Madsen LH, Darbani B, Noeparvar S, Holm PB (2012) Wheat ferritins: improving the iron content of the wheat grain. J Cereal Sci 56(2):204–213
Boyd L, Thummel CS (1993) Selection of CUG and AUG initiator codons for Drosophila E74A translation depends on downstream sequences. Proc Natl Acad Sci U S A 90:9164–9167
Brooks SA (2006) Protein glycosylation in diverse cell systems: implications for modification and analysis of recombinant proteins. Exp Rev Proteomics 3:345–359
Brown JWS, Simpson CG (1998) Splice site selection in plant pre-mRNA splicing. Annu Rev Plant Physiol Plant Mol Biol 49:77–95
Bruyns A-M, de Jaeger G, de Neve M, de Wilde C, van Montagu M, Depicker A (1996) Bacterial and plant-produced scFv proteins have similar antigen-binding properties. FEBS Lett 386:5–10
Butaye KM, Cammune BPA, Delaure SL, De Bolle MF (2005) Approaches to minimize variation of transgene expression in plants. Mol Breed 16:79–91
Castilho A, Strasser R, Stadlmann J, Grass J, Jez J, Gattinger P, Kunert R, Quendler H, Pabst M, Leonard R, Altmann F, Steinkellner H (2010) In planta protein sialylation through overexpression of the respective mammalian pathway. J Biol Chem 285:15923–15930
Chakravarthi M, Narayan JA, Subramonian N, Appunu C (2015) Development of novel synthetic promoters for gene expression in transgenic sugarcane. J Sugarcane Res 5(2):42–52
Chan MT, Chang HH, Ho SL, Tong WF, Yu SM (1993) Agrobacterium-mediated production of transgenic rice plants expressing a chimeric α-amylase promoter/β-glucuronidase gene. Plant Mol Biol 22(3):491–506
Chan MT, Yu SM (1998) The 3′ untranslated region of a rice alpha-amylase gene functions as a sugar-dependent mRNA stability determinant. Proc Natl Acad Sci U S A 95:6543–6547
Chaturvedi CP, Sawant SV, Kiran K, Mehrotra R, Lodhi N, Ansari SA, Tuli R (2006) Analysis of polarity in the expression from a multifactorial bidirectional promoter designed for high-level expression of transgenes in plants. J Biotechnol 123:1–12
Chen M, Liu X, Wang Z, Song J, Qi Q, Wang PG (2005) Modification of plant N-glycans processing: the future of producing therapeutic protein by transgenic plants. Med Res Rev 25: 343–360
Chen ZL, Schuler MA, Beachy RN (1986) Functional analysis of regulatory elements in a plant embryo-specific gene. Proc Natl Acad Sci 83(22):8560–8564
Cheung SCK, Sun SSM, Chan JCN, Tong PCY (2009) Expression and subcellular targeting of human insulin-like growth factor binding protein −3 in transgenic tobacco plants. Transgenic Res 18(6):943–951
Chikwamba RK, Scott MP, Mejia LB, Mason HS, Wang K (2003) Localization of a bacterial protein in starch granules of transgenic maize kernels. Proc Natl Acad Sci U S A 100:11127–11132
Christensen AH, Sharrock RA, Quail PH (1992) Maize polyubiquitin genes: structure, thermal perturbation of expression and transcript splicing, and promoter activity following transfer to protoplasts by electroporation. Plant Mol Biol 18:675–689
Comai L, Moran P, Maslyar D (1990) Novel and useful properties of a chimeric plant promoter combining CaMV35S and MAS elements. Plant Mol Biol 15:373–381
Cominelli E, Galbiati M, Albertini A, Fornara F, Conti L, Coupland G, Tonelli C (2011) DOF–binding sites additively contribute to guard cell–specificity of *AtMYB60* promoter. BMC Plant Biol 11(1):1–3
Conrad U, Fiedler U (1998) Compartment-specific accumulation of recombinant immunoglobulins in plant cells: an essential tool for antibody production and immunomodulation of physiological functions and pathogen activity. Plant Mol Biol 38:101–109

Conrad U, Fiedler U, Artsaenko O, Phillips J (1998) High-level and stable accumulation of singlechain Fv antibodies in plant storage organs. J Plant Physiol 152:708–711

Copenhaver GP, Preuss D (2000) Plant artificial chromosome compositions and methods, United States Patent, 6156953

Cornejo MJ, Luth D, Blankenship KM, Anderson OD, Blechl AE (1993) Activity of a maize ubiquitin promoter in transgenic rice. Plant Mol Biol 23(3):567–581

Coussens G, Aesaert S, Verelst W, Demeulenaere M, De Buck S, Njuguna E, Inzé D, Van Lijsebettens M (2012) *Brachypodium distachyon* promoters as efficient building blocks for transgenic research in maize. J Exp Bot 63(11):4263–4273

Cox KM, Sterling JD, Regan JT, Gasdaska JR, Frantz KK, Peele CG, Black A, Passmore D, Moldovan-Loomis C, Srinivasan M, Cuison S, Cardarelli PM, Dickey LF (2006) Glycan optimization of a human monoclonal antibody in the aquatic plant *Lemna minor*. Nat Biotechnol 24:12

Curtis IS, Power JB, Hedden P, Phillip A, Lowe KC, Ward DA, Davey MR (2000) Transformation and characterization of transgenic plants of Solanum dulcamara L. –incidence of transgene silencing. Ann Bot 86:63–71

Dai Z, Hooker BS, Quesenberry RD, Thomas SR (2005) Optimization of *Acidothermus cellulolyticus* (E1) production in transgenic tobacco plants by transcriptional, post-transcription and post-translational modification. Transgenic Res 14:627–643

Daniell H, Ruiz G, Denes B, Sandberg L, Langridge W (2009) Optimization of codon composition and regulatory elements for expression of human insulin like growth factor-1 in transgenic chloroplasts and evaluation of structural identity and function. BMC Biotechnol 9(1):1–16

Datla RS, Bekkaoui F, Hammerlindl JK, Pilate G, Dunstan DI, Crosby WL (1993) Improved high level constitutive foreign gene expression in plants using an AMV RNA4 untranslated leader sequence. Plant Sci 94:139–149

De Paepe A, De Buck S, Hoorelbeke K, Nolf J, Peck I, Depicker A (2009) High frequency of single copy T-DNA transformants by floral dip in CRE-expressing *Arabidopsis* plants. Plant J 59(4): 517–527

Dey N, Maiti IB (1999) Structure and promoter/leader deletion analysis of *mirabilis mosaic virus* (*MMV*) full-length transcript promoter in transgenic plants. Plant Mol Biol 40(5):771–782

Di Rita VJ, Gelvin SB (1987) Deletion analysis of mannopine synthase gene promoter in sunflower crown gall tumors and *Agrobacterium tumefaciens*. Mol Gen Genet 207:233–241

Dong JL, Liang BG, Jin YS, Zhang WJ, Wang T (2005) Oral immunization with pBsVP6-transgenic alfalfa protects mice against rotavirus infection. Virology 339:153–163

Doran PM (2006) Foreign protein degradation and instability in plants and tissue cultures. Trends Biotech 24:426–432

Dowson Day MJ, Ashurst JA, Mathias SF, Watts JW, Wilson TM, Dixon RA (1993) Plant viral leaders influence expression of a reporter gene in tobacco. Plant Mol Biol 23(1):97–109

Du L, Lou Q, Zhang X, Jiao S, Liu Y, Wang Y (2014) Construction of flower-specific chimeric promoters and analysis of their activities in transgenic torenia. Plant Mol Biol Rep 32(1): 234–245

Ducreux LJ, Morris WL, Hedley PE, Shepherd T, Davies HV, Millam S, Taylor MA (2005) Metabolic engineering of high carotenoid potato tubers containing enhanced levels of β-carotene and lutein. J Exp Bot 56(409):81–89

Dutta I, Majumder P, Saha P, Ray K, Das S (2005) Constitutive and phloem specific expression of *Allium sativum* leaf agglutinin (ASAL) to engineer aphid (*Lipaphis erysimi*) resistance in transgenic Indian mustard (*Brassica juncea*). Plant Sci 169(6):996–1007

Dwek RA (1996) Glycobiology: toward understanding the function of sugars. Chem Rev 96:683–720

Dwivedi KK, Roche D, Carman JG (2010) Expression in *Arabidopsis* of a nucellus-specific promoter from watermelon (*Citrullus lanatus*). J Plant Sci 179(5):549–552

Elkholy SF, Ismail RM, Bahieldin A, Sadik AS, Madkour MA (2009) Expression of hepatitis B surface antigen (HBsAg) gene in transgenic banana (Musa Sp.). Arab J Biotechnol 12:291–302

Erpen-Dalla Corte L, Mendes BM, Mourão Filho FA, Grosser JW, Dutt M (2020) Functional characterization of full-length and 5′ deletion fragments of *Citrus sinensis*-derived constitutive promoters in *Nicotiana benthamiana*. In Vitro Cell Dev Biol Plant 8:1

Fagard M, Vaucheret H (2000) (Trans) gene silencing in plants: how many mechanisms? Plant Mol Biol 51:167–194

Falco SC, Guida T, Locke M, Mauvais J, Sanders C, Ward RT, Webber P (1995) Transgenic canola and soybean seeds with increased lysine. Bio/Technol 13(6):577–582

Fang RX, Nagy F, Sivasubramaniam S, Chua NH (1989) Multiple cis-regulatory elements for maximal expression of the cauliflower mosaic virus 35S promoter in transgenic plants. Plant Cell 1:141–150

Fauser F, Schiml S, Puchta H (2014) Both CRISPR/Cas-based nucleases and nickases can be used efficiently for genome engineering in *Arabidopsis thaliana*. Plant J 79(2):348–359

Fecker LF, Kaufmann A, Commandeur U, Commandeur J, Koenig R, Burgermeister W (1996) Expression of a single chain antibody fragment scFv specific for beet necrotic yellow vein virus coat protein or 25 kDa protein in *Escherichia coli* and *Nicotiana benthamiana*. Plant Mol Biol 32:979–986

Feng C, Su H, Bai H, Wang R, Liu Y, Guo X, Liu C, Zhang J, Yuan J, Birchler JA, Han F (2018) High-efficiency genome editing using a dmc1 promoter-controlled CRISPR/Cas9 system in maize. Plant Biotechnol J 16(11):1848–1857

Feng Z, Zhang B, Ding W, Liu X, Yang DL, Wei P, Cao F, Zhu S, Zhang F, Mao Y, Zhu JK (2013) Efficient genome editing in plants using a CRISPR/Cas system. Cell Res 23(10):1229–1232

Fennoy SL, Bailey-serres J (1993) Synonymous codon usage in *Zea mays* L nuclear genes is varied by levels of C-ending and G-ending codons. Nucl Acid Res 21:5294–5300

Fiedler U, Phillips J, Artsaenko O, Conrad U (1997) Optimization of scFv antibody production in transgenic plants. Immunotechnology 3:205–216

Firek S, Draper J, Owen MR, Gandecha A, Cockburn B, Whitelam GC (1993) Secretion of a functional single-chain Fv protein in transgenic tobacco plants and cell suspension cultures. Plant Mol Biol 23:861–870

Fischer R, Schumann D, Zimmermann S, Drossard J, Sack M, Schillberg S (1999) Expression and characterization of bispecific single-chain Fv fragments produced in transgenic plants. Eur J Biochem 262:810–816

Fischer R, Stoger E, Schillberg S, Christou P, Twyman RM (2004) Plant-based production of biopharmaceuticals. Curr Opin Plant Biol 7:152–158

Fiume E, Christou P, Giani S, Breviario D (2004) Introns are key regulatory elements of rice tubulin expression. Planta 218:693–703

Frey AD, Karg SR, Kallio PT (2009) Expression of rat β(1,4)-N-acetylglucosaminyltransferase III in *Nicotiana tabacum* remodels the plant-specific N-glycosylation. Plant Biotechnol J 7:33–48

Gallie DR, Tanguay RL, Leathers V (1995) The tobacco etch viral 5′ leader and poly(A) tail are functionally synergistic regulators of translation. Gene 165:233–238

Gao SQ, Chen M, Xia LQ, Xiu HJ, Xu ZS, Li LC et al (2009) A cotton (*Gossypium hirsutum*) DRE-binding transcription factor gene, GhDREB, confers enhanced tolerance to drought, high salt, and freezing stresses in transgenic wheat. Plant Cell Rep 28(2):301–311

Gao T, Wu Y, Zhang Y, Liu L, Ning Y, Wang D et al (2011) OsSDIR1 overexpression greatly improves drought tolerance in transgenic rice. Plant Mol Biol 76(2):145–156

Ge H, Li X, Chen S, Zhang M, Liu Z, Wang J, Li X, Yang Y (2018) The expression of *CARK1* or *RCAR11* driven by synthetic promoters increases drought tolerance in *Arabidopsis thaliana*. Int J Mol Sci 19(7):1945

Gerasymenko IM, Sheludko YV (2017) Synthetic cold-inducible promoter enhances recombinant protein accumulation during Agrobacterium-mediated transient expression in *Nicotiana* excelsior at chilling temperatures. Biotechnol Lett 39(7):1059–1067

Gerngross TU (2004) Advances in the production of human therapeutic proteins in yeasts and filamentous fungi. Nat Biotechnol 22:1409–1414

Gilissen LJW, Metz PLJ, Stiekema WJ, Nap J-P (1998) Biosafety of *E. coli* β-glucuronidase (GUS) in plants. Transgenic Res 7(3):157–163
Gils M, Kandzia R, Marillonnet S, Klimyuk V, Gleba Y (2005) High-yield production of authentic human growth hormone using a plant virus-based expression system. Plant Biotechnol J 3:613–620
Gomord V, Wee E, Faye L (1999) Protein retention and localization in the endoplasmic reticulum and the Golgi apparatus. Biochimie 81:1–12
Grichko VP, Filby B, Glick BR (2000) Increased ability of transgenic plants expressing the bacterial enzyme ACC deaminase to accumulate Cd, Co, Cu, Ni, Pb, and Zn. J Biotechnol 81(1):45–53
Guerineau F, Lucy A, Mullineaux P (1992) Effect of two consensus sequences preceding the translation initiator codon on gene expression in plant protoplasts. Plant Mol Biol 18:815–818
Gustafsson C, Govindarajan S, Minshull J (2004) Codon bias and heterologous protein expression. Trends Biotechnol 22:346–353
Gutiérrez-Ortega A, Sandoval-Montes C, Olivera-Flores TJ, Santos-Argumedo L, Gómez-Lim MÁ (2005) Expression of functional interleukin-12 from mouse in transgenic tomato plants. Transgenic Res 14:877–885
Halpin C (2005) Gene stacking in transgenic plants-the challenge for 21st century plant biotechnology. Plant Biotechnol J 3:141–155
Hampsey M (1998) Molecular genetics of the RNA polymerase II general transcriptional machinery. Microbiol Mol Biol Rev 62(2):465–503
Han YJ, Kim YM, Hwang OJ, Kim JI (2015) Characterization of a small constitutive promoter from *Arabidopsis* translationally controlled tumor protein (*AtTCTP*) gene for plant transformation. Plant Cell Rep 34(2):265–275
Hanumappa M, Choi G, Ryu S, Choi G (2007) Modulation of flower colour by rationally designed dominant-negative chalcone synthase. J Exp Bot 58(10):2471–2478
Haq TA, Mason HS, Clements JD, Arntzen CJ (1995) Oral immunization with a recombinant bacterial antigen produced in transgenic plants. Science 268:714–716
Haseloff J, Siemering KR, Prasher DC, Hodge S (1997) Removal of a cryptic intron and subcellular localization of green fluorescent protein are required to mark *Arabidopsis* plants brightly. Proc Natl Acad Sci U S A 94:2122–2127
Hashimoto R, Ueta R, Abe C, Osakabe Y, Osakabe K (2018) Efficient multiplex genome editing induces precise, and self-ligated type mutations in tomato plants. Front Plant Sci 3(9):916
He X, Hou X, Shen Y, Huang Z (2011) TaSRG, a wheat transcription factor, significantly affects salt tolerance in transgenic rice and Arabidopsis. FEBS Lett 585(8):1231–1237
He H, Ke H, Keting H, Qiaoyan X, Silan D (2013) Flower colour modification of chrysanthemum by suppression of F3′H and overexpression of the exogenous Senecio cruentus F3′5′H gene. PLoS One 8(11):e74395
Herman EM, Tague BW, Hoffman LM, Kjemtrup SE, Chrispeels MJ (1990) Retention of phytohemagglutinin with carboxy terminal tetrapeptide KDEL in the nuclear envelope and the endoplasmic reticulum. Planta 182:305–312
Hernandez-Garcia CM, Martinelli AP, Bouchard RA, Finer JJ (2009) A soybean (*Glycine max*) polyubiquitin promoter gives strong constitutive expression in transgenic soybean. Plant Cell Rep 28:837–849
Hiatt A, Cafferkey R, Bowdish K (1989) Production of antibodies in transgenic plants. Nature 342: 76–78
Hood E, Witcher D, Maddock S, Meyer T, Baszczynski C, Bailey M (1997) Commercial production of avidin from transgenic maize: characterization of transformant, production, processing, extraction and purification. Mol Breed 3:291–306
Horvath H, Huang J, WongOi Kohl E, Okita T, Kannangara CG (2000) The production of recombinant proteins in transgenic barley grains. Proc Natl Acad Sci U S A 97:1914–1919
Hull AK, Yusibov V, Mett V (2005) Inducible expression in plants by virus-mediated transgene activation. Transgenic Res 14:407–416

Hyun Y, Kim J, Cho SW, Choi Y, Kim JS, Coupland G (2015) Site-directed mutagenesis in *Arabidopsis thaliana* using dividing tissue-targeted RGEN of the CRISPR/Cas system to generate heritable null alleles. Planta 241(1):271–284

Im Choi Y, Noh EW, Lee HS, Han MS, Lee JS, Choi KS (2007) Mercury-tolerant transgenic poplars expressing two bacterial mercury-metabolizing genes. J Plant Biol 50(6):658–662

Jani D, Singh NK, Bhattacharya S, Meena LS, Singh Y, Upadhyay SN, Sharma AK, Tyagi AK (2004) Studies on the immunogenic potential of plant-expressed cholera toxin B subunit. Plant Cell Rep 22:471–477

Jarvis DL (2003) Developing baculovirus-insect cell expression systems for humanized recombinant glycoprotein production. Virology 310:1–7

Jarvis P, Belzile F, Dean C (1997) Insufficient and incorrect processing of the Ac transposase transcript in iae1 and wild-type *Arabidopsis thaliana*. Plant J 11:921–931

Jefferson RA (1987) Assaying chimeric genes in plants: the GUS gene fusion system. Plant Mol Biol Rep 5:387–405

Jiang P, Zhang K, Ding Z, He Q, Li W, Zhu S, Cheng W, Zhang K, Li K (2018) Characterization of a strong and constitutive promoter from the *Arabidopsis* serine carboxypeptidase-like gene *AtSCPL30* as a potential tool for crop transgenic breeding. BMC Biotechnol 18(1):1–3

Jin RG, Liu YB, Tabashnik BE, Borthakur D (2000) Development of transgenic cabbage (*Brassica oleracea* var. capitata) for insect resistance by *Agrobacterium tumefaciens*-mediated transformation. In Vitro Cell Dev Biol Plant 36(4):231–237

Joshi CP, Zhou H, Huang X, Chiang VL (1997) Context sequences of translation initiation codon in plants. Plant Mol Biol 35:993–1001

Kane JF (1995) Effects of rare codon clusters on high-level expression of heterologous proteins in *Escherichia coli*. Curr Opin Biotechnol 6:494–500

Kang TJ, Loc NH, Jang MO, Yang MS (2004) Modification of the cholera toxin B subunit coding sequence to enhance expression in plants. Mol Br 13:143–153

Kanofsky K, Lehmeyer M, Schulze J, Hehl R (2016) Analysis of microbe-associated molecular pattern-responsive synthetic promoters with the parsley protoplast system. In: Plant synthetic promoters. Humana Press, New York, NY, pp 163–174

Kanrar S, Venkateswari J, Kirti P, Chopra V (2002) Transgenic Indian mustard (Brassica juncea) with resistance to the mustard aphid (*Lipaphis erysimi* Kalt.). Plant Cell Rep 20(10):976–981

Karg SR, Frey AD, Ferrara C, Streich DK, Umaña P, Kallio PT (2009) A small-scale method for the preparation of plant N-linked glycans from soluble proteins for analysis by MALDITOF mass spectrometry. Plant Physiol Biochem 47:160–166

Katsumoto Y, Fukuchi-Mizutani M, Fukui Y, Brugliera F, Holton TA, Karan M et al (2007) Engineering of the rose flavonoid biosynthetic pathway successfully generated blue-hued flowers accumulating delphinidin. Plant Cell Physiol 48(11):1589–1600

Kay R, Chan A, Daly M, McPherson J (1987) Duplication of CaMV35S promoter sequence creates a strong enhancer for plant genes. Science 236:1299–1302

Khan ZA, Abdin MZ, Khan JA (2015) Functional characterization of a strong bi-directional constitutive plant promoter isolated from cotton leaf curl *Burewala virus*. PLoS One 10(3): e0121656

Kim T-G, Baek M-Y, Lee E-K, Kwon T-H, Yang M-S (2008) Expression of human growth hormone in transgenic rice cell suspension culture. Plant Cell Rep 27:885–891

Kinney AJ, Knowlton S (1998) Designer oils: the high oleic acid soybean. Genetic modification in the food industry. Springer, Boston, MA, pp 193–213

Ko K, Tekoah Y, Rudd PM, Harvey DJ, Dwek RA, Spitsin S (2003) Function and glycosylation of plant-derived antiviral monoclonal antibody. Proc Natl Acad Sci U S A 100:8013–8018

Koia J, Moyle R, Hendry C, Lim L, Botella JR (2013) Pineapple translation factor SUI1 and ribosomal protein L36 promoters drive constitutive transgene expression patterns in *Arabidopsis thaliana*. Plant Mol Biol 81(4–5):327–336

Koul B, Yadav R, Sanyal I, Sawant S, Sharma V, Amla DV (2012) Cis–acting motifs in artificially synthesized expression cassette leads to enhanced transgene expression in tomato (*Solanum lycopersicum* L.). Plant Physiol Bioch 61:131–141

Kozak M (1986) Point mutations define a sequence flanking the AUG initiator codon that modulates translation by eukaryotic ribosomes. Cell 44:283–292

Kozak M (1987) An analysis of 5′-noncoding sequences from 699 vertebrate messenger RNAs. Nucleic Acids Res 15:8125–8148

Kozak M (1999) Initiation of translation in prokaryotes and eukaryotes. Gene 234:187–208

Kozak M (2002) Pushing the limits of the scanning mechanism for initiation of translation. Gene 299:1–34

Kumar GS, Ganapathi TR, Revathi CJ, Srinivas L, Bapat VA (2005) Expression of hepatitis B surface antigen in transgenic banana plants. Planta 222(3):484–493

Kumar D, Patro S, Ranjan R, Sahoo DK, Maiti IB, Dey N (2011) Development of useful recombinant promoter and its expression analysis in different plant cells using confocal laser scanning microscopy. PLoS One 6(9):e24627

Kummari D, Palakolanu SR, Kishor PK, Bhatnagar-Mathur P, Singam P, Vadez V, Sharma KK (2020) An update and perspectives on the use of promoters in plant genetic engineering. J Biosci 45(1):1–24

Kusnadi AR, Nikolov ZL, Howard JA (1997) Production of recombinant proteins in transgenic plants: practical considerations. Biotechnol Bioeng 56:473–484

La Frano MR, Woodhouse LR, Burnett DJ, Burri BJ (2013) Biofortified cassava increases β-carotene and vitamin A concentrations in the TAG-rich plasma layer of American women. Br J Nutr 110(2):310–320

Lawton MA, Tierney MA, Nakamura I, Anderson E, Komeda Y, Dubé P et al (1987) Expression of a soybean β-conclycinin gene under the control of the *cauliflower mosaic virus* 35S and 19S promoters in transformed petunia tissues. Plant Mol Biol 9(4):315–324

Leisner SM, Gelvin SB (1988) Structure of the octopine synthase upstream activator sequence. Proc Natl Acad Sci U S A 85:2533–2557

Leite A, Kemper E, da Silva M, Luchessi A, Siloto R, Bonaccorsi E (2000) Expression of correctly processed human growth hormone in seeds of transgenic tobacco plants. Mol Breeding 6:47–53

Lerouge P, Cabanes-Macheteau M, Rayon C, Fischette-Laine AC, Gomord V, Faye L (1998) N-glycoprotein biosynthesis in plants: recent developments and future trends. Plant Mol Biol 38:31–48

Li D, Leary JO, Huang Y, Huner NPA, Jevnikar AM, Ma S (2006) Expression of cholera toxin B subunit and the B chain of human insulin as a fusion protein in transgenic tobacco plants. Plant Cell Rep 25:417–424

Li M, Song B, Zhang Q, Liu X, Lin Y, Ou Y, Zhang H, Liu J (2013) A synthetic tuber-specific and cold–induced promoter is applicable in controlling potato cold–induced sweetening. Plant Physiol Bioch 67:41–47

Li M, Xie C, Song B, Ou Y, Lin Y, Liu X, Zhang H, Liu J (2015) Construction of efficient, tuber–specific, and cold-inducible promoters in potato. J Plant Sci 235:14–24

Liu S, Bao Y (2009) Effects of copy number of an octopine synthase enhancer element and its distance from the TATA box on heterologous expression in transgenic tobacco. Acta Physiol Plant 31(4):705–710

Liu X, Li S, Yang W, Mu B, Jiao Y, Zhou X, Zhang C, Fan Y, Chen R (2018) Synthesis of seed–specific bidirectional promoters for metabolic engineering of anthocyanin-rich maize. Plant Cell Physiol 59(10):1942–1955

Liu W, Mazarei M, Peng Y, Fethe MH, Rudis MR, Lin J, Millwood RJ, Arelli PR, Stewart CN Jr (2014) Computational discovery of soybean promoter cis-regulatory elements for the construction of soybean cyst nematode-inducible synthetic promoters. Plant Biotechnol J 12(8): 1015–1026

Liu W, Mazarei M, Rudis MR, Fethe MH, Stewart CN (2011) Rapid in vivo analysis of synthetic promoters for plant pathogen phytosensing. BMC Biotechnol 11(1):1–9

Liu JW, Tabe LM (1998) The influences of two plant nuclear matrix attachment regions (MARs) on gene expression in transgenic plants. Plant Cell Physiol 39:115–123
Liu L, White MJ, MacRae TH (1999) Transcription factors and their genes in higher plants functional domains, evolution and regulation. Eur J Biochem 262:247–257
Long L, Guo DD, Gao W, Yang WW, Hou LP, Ma XN, Miao YC, Botella JR, Song CP (2018) Optimization of CRISPR/Cas9 genome editing in cotton by improved sgRNA expression. Plant Methods 14(1):1–9
Loos A, Van Droogenbroeck B, Hillmer S, Grass J, Kunert Cao J, Robinson DG, Depicker A, Steinkellner H (2011) Production of monoclonal antibodies with a controlled N-glycosylation pattern in seeds of *Arabidopsis thaliana*. Plant Biotechnol J 9(2):179–192
Lota F, Wegmüller S, Buer B, Sato S, Bräutigam A, Hanf B, Bucher M (2013) The cis-acting CTTC–P1 BS module is indicative for gene function of L j VTI 12, a Q b-SNARE protein gene that is required for arbuscule formation in *Lotus japonicus*. Plant J 74(2):280–293
Lu J, Sivamani E, Azhakanandam K, Samadder P, Li X, Qu R (2008) Gene expression enhancement mediated by the 5′ UTR intron of the rice rubi3 gene varied remarkably among tissues in transgenic rice plants. Mol Gen Genomics 279:563–572
Lv X, Song X, Rao G, Pan X, Guan L, Jiang X (2009) Construction vascular-specific expression bi-directional promoters in plants. J Biotechnol 141:104–108
Ma JK, Hiatt A, Hein M, Vine ND, Wang F, Stabila P, van Dolleweerd C, Mostov K, Lehner T (1995) Generation and assembly of secretory antibodies in plants. Science 268:716–719
Ma S, Huang Y, Davis A, Yin Z, Mi Q, Menassa R, Brandle JE, Jevnikar AM (2005) Production of biologically active human interleukin-4 in transgenic tobacco and potato. Plant Biotechnol J 3: 309–318
Maclean J, Koekemoer M, Olivier AJ, Stewart D, Hitzeroth II, Rademacher T, Fischer R, Williamson AL, Rybicki EP (2007) Optimization of human papillomavirus type 16 (HPV-16) L1 expression in plants: comparison of the suitability of different HPV-16 L1 gene variants and different cell-compartment localization. J Gen Virol 88(Pt 5):1460–1469
Maiti IB, Gowda S, Kiernan J, Ghosh SK, Shepherd RJ (1997) Promoter/leader deletion analysis and plant expression vectors with the *figwort mosaic virus* (*FMV*) full length transcript (*FLt*) promoter containing single or double enhancer domains. Transgenic Res 6(2):143–156
Maiti IB, Shepherd RJ (1998) Isolation and expression analysis of peanut *chlorotic streak caulimovirus* (*PClSV*) full-length transcript (*FLt*) promoter in transgenic plants. Biochem Biophs Res Commun 244(2):440–444
Mao Y, Zhang Z, Feng Z, Wei P, Zhang H, Botella JR, Zhu JK (2016) Development of germ-line-specific CRISPR-Cas9 systems to improve the production of heritable gene modifications in *Arabidopsis*. Plant Biotechnol J 14(2):519–532
Marusic C, Nuttall J, Buriani G, Lico C, Lombardi R, Baschieri S, Benvenuto E, Frigerio L (2007) Expression intracellular targeting and purification of HIV Nef variants in tobacco cells. BMC Biotechnol 7:12
Mason HS, Haq TA, Clements JD, Arntzen CJ (1998) Edible vaccine protects mice against *Escherichia coli* heat-labile enterotoxin LT: potatoes expressing a synthetic LT-B gene. Vaccine 16:1336–1343
Matsuo K, Matsumura T (2010) Deletion of fucose residues in plant N-glycans by repression of the GDP-mannose 4,6-dehydratase gene using virus-induced gene silencing and RNA interference. Plant Biotechnol J 1–18:2010
Mavituna AM (2005) Production of recombinant human serum albumin in transgenic plants and plant cells. Master of Science dissertation, Rheinisch-Westfälischen Technischen, Hochschule, Aachen
McElroy D, Blowers AD, Jenes B, Wu R (1991) Construction of expression vectors based on the rice actin 1 (*Act1*) 5′ region for use in monocot transformation. Mol Gen Genet MGG 231(1): 150–160

Mishra S, Yadav DK, Tuli R (2006) Ubiquitin fusion enhances cholera toxin B subunit expression in transgenic plants and the plant-expressed protein binds GM1 receptors more efficiently. J Biotechnol 127:95–108

Mitsuhara I, Ugaki M, Hirochika H, Ohshima M, Murakami T, Gotoh Y, Katayose Y, Nakamura S, Honkura R, Nishimiya S, Ueno K, Mochizuki A, Tanimoto H, Tsugawa H, Otsuki Y, Ohashi Y (1996) Efficient promoter cassettes for enhanced expression of foreign genes in dicotyledonous and monocotyledonous plants. Plant Cell Physiol 37:49–59

Miyasaka H, Kanai S, Tanaka S, Akiyama H, Hirano M (2002) Statistical analysis of the relationship between translation initiation AUG context and gene expression level in humans. Biosci Biotechnol Biochem 66:667–669

Mohan C, Jayanarayanan AN, Narayanan S (2017) Construction of a novel synthetic root-specific promoter and its characterization in transgenic tobacco plants. Biotechnol J 7(4):1–9

Morishita K, Johnson DE, Williams LT (1995) A novel promoter for vascular endothelial growth factor receptor (flt–1) that confers endothelial-specific gene expression. J Biol Chem 270(46): 27948–27953

Mosa KA, Saadoun I, Kumar K, Helmy M, Dhankher OP (2016) Potential biotechnological strategies for the cleanup of heavy metals and metalloids. Front Plant Sci 7:303

Murray EE, Lotzer J, Eberle M (1989) Codon usage in plant genes. Nucl Acid Res 17(2):477–498

Murray EE, Rocheleau T, Eberle M, Stock C, Sekar V, Adang M (1991) Analysis of unstable RNA transcripts of insecticidal crystal protein genes of *Bacillus thuringiensis* in transgenic plants and electroporated protoplasts. Plant Mol Biol 16:1035–1050

Murray C, Sutherland PW, Phung MM, Lester MT, Marshall RK, Christeller JT (2002) Expression of biotin-binding proteins, avidin and streptavidin, in plant tissues using plant vacuolar targeting sequences. Transgenic Res 11:199–214

Nagamiya K, Motohashi T, Nakao K, Prodhan SH, Hattori E, Hirose S et al (2007) Enhancement of salt tolerance in transgenic rice expressing an *Escherichia coli* catalase gene, katE. Plant Biotechnol Rep 1(1):49–55

Nakamura N, Suzuki T, Shinbo Y, Chandler S, Tanaka Y (2020) Development of violet transgenic carnations and analysis of inserted transgenes. In: The carnation genome. Springer, Singapore, pp 135–146

Ni M, Chui D, Einstein J, Narasimhulu S, Vergara CE, Gelvin SB (1995) Strength and tissue specificity of chimeric promoter derived from octopine and mannopine synthase gene. Plant J 7: 661–676

Ni M, Cui D, Einstein J, Narasimhulu S, Vergara CE, Gelvin SB (1995) Strength and tissue specificity of chimeric promoters derived from the octopine and mannopine synthase genes. Plant J 7:661–676

Noman A, Aqeel M, Deng J, Khalid N, Sanaullah T, Shuilin H (2017) Biotechnological advancements for improving floral attributes in ornamental plants. Front Plant Sci 8:530

Nowak W, Gawlowska M, Jarmolowski A, Augustyniak J (2001) Effect of nuclear matrix attachment regions on transgene expression in tobacco plants. Acta Biochem Pol 48:637–646

Odell JT, Nagy F, Chua NH (1985) Identification of DNA sequences required for activity of the *cauliflower mosaic virus 35S* promoter. Nature 313(6005):810–812

Ordon J, Bressan M, Kretschmer C, Dall'Osto L, Marillonnet S, Bassi R, Stuttmann J (2020) Optimized Cas9 expression systems for highly efficient Arabidopsis genome editing facilitate isolation of complex alleles in a single generation. Funct Integr Genomic 20(1):151–162

Outchkourov NS, Rogelj B, Strukelj B, Jongsma MA (2003) Expression of sea anemone equistatin in potato. Effect of plant proteases on heterologous protein production. Plant Physiol 133:379–390

Owen M, Gandecha A, Cockburn B, Whitelam GC (1992) Synthesis of a functional anti-phytochrome single-chain Fv protein in transgenic tobacco. Biotechnology (N Y) 10:790–794

Panahi M, Alli Z, Cheng X, Belbaraka L, Belgoudi J, Sardana R (2004) Recombinant protein expression plasmids optimized for industrial *E. coli* fermentation and plant systems produce biologically active human insulin-like growth factor-1 in transgenic rice and tobacco plants. Transgenic Res 13:245–259

Park SH, Yi N, Kim YS, Jeong MH, Bang SW, Choi YD, Kim JK (2010) Analysis of five novel putative constitutive gene promoters in transgenic rice plants. J Exp Bot 61(9):2459–2467

Patro S, Maiti IB, Dey N (2013) Development of an efficient bidirectional promoter with tripartite enhancer employing three viral promoters. Biotechnol J 163(3):311–317

Peeters K, De Wilde C, Depicker A (2001) Highly efficient targeting and accumulation of a F (ab) fragment within the secretory pathway and apoplast of *Arabidopsis thaliana*. Eur J Biochem 268:4251–4260

Pelham HR (1989) Control of protein exit from the endoplasmic reticulum. Annu Rev Cell Biol 5: 1–23

Peng RH, Yao Q-H, Xiong A-S, Cheng Z-M, Li Y (2006) Codon-motifications and an endoplasmic reticulum-targeting sequence additively enhance expression of an *Aspergillus* phytase gene in transgenic canola. Plant Cell Rep 25:124–132

Perlak FJ, Fuchs RL, Dean DA, McPherson SL, Fischhoff DA (1991) Modification of the coding sequences enhances plant expression of insect control protein genes. Proc Natl Acad Sci U S A 88:3324–3328

Petruccelli S, Otegui MS, Lareu F, Dinh OT, Fitchette AC, Circosta A (2006) A KDEL-tagged monoclonal antibody is efficiently retained in the endoplasmic reticulum in leaves, but is both partially secreted and stored to protein storage vacuoles in seed. Plant Biotech J 4:511–527

Pueyo JJ, Chrispeels MJ, Herman EM (1995) Degradation of transport-competent destabilized phaseolin with a signal for retention in the endoplasmic reticulum occurs in the vacuole. Planta 196:586–596

Ramírez N, Ayala M, Lorenzo D, Palenzuela D, Herrera L, Doreste V, Pérez M, Gavilondo JV, Oramas P (2002) Expression of a single-chain Fv antibody fragment specific for the hepatitis B surface antigen in transgenic tobacco plants. Trans Res 11:61–64

Ranjan R, Dey N (2012) Development of vascular tissue and stress inducible hybrid–synthetic promoters through DOF–1 motifs rearrangement. Cell Biochem Biophys 63(3):235–245

Ranjan R, Patro S, Pradhan B, Kumar A, Maiti IB, Dey N (2012) Development and functional analysis of novel genetic promoters using DNA shuffling, hybridization and a combination thereof. PLoS One 7(3):e31931

Rao KV, Rathore KS, Hodges TK, Fu X, Stoger E, Sudhakar D et al (1998) Expression of snowdrop lectin (GNA) in transgenic rice plants confers resistance to rice brown planthopper. Plant J 15(4):469–477

Rawat P, Singh AK, Ray K, Chaudhary B, Kumar S, Gautam T, Kanoria S, Kaur G, Kumar P, Pental D, Burma PK (2011) Detrimental effect of expression of Bt endotoxin Cry1Ac on in vitro regeneration, in vivo growth and development of tobacco and cotton transgenics. J Biosci 36: 363–376

Rethmeier N, Seurinck J, Van Montagu M, Cornelissen M (1997) Intron-mediated enhancement of transgene expression in maize is a nuclear, gene-dependent process. Plant J 12:895–899

Rose AB (2004) The effect of intron location on intron-mediated enhancement of gene expression in Arabidopsis. Plant J 40:744–751

Rouwendal GJ, Mendes O, Wolbert EJ, Douwe de Boer A (1997) Enhanced expression in tobacco of the gene encoding green fluorescent protein by modification of its codon usage. Plant Mol Biol 33:989–999

Ruggiero F, Exposito JY, Bournat P, Gruber V, Perret S, Comte J, Olagnier B, Garrone R, Theisen M (2000) Triple helix assembly and processing of human collagen produced in transgenic tobacco plants. FEBS Lett 469(1):132–136

Rugh CL, Senecoff JF, Meagher RB, Merkle SA (1998) Development of transgenic yellow poplar for mercury phytoremediation. Nature Biotechnol 16(10):925–928

Rushton PJ, Reinstadler A, Lipka V, Lippok B, Somssich IE (2002) Synthetic plant promoters containing defined regulatory elements provide novel insights into pathogen- and wound-induced signalling. Plant Cell 14:749–762

Rushton PJ, Reinstädler A, Lipka V, Lippok B, Somssich IE (2002) Synthetic plant promoters containing defined regulatory elements provide novel insights into pathogen-and wound-induced signaling. Plant Cell 14(4):749–762

Sardana RK, Alli Z, Dudani A, Tackaberry E, Panahi M, Narayanan M (2002) Biological activity of human granulocyte-macrophage colony stimulating factor is maintained in a fusion with seed glutelin peptide. Transgenic Res 11:521–531

Sawant SV, Singh PK, Gupta SK, Madanala R, Tuli R (1999) Conserved nucleotide sequence in highly expressed genes in plants. J Genet 78:1–8

Sawant SV, Kiran K, Singh PK, Tuli R (2001) Sequence architecture downstream of the initiator codon enhances gene expression and protein stability in plants. Plant Physiol 136:1630–1636

Sawant SV, Kiran K, Mehrotra R, Chaturvedi CP, Ansari SA, Singh P, Lodhi N, Tuli R (2005) A variety of synergistic and antagonistic interactions mediated by cis-acting DNA motifs regulate gene expression in plant cells and modulate stability of the transcription complex formed on a basal promoter. J Exp Bot 56(419):2345–2353

Schillberg S, Zimmermann S, Voss A, Fischer R (1999) Apoplastic and cytosolic expression of full-size antibodies and antibody fragments in Nicotiana tabacum. Transgenic Res 8:255–263

Schillberg S, Emans N, Fischer R (2002) Antibody molecular farming in plants and plant cells. Phytochem Rev 1:45–54

Schillberg S, Fischer R, Emans N (2003) Molecular farming of recombinant antibodies in plants. Cell Mol Life Sci 60:433–445

Schiml S, Fauser F, Puchta H (2014) The CRISPR/C as system can be used as nuclease for in planta gene targeting and as paired nickases for directed mutagenesis in *Arabidopsis* resulting in heritable progeny. Plant J 80(6):1139–1150

Schmülling T, Schell J, Spena A (1989) Promoters of the *rol A*, *B*, and *C* genes of Agrobacterium rhizogenes are differentially regulated in transgenic. Plant Cell 1(7):665–670

Schouten A, Roosien J, de Boer JM, Wilmink A, Rosso MN, Bosch D, Stiekema WJ, Gommers FJ, Bakker J, Schots A (1997) Improving scFv antibody expression levels in the plant cytosol. FEBS Lett 415:235–241

Schouten A, Roosien J, van Engelen FA, de Jong GA, Borst-Vrenssen AW, Zilverentant JF, Bosch D, Stiekema WJ, Gommers FJ, Schots A, Bakker J (1996) The C-terminal KDEL sequence increases the expression level of a single-chain antibody designed to be targeted to both the cytosol and the secretory pathway in transgenic tobacco. Plant Mol Biol 30:781–793

Scranton MA, Ostrand JT, Georgianna DR, Lofgren SM, Li D, Ellis RC, Carruthers DN, Dräger A, Masica DL, Mayfield SP (2016) Synthetic promoters capable of driving robust nuclear gene expression in the green alga *Chlamydomonas reinhardtii*. Algal Res 15:135–142

Shan Q, Wang Y, Chen K, Liang Z, Li J, Zhang Y, Zhang K, Liu J, Voytas DF, Zheng X, Zhang Y (2013) Rapid and efficient gene modification in rice and *Brachypodium* using TALENs. Mol Plant 6(4):1365

Shani N, Rosenberg N, Kasarda DD, Galili G (1994) Mechanism of assembly of wheat high molecular weight glutenins inferred from expression of wild-type and mutant subunits in transgenic tobacco. J Biol Chem 269:8924–8930

Sharma AK, Sharma MK (2009) Plants as bioreactors: recent developments and emerging opportunities. Biotechnol Adv 27(16):811–832

Sharma MK, Jani D, Thungapathra M, Gautam JK, Meena LS, Singh Y (2008) Expression of accessory colonization factor subunit A (ACFA) of Vibrio cholerae and ACFA fused to cholera toxin B subunit in transgenic tomato (Solanum lycopersicum). J Biotechnol 135:22–27

Shepherd CT, Scott MP (2009) Construction and evaluation of a maize (*Zea mays*) chimaeric promoter with activity in kernel endosperm and embryo. Biotechnol Appl Biochem 52(3): 233–243

Shokouhifar F, Zamani MR, Motallebi M, Mousavi A, Malboobi MA (2011) Construction and functional analysis of pathogen-inducible synthetic promoters in *Brassica napus*. Biol Plant 55(4):689

Shu Q, Ye G, Cui H, Cheng X, Xiang Y, Wu D et al (2000) Transgenic rice plants with a synthetic cry1Ab gene from *Bacillus thuringiensis* were highly resistant to eight lepidopteran rice pest species. Mol Breed 6(4):433–439

Sijmons PC, Dekker BM, Schrammeijer B, Verwoerd TC, van den Elzen PJ, Hoekema A (1990) Production of correctly processed human serum albumin in transgenic plants. Biotechnology 8: 217–221

Singh SP, Gruissem W, Bhullar NK (2017) Single genetic locus improvement of iron, zinc and β-carotene content in rice grains. Sci Rep 7(1):1–11

Smale ST, Kadonaga JT (2003) The RNA polymerase II core promoter. Annu Rev Biochem 72: 449–479

Smith ML, Mason HS, Shuler ML (2002) Hepatitis B surface antigen (HBsAg) expression in plant cell culture: kinetics of antigen accumulation in batch culture and its intracellular form. Biotechnol Bioeng 80:812–822

Sojikul P, Buehner N, Mason HS (2003) A plant signal peptide-hepatitis B surface antigen fusion protein with enhanced stability and immunogenicity expressed in plant cells. Proc Natl Acad Sci 100(5):2209–2214

Somerville C, Somerville S (1999) Plant functional genomics. Science 285:380–383

Song XY, Zhu WJ, Tang RM, Cai JH, Chen M, Yang Q (2016) Over-expression of StLCYb increases β-carotene accumulation in potato tubers. Plant Biotechnol Rep 10(2):95–104

Staub J, Garcia B, Graves J, Hajdukiewicz P, Hunter P, Nehra N, Paradkar V, Schlittler M, Carroll J, Spatola L, Ward D, Ye G, Russell D (2000) High-yield production of a human therapeutic protein in tobacco chloroplasts. Nat Biotechnol 18:333–338

Stoger E, Vaquero C, Torres E, Sack M, Nicholson L, Drossard J (2000) Cereal crops as viable production and storage systems for pharmaceutical scFv antibodies. Plant Mol Biol 42:583–590

Stoger E, Sack M, Perrin Y, Vaquero C, Torres E, Twyman RM, Christou P, Fischer R (2002) Practical considerations for pharmaceutical antibody production in different crop systems. Mol Breed 9(3):149–158

Storozhenko S, De Brouwer V, Volckaert M, Navarrete O, Blancquaert D, Zhang GF et al (2007) Folate fortification of rice by metabolic engineering. Nat Biotechnol 25(11):1277–1279

Streatfield SJ (2007) Approaches to achieve high-level heterologous protein production in plants. Plant Biotechnol J 5:2–15

Sunkara S, Bhatnagar-Mathur P, Sharma KK (2014) Isolation and functional characterization of a novel seed-specific promoter region from peanut. Appl Biochem Biotechnol 172(1):325–339

Suo G, Chen B, Zhang J, Duan Z, He Z, Yao W, Yue C, Dai J (2006) Effects of codon modification on human BMP2 gene expression in tobacco plants. Plant Cell Rep 25:689–697

Tabashnik BE, Dennehy TJ, Sims MA, Larkin K, Head GP, Moar WJ, Carriere Y (2002) Control of resistant pink bollworm (*Pectinophora gossypiella*) by transgenic cotton that produces *Bacillus thuringiensis* toxin Cry2Ab. Appl Environ Microbiol 68(8):3790–3794

Tao YB, He LL, Niu LJ, Xu ZF (2015) Isolation and characterization of an ubiquitin extension protein gene (*JcUEP*) promoter from *Jatropha curcas*. Planta 241(4):823–836

Tashkandi M, Ali Z, Aljedaani F, Shami A, Mahfouz MM (2018) Engineering resistance against tomato *yellow leaf curl virus* via the CRISPR/Cas9 system in tomato. Plant Signal Behav 13(10):e1525996

Tavladoraki P, Benvenuto E, Trinca S, De Martinis D, Cattaneo A, Galeffi P (1993) Transgenic plants expressing a functional single-chain Fv antibody are specifically protected from virus attack. Nature 366:469–472

Teli NP, Timko MP (2004) Recent developments in the use of transgenic plants for the production of human therapeutics and biopharmaceuticals. Plant Cell Tissue Organ Cult 79:125–145

Torres JB, Ruberson, JR, Whitehouse M (2009) Transgenic cotton for sustainable pest management: a review. Organic Farming, Pest Control and Remediation of Soil Pollutants 15–53

Tsuboi T, Takeo S, Iriko H, Jin L, Tsuchimochi M, Matsuda S, Han ET, Otsuki H, Kaneko O, Sattabongkot J, Udomsangpetch R, Sawasaki T, Torii M, Endo Y (2008) Wheat germ cell-free system-based production of malaria proteins for discovery of novel vaccine candidates. Infect Immun 76(4):1702–1708

Twyman RM, Stoger E, Schillberg S, Christou P, Fischer R (2003) Molecular farming in plants: host systems and expression technology. Trends Biotechnol 21:570–578

Tyc K, Konarska M, Gross HJ, Filipowicz W (1984) Multiple ribosome binding to the 5′-terminal leader sequence of tobacco mosaic virus RNA. Assembly of an 80S ribosome X mRNA complex at the AUU codon. Eur J Biochem 140:503–511

Tzafrir I, Torbert KA, Lockhart BE, Somers DA, Olszewski NE (1998) The sugarcane *bacilliform badnavirus* promoter is active in both monocots and dicots. Plant Mol Biol 38(3):347–356

Van der Does D, Leon-Reyes A, Koornneef A, Van Verk MC, Rodenburg N, Pauwels L et al (2013) Salicylic acid suppresses jasmonic acid signaling downstream of SCFCOI1–JAZ by targeting GCC promoter motifs via transcription factor ORA59. Plant Cell 25(2):744–761

Van der Geest AH, Hall TC (1997) The β–phaseolin 5′ matrix attachment region acts as an enhancer facilitator. Plant Mol Biol 33(3):553–557

Van Engelen FA, Schouten A, Molthoff JW, Roosien J, Salinas J, Dirkse WG, Schots A, Bakker J, Gommers FJ, Jongsma MA (1994) Coordinate expression of antibody subunit genes yields high levels of functional antibodies in roots of transgenic tobacco. Plant Mol Biol 26:1701–1710

Vaquero C, Sack M, Schuster F, Finnern R, Drossard J, Schumann D, Reimann A, Fischer R (2002) A carcinoembryonic antigen-specific diabody produced in tobacco. FASEB J 16:408–410

Venter M (2007) Synthetic promoters: genetic control through cis engineering. Trends Plant Sci 12: 118–124

Vitale A, Ceriotti A, Denecke J (1993) The role of the endoplasmic reticulum in protein synthesis, modification and intracellular transport. J Exp Bot 44:1417–1444

Wandelt CI, Khan MRI, Craig S, Schroeder HE, Spencer D, Higgens TJV (1992) Vicilin with carboxy-terminal KDEL is retained in the endoplasmic reticulum and accumulates to high levels in the leaves of transgenic plants. Plant J 2:181–192

Wang ET, Sandberg R, Luo S, Khrebtukova I, Zhang L, Mayr C, Kingsmore SF, Schroth GP, Burge CB (2008) Alternative isoform regulation in human tissue transcriptomes. Nature 456 (7221):470–476

Wang P, Du Y, Zhao X, Miao Y, Song CP (2013) The MPK6–ERF6–ROS-responsive cis-acting Element7/GCC box complex modulates oxidative gene transcription and the oxidative response in *Arabidopsis*. Plant Physiol 161(3):1392–1408

Wang WC, Wu CY, Lai YC, Lin NS, Hsu YH, Hu CC (2014) Characterization of the cryptic AV3 promoter of *Ageratum yellow vein virus* in prokaryotic and eukaryotic systems. PLoS One 9(9): e108608

Wang XG, Zhang GH, Liu CX, Zhang YH, Xiao CZ, Fang RX (2001) Purified cholera toxin B subunit from transgenic tobacco plants possesses authentic antigenicity. Biotechnol Bioeng 72: 490–494

Wang R, Zhu M, Ye R, Liu Z, Zhou F, Chen H, Lin Y (2015) Novel green tissue-specific synthetic promoters and cis–regulatory elements in rice. Sci Rep 5(1):1–1

Wever W, McCallum EJ, Chakravorty D, Cazzonelli CI, Botella JR (2010) The 5′ untranslated region of the VR–ACS1 mRNA acts as a strong translational enhancer in plants. Transgenic Res 19(4):667–674

Wilde CD, Houdt HV, Buck SD, Angenon G, Jaeger GD, Depicker A (2000) Plant as bioreactors for protein production: avoiding the problem of transgenic silencing. Plant Mol Biol 43:347–359

Wilson IB (2002) Glycosylation of proteins in plants and invertebrates. Curr Opin Struct Biol 12 (5):569–577

Wirth S, Calamante G, Mentaberry A, Bussmann L, Lattanzi M, Baranao L, Bravo-Almonacid F (2004) Expression of active human epidermal growth factor (hEGF) in tobacco plants by integrative and non-integrative systems. Mol Breed 13:23–35

Wong EY, Hironaka CM, Fischhoff DA (1992) *Arabidopsis thaliana* small subunit leader and transit peptide enhance the expression of *Bacillus thuringiensis* proteins in transgenic plants. Plant Mol Biol 20:81–93

Xiao K, Zhang C, Harrison M, Wang ZY (2005) Isolation and characterization of a novel plant promoter that directs strong constitutive expression of transgenes in plants. Mol Plant Breed 15(2):221–231

Xu D, Duan X, Wang B, Hong B, Ho TH, Wu R (1996) Expression of a late embryogenesis abundant protein gene, HVA1, from barley confers tolerance to water deficit and salt stress in transgenic rice. Plant Physiol 110(1):249–257
Xu D, Xue Q, McElroy D, Mawal Y, Hilder VA, Wu R (1996) Constitutive expression of a cowpea trypsin inhibitor gene, CpTi, in transgenic rice plants confers resistance to two major rice insect pests. Mol Breed 2(2):167–173
Yan J, Kandianis CB, Harjes CE, Bai L, Kim EH, Yang X et al (2010) Rare genetic variation at *Zea mays* crtRB1 increases β-carotene in maize grain. Nat Gen 42(4):322–327
Yang J, Barr LA, Fahnestock SR, Liu ZB (2005) High yield recombinant silk-like protein production in transgenic plants through protein targeting. Transgenic Res 14:313–324
Yang Z, Chen H, Tang W, Hua H, Lin Y (2011) Development and characterization of transgenic rice expressing two *Bacillus thuringiensis* genes. Pest Manag Sci 67(4):414–422
Yang D, Wu L, Hwang YS, Chen L, Huang N (2001) Expression of the REB transcriptional activator in rice grains improves the yield of recombinant proteins whose genes are controlled by a Reb-responsive promoter. Proc Natl Acad Sci U S A 98:11438–11443
Yang QQ, Zhang CQ, Chan ML, Zhao DS, Chen JZ, Wang Q et al (2016) Biofortification of rice with the essential amino acid lysine: molecular characterization, nutritional evaluation, and field performance. J Exp Bot 67(14):4285–4296
Yi D, Yang W, Tang J, Wang L, Fang Z, Liu Y et al (2016) High resistance of transgenic cabbage plants with a synthetic cry1Ia8 gene from *Bacillus thuringiensis* against two lepidopteran species under field conditions. Pest Manag Sci 72(2):315–321
Yoshida K, Shinmyo A (2000) Transgene expression systems in plant, a natural bioreactor. J Biosci Bioeng 90:353–362
Yunus AM, Parveez GKA, Ho C (2008) Transgenic plants producing polyhydroxyalkanoates. Asia Pac J Mol Biol Biotechnol 16:1–10
Zavallo D, Bilbao ML, Hopp HE, Heinz R (2010) Isolation and functional characterization of two novel seed-specific promoters from sunflower (*Helianthus annuus* L.). Plant Cell Rep 29(3): 239–248
Zhang C, Gai Y, Wang W, Zhu Y, Chen X, Jiang X (2008) Construction and analysis of a plant transformation binary vector pBDGG harboring a bi-directional promoter fusing dual visible reporter genes. J Genet Genomics 35:245–249
Zhang W, Wu R (1988) Efficient regeneration of transgenic plants from rice protoplasts and correctly regulated expression of the foreign gene in the plants. Theor Appl Genet 76(6): 835–840
Zheng GG, Yang YH, Rao Q, Lin YM, Zhang B, Wu KF (2006) Expression of bioactive human M-CSF soluble receptor in transgenic tobacco plants. Protein Expr Purif 46(2):367–373
Zhou J, Liu WJ, Peng SW, Sun XY, Frazer I (1999) Papillomavirus capsid protein expression level depends on the match between codon usage and tRNA availability. J Virol 73:4972–4982
Zhou J, Yang Y, Wang X, Yu F, Yu C, Chen J, Cheng Y, Yan C, Chen J (2013) Enhanced transgene expression in rice following selection controlled by weak promoters. BMC Biotechnol 13(1): 1–2
Ziegelhoffer T, Raasch JA, Austin-Phillips S (2001) Dramatic effects of truncation and sub-cellular targeting on the accumulation of recombinant microbial cellulase in tobacco. Mol Breeding 8: 147–158
Zimmermann S, Schillberg S, Liao YC, Fischer R (1998) Intracellular expression of a TMV-specific single chain Fv fragment leads to improved virus resistance in *Nicotiana tabacum*. Mol Breed 4:369–379
Zuo J, Chua NH (2000) Chemical-inducible systems for regulated expression of plant genes. Curr Opin Biotechnol 11:146–151

Cisgenics and Crop Improvement 3

Abstract

The word "cis" in cisgenics means "on the same side/same kingdom" and "genics" means "pertaining to genes." Cisgenics involves genetic transformation of a plant species with a natural gene derived from a crossable sexually compatible plant. Cisgenics and intragenics are eco-friendly crop improvement techniques that are alternative to transgenics, which is the genetic modification of a plant with one or even more genes from any non-plant organism, or from a plant that is sexually incompatible with the recipient plant. In cisgenesis, the inserted gene remains unchanged and has its own introns and regulatory elements. The concept of cisgenesis was brought by Dutch researchers Schouten, Krens, and Jacobsen in the year 2006. Cisgenic crops are accepted by the general public; they do not involve hectic and costly time-consuming procedure of approvals; are not a threat to the biodiversity and are not a potential risk to the health and the ecosystem; are safer than those raised by conventional breeding; and preclude linkage drag. This chapter encompasses the advantages and drawbacks of cisgenesis, their implications towards sustainable crop improvement, and their future prospects.

Keywords

Cisgenics · Intragenics · Genetic modification · Transformation · Crop improvement

3.1 Introduction

Phyto-biotechnology is the development of technological techniques relating to living systems to improve agricultural production. The genetic modification of plant species is the process of introducing foreign genetic material into the genome of a plant with the aim of incorporating novel desirable traits either by inhibition of

B. Koul, *Cisgenics and Transgenics*, https://doi.org/10.1007/978-981-19-2119-3_3

an endogenous protein or by expression of a foreign gene to modify a particular function (Telem et al. 2013; Noor et al. 2019). This genetic transformation of living organisms through biotechnological methods leads to the development of genetically modified organisms, which is the unique feature of biotechnology (Aaron et al. 2018).

The Food and Agricultural Organization reported that the global decline in hunger for several years as assessed using the prevalence of undernourishment (PoU) had regrettably stopped (FAO et al. 2020). More information and several data updates incorporating reviews of the entire PoU progressions back to the year 2000 demonstrate that, virtually, 690 million people representing 8.9% of the world population were estimated to have been malnourished in 2019. This underscored the serious challenge of achieving the zero-hunger target (sustainable development goal 2) by 2030. It is estimated that the food supply must increase immensely (about 70%) before 2050 so as to feed the increasing global population (Delwaide et al. 2015). Traditional plant breeding methods have been used to effectively manage the abiotic and biotic constraints and to introduce desirable characters in the cultivated crops, to enhance the agricultural yield and profitability of farmers (Gaetano et al. 2021). However, the emerging biotic constraints (bacterial, fungal, insect, nematode attack, etc.), occasional climatic fluctuations (drought, high temperature, heavy rainfall, etc.), and dwindling agriculture globally have forced scientists to deliberate on alternatives to overcome food and related health security crises (Gaetano et al. 2021). Green biotechnological approaches like cisgenesis, genome editing, intragenesis, and production of marker-free transgenic plants are among the current strategies deployed to produce crop plants that are resistant to the aforementioned biotic and abiotic stresses (Thapa et al. 2015; Paula et al. 2019; Gaetano et al. 2021), which results in precise and faster genetic modification of crops. To achieve this, molecular cloning technique is used to manipulate the plant genome by gene insertion from another plant species (Mariam 2015).

Genetic improvement in crop plants through conventional breeding methods is a sluggish and repetitive process, characterized with drawbacks generated by high heterozygosity, lengthy juvenile periods, and auto-incompatibility (Cecilia et al. 2017; Ahmar et al. 2020). To overcome this issue, two alternative transformation concepts (cisgenesis/intragenesis) were developed, and in both of the concepts, the plants involved are modified using alleles that are obtained from the species itself or from the wild relatives that are sexually crossable plants (Aaron et al. 2018; Noor et al. 2019). These indigenous natural genes isolated from closely related crops or from species itself are classified as cisgenes, to differentiate them from transgenes (Shiva et al. 2013).

3.2 Difference Between Cisgenics and Transgenics

The advances in genetic engineering and genome sequencing of different crop plants have facilitated the isolation, identification, and characterization of genes from wild relatives and auto-compatible or crossable species (Thapa et al. 2015). Cisgenesis as a biotechnological terminology was first coined by Jochemsen and Schouten (2000), and they described it as a genetic transformation technique that is used to genetically modify plants using alleles or genes derived from sexually compatible organisms. The genes from sexually crossable species contain introns and are flanked by its native promoter and terminator in the normal sense orientation (Mansoor et al. 2019). Cisgenics are developed through cisgenesis, and they may be similar to those plants developed through the classical breeding technique as they share the same gene pool. This is because the required genetic material (genes) is acquired from crop plants capable of undergoing traditional breeding, which are transferred conserving the endogenous characters (Gaetano et al. 2021).

Transgenesis, on the other hand, is the genetic modification of the crop plant where the gene(s) of interest is/are sourced from either non-plant (interkingdom transfer) or from sexually incompatible species. A transgenic plant can thus be defined as a plant modified genetically by incorporation of gene(s), coding for a desirable character(s) that did not occur naturally in the modified crop (Jhansi and Usha 2013). The use of transgenic technology to produce crop through the transfer of hereditary materials (genes) from varied sources would be useful but it is associated with hidden probable risks of strict regulatory response (Shiva et al. 2013).

Table 3.1 presents the distinguishing features of cisgenic and transgenic plants.

Cisgenics and transgenics also share some common characteristics as enumerated below:

(a) Both are biotechnological techniques that involve the introduction of genes into the host plants so as to modify their genetic constitution for a better desirable trait. This helps a lot in increased crop yield, biofortification of crops (e.g., golden rice), antibiotic production, and synthesis of vitamin A.
(b) One or more genes can be introduced into the host species using both techniques.
(c) Both techniques involve the use of regulatory sequences to insert novel genes into the recipient species.

3.3 The Limitations of Cisgenesis

1. In contrast to transgenesis, the gene(s) with desirable traits from sexually non-crossable species cannot be introduced into the recipient species (Thapa et al. 2015).
2. The development of cisgenics requires outstanding proficiency compared to the generation of transgenic crops. This is because the genes/fragments of genes needed for the process may not be readily available but must be isolated from the sexually crossable gene bank/pool (Telem et al. 2013). This may involve the development of marker-free plants through the formulation of inventive protocols, which may not be readily available for the crop to be transformed or engineered.

Table 3.1 Differences between cisgenic and transgenic plants

S. no.	Cisgenics	Transgenics
1	In cisgenics, the plants are genetically modified using sexually crossable species	In transgenics, the plants are genetically modified using genes from sexually incompatible plant species
2	The transferred gene together with its promoter is obtained from native species or from its sexually compatible relatives	The transferred gene is usually obtained from foreign species that is neither closely related, nor sexually crossable
3	There is no additional trait in cisgenic plants because there are no changes in the gene pool of cisgenically transformed plants	The gene pool is altered, thereby producing new additional characteristics that do not occur naturally in the transgenic plants
4	The fitness of the recipient plant does not change. The effects associated with GMOs like the effects on soil ecosystems, nontarget organisms, and possible allergy other than those produced by traditional breeding do not occur in cisgenics	The fitness of the recipient species might be affected by the novel/new trait in different ways. Changes in fitness can spread through gene flow (horizontal gene transfer) creating potential shifts in natural vegetation between GM crops and their wild relatives
5	The release of these plants in markets is risk-free and normal. With regard to the safety issue, cisgenics could be treated by regulatory authorities the same way as conventionally bred plants	Due to the risk associated with these plants, their release into the markets or environment has received much attention by the regulatory agencies
6	There is no risk associated with the cisgenics; as a result, they are considered safe for human consumption	They may be unsafe for consumption; thus, they are controlled very strictly

Redrawn from Shiva et al. (2013), Telem et al. (2013), Thapa et al. (2015)

3. Many important traits in crop plants are predominantly complex, regulated by several genes and their interactions, and the application of cisgenesis may not be possible for polygenic traits (Kronberger et al. 2015; De Steur et al. 2019).
4. Insertion of cisgenes at unknown points in the plant genome is one of the major problems of this technique as it may affect the DNA methylation leading to unexpected consequences like mutation at the insertion site, knocking out genes, and creating new reading frames, thereby introducing undesirable phenotypic characteristics (Thapa et al. 2015; Rameshraddy et al. 2017).
5. The experimental procedures required to identify and characterize a particular gene that specifically encodes for the required traits and to utilize the identified gene with maximum efficiency are very costly and time consuming (Vaibhav et al. 2018).

3.4 Cisgenesis and Sustainable Crop Improvement

One of the major problems for the growing population across the globe is food insecurity. Crop improvement through modern biotechnological methods like "genetic modification strategies" is one of the potential solutions to cope with the challenges of agricultural productivity (Giraldo et al. 2019). Cisgenesis as a genetic engineering technique has been used to generate improved crop cultivars that are tolerant to biotic and abiotic stresses, thereby decreasing the "yield penalty" caused due to these stresses (Mansoor et al. 2019; Ahmar et al. 2020). In cisgenesis, the genes involved (cisgenes) must be identical to the host's native genes, with their regulatory sequences being integrated in the plant species being improved in the normal sense orientation (Mariam 2015).

3.4.1 Techniques Involved in the Development of Cisgenic/Intragenic Crop

Cisgenesis and intragenesis normally follow the same/similar procedure as applied in transgenesis, to generate cisgenics, in addition to a few extra measures (Ankita et al. 2015). They are developed by insertion of DNA fragment derived from the species itself or from sexually crossable species into the host or recipient species so as to genetically modify a particular function (Moradpour and Abdullahi 2017). For a cisgenic or transgenic crop to be developed, the following requirements must be met: (a) the sequence information of the plant under study must be known, (b) the plant-derived T-DNA (P-DNA) borders containing the cisgenic/intragenic vectors should be well understood, (c) the desirable genes from the compatible species must be isolated and characterized, (d) the isolated genome is inserted into a plasmid vector, (e) suitable method for nuclear or chloroplast transformation, (f) the presence of suitable selection marker or procedure for the selection of putative transformant, and (g) selection and regeneration of the transformants that receive the insert (s) (Hunter et al. 2014; Koul et al. 2014; Mansoor et al. 2019; Taak et al. 2020). Holme et al. (2013) mentioned that for raising an ecologically, ethically, and socially acceptable cisgenic plant, eradication of selection marker genes is crucial. The plant developed through this technique (cisgenesis) will finally possess the gene of interest available in cross-compatible donor plant. These techniques have been used to develop a variety of crops, and some are currently under production as highlighted (Sheikh Mansoor et al. 2019). The use of cisgenesis to improve crop plants is considered a crucial and efficient biotechnological technique due to the fact that the linkage drag is avoided; as a result, the threat from the complications of unknown genes (possible contamination by undesirable genes) is thwarted (Holme et al. 2013).

Intragenesis is a biotechnological method deployed to insert gene cassettes comprising distinct genetic sequences derived from plants that belong to the same sexually compatible gene bank into the genome of the host species (Mariam 2015). In this regard, the coding regions of a single gene with or without introns can be controlled by promoters and terminators from varied genes found from the same sexually crossable gene pool (Vaibhav et al. 2018). Furthermore, the silencing constructs/protocols can be formulated by integrating numerous genetic elements from cross-compatible species, and if Agrobacterium-mediated modification is used, the sequences of the T-DNA border have to arise from sexually compatible gene pool (P-DNA borders). Foreign genes like selectable marker genes or insecticidal genetic materials do not exist in the intragenic crop plants. The main disparity between cisgenesis and intragenesis is that the former employs innate genes encompassing their native promoter, while the latter (intragenesis) enables the new combination of promoters and coding sequences available in the crop plant that can be used in a classical breeding method (Jacobsen and Schouten, 2010).

Currently, a number of variable traits have been introduced using intragenesis/cisgenesis approaches in different crop species, as presented in Table 3.2.

3.4.2 Sources of Genes for Cisgenic/Intragenic Technology

The genetic materials to be used for cisgenesis or intragenesis can be derived from gene pools used for conventional plant breeding. Majhi (2020) defined gene pool as a set of genetic information (genes) possessed by a specific species in any population. He further described the classes of gene pools as primary, secondary, and tertiary gene pools. The primary gene pool (known as GP1) is one in which intermating is simply achieved leading to the development of productive/fertile hybrids referred to as primary gene pool. This consists of plants that are of the same species or closely relevant species that generate entirely fertile progeny when interbred. A cross between improved (cultivated) and wild relatives of sunflower is an example. In this type, genes can be swapped between lines through normal crosses with normal seed set segregation and recombination. It is the material of high breeding significance that includes cultivated races and wild or woody (spontaneous) species. Secondary gene pool, otherwise called gene pool two (GP2), consists of genes that result to be partial or incomplete fertile offspring when crossed with GP1. They constitute plants belonging to related species characterized with some drawbacks of crossability with other crops. GP2 material can be crossed with GP1, but the hybrids generated are usually sterile or fragile, although some are fertile to some extent. In tertiary gene pools (GP3), the species are less closely related to the GP1 species, and the hybrids produced when intermated are sterile or lethal due to anomaly in the development of embryo. Thus, the two gene pools can be crossbred, but for the gene transfer to occur, some revolutionary measures have to be employed. Such measures include embryo rescue, chromosome doubling (induced polyploidy), and bridging crosses for instance using the GP2 species. The genes for cisgenesis or intragenesis can be obtained from the following:

(a) The local cultivars or wild allied species with well-known history of production and consumption by man: for example, the efforts by researchers to transfer genes for resistance to fusarium wilt disease caused by *Fusarium oxysporum* from cultivated to wild banana species with a history of utilization by humans.

Table 3.2 Crop plants with variable traits developed through cis/intragenesis

Crop name	Technique involved	Modification type	Gene	Traits improved	References
Wheat (*Triticum turgidum* var. durum)	Cisgenesis	Gene introduced from allied species	*1Dy10*	Baking quality	
Barley (*Hordeum vulgare*)	Cisgenesis	Expression	*HvPAPhy*	Improvement in the activity of enzymes (phytase) in grains	Holme et al. (2012)
Wheat (*Triticum aestivum*)	Cisgenesis	Insertion of gene from only crossable species	Wheat class 1 chitinase gene	Resistance to fungal attack	Maltseva et al. (2014)
Poplar (poplar species)	Cisgenesis	Overexpression	Genes involved in growth	Enhancement of growth during drought	Han et al. (2011)
Perennial rye grass (*Lolium perenne*)	Intragenesis	Overexpression	*Lpvp1*	Increased resistance to drought	Bajaj et al. (2008)
Alfalfa (*Medicago sativa*)	Intragenesis	RNA interference	*Comt*	Decreased polymer (lignin) level	Weeks et al. (2008)
Potato (*Solanum tuberosum*)	Intragenesis	Silencing	*GBSS*	Increased amylopectin level	de Vetten et al. (2003)
Potato (*Solanum tuberosum*)	Intragenesis	Silencing	*Ppo, R1,PhL*	Reduced starch degradation; acryl amide formation restricted	Rommens et al. (2006)
Strawberry (*Fragaria spp.*)	Intragenesis	Expression	*PGIP*	Fungal (gray mold) tolerance	Schaart, (2004)
Potato (*Solanum tuberosum*)	Intragenesis	RNA interference	*StAs1, StAs2*	Decreased formation of acrylamide	Rommens et al. (2008)
Melon (*Cucumis melo* L.)	Cisgenesis	Introduction of genes from allied species	*At1/At2*-glyoxylate aminotransferase	Resistance to downy mildew attack	Benjamin et al. (2009)

(continued)

Table 3.2 (continued)

Crop name	Technique involved	Modification type	Gene	Traits improved	References
Grapevine (*Vitis vinifera*)	Cisgenesis	Introduction of gene from allied species	*VVTL-1*	Resistance to fungal infection	Dhekney et al. (2011)
Apple (*Malus domestica*)	Cisgenesis	Introduction of gene from allied species	*HcrVf2*	Resistance to fungal scab disease	Vanblaere et al. (2011), Krens et al. (2015), Chizzali et al. (2016)
Potato (*Solanum tuberosum*)	Cisgenesis	Introduction of gene from relevant species	R-genes	Tolerance to late blight disease	Huang et al. (2004)
Potato (*Solanum tuberosum*)	Cisgenesis	RNA interference	*StAs1*	Improved acrylamide formation	Chawla et al. (2012)
Barley	Cisgenesis	Expression	*Gtip2*, *Ggs1a*	Nitrogen-use efficiency	Lutken et al. (2011)

Redrawn from Telem et al. (2013), Moradpour and Abdullahi (2017), Noor et al. (2018)

Since banana is sterile, it is not possible to practically insert resistance genes by a classical or conventional technique, but cisgenic technique can solve this problem.

(b) Cisgenes/intragenes can also be sourced from the donor crop plant without the history of human consumption but with a history of being used for conventional breeding in its production.

(c) Cisgenes/intragenes can also be obtained from the donor plants whose family genetic information concerning the protein structure and function is well understood regardless of its exploitations in crop development.

3.4.3 Cisgenesis/Intragenesis as a Novel Biotechnology in Plant Breeding

Crop improvement using cisgenic technology depends strongly on the full knowledge of the specific alleles that produce the desirable trait. The availability of the molecular markers that will be used to identify these alleles or genes is highly recommended being essential tools that help in plant breeding process (Moradpour and Abdullahi 2017). The introduction of sequencing in plant genome rendered the isolation and subsequent identification of alleles much easier. Large number of sequences that can be utilized in transgenic/cisgenic transformation are now available, and depending on the sequence analogy, many of these genes may be related to a specified function. This biotechnological approach has been used to determine the sequences of similar activity to get DNAs derived from plants (P-DNAs), which are used to develop classical T-DNAs in Agrobacterium-mediated modification in some specific plant genomic base. Genome editing methods in collaboration with other biotechnology techniques like cisgenesis can serve as a way forward in harnessing the plant genetic materials in the improvement of the cultivated crop (Cardi 2016).

3.5 Merits of Cisgenesis Over Conventional Breeding Methods

Cisgenesis as a novel biotechnological technique has the following advantages over other conventional breeding techniques:

3.5.1 Time-Saving Technique

The conventional breeding methods are associated with linkage drag, which leads to the inheritance of several undesirable genes by the offspring. To overcome this challenge, several generations need to be produced through backcrossing, which is time consuming. Cisgenesis introduces desirable alleles (gene) into the genome of plant of interest (recipient species) in a short time period, leaving behind any undesirable gene, thus averting the issue of linkage drag. For instance, the introgression of disease resistance gene in apple plant took almost 40 years through conventional breeding technique, while the insertion of scab (fungal disease) resistance gene Vf in apple into new crop varieties through cisgenic technique could provide better results in a very short time period. Therefore, cisgenesis can be used to rapidly introduce the gene of interest into the host plants in a relatively lesser time compared to traditional breeding methods (Telem et al. 2013).

3.5.2 Maintenance of Plant Genetic Constitution

Plants developed through hybridization techniques possess a different genetic makeup from their parents due to mixture of genes from both parents. Despite this, there is need to preserve some parts of the genome (parents' DNA), which will reveal a particular formative or constructive trait in the offspring. This approach is impossible through traditional breeding because of auto-incompatibility in crop plants that are vegetatively propagated (e.g., apple, grape, tomatoes). For example, when crossing is done in a grape variety with disease-resistant cultivar, the offspring's genetic makeup will be entirely dissimilar to the parents' species. Thus, conventional breeding methods will not grant disease and pest tolerance in notable parent species (EFSA 2012). Cisgenic approach has been used to transfer up to four genes with disease resistance into a single crop cultivar without altering the natural traits of the cisgenic cultivar (Haverkort et al. 2009).

3.5.3 Overcomes the Problem of Linkage Drag

Linkage drag is a problem related to the gene with desirable traits when it is linked with genetic material encoding for inferior/undesirable characteristics. The recombination between the gene of interest and unwanted genes is nearly impossible. As a result, the backcrossed line may turn out useless due to the linkage drag issue. The genes with inferior traits affect the natural attributes of the crops being engineered as they may influence the synthesis of different types of toxins and allergens (Telem et al. 2013). Cis/intragenic approach prevents the possible linkage drag impediments linked to traditional breeding technique by integrated application of specific genes (cisgenes). As a result, cisgenesis averts this issue (linkage drag) and hence deters the dangers that may arise from the unknown hitchhiking genes (Thapa et al. 2015). In crop plants that are vegetatively propagated like apples and potatoes, their heterozygous characters also result in the limitation of transferring the desirable traits successfully (Noor et al. 2018). Therefore, cisgenesis precludes linkage drag, and hence prevents hazards from unidentified hitchhiking genes.

3.5.4 Traits with Limited Allelic Variability Are Improved

The plant desirable traits with higher level of expression are produced by reintroduction of the genetic materials (genes) with that particular feature using both the promoter and the terminator from the crop itself (i.e., cisgenesis) or it may be obtained from the gene pool of plants that are sexually crossable (i.e., intragenesis).

3.5.5 The Decreased Use of Pesticides

Cisgenesis provides a means of transferring disease resistance genes to susceptible cultivars, which minimizes the rate of pesticide usage in managing disease-causing organisms. It therefore results in decline in the farmers' input cost and pesticide residues in the plant themselves and the environment, which increases consumers' acceptance of the crops and decreases environmental pollution. This will ultimately enhance sustainable crop production (Lusser et al. 2011).

3.6 The Potential Roles of Cisgenesis in Other Breeding Techniques

Cisgenesis can serve as an alternative for other breeding techniques such as gene introgression, induced translocation, and crossbreeding after cloning of the genes that code for desirable characters. This technique is basically required when a number of traits arising from different plant species need to be merged together at the same time. Initially, the plant species to be improved is required, and if the cisgenes cassette is used for the introgression of the desired genes into the plant, the cisgenes will be genetically linked together so tightly in the next crossbreeding. This aids in giving more attention to quantitative characters that have polygenic basis to be handled well in crossbreeding. Table 3.3 simplifies the breeding features of some crops as described by Jacobsen and Schouten (2010).

Table 3.3 Breeding features of some important crop plants

Crop name	Reproduction type	Cultivar	Homogeneity	Zygosity
Barley Wheat	Autogamous	Landrace Modern variety	Heterogeneous Homogeneous	Homozygous Homozygous
Potato	Vegetative Sexual	Varieties Potato seed (true)	Homogeneous	Heterozygous
Apple Asparagus	Vegetative Allogamous	Cultivars Pollination (open)	Homogeneous Heterogeneous	Heterozygous Heterozygous
Maize	Allogamous (monoecious)	Landrace and open pollination F1 hybrid Inbred lines	Heterogenous	Heterozygous
Rye Tomato Rice	Allogamous Allogamous	Open pollination Modern/old cultivars F1 hybrid	Heterogeneous Heterogeneous Homogeneous Homogeneous	Heterozygote Heterozygote Homozygote Homozygote
Cucumber	Allogamous (monoecious)	F1 hybrids	Homogeneous	Heterozygote

3.7 Rules and Regulations on the Use of Cis/Intragenic Plants

The global regulations on genetically modified organisms (GMOs) do not differentiate cisgenic from transgenic plants indicating that the rules guiding the development and the release of transgenes are also applicable to cisgenic crop plants (Jacobsen and Schouten 2009). The directive by the EU (2001/18/EC) described genetic modification in a broader term as breeding practices such as polyploidization, protoplast fusion, in vitro fertilization, and induced mutation, all of which are known to be safe for use long time ago, with the exception of protoplast fusion. As a result, plant species and their products developed through these techniques are not regarded as GMOs. The exemption of cisgenic/intragenic crops from the legislation guiding the deliberate release of GMOs into the public domain (environment) has been suggested (Schouten et al. 2006; Jacobsen and Schouten 2007; Rommens et al. 2007). This is because the features related to cisgenic/intragenic crop plants are safe and are very much similar to those plants developed through traditional breeding techniques (Table 3.4).

3.7.1 Rejections of Cisgenics Exemption from GMOs

Despite the recommendations to exempt cisgenic crops from GMO regulations, few objections were raised based on the following reasons: (a) introduction of the desirable genes or alleles into the genome of a plant to be modified is done randomly and (b) sudden change in gene (mutation) may occur in plant genome due to random insertion of the required genes. However, such things are not new in plant breeding practices. It was highlighted that about 2500 mutant cultivars from different crop species are developed through induced mutation and they are safe for use. The new GM cultivars with transgenes constitute a clear example of random insertion devoid of any issue, provided that the normal selection methods are employed (Jacobsen and Schouten 2010).

Vaibhav et al. (2018) stated that many crop plants including fruits, vegetables, and ornamental species were genetically modified through cisgenesis and

Table 3.4 GM regulations with respect to crop improvement strategies

		GM regulations	
S. no.	Gene types	A	B
1	Cisgenes	Exempted	Exempted
2	New transgenes	Full	Full
3	New intragenes	Partial	Exempted
4	New changes within the current gene in crop plant collection	Partial	Partial

A: The basis is a perfect endogenous gene. B: The basis is an endogenous sequence that is functional in activities like coding parts, promoters, and terminators
Redrawn from Jacobsen and Schouten (2010), Shiva et al. (2013), EFSA (2012)

intragenesis. The generation of cisgenic melon species using At1 and At2 as genetic materials as obtained from wild allied melon species has resulted in increased resistance of the modified species against a fungal disease (powdery mildew) caused by *Podosphaera cubensis* (Moradpour and Abdullahi 2017). Similarly, cisgenesis was employed to generate cabbage plant using an enzyme acetolactate synthase as selectable marker for *Agrobacterium*-mediated gene transfer regime. The cisgenic apple resistant to scab disease was developed by incorporation of Rvi6 gene (a gene resistant to scab of *Malus floribunda* 821), which led to an increase in the production of the crop (Würdig et al. 2015).

Previous studies (Holme et al. 2012; Shiva et al. 2013) applied cisgenic technology to obtain polygenic durable resistance to *Phytophthora infestans* and apple scab in potato and apple plants, respectively. It helped in minimizing the devastative effects of disease and increased the fruit production.

Thapa et al. (2015) explained cisgenesis as a rapid biotechnological technique that aids the gene transfer together with its promoter in a single step without unnecessary several backcrosses to overcome linkage drag problems. This method produces clean genetically transformed plant or organism without leaving any selective marker gene such as herbicide or antibiotic resistance behind. It therefore transforms the target plant without changing the genetic basis of the cisgenic plant.

As a result, cisgenesis is considered normal and safe than traditional breeding methods, and also resistance to a variety of stresses (biotic and abiotic) can be a useful tool in providing extensive and durable resistance (Rameshraddy et al. 2017). Kronberger et al. (2015) stated that cisgenic transformation has wider acceptance than transgenic, but it is unlikely to be regarded "natural" as this can be well elaborated through various theories that emphasized the role of the technique, human involvement, as well as role of symbolic orders and essential reasoning (Wagner et al. 2010). Currently, in the debates on new technologies of crop improvement, much attention has been paid to the regulation of these methods as well as their public acceptance. Both cisgenics and transgenics are presently placed under GMO regulation as both are necessarily labeled as GMOs in Europe (Van Hove and Gillund 2017) as previous studies on the early development of genetically modified cisgenic plants showed that the public perceptions on these transformed plants will be negatively affected if the GMO regulations remained unchanged.

Recent research on cisgenesis, a step towards the development of a new efficient crop by Sheikh Mansoor et al. (2019), revealed that cisgenesis is a technique with good potential of developing crops with economic, environmental, and health benefits and they can help in meeting the requirement of sustainable crop production more efficiently. They described transgenesis as a major problem to common people as it deals with manipulating genes of crop plant that are unable to crossbreed naturally. Cisgenic technique was developed as a replacement to transgenesis to overcome this public concern as it depends on genetic material derived from sexually crossable species. In this technique, the limitations of classical breeding are eliminated, while some foreign gene sequences and vector backbone sequences may not be present. Apple species that are resistant to scab, being the most destructive disease caused by fungal pathogen (Ascomycete, *Venturia inaequalis*), have been developed through cisgenesis by transferring a specific gene (*HcrVf2*) clone to improve the apple cv Gala plant (Vanblaere et al. 2011).

The work of Francis et al. (2017) on Ghanaian consumers' attitudes towards cisgenic rice describes rice blast as a fungal disease that leads to increased use of fungicides and costs of production and decrease in the yield potential of the crop. Blast-resistant rice can therefore lessen toxicity and increase productivity, thereby raising its potential to be labeled a good and beneficial crop environmentally. The blast disease of rice is among the major problems that cause food scarcity with up to 30% yield losses in rice production, which if resolved is assumed to take more than 60 million people out of hunger worldwide. They mentioned that rice varieties that are cisgenically generated to be blast resistant have the potential of increasing rice production with a less cost of production and eradicating the application of fungicides, which are toxic to the environment. The total success of cisgenic rice is however hindered by the level of its acceptance by the farmers, safety assessment, legal status, and labeling that need to be done on genetically engineered crops in a given nation. In barley crop, phytase activity was improved through cisgenesis. Similarly, rice pathogenic fungus (*Magnaporthe grisea*), one of the most destructive pathogens of rice, has been managed through cisgenesis by means of co-retrogression strategy to put in Pi9, a blast-resistant gene into the rice cultivars (Tamang 2018).

Gaetano et al. (2021) in their study "Novel and emerging biotechnological crop protection approaches" highlighted that cisgenic techniques were severally used in different crops like poplar, barley, potato, melon, apple, grapevine, wheat, strawberry, and rice, and in most of the studies, increasing resistance to biotic constraint related to pathogenic infection and improvement in the qualitative features of the crop were the major aspects considered (Krens et al. 2015; Maltseva et al. 2018). Haverkort and coworkers (Haverkort et al. 2016) followed a marker-free technique to get four genetically modified potato-resistant cultivars to *Phytophthora* species through the transfer of 1–3 resistance genes. Many genes resistant to pathogens including PR1 variants, VvA1b1, VvTL1, VvA1b1 homologics of VvAMP2, and orthologue to Snakin-1 were obtained from compatible species and expressed through transgenic methods and are now being evaluated in the field (Gray et al. 2014). Costa et al. (2010) also use heat shock as a recombination method for a selectable marker excision in the grapevine species.

Abdullah (2016) in his research on oilseed production in relation to climate change, explained cisgenesis as a means of modifying useful alleles derived from a sexually compatible plant into a host plant without any risk of it being released to the environment, and this feature distinguished them from transgenics. Erickson et al. (2014) in a debate on the "slippery slope of cisgenesis" warned that "the expertise and time required to create case-specific genomic clones that include endogenous promoters and terminators, free from selectable marker genes and vector backbone, should not be underestimated." Cisgenesis utilizes modern biotechnology and traditional breeding methods to hasten the breeding process while simultaneously avoiding the linkage drag, thereby enhancing the use of existing gene alleles (Hou et al. 2014). The introgression of extra copies of local genetic material through cisgenesis will practically establish a means of modifying plant genetic makeup, thereby expanding the variability in the structure of plants that are accessible to breeders and also enhancing the transfer of alleles in plant species that do not easily crossbreed. Advancement in genetic engineering has simplified the process of gene isolation and its introgression between plant species that are sexually crossable to improve a particular trait of interest (Lusser et al. 2012).

Noor et al. (2019) explained the techniques involved in the generation of cisgenic crops. They described that cisgenics and intragenics are produced using similar modification methods used to produce transgenic plants. Initially, the genes of interest must be isolated, synthesized, or cloned and then transferred to the recipient species where there is stability in the integration and expression of genes. Moreover, the generation of cisgenic/intragenic plants needs extra/additional investigations compared to the generation of transgenic crop species. They stated that marker gene elimination is achieved through the following strategies:

3.7.1.1 Co-Transformation

This is a highly effective and simple procedure of eliminating genetic markers from the genome of transgenic plant, which involves the transformation of crop cells using two plasmids that direct the gene insertion at two different loci in the DNA of a plant. The selection of marker gene is achieved by one plasmid, while the gene of interest is carried by the other plasmid. The major concepts involved in co-transformation strategy are three in number. Initially, there should be a particular plasmid carrying two T-DNAs in a single Agrobacterium strain (Ling et al. 2016); the next concept consists of two T-DNAs situated on two dissimilar plasmids in identical Agrobacterium strain, while the last avenue deals with two T-DNAs present in a variable *Agrobacterium* strain (Dutt et al. 2012). The disintegrated consolidation may be due to the presence of two T-DNA vectors with one of the vectors carrying the marker gene while the other one carrying the required gene that can separate in the offspring. To avoid the stage of segregation, the two T-DNAs can be co-transferred with one of them harboring the selection gene while the other one carrying the gene of interest, which will be followed by selection stage. Wheat and barley plants (cisgenics) have been developed successfully using this procedure (Gadaleta et al. 2008; Holme et al. 2012).

The rules in plant transformation (Agrobacterium mediated) can be altered to allow the mix-up of transgenes and selectable marker genes into two dissimilar chromosomes or genome loci. The co-transformation strategy can be considered an efficiently matured technology when the segregation of the entire cell lines that were co-transformed attains 25%. Figure 3.1 represents the steps involved in co-transformation technology.

3.7.1.2 Marker-Free Transformation

Another strategy to eliminate marker genes from the transgenic plant is to avert their use in plant transformation (Fig. 3.2). The transformation of potato cv. Kanico was first reported by de Vetten et al. (2003), where *Agrobacterium tumefaciens* strain AGLO was used in the study; the result shows high transformation efficiency as it

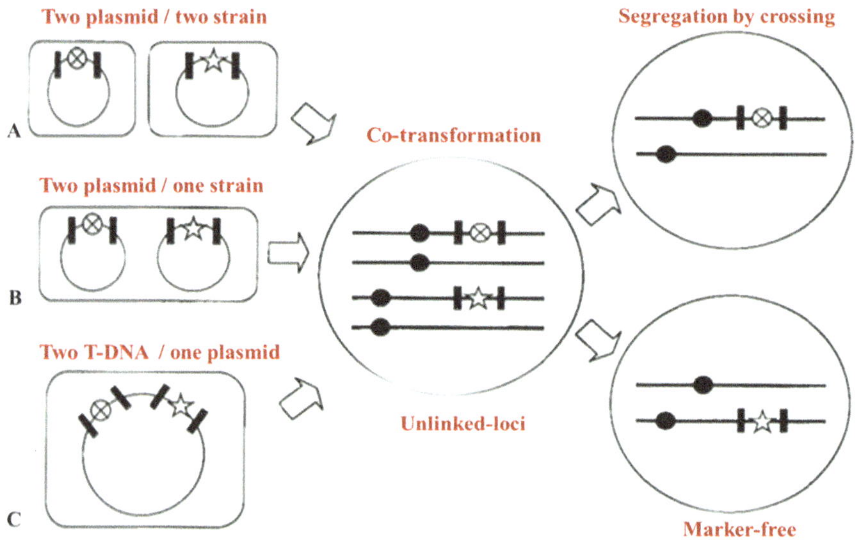

Fig. 3.1 Co-transformation technique with two plasmids targeting insertion site at two dissimilar loci in the genome of a plant. ☆ = Desirable gene. ▬ = T-DNA border. ⊗ = Selectable marker

has a DNA region with its origin from a supervirulent *A. tumefaciens* strain. Polymerase chain reaction (PCR) was used to analyze the regenerated shoots isolated from approximately 5000 shoots in the experiment. The method to be used for the development of marker-free crop species will depend on the mode of propagation and modification efficiencies of the target plant. A potato species with abundant amylopectin was developed through transformation system that was devoid of marker gene, and a cisgenic potato plant tolerant to late blight disease was generated using similar method. Similarly, intragenic marker-free potato with low acrylamide was developed using less complex method (Richael et al. 2008). The ipt alleles from bacterial cell are used to encode an enzyme, isopentenyl transferase, which helps in catalyzing the cytokinin isopentenyl adenosine formation in crops. The injection of this gene into the vector backbone containing the desirable gene renders the identification of transformants integrated with vector backbone so simple due to abnormality in their cytokinin-induced structure (Holme et al. 2013).

3.7.1.3 Recombinase-Induced Excision

In this strategy, specific recombination site is required where the exchange of DNA strand occurs between segments having a restricted level of sequence homology only. A number of specific recombination sites are well known and explained for the elimination of selection of marker genes. The first system is the Cre/1ox site-specific recombination using bacteriophage P1, the second is the FLP/FRT system using *Saccharomyces cerevisiae*, while the third system is the R/RS recombination from *Zygosaccharomyces rouxii*. The length of the recombination site is typically between

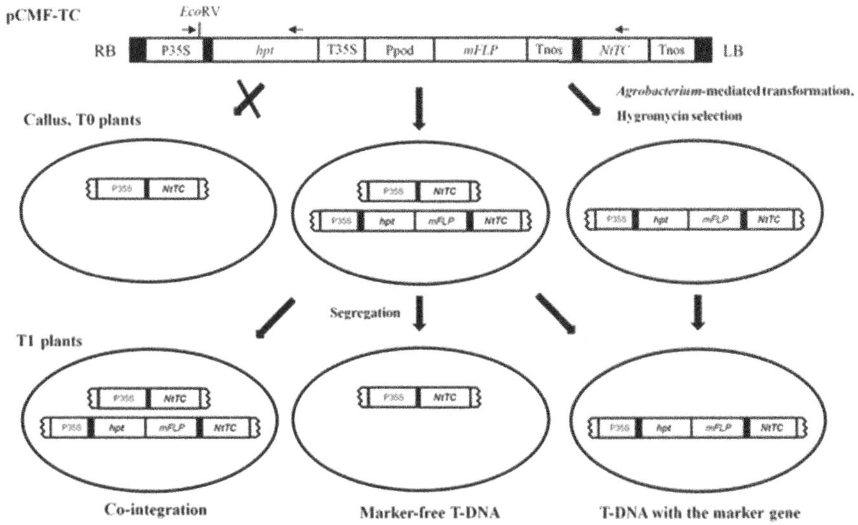

Fig. 3.2 Schematic representation for the development of selectable marker-free transgenic crop. (Adapted from Woo et al. (2015))

30 and 200 nucleotides with two motifs having a repeated symmetry that is partially inverted. The motifs will be bound by enzyme recombinase during the recombination process, which flank a central crossover sequence. Despite many merits, these methods are associated with some limitations including undesirable changes in the plant genome at the excision site due to prolonged presence of enzyme recombinase in plants. The self-excision method is also successfully applicable to only flowering plants. The method is not applicable to vegetatively propagated plants like banana, grapes, and potato. Diagrammatic representation of the removal of selectable marker approach through an inducible recombinase mediation is given in Fig. 3.3. The recombinase genes and selectable marker are directly flanked by recombinase-oriented recognition points (indicated by red arrows). The marker gene and recombinase excision are caused by the activation of the enzyme recombinase, where only gene of interest (GOI), nonreplicating globular fragment of DNA, and single recognition sites are left behind.

3.7.1.4 Transposon-Based Excision

The transposon excision, also known as jumping genes, can be as used as a useful tool to excise the marker sequence from the gene of interest. In this method, Ac/Ds transposition system is used based on the sequences of the genome present in the Ds repeat, which can be easily translocated to excise together with the Ds elements. It involves *Agrobacterium*-mediated modification, which is followed by relocation of desirable intragenic alleles.

Yajie et al. (2019) demonstrated the transformation of barley (*Hordeum vulgaris* L.) through cisgenesis using *HvGS1–1*, which results in the expression of high

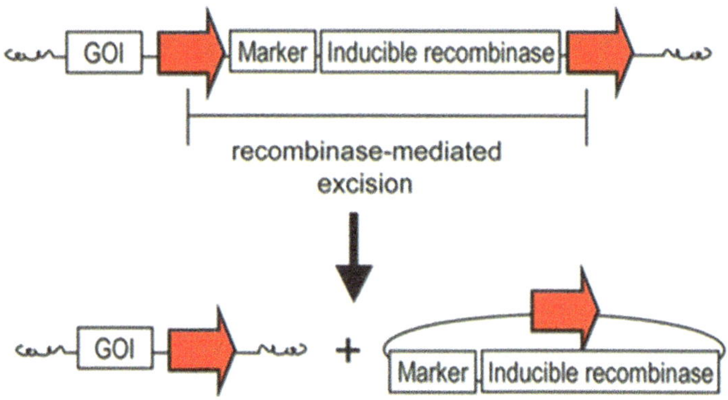

Fig. 3.3 Diagrammatic representation of recombinase-induced excision strategy. (Adapted from Yueju et al. (2011))

HvGS1–1 and cytosolic glutamate synthesis (GSI) enzyme activity. They showed that the overexpression of GSI lines has exhibited grain yields and nitrogen utilization efficiency compared to wild plant species grown under different supplies of nitrogen and two levels of CO_2 in the atmosphere. The protein content of the grain did not decrease in the GS1 overexpressed lines in domestic species when the plants are exposed to elevated CO_2 (about 800–900 μl L^{-1}). They concluded that through cisgenesis, the activity of GS1 can be increased leading to overexpression of HvGS1–1, which subsequently helped in the improvement of grain yield and nitrogen utilization in barley plants. The declining grain protein levels may also be prevented under elevated CO_2 in the atmosphere by the extra nitrogen utilization through GSI cisgenic line strategy. Similarly, there were positive correlation reports between protein concentrations of grains and GS activity in *Triticum durum* L. (Nigro et al. 2016; Zhang et al. 2017).

Jacobsen and Karaba (2008) described cisgenesis as a better breeding method than traditional introgression techniques because it requires less time, decreased steps, and lack of linkage drag. Cisgenesis leads to the elimination of undesirable traits or toxic substances that may be present in wild plant species, thereby maintaining the desirable traits by gene stacking. This method is used to develop potatoes that are resistant to *Phytophthora infestans* by stacking of many R-genes. This technique can similarly be applied to develop cisgenic plant species for qualitative traits as well as production of biotic and abiotic tolerant crop plants. Cisgenesis is believed to be used in providing new options for crop development as it can be applied directly in the improvement of existing varieties, which were considered to be safe for use in our environment (Rameshraddy et al. 2017).

Cecilia et al. (2017) explained cisgenesis and intragenesis as two recent biotechnological methods of transforming plant species, and they defined cisgenesis as a genetic modification of living species using genetic materials originating from the same species or from sexually crossable species. The gene inserted to generate cisgenic crop is an additional copy of incumbent/native DNA, which is considered a natural copy with its introns, native promoter, and terminator in a normal sense orientation (Giraldo et al. 2019). Intragenesis on the other hand was described as a biotechnological approach in which the genes with required traits come from species that are sexually crossable or from species itself. Studies on genomic sequencing provide information on genetic materials (cisgenes) that could be used genetically to improve specific crops, but the availability of cisgenic efficient marker genes and promoters is the main limitation. The two concepts have been used in various plant species to provide solutions for different diseases and environmental stress that affect crop plants leading to low agricultural productivity (Lamalakshhmi et al. 2013; Kost et al. 2015; Teodora 2016).

3.8 Issues Associated with Genetic Modification

Plant species with lengthy reproductive cycle can only be improved through conventional breeding in a very long time. Cisgenic technique is a special efficient method that can be used to produce crop plants with such lengthy reproductive cycle in a short time period (Schaart and Visser 2009). Maintaining the genetic constitution of verified/proven varieties unaltered is among the significance of genetic modification techniques of plant breeding; for instance, in a heterozygous and self-incompatible crop like apple, restoring the genetic composition of the proven cultivars in the subsequent offspring or progeny cannot be fully achieved by conventional breeding methods. Genetic modification is the technique of introducing the desirable alleles or genes of interest to improve a particular plant function without the co-insertion of genes that are not required (Joshi et al. 2009). The techniques are costly and time demanding because the alteration of the individual crop needs a repeated breeding, selection of the required features, and confirmation. However, a single gene or few alleles with desirable traits can be inserted to an existing crop variety through cisgenic and intragenic methods (Mans et al. 2015).

3.9 Conclusions

Traditional breeding techniques have influenced agricultural productivity by developing plants with excellent desirable traits that have added quality and quantity to cultivated food crops. However, these methods need lengthy time for the backcrosses and synchronic selection stages to subdue the problem of linkage drag. Biotechnological methods have rapidly added novel and valuable tools for plant breeders, which has helped a lot in developing crop cultivars with desirable traits in a very fast and more efficient manner, to overcome the challenges of food security by maintaining sustainable crop production to satisfy the needs of increasing global population. Novel techniques in biotechnology including cisgenesis/intragenesis can upgrade the entire breeding program excellently using the existing promising varieties. Cisgenesis is believed to be the basis of the evergreen revolution required in the classical breeding. Cisgenic crops have a wider acceptance level than transgenic crops as they were to be exempted from GMO regulations. It is hoped that cisgenesis may remove the possible undesirable breeding results as well as social considerations and beliefs with regard to GM technology. A significant role can therefore be played by cisgenesis for sustainable crop improvement.

References

Aaron M, Shewa L, Lanier N, Heather AS, Rodolfa M, Bruce LD (2018) CRIPR versus GMOS: public acceptance and valuation. Glob Food Sec 19:71–80

Abdullah AJ (2016) Breeding oilseed crops for climate change, pp 421–472

Ahmar S, Gill RA, Jung K-H, Faheem A, Qasim MU, Mustansar M, Zhou W (2020) Conventional and molecular techniques from simple breeding to speed breeding in crop plants: recent advances and future outlook. Int J Mol Sci 21:2590

Ankita S, Joshi M, Devi EL (2015) Alternative to transgenesis: cisgenesis and intragenesis. In: Al-Khayri JM, Jain SM, Johnson DV (eds) Advances in plant breeding strategies: breeding, biotechnology and molecular tools, pp 345–367

Bajaj S, Puthigae S, Templeton K, Bryant C, Gill G, Lomba P, Zhang H, Altpeter F, Hanley Z (2008) Towards engineering drought tolerance in perennial ryegrass using its own genome. In: 6th Canadian plant genomics workshop, p 62

Benjamin I, Kenigsbuch D, Galperin M, Abrameto JA, Cohen Y (2009) Cisgenic melons over expressing glyoxylate-aminotransferase are resistant to downy mildew. Eur J Plant Pathol 125:355–365

Cardi T (2016) Cisgenesis and genome editing: combining concepts and efforts for a smarter use of genetic resources in crop breeding. Plant Breed 135:139–147

Cecilia L, Silvia S, Jeremy BS, Bruno M (2017) New biotechnological tools for the genetic improvement of major woody fruit species. Front Plant Sci 8:1418

Chawla R, Shakya R, Rommens CM (2012) Tuber-specific silencing of asparagine synthetase-1 reduces the acrylamide-forming potential of potatoes grown in the field without affecting tuber shape and yield. Plant Biotechnol J 10:913–924

Chizzali C, Gusberti M, Schouten HJ, Gessler C, Broggini GAL (2016) Cisgenic Rvi6 scabresistant apple lines show no differences in Rvi6 transcription when compared with conventionally bred cultivars. Planta 243:635–644

Costa L, Mandolini M, Poletti V, Martinelli L (2010) Comparing 17-β-estradiol supply strategies for applying the XVE-Cre/loxP system in grape gene transfer (Vitis vinifera L.). VITIS J Grapevine Res 49:201–208

De Steur H, Van Loo EJ, Maes J, Gheysen G, Verbeke W (2019) Farmers' willingness to adopt late blight–resistant genetically modified potatoes. Agronomy 9:1–17

de Vetten N, Wolters A, Raemakers K, van der Meer I, ter Stege R, Heeres E, Heeres P, Visser R (2003) A transformation method for obtaining marker-free plants of a cross-pollinating and vegetatively propagated crop. Nat Biotechnol 21:439–442

Delwaide AC, Nalley LL, Dixon BL, Danforth DM, Nayga RM, Van Loo EJ, Verbeke W (2015) Revisiting GMOs: are there differences in European consumers' acceptance and valuation of cisgenically vs. transgenically bred rice? PLoS One 10(5):1–16

Dhekney SA, Li ZT, Gray DJ (2011) Grapevines engineered to express cisgenic *Vitis vinifera* thaumatin–like protein exhibit fungal disease resistance. In vitro cellular development. Biol Plant 47:458–466

Dutt M, Li ZT, Dhekney SA, Gray DJ (2012) Co-transformation of grapevine somatic embryos to produce transgenic plants free of marker genes. In: Transgenic plants: methods and protocols, pp 201–203

Erickson D, Stymne S, Schjoeerring JK (2014) The slippery slope of cisgenesis. Nat Biotechnol 32(8):727

FAO, IFAD, UNICEF, WFP and WHO (2020) The state of food security and nutrition in the world. p 20

Francis T, Lawton LN, Bruce LD, Dianna D, Anne CD, Rodolfa MN (2017) Ghanaian consumer's attitudes towards cisgenic rice: are all genetically modified rice the same? GJDS 14(1):1–18

Gadaleta A, Giancaspro A, Blechl AE, Blanco A (2008) A transgenic durum wheat line that is free of marker genes and expresses 1Dy10. J Cereal Sci 48:439–445

Gaetano G, Moffa L, Varotto S, Cardone MF, Bergamini C, De Lorenzis G, Velasco R, Nerva L, Chitarra W (2021) Novel and emerging biotechnological crop protection approaches. Plant Biotechnol J 19(8):1495–1510

Giraldo PA, Shinizuka H, Spangenberg GC, Cogan NOI, Smith KF (2019) Safety assessment of genetically modified feed: is there any difference from food? Front Plant Sci 10:1592

Gray DJ, Li ZT, Dhekney SA (2014) Precision breeding of grapevine (*Vitis vinifera* L.) for improved traits. Plant Sci 228:3–10

Han KM, Dharmwardhana P, Arias RS, Ma C, Busov V, Straus SH (2011) Gibberellin–associated cisgenes modify growth, stature and wood properties in populous. Plant Biotechnol J 9:162–178

Haverkort J, Struik PC, Visser RGF, Jacobsen E (2009) Applied biotechnology to combat late blight in potato caused by Phytophthora infestans. Potato Res 52:249–264

Haverkort AJ, Boonekamp PM, Hutten R (2016) Durable late blight resistance in potato through dynamic varieties obtained by cisgenesis: scientific and societal advances in the DuRPh Project https://doi.org/10.1007/s11540-015-9312-6

Holme IB, Dionisio G, Brinch-Pedersen H, Wendt T, Madsen CK, Vincze E, Holm PB (2012) Cisgenic barley with improved phytase activity. Plant Biotechnol J 10:237–247

Holme IB, Wendt T, Holm PB (2013) Intragenesis and cisgenesis as alternatives to transgenic crop development. Plant Biotechnol J 11:395–407

Hou H, Neslihan A, Zhen-Xiang L (2014) New biotechnology enhances the application of cisgenesis in plant breeding. Front Plant Sci 5:389

Huang S, Vleeshouwers VGAA, Werij JS, Hutten RCB, van Eck HJ, Visser RGF, Jacobsen E (2004) The R3 resistance to Phytophthora infestans in potato is conferred by two closely linked R genes with distinct specificities. Mol Plant Microbe Interact 17:428–435

Hunter CT, Suzuki M, Saunders J, Wu S (2014) Phenotype to genotype using forward-genetic Mu-seq for identification and functional classification of maize mutants. Front Plant Sci 4:545

Jacobsen E, Schouten HJ (2007) Cisgenesis, a new tool for traditional plant breeding, should be exempted from the regulation on genetically modified organisms in a step-by-step approach. Potato Res 51:75–88

Jacobsen E, Karaba NN (2008) Cisgenics - facilitating the second green revolution in India by improved traditional plant breeding. Curr Sci 94(11):10

Jacobsen E, Schouten HJ (2009) Cisgenesis: an important sub-invention for traditional plant breeding companies. Euphytica 170(1):235–247

Jacobsen E, Schouten HJ (2010) Cisgenesis: an important sub-invention for traditional plant breeding companies. Euphytica 170:235–247

Jochemsen H, Schouten HJ (2000) Ethische beoordeling van genetische modificatie. In: Jochemsen H (ed) Toetsen en Begrenzen. Een Ethische en Politieke Beoordeling van de Moderne Biotechnologie. Buijten and Schipperheijn, Amsterdam, pp 88–95

Joshi S, Soriano MJ, Schaart J, Broggini GAL, Szankowski I, Jacobsen E, Krens F, Schouten H (2009) Approaches for development of Cisgenic apples. Transgenic Plant J 3:40–46

Jhansi SR, Usha R (2013) Transgenic plants: types, benefits, public concerns and future. J Pharm Res 6:879–883

Kost TD, Gessler C, Jänsch M, Flachowsky H, Patocchi A, Broggini GAL (2015) Development of the first cisgenic apple with increased resistance to fire blight. PLoS One 10:980

Koul B, Srivastava S, Amla DV, Sanyal I (2014) Establishment and optimization of *Agrobacterium*-mediated transformation and regeneration of tomato (*Solanum lycopersicum* L.). Int J Biosci 4(10):51–69

Krens FA, Schaart JG, Van der Burgh AM, Tinnenbroek-Capel IE, Groenwold MR, Kodde LP (2015) Cisgenic apple trees; development, characterization, and performance. Front Plant Sci 6: 286

Kronberger N, Wolfgang W, Motohiko N (2015) How natural is "more natural"? The role of method, type of transfer, and familiarity for public perceptions of cisgenic and transgenic modification. Sci Commun:1–25

Lamalakshhmi DE, Chongtham SK, Holyachi P, Kuosar N, Sinh M, Behera RS, Telem NB, Wani SH (2013) Cisgenesis and intragenesis: twin sisters for crop improvement. Res J Agric Forestry Sci 1:22–26

Ling F, Zhou F, Chen H, Lin Y (2016) Development of marker-free insect-resistant indica rice by Agrobacterium tumefaciens-mediated co-transformation. Front Plant Sci 7:1–10

Lusser M, Parisi C, Plan D, Rodríguez-Cerezo E (2011) Deployment of new biotechnologies in plant breeding. Nat Biotechnol 30:231–239

Lusser M, Parisi C, Plan D, Rodríguez-Cerezo E (2012) Deployment of new biotechnologies in plant breeding. Nat Biotechnol 30:231–239

Lutken H, Laura M, Borghi C, Orgaard M, Allavena A, Rasmussen SK (2011) Expression of KxhKN4 and KxhKN5 genes in Kalanchoe blossfeldiana 'Molly' results in novel compact plant phenotypes: towards a cisgenesis alternative to growth retardants. Plant Cell Rep 30(12):2267–2279

Majhi PK (2020) Chapter-4 Gene pool concept in plant breeding. In: Current research and innovations in plant pathology, p 85

Maltseva E, Ismagul A, Iskakova G, Chirkin A, Skiba Y, Ismagulova G, Eliby S, Aitkhozhina N (2014) Wheat cisgenic transformation with class I chitinase gene. J Biotechnol 185:S116–S117

Maltseva ER, Iskakova GA, Rsaliev AS, Skiba YA, Naizabaeva DA, Ismagulova GA (2018) Assessment of cisgenic bread wheat lines carrying class I chitinase gene to leaf rust. J Biotechnol 280:S80–S81

Mans R, van Rossum HM, Wijsman M, Backx A, Kuijpers NGA, van den Broek M, DaranLapujade P, Pronk JT, van Maris AJA, Daran JMG (2015) CRISPR/Cas9: a molecular Swiss army knife for simultaneous introduction of multiple genetic modifications in Saccharomyces Cerevisiae. FEMS Yeast Res 15:1–15

Mansoor SS, Nazeer A, Khalid ZM (2019) Cisgenesis: a step towards development of new efficient crops. EC Agric 5(3):165–167

Mariam S (2015) Transgenic, cisgenic, intragenic and subgenic crops. Adv Crop Sci Tech 3:123

Moradpour M, Abdullahi SNA (2017) Cisgenesis and intragenesis as new strategies for crop improvement. Springer International Publishing AG, Cham, pp 191–209

Nigro D, Fortunato S, Giove SL, Paradiso A, Gu YQ, Blanco A, Pinto MCD (2016) Glutamine synthetase in durum wheat: genotypic variation and relationship with grain protein content. Front Plant Sci 7:1–11

Noor EM, Mallik MN, Krishna KR, Nitesh SD (2018) Cisgenesis and intragenesis: a new tool for conventional plant breeding. J Pharmacogn Phytochem 8(1):2485–2489

Noor EM, Mallik MN, Krishna KR, Nitesh SD (2019) Cisgenesis and intragenesis: a new tool for conventional plant breeding. J Pharmacogn Phytochem 8(1):2485–2489

Paula AG, Hiroshi S, German CS, Noel OIC, Kevin FS (2019) Safety assessment of genetically modified feed: is there any difference from food? Front Plant Sci 10:1592

Rameshraddy, Khan JA, Mahesh S, Shankar AG (2017) Cisgenesis: an approach for crop improvement. Int J Pure Appl Biosci 5(3):245–250

Richael CM, Kalyaeva M, Chretien RC, Yan H, Adimulam S, Stivison A, Weeks JT, Rommens CM (2008) Cytokinin vectors mediate marker-free and backbonefree plant transformation. Transgenic Res 17:905–917

Rommens CM, Ye J, Richael C (2006) Improving potato storage and processing characteristics through all-native DNA transformation. J Agric Food Chem 54:9882–9887

Rommens CM, Haring MA, Swords K, Davies H (2007) The intragenic approach as a new extension to traditional plant breeding. Trends Plant Sci 12:397–340

Rommens CM, Yan H, Swords K (2008) Low-acrylamide French fries and potato chips. Plant Biotechnol J 6:843–853
Schaart JG (2004) Towards consumer-friendly cisgenic strawberries which are less susceptible to Botrytis cinerea. Ph.D. thesis. Wageningen University, Wageningen
Schaart JG, Visser RGF (2009) Novel plant breeding techniques consequences of new genetic modification-based techniques in comparison to conventional plant breeding. The Netherlands commission on genetic modification. COGEM Research Report number 2009–02
Schouten HJ, Krens FA, Jacobsen E (2006) Cisgenic plants are similar to traditionally bred plants. EMBO Rep 7:750–753
Shiva GP, Raju B, Kumar RK (2013) CISGENESIS – an alternate approach for development of genetically modified crops. Ann Biol Res 4(9):109–115
Taak P, Tiwari S, Koul B (2020) Optimization of regeneration and *Agrobacterium*-mediated transformation of Stevia (*Stevia rebaudiana* Bertoni): a commercially important natural sweetener plant. Sci Rep 10(1):1–12
Tamang T (2018) Development of selectable marker-free cisgenic rice plants expressing a blast resistance gene Pi9. World Cong In Vitro Biol 54:544
Telem RS, Shabir HW, Singh NB, Nandini R, Sadhukhan R, Bhattacharya S, Mandal N (2013) Cisgenesis: a sustainable approach for crop improvement. Curr Genomics 14:468–476
Teodora CA (2016) Cisgenesis and genome editing: combining concepts and efforts for a smarter use of genetic resources in crop breeding. Plant Breed 135:139–147
Thapa B, Joshi T, Krishna KJ, Pawan B (2015) Cisgenesis: genetic engineering introduced spark to traditional breeding methods. Res Environ Life Sci 8(4):757–760
Vaibhav S, Saurabh S, Kamuri S, Ashutosh K (2018) Cisgenesis: a sustainable approach to gene intogession and its utilization in horticultural crops. Int J Curr Microbiol Appl Sci 7:5002–5009
Van Hove L, Gillund F (2017) Is it only the regulatory status? Broadening the debate on cisgenic plants. Environ Sci Eur 29:22
Vanblaere T, Szankowski I, Schaart J, Schouten H, Flachowsky H, Giovanni ALB, Gessler C (2011) The development of cisgenic apple plant. J Biotechnol 154:304–311
Wagner W, Kronberger N, Nagata M, Sen R, Holtz P, Palacios FF (2010) Essentialist theory of "hybrids": from animal kinds to ethnic categories and race. Asian J Soc Psychol 13:232–246
Weeks JT, Ye J, Rommens CM (2008) Development of an in-planta method for transformation of alfalfa (Medicago sativa). Transgenic Res 17:587–597
Woo HJ, Qin Y, Park SY, Park SK, Cho YG, Shin KS (2015) Development of selectable marker-free transgenic rice plants with enhanced seed tocopherol content through FLP/FRT– mediated spontaneous auto-excision. PLoS One 10(7):132667
Würdig J, Flachowsky H, Sa A, Peil A, Hanke MV (2015) Improving resistance of different apple cultivars using the Rvi6 scab resistance gene in a cisgenic approach based on the Flp/FRT recombinase system. Mol Breed 35:95
Yajie G, Thomas C, De Bang D, Schjoerring KJ (2019) Cisgenic overexpression of cytosolic glutamine synthetase improves nitrogen utilization efficiency in barley and prevents grain protein decline under elevated. Plant Biotechnol J 17:1209–1221
Yueju W, Yuan-Yeu Y, Donna PB, James GT (2011) Recombinase technology: applications and possibilities. Plant Cell Rep 30:267–285
Zhang H, Zhang J, Lang Z, Botella JR, Zhu JK (2017) Genome editing-principles and applications for functional genomics research and crop improvement. Crit Rev Plant Sci 36:291–309

Transgenics and Crop Improvement 4

Abstract

In developing nations, where arable land per capita is declining but human and animal populations are constantly expanding, the key constraint for food and nutritional security for the human population in the next years will be sustained plant productivity and crop yield(s). Apart from the genetic potential of plant species, agricultural plant output is quite variable and is impacted by a variety of physical, abiotic, and biotic factors. Transgenic technology has the potential to cope with these situations and to feed the teeming millions through crop improvement strategies. The basic objective of any transformation technique is to get the desired gene into the cell's nucleus without compromising the cell's capacity to live. The plant is considered to be transformed if the inserted gene is functional and the gene product is produced. The plant is called transgenic after the gene introduced is stable, inherited, and expressed in following generations. Therefore, transgenic plants are plants that have been genetically engineered with novel traits and are identified as a class of genetically modified organism (GMO). Several GM crops such as corn, cassava, soybean, canola, squash, tobacco, mustard, tomato, rice, papaya, cotton, alfalfa, sugar beet, and brinjal have been commercialized worldwide, and some are under pipeline. These crops and their products have now gained acceptance in several countries across the world. Previously, the focus of development of transgenic plants was to develop water, salinity, temperature, insect, and disease tolerance, but now the focus of this technology is enhancement of nutritional components in edible crops so as to improve people's health. Transgenic technology is an indispensable tool for the biotechnologists and has a bright future towards sustainable development goals.

Keywords

Transgenic technology · GMO · Genetic modification · Biosafety · Crop improvement · Bt technology

B. Koul, *Cisgenics and Transgenics*, https://doi.org/10.1007/978-981-19-2119-3_4

4.1 Introduction

To satisfy the growing global community, particularly the underfed citizens in some parts of the universe (e.g., Africa, Latin America, and Asia), crop cultivation should be improved by steadily raising staple agricultural output (Saeed et al. 2020). Humans have been choosing crops from obtainable collection to produce novel crop varieties for hundreds of years across the globe. This selection procedure laid the groundwork for the development of food plants or crops, as we now recognize them. Plant breeding is the process of identifying and selecting desirable attributes in plants and then incorporating them into such a single species. The production of genetically modified plants has substantially improved global crop production throughout the recent decades (Kumar et al. 2020).

The following steps are crucial towards the construction of genetically modified plants: tissue culture and transformation technologies that are effective, preparing gene sequences and converting them using appropriate vectors, mechanisms of introducing genes into agricultural crops that are both efficient and effective, genetically modified plants' recovery and proliferation, genomic characterization of the modified plants for accurate and fruitful expression of genes, transferring the genetic material (genes) into superior crop varieties via traditional breeding techniques, as well as assessment of the modified crops regarding their ability to mitigate biotic and abiotic pressures while posing no danger to the public (Birch 1997; Tarafdar et al. 2014). According to an international meta-analysis of the influence of GM crops' acceptance, transgenesis boosted crop production by 22% on average, resulting in a 68% rise in farmer profitability (Klumper and Qaim 2014). At the same time, the initial modified crops have been created, namely antibiotic-tolerant tobacco as well as petunia (Fraley et al. 1983; Herrera-Estrella et al. 1983). Moreover, it was discovered that the bean gene "phaseolin" was expressed in sunflower, illustrating that a crop genetic material can be manifested/expressed even when it is added to a botanically different species (Murai et al. 1983). A genetically modified tomato "Flavr Savr" from Calgene (Monsanto) was authorized for sale by the FDA in the USA in 1994, having the property of an extended shelf life or deferred fruiting (Kumar et al. 2020). Afterward, other transgenic crops were certified for commercialization. Such crops include canola characterized with improved oil properties and potato, maize, and soybean that are tolerant to glyphosate among others (James 1997). Microbial genomes and other genetic elements have been used extensively in commercialized genetically engineered crops (Kumar et al. 2018).

4.2 Crop Improvement Through Transgenic Technology

Although various ways are available for genetic modification, four are commonly deployed and have facilitated researchers to insert genes into a variety of agricultural crops (Dale et al. 1993). Genomic transformation mediated by *Agrobacterium* species, biolistic techniques (particle bombardment), microinjection of the modified

DNA directly into the plant cells, and protoplast modification/transformation method are a few examples. These techniques are briefly mentioned below:

1. The bacterium *Agrobacterium tumefaciens* has been commonly deployed to transfer appropriate genetic material into agricultural crop plants. This bacterium species dwells in the soil, and it is linked to gall development in dicots plants' infection site. The involvement of a high tumor-inducing (Ti) plasmid in pathogenic variants of *Agrobacterium* is responsible for this tumor-causing potential. Similarly, aggressive species of *Agrobacterium rhizogenes*, the pathogen that causes "hairy root" malady, contain root-inducing (Ri) megaplasmids. The genomic biochemistry of crown gall and hairy root activation, as well as the Ti and Ri plasmids, was extensively researched (Zambryski et al. 1983). This technique is achieved by the integration of desired genes obtained from tumor-inducing (Ti) plasmid that replicate independently within the cell of *A. tumefaciens*. It will subsequently infect the plant tissue, thereby transferring the T-DNA incorporating the required genetic material (gene) into chromosomes of the host plant cells that are energetically dividing (Tzfira and Citovsky 2006).
2. Direct injection of transgenic genetic material into the nucleus of somatic embryo cells could be used to elicit plant regeneration in cultured cells (Karesch et al. 1991). Using microscopy, individual cells or minute clusters of cells must be micromanipulated, and small volumes of DNA solution must be precisely injected with an extremely fine minute pipette. Infused tissues or bunches of cells are developed via in vitro technique and reproduced into new seedlings.
3. Particulate explosion (biolistics) method involves coating tungsten or gold particle (tiny microprojectiles) mostly with DNA to be incorporated and then inundating them into cells/tissues that have the capacity of plant restoration. Massive microprojectiles (tungsten or gold granules of 0.5–5.0 m diameter coated with DNA) are accelerated and carry genes into nearly all forms of tissues or cells (Sanford 1988). The genome components access the cells of the plant, the genome DNA is integrated within a little fraction of the treated cells, and the transformants are chosen for genetic transformation.
4. The intended/targeted cell walls are disrupted by catalytic activity, and the plasma membrane acts as a barrier between the cells in protoplast transformation technique. DNA could be inserted to a cell suspension, which could be delivered by putting polyethylene glycol on the cell membrane or conducting an electrical current using protoplast suspension. Only a few cells' genomes are integrated with the DNA. To choose the protoplasts that have been changed and the cellular clusters that resulted through them, an appropriate marker should be added (Shimamoto et al. 1989).

4.3 Advantages of Transgenic Techniques in Crop Improvement

4.3.1 Improved Crop Yields

Zhang et al. (2012) examined the QTL, qGL3 that codes for a phosphate-containing putative protein through a kelch-like region/domain (OsPPKL1) for a major grain length in rice, and the result showed that the qgl3 gene has led to a long grain genotype in the rice crop. Overexpression of TYFY11b allele/gene was also shown to increase the grain size in rice as reported by Hakata et al. (2012). Similarly, improvement in grain quantity/number in same crop has been achieved via the expression of zinc finger transcription element that controls an enzyme (cytokinin oxidase) expression (Li et al. 2013). Bednarek et al. (2012) in another study on *Triticum aestivum* has determined the function of TaGW2-A in regulating the size of the wheat grains. An attempt to introduce C4 photosynthetic attribute, which is seen in maize, into C3 crops like rice is one of the radical efforts made to improve crop yield via transgenic techniques. The expression of a maize-gene (phosphoenolpyruvate carboxylate) into wheat crop resulted in an increase in photosynthetic rate, 26% higher than that in non-transgenic crops (Qin et al. 2016; Cordeiro et al. 2022). Crop plants with traits like tolerance to disease and greater grain quality are the benefits of plant biotechnology. Now, scientists may choose disease-resistant genes obtainable through some other animal species and introduce them to some important crops. Furthermore, researches from the University of Hawaii and Cornell University developed two pawpaw cultivars conferring resistance to pawpaw virus infection by delivering the virus' genomes to the crop plants. Since 1998, papaya producers were able to obtain seedlings of the two cultivars, "SunUp" and "Rainbow," distributed under licensing arrangements. Several abiotic factors like drought, cold, salinity, heat, and flooding render an adverse effect on the agricultural crop yield resulting in decreased crop output (Suzuki et al. 2014; Kumar et al. 2020). To overcome this issue and improve crop yield, seven, three, and two events associated with ecological constraint have been released and marketed in crops like maize, sugarcane, and soybean, respectively (ISAAA 2019). A cold-shock protein from bacteria was deployed to mitigate the impacts of cold in *Arabidopsis* species, cold, and moisture deficiency in rice as well as shortage of water in *Zea mays* crop resulting in improved yield in the crop (Castiglioni et al. 2008). The modified maize showed proper phenotype with better adaptation during water-limited situations. Furthermore, about six other events were generated in maize by herbicide tolerance stacking of insect resistance events relating to abiotic stress resistance (ISAAA Database 2019). Diseases caused by pathogenic organisms lead to decrease in the quantity of crop plants, and through transgenic techniques such problems have been minimized. For example, Tricoli et al. (1995) and Yan et al. (2007) developed virus-resistant modified crops using gene silencing strategies like co-suppression or RNAi and antisense RNA methods. Similarly, tolerance from cucumber virus disease was demonstrated in sweet pepper and tomato through the expression of viral coat proteins (Zhu et al. 1996; Yang et al. 1995). Genetic modification technique employing defective replicase genes was used to generate improved papaya (Huanong No. 1 papaya) that is tolerant to virus infection as reported by Guo et al. (2009).

4.3.2 Enhancement in Crop Protection

Agricultural methods are used by farmers as they offer cost-effective remedies to insect pest infestations that, when left untreated, would result in significantly decreased yields. As previously indicated, crops like corn and potatoes were successfully modified via transgenic technique to produce a biomolecule (proteins) that kills insect pests that consume such crop plants. The protein is derived from a soil bacterium species (*Bacillus thuringiensis*), which is being used as a natural insecticide for decades (Kamthan et al. 2016). When Bt was built into a maize crop, for example, the whole crop turned to be insect tolerant, not just the section where the Bt toxin was applied. Organic farmers, for example, use Bt as an insecticide to keep insect pests out of their farms, but transgenic Bt crops may be objectionable to them. Wally and Punja (2010) mentioned that transgenic crops were severally utilized to successfully increase resistance to disease caused by bacterial and fungal pathogens without posing negative effects beneficial to soil microbes. The work of Kesarwani et al. (2000) and Kumar et al. (2016a, b) showed that expressing oxalate-disintegrating enzyme (OXDC) in crops like tomato, soybean, tobacco, and lathyrus has led to high tolerance to pathogenic fungi (*Sclerotinia sclerotiorum*) that employ oxalic acid while colonizing the host. Similarly, as a substitute to synthesized chemicals, transgenic crops like cotton and corn have been developed to provide inherent insecticidal proteins that are coded by genetic material from *B. thuringiensis* as reported (Sanahuja et al. 2011). Tabashnik et al. (2013) and Sanahuja et al. (2011) reported that Bt substances are poisonous to the ravaging insect species but not harmful to other living beings, human beings inclusive. Bt crops are considered environment friendly in addition to the tremendous improvement in the yields of the crop and resulting benefits to the crop producers (Carpenter 2010; Lu et al. 2012). Several gene silencing techniques utilizing small noncoding RNAs were used to increase protection mechanisms in agricultural crops against the attack of pathogenic microbes like bacteria, viruses, fungi nematodes, and insect pests as reported by Kamthan et al. (2015). The modification of LEA (late embryonic protein) genes has been done in several crop plants so as to improve resistance to drought (Duan and Cai 2012). For instance, transgenic species of *Arabidopsis* expressing a wheat protein (dehydrin, DHN-5) exhibited increased growth, seed germination capacity, moisture retention, as well as improved H-proline volume (Brini et al. 2011). Antibody transformation involving the expression of antibodies or a segment of rAbs that are capable to deactivate pathogens directly or deactivate the protein responsible for the pathogenesis is another technique used to protect plants against pathogens (Cardoso 2014). Pokeweed antiviral protein has been employed by Thamizhmani and Vijayachari (2014) to generate plants that are tolerant to viral infection.

4.3.3 Improvement in Food Processing

Transgenic technology plays a significant role in providing biomolecules that help in food purification process. Chymosin, for example, is an enzyme secreted by bioengineered bacterium that was the pioneer dietary product originating from biotechnological methods to get legal authorization in 1990. It is now used in 60% of all cheese production, replacing calf rennet. Enhanced purity, consistent production, 50% cost decrease, and great cheese yield performance are just a few of the advantages (Deepak et al. 2018). Production process has been increased by researchers via genetic transformation techniques. The development of rDNA-mediated modification of *Aspergillus nidulans* (Campbell et al. 1989) and its subsequent development in *A. niger* have been achieved successfully (Ward et al. 1988). The acceptance of various enzymes in food making and processing by the US FDA in the early 1960s and addressing the potential of *Aspergillus niger* in food production have helped a lot in increasing industrialization in non-toxigenic and nonpathogenic circumstances (Deepak et al. 2018). Catalase enzyme is another form of enzyme developed via genetic engineering methods in microorganisms like *A. niger* and *Vibrio cholera*. These enzymes facilitate in maintaining reactive oxygen species (ROS) level, therefore making a substantial contribution in ROS homoeostatic activities (Goulart et al. 2016). *Geobacillus* spp. are known to possess a unique combination of numerous vital extremophilic industrial attributes, which aid in the higher production of catalase enzyme (Kauldhar and Sooch 2016).

4.3.4 Improved Nutritive Value

Transgenic methods have unlocked novel possibilities of enhancing nutritive value, flavor, and composition of agricultural crops. Soybeans with relatively better protein levels, potatoes with much more nutritious usable carbohydrate and optimized composition of amino acid, beans with improved micronutrients, as well as rice with the potential to provide a precursor to vitamin A synthesis (beta-carotene) are among the genetically engineered crops in development that help individuals who are nutritionally deficient. Rice, being a vital crop plant, is wanting in beta-carotene that serves as a precursor of providing vitamin A. Using transgenic methods, scientists have produced nutritionally enhanced rice (golden rice) having beta-carotene being expressed in the crop seeds (Ye et al. 2000; Kamthan et al. 2016). A superior banana variety with additional level of beta-carotene has also been generated by altering an enzyme, phytoene synthase gene extracted from a banana cultivar (Asupina) species (Mlalazi et al. 2012). The benefit of these bananas compared to rice is that there is no issue of transgene movement in bananas because they are sterile (Kamthan et al. 2016). Transformed potato tuber (expressing AmA1) has been developed with substantial increment in the production of tubers with high protein content and improvement in several crucial amino acids (Chakraborty et al. 2000). Kamthan et al. (2012) stated that various preferable traits such as resistance to various constraints (biotic and environmental), as well as enhancement in iron and other useful fatty acid quantity, were achieved in tomatoes by conveying a particular gene from fungi (C-5 sterol desaturase). This method can be deployed for improving the nutritional content of other vital crops like maize, brinjal, and cowpea.

4.3.5 Improved Shelf Life

Increase in softening of fruit crops while ripening has led to the damages and loss of various fruit (climacteric and non-climacteric) species (Kamthan et al. 2016). There are two approaches for improving the shelf life of tomato (crop) as stated by Tarafdar et al. (2014), which deal with antisense RNA approach and employing the ACC gene (1-aminocyclopropane-1-carboxylic acid) deaminase enzyme that breaks the ACC to an alkene compound (ethylene) leading to fruit ripening. The main way of improving the storage quality of fruits while maturing is to enhance the required texture of the fruits (Chapple and Carpita 1998). For this purpose, several techniques including RNAi were utilized (Matas et al. 2009). Reduction in the development of autocatalytic ethylene by repression of genetic material (genes) from ethylene biosynthetic route such as 1-aminocyclopropane-1-carboxylic acid synthase (ACS) or ACC oxidase (ACO) is one of such strategies (Hamilton and Baulcombe 1999). Numerous researches were conducted to lengthen the fruit storage quality by modification via ACS or ACO alleles in antisense orientation (Bapat et al. 2010; Litz and Padilla 2012). The work of Schaffer et al. (2013) showed that the repression/suppression of SEP1/2 category of gene (MADS8/9) prevented quick ripening in apple fruits. Atkinson et al. (2012) demonstrated the use of an enzyme (polygalacturonase, PG) that helps greatly in the breakdown of plant cell wall, and softening of fruits was suppressed using RNAi approach with decreased pectin-polymerizing process without alteration of some other fruit attributes like weight, color, and soluble solids. Several earlier studies (Priem et al. 1993; Meli et al. 2010) showed that the breakdown of N-glycoprotein cell wall and levels of free N-glycan has substantially improved softening of tomato fruits, which is a climacteric form of fruits. Other enzymes (α-Man and β-Hex) utilized in the repression of N-glycan were reported to improve the fruit shelf life in tomato plants as they minimized the softening rate of fruits (Meli et al. 2010). Furthermore, the elimination/suppression of α-Man and β-Hex reported by Ghosh et al. (2011) using non-climacteric fruit (*Capsicum* sp.) has led to delay in the deterioration of fruits by 7 days, and the fruits were two times stronger than the control. This shows that genetic modification of N-glycan enzymes might be useful in minimizing postharvest damages in many vital fruit crop (climacteric and non-climacteric) species. Recently, β-Hex and α-Man promoters have been determined and characterized to shed light on transcriptional regulation of such genes when fruits ripen (Irfan et al. 2014, 2016).

4.3.6 Environmental Benefits

Pesticide remains on foodstuffs are lowered, pesticide leakage into waterways is minimized, and farm laborer exposure to toxic materials is curtailed because genetic engineering reduces pesticide reliance. Because Bt cotton is resistant to three important pests, it now accounts for half of the US cotton crop, resulting in a 15% reduction in global insecticide use. Increase in the acceptance of soybean crop resistant to herbicide was related with minor increase in yields and variable profits but significant decrease in herbicide utilization (Ervin and Welsh 2006).

Pesticide utilization reduction is a major direct benefit of GM agricultural production, as it reduces farmers' chemical exposure and lowers pesticide contaminants in agricultural crops, releasing fewer chemicals in the atmosphere and possibly increasing on-farm pollinators (Hossain et al. 2004; Huang et al. 2005; Nickson 2008). Furthermore, insect resistance conferred by genetic engineering can minimize the concentration of aflatoxins in agricultural crops. Although reduction in the application of pesticide has appeared to be among the most vital direct impacts of transgenic crops on the ecosystem, the issue of how to determine the effect is still unsolved (Kleter and Kuiper 2005). Ecological impact quotient (EIQ) is the most commonly applied parameter in such studies, which deals with the impacts of herbicides on the ecosystem, farm laborers, and consumers. A positive ecological benefit has been indicated using EIQ to herbicide-resistant soybeans compared to non-HR crops. A study to compare the impacts of GM crops to those of conventional ones using oilseed rape showed that applying pesticide active ingredients has reduced by 30% and environmental impact per hectare reduced by 42%, with 54% reduction of the herbicide effects on farm workers (Kleter et al. 2007). Similar research by Brookes and Barfoot (2008) reported 15.4% reduction in the impacts of herbicides on the ecosystem. There was also a report on the decrease (12.6%) in the use of chemicals on HR canola due to the presence of herbicide resistant transgene (Smyth et al. 2011a). Gusta et al. (2011) and Smyth et al. (2011a, b) examined the ecological aftermaths of herbicide-tolerant canola in Canada, the acceptance of which has transformed the weed control practices.

4.3.7 Benefits for Developing Countries

Transgenic crop plants can lead to poverty mitigation in developing nations by enhancing agricultural output leading to high profits to farmers and at the same time dealing with the persistent issues of starvation in the said countries (Juma 2011; Brookes and Barfoot 2009, 2013). Transgenic technology can aid in the development of genetically transformed crops in developing nations, provided that the problems related to safety of the products, ecological issues, and ethics are sufficiently addressed (Anderson 2010). Furthermore, the therapeutic proteins (antibodies, enzymes and vaccines) can be produced safely and economically in large quantities using transgenic plants as living bioreactors (Jhansi and Usha 2013; Agarwal et al. 2008; Sharma and Sharma 2009; Tiwari et al. 2009). This will lead to mass production of drugs that can be used in the medication and prevention of different ailments in developing nations. Therefore, the significance of this technique lies in improving the quality and quantity of food supply as well as provision of therapeutic substances to the increasing population of the globe, especially the underdeveloped countries, in the coming years (Jaiwa et al. 2014).

4.4 Disadvantages of Transgenic Crops

Previous researches (Zhang et al. 2016; ISAAA 2018; Brookes and Barfoot 2018) demonstrated the benefits provided by transgenic crops in the last 20 years in terms of yield improvement, decreased application of herbicides and pesticides, minimizing CO_2 emission, farmer income enhancement, as well as consumer health improvement. Despite the positive impacts delivered by these crops, a number of problems are known to be associated with the adoption of the crops as briefly explained below:

4.4.1 Biosafety-Related Issues

Among the issues of adopting transgenic crops, biosafety concerns are related to human health menace such as allergenicity and toxicity (Kumar et al. 2020). The possible health threat associated with allergenicity and toxicity of engineered crops had been contentious for long. For example, cry9c expressed in maize was accepted as animal feed and commercial application but not allowed for human utilization due to expectation of its allergenic and toxigenic effects on human health as a result of high protein stability and its potential in interacting with human immune system (Bucchini and Goldman 2002; Carzoli et al. 2018). Séralini et al. (2012) reported that transgenic maize caused potential health threats including chronic kidney disorder, high tumor occurrence, and high liver congestion as well as increased death rate of the experimental animals (rats), although the report was later retracted due to some reason. However, subsequent studies related to health hazards of modified crops like rice, soybean, maize, and soybean on poultry, rodents, frogs, monkeys, and cows did not report any negative impacts of transgenic crops on the status of animal's fitness as reviewed (Domingo 2016; Tsatsakis et al. 2017; de Vos and Swanenburg 2018; Kumar et al. 2020). The danger of introducing allergens and poisons into otherwise acceptable meals has been mentioned as a serious safety problem using genetic engineering technologies. The FDA monitors naturally existing allergen concentrations in foodstuffs derived from GMO crops to confirm that they do not exceed the ideal range seen in traditional foods. Groundnut, among the most prominent causes of allergies in food, is also being treated with transgenic technology to eliminate allergies.

4.4.2 Antibiotic Resistance

Another problem related to adoption of transgenic crops is the likelihood of transferring antibiotic-resistant genetic material (genes) from genetically modified crops to human and other animal gut microorganisms that can lead to antimicrobial tolerance in the abdominal microorganisms (Netherwood et al. 2004; Keese 2008). Tuteja et al. (2012) stated that the potentiality of this kind of gene transfer is, however, remarkably low. Furthermore, to solve this problem, marker-free transgenic crops are being produced (Tuteja et al. 2012). Genetic materials (genes) for antibiotic tolerance are utilized to discover and track a characteristic of concern in plant cells. This method leads to the improvement of a gene transfer during genetic alteration. These indicators have sparked fears that novel antibiotic-resistant bacterium strains would develop as a result of their use. Several critics of transgenic technology are concerned about the advent of infections that are difficult to treat using the standard antibiotics. The prospective hazard of transferring such materials from crop plants to microbes is significantly much less considering the danger of regular movement among the microbes, or between humans and the microbes that typically reside in our digestive tracts. However, to remain on the right track, the FDA has encouraged food producers to avoid utilizing marker genes that encode tolerance to therapeutically essential medicines (Kumar et al. 2020).

4.4.3 Environmental Effects

The environmental effects of accepting transgenic crops may include the pollen-interceded transgene flow that was reported from genetically modified crops, traditional crop varieties, and also numerous wild-related crop species. Reports by Chen et al. (2004), Han et al. (2015), and Yan et al. (2015) showed that such transgene flow has been observed in crops like barley, rice, cotton, beans, maize, and bent grass. Heap (2014) explained that the transfer of a modified gene from transgenic crops to wild closely related species may lead to loss of biodiversity whereas, transgene movement to weedy species that are closely related may cause negative effects like appearance/emergence of superweeds that are resistant to herbicides. For example, 16 out of 24 glyphosate-tolerant weedy species were reported to emerge from crop cultivation via transgenic techniques (Heap 2014). Among the species, horseweed (*Conyza canadensis*) is regarded as the most prominent species of weeds, whereas palmer amaranth and rough-fruited water hemp (*Amaranthus tuberculatus*) are the most important weed species that are resistant to glyphosate world over (Heap and Duke 2018). Various critics of transgenesis believe that GM crops may cross-pollinate with allied weed species, leading to the generation of "superweeds" that are not easy to manage. Many crops possess substantial development and seed dissemination limits, making them unable to flourish in the wild like the unwanted species (weeds).

4.4.4 Impacts on Nontarget Organism

Numerous analysts believed that after transgenic crops are introduced into the ecosystem, they may have unanticipated and harmful consequences. Despite the fact that transgenic crops are thoroughly evaluated before hitting the market, not all possible effects could be predicted. For example, Bt maize (corn) generates a pesticide that is exclusively effective against insects that consume grains of corn. Nonetheless, a study by researchers in Cornell University around 1999 discovered that microspores from Bt maize might kill the caterpillars of the monarch butterfly, which were previously thought to be safe. In the lab, 50% of the monarch butterflies perished after being administered plant sprinkled with Bt corn particles. However, field investigation indicates that under actual situation, butterfly larvae are extremely rare to interact with microspores of Bt corn, which has fallen on the leaf surface of milkweed, much less eat sufficient of it to destroy them. Literature has also looked into the possibility of GM crops having unexpected impacts on non-intended living species. Losey et al. (1999) found that monarch butterfly caterpillars fed on milk-weed leaves sprinkled with Bt maize died at a greater rate than the control under laboratory setting. Due to flaws in the study structure and extension of the laboratory assessments to the field, this study created a firestorm of criticism. Dively et al. (2004) and Sears et al. (2001) in their experiments found that Bt maize pollen had little or no effect on the larvae of monarch butterfly. However, pesticide-producing transgenic plants have been linked to a decrease in monarch butterfly populations in Mexico and the USA. Brower et al. (2012) discovered that the drop in the monarch butterfly abundance was due to the decline in milkweed population due to the increased use of glyphosate sprays. According to Lu et al. (2010), eliminating primary insect pests may allow subordinate insects to take over as significant crop pests. In China, for example, widespread use of Bt cotton has resulted in an upsurge in the number of hitherto uncontrolled insect pests during the last 10 years.

4.4.5 Cost for Commercialization

The hefty costs of safety assessment and the long and convoluted legislative procedure required for legalizing and commercializing biotech crops are only a few of the problems surrounding their development and dissemination (Davison 2010; Miller and Bradford 2010). The cost of legislative safety review, certification, and authorization is projected to be around USD35.01 million on average (Kumar et al. 2020). Thirteen years was estimated to be the length of time required for the development of transgenic crops from beginning to its commercial establishment (McDougall 2011). Similarly, it was estimated that the length of time needed for GM crops to go through regulatory channels of the EU and the USA was around 5 years and 7 days, respectively (Smart et al. 2017).

4.5 Safety and Regulations

The United States Department of Agriculture (USDA), the United States Environmental Protection Agency (EPA), and the United States Food and Drug Administration (FDA) are three governmental entities that perform detailed investigations and evaluations of GMOs and the foods they develop (FDA). Each entity is responsible for a particular aspect of the investigation. The USDA determines whether a unique food is healthy to manufacture, while the EPA evaluates the item's ecological impact. The FDA is focused on the consumer safety and has the ultimate say on whether an item is acceptable for consumption (Kramkowska et al. 2013). Shukla et al. (2018) mentioned that in India, legislation related to transgenic products commenced in 1982 with the formation of the National Biotechnology Board and the introduction of biotechnology safety policies to carry out biotech experiment in the laboratory (Chaturvedi 2004). The board was later on modified to the Department of Biotechnology since 1986 under the Ministry of Science and Technology (Sharma et al. 2003).

4.6 Transgenics for Herbicide Resistance

Weeds struggle to get nutrients, moisture, sunshine, and space as they compete alongside agricultural crops, resulting in severe output reductions (Nandula 2010). Weed control, including the use of chemicals, is required to prevent crop production losses incurred by weeds. Since the majority of weeds are grassy plants, selective weed control while safeguarding the crop plant is often not practicable. As a result, establishing herbicide-tolerant attributes in the major crop could be a viable approach for allowing the utilization of non-discriminating and wide-range herbicides with greater flexibility. Glyphosate and glufosinate are by far the highest widely utilized nonselective herbicides. Most glyphosate (HT) transgenic plants, for example, were established to endure glyphosate and glufosinate. Glyphosate hinders the main enzyme in the biosynthetic (shikimic) acid platform of heterocyclic (aromatic) amino acid syntheses, 5-enolpyruvylshikimate-3-phosphate synthase (Kumar et al. 2020).

4.6.1 Herbicide-Tolerant Crops

Herbicide resilience is and will continue to be the most ubiquitous distinctive feature of developed transgenic crops in the nearest future. GM crops durable to the expansive chemical pesticides (glyphosate and glufosinate) were initially profitably cultivated in the 1990s (Nandula 2010), and GMOs impervious to some weed killers (herbicides) are in progression (Green 2014) or already on the public domain, with various HR traits progressively blended in a single crop (Green 2014; USDA 2015). Another, more recent method is the creation of crops that are tolerant to high-quantity glyphosate without displaying any productivity loss/drag (Dun et al. 2014; Guo et al. 2015). In crop plants and microbes, glyphosate hinders 5-enolpyruvylshikimate-3-phosphate synthase (EPSPS), a protein of the metabolic pathway for the biogenesis of branched-chain amino acids and polyphenol compounds. This enzyme is absent from both human and animal cell lines (OECD 1999a, b).

4.6.2 Yields of Herbicide-Resistant (HR) Crops

HR crops are not always better in terms of yield than traditional crops, contradicting to public belief. Farmers are not solely motivated to use HR crops because of higher yield. If output variations arise for both herbicide-tolerant and ordinary crops, they could be linked to a variety of elements, including cultivation dimension and geographic area, farm size and location, soil, weather, cultivation system, weed availability, weed control practice, farmer expertise, and farm operator training (Brookes and Barfoot 2018). Areal et al. (2013) reviewed data on GM plant agricultural productivity and reported that while GM crops outperform traditional equivalents in agricultural and socioeconomic parameters, the results on herbicide-tolerant crop yield efficiency varied. It was not easy to present that HR crops have a constant production benefit over existing systems (Gurian et al. 2009; Heinemann et al. 2014; Khan 2015).

4.6.3 Methods for Developing Herbicide-Resistant (HR) Crops

A variety of methods are applicable in generating plants that are resilient to herbicides as well as other phytotoxins. The herbicide's genomic binding region could be altered to make it unable to bind to it, making it tolerant. A plant can be offered or enhanced with one or even more herbicide-inactivating or antibiotic enzymes. It is possible to modify the plant so that it has a mechanism that stops the herbicide from accessing the specific binding sites (Stephen and Antonio 2010). In weeds that have developed herbicide resistance, all three techniques have been identified. In most examples of inherent crop tolerance to preferential herbicides, metabolic deactivation or disintegration is the major strategy at work. In the production of commercial HRCs, the first two procedures have proven to be effective. An enzyme in the heterocyclic amino acid pathway is glyphosate's molecular target location (the shikimate pathway). An enzyme is the genomic target site of glyphosate in the heterocyclic amino acid pathway (the shikimate pathway). EPSPS, or 5-enolpyruvylshikimate-3-phosphate synthase, is highly vulnerable to herbicides (Duke 1988), and no other inhibitors seem to be impactful. Because it is particularly prone to every plant EPSPS, glyphosate is a broad-spectrum herbicide that can decapitate virtually all weed species. EPSPS is found in fungi and bacteria, and bacterial variants of the enzyme are glyphosate resistant.

4.6.4 Advantages of Herbicide-Resistant (HR) Crops

Farmers have benefited from the broad production of HT transgenic crops in a number of ways, including higher yield resulted from effective and facilitated system of controlling weeds as well as lower cost of managing the weeds (Brookes and Barfoot 2018). According to estimates, 38% of the economic gain in glyphosate-resistant soybean comes from increased yield, while the other 62% comes from lower weed control costs (Green 2012; Brookes and Barfoot 2018). Furthermore, the introduction of herbicide-tolerant crops has lowered weed management's ecological consequences. Herbicide usage has switched to much more eco-friendly herbicides like glyphosate and glufosinate, which break down quickly after administration. Additionally, the deployment of HT technology encouraged the transition from traditional plough-based agricultural strategies to decreased, minimal, or no tillage production processes, resulting in fewer emissions of greenhouse gases due to limited utilization of machines like tractors (Brookes and Barfoot 2017).

4.6.5 Disadvantages of Herbicide-Resistant (HR) Crops

One of the worries expressed by individuals opposed to GMOs (transgenic) is the potentiality of foreign genes (transgenes) in affecting the virtue and safety of the plant's edible parts. This could be due to toxicity of the transgene's protein, toxicity of a biochemical output of the protein (enzyme) transcribed by the mutant gene, pleiotropic effects of the transgene, modification of nonmutant gene expression due to the transgene's location in the DNA molecule, or indirect effects (Stephen and Antonio 2010). Transgenic crops are tested more thoroughly than traditional crops for FDA approval, using technical, dietary, and pharmacological methods (Konig et al. 2004), while some have suggested much more rigorous studies employing metagenomic, proteomic, and genomic analysis to pinpoint the prospective unforeseen consequences of the transgene and its products (Atherton 2002; Malarkey 2003). However, Harrison et al. (1996) discovered CP4 EPSPS gene, expressed in soybean to generate glyphosate tolerance to be (1) harmless to rodents if ingested at dose levels many times extremely high than prospective human exposure; (2) easily disintegrate by gastric juices; and (3) architecturally or physiologically unrelated to all other documented enzymatic allergens or carcinogens, depending on the nucleotide sequence evaluation. Before a transgene's protein products are approved, their possible allergenic characteristics must be identified. Although regulatory bodies get these data, there are few publications on the subject. Nonetheless, a few investigations have shown that transgenic products linked with HRCs had no allergenic characteristics. Sten et al. (2004) observed that the genotoxicity of ten GR and eight non-GR soybean varieties was similar in a trial with soybean-sensitized patients. Chang et al. (2003) discovered that the glyphosate resistance gene product CP4 EPSPS had no significant allergenicity in rats. Herouet et al. (2005) determined that there is still a considerable confidence that the addition of the glufosinate resistance gene in human food or animal diet will cause no harm based on a study of existing literature and experimental investigations.

4.6.6 Herbicide-Resistant Crops (HRCs) and Crop Disease

GR crops, as well as possibly glufosinate-resistant crops, have some untapped benefits. Glufosinate and glyphosate are both fungicides (Duke and Cerdeira 2007). As a result, if ample treatment rates are applied, these glyphosates may provide enough defense against plant pathogens to avert crop losses or eliminate the need for fungicide treatment in some cases. The glyphosate impacts of Asian rust in GR soybeans (Feng et al. 2005, 2008) may be the well-studied example, where herbicide (glyphosate) sprays to soybeans curtailed infestation by rust and injury as a prophylactic and therapeutic treatment. In many field circumstances, meanwhile, the best time to use glyphosate for weed control is doubtful to match the best time for rust suppression (Bradley and Sweets 2008).

4.6.7 Glyphosate-Resistant Crops

Given the immense performance of GR plants and the reality that glyphosate became a universal herbicide, other firms have developed or synthesized additional herbicide-tolerant transgenes and, in at least one instance, are pushing forward with the industrialization stage. This crop was already genetically modified to be resistant to glyphosate (Castle et al. 2004; Siehl et al. 2005). The efficiency of a weak glyphosate *N*-acetyltransferase (GAT) gene from *Bacillus licheniformis* was increased by nearly four orders of magnitude after 11 rounds of gene shuffling. The features of the GAT that resulted were described by Siehl et al. (2005, 2007). Plants with this mutant gene were 100 times stronger and more resilient to herbicides compared to non-transgenic lines (Green et al. 2008). Several glyphosate-deactivating enzymes tend to be expressed by soil microorganism genes, as there are multiple paths of degradation or inactivation. C-P lyase is found in *Arthrobacter* spp., *Rhizobium* spp., and *Pseudomonas* spp.; for example, it transforms glyphosate to inorganic phosphate and sarcosine (Kishore and Jacob 1987; Liu et al. 1991; Dick and Quinn 1995). Glyphosate tolerant maize was raised with bacteria-derived EPSP synthase (aroA1398) gene which showed significant resistance to glyphosate sprays. The E. coli expressed EPSPS enzyme offered 800 times more glyphosate resistance than the maize- derived EPSPS (Vande Berg et al. 2008).

4.6.8 The Future of HRCs

The exploration of innovative weed control measures will have an impact on HRCs' future. The emergence of appropriate emerging techniques that are cost competitive with HRCs might ultimately halt their widespread acceptance. Other options for employing transgenes for weed control include improving crop competition and allelopathy, as well as improving weed biocontrol agents (Duke et al. 2003, 2006). Although these measures are unlikely to make a significant influence in the next decade, they may have a significant impact in the long run.

4.7 The Story of Transgenic Mustard

Mustard crop (*Brassica juncea* L. (Czern & Coss.)) is a prominent dietary oil plant of the oilseed rape (oilseed Brassica) subgroup that is currently produced mostly in China, India, Russia, Canada, and Australia. It represents India's most important oilseed crop, but *B. napus* is a crucial oilseed crop in other nations. Although it is mostly a self-pollinating crop, cross-pollination can occur up to 12% of the time in this species. *B. juncea* is an amphidiploid ($2n = 36$, AABB) crop that was created thousands of years ago by mating its diploid parents *B. rapa* ($2n = 20$, AA) with *B. nigra* ($2n = 16$, BB) (Redden et al. 2009). Indian mustard has a variety of uses in the culinary and chemical industries, as well as being a biofertilizer. Mustard seed meal is an excellent chicken feed, and India is the leading exporter of the crop (Thakur et al. 2020).

4.7.1 Genetic Modification of Mustard

The genetic modification procedure relies on the maturity level of the plantlets, the influence of preculture, *Agrobacterium* cell-density, Agro-inoculation time, the use of antibacterial agents to destroy the surplus agrobacterial cells, and the choice of selection medium utilized to pick transformed cells/explants. These collectively play a major function in the achievement of any genetic manipulation procedure. Plant regeneration success greatly depends on the genotype. However, in vitro plant regeneration protocols differ from one genotype to another. Using cotyledon explants, Yadav et al. (1991) found varying regeneration rates in three genotypes of *B. juncea*. The overall performance of any genomic modification procedure is also determined by the age of the explants. Younger explants seem to produce superior outcomes compared to older explants. Explants removed from 4–6-day-old in vitro-grown plantlets provide a greater sensitivity to regeneration in *B. juncea* (Bose et al. 2019). Hypocotyl is often used as an explant in mustard genetic manipulation investigations because it can withstand the transformation stress more than other types of explants (Arora et al. 2019a, b). Prior to cocultivation using the bacterium strain, majority of the researchers recommended giving the explants a 2-day, i.e., 48-h, preculture incubating period (Ahmed et al. 2017; Rani et al. 2017a). This range of time was indicated as ideal for boosting transformation intensity because plant cells are more prepared to withstand the conversion stress during this time. Plant cells continue to regenerate after 2 days, rendering them unsuited for mutation, leading to lower rate of transformation. One of the crucial elements influencing transformation frequency is the Agrobacterium cell-density. At 600 nm wavelength, the agrobacterial cell-density required for *B. juncea* transformation is commonly between 0.01 and 0.1 (Pental et al. 1993; Das et al. 2006; Bose et al. 2019). The length of time it would take for the Agrobacterium strain to infiltrate explants differs from 1 to 30 min. Singh et al. (2010) found that a 30-min infection period using cotyledonary petiole improved the transformation of mustard species. In another investigation, Bose et al. (2019) immersed *B. juncea* hypocotyl and cotyledon explants in agrobacterial for 20–30 min to generate a significant conversion rate. To choose the transformed cells/tissues, the modified explants are deposited on selective media comprising selection agent following cocultivation. The type of marker used and the concentration of selection agent have an influence on the selection and regeneration response. When NPT-II gene is used in the gene construct, then the antibiotic kanamycin can be used as a selection agent (Taj et al. 2004; Mondal et al. 2006; Ahmed et al. 2017). Various additional selective agents, such as hygromycin (Bisht et al. 2007; Dutta et al. 2008; Bhuiyan et al. 2011; Das et al. 2018), 2,4-D (Bisht et al. 2004a, b), and phosphinothricin (Mehra et al. 2000; Arora et al. 2019a, b), have also been used in mustard-transformation.

Marker genes could be scored and their activity checked visually, or they can be chosen to allow the identification of altered cells among the non-modified cells. Barfield and Pua (1991) used the GUS and NPT-II genes in Indian mustard to design a strong transformation procedure. The GUS enzyme activity varied in all T0 plants. Pental et al. (1993) improved the GUS and HPT genes in Indian mustard using an *A. tumefaciens*-mediated procedure. Using the *Agrobacterium*-mediated gene transfer approach, Nirupa et al. (2007) genetically transformed mustard using the GUS gene. Transgenic mustard containing the marker genes (GUS and NPT-II) was also developed by Akter et al. (2016). With 30 min of preincubation, 72 h of coculturing with the explants of petiole and hypocotyl, the maximum transformation percentage was observed with 0.08 optical density at a wavelength of 600 nm. PCR testing and a GUS histopathological assay were performed to validate gene incorporation.

4.7.2 Aphid/Insect Pest-Resistant Mustard

Aphids are among the most common insect pests attacking mustard crops, thereby causing detrimental impacts on the crop yields. They are sap-sucking species of insects belonging to the order Hemiptera. It results in significant crop damage and yield losses in India when described in relation to mustard-cultivating regions of the world (Rao et al. 2014). It infects the whole mustard crop's body, including the foliage, branches, thorns, and flower buds (Das et al. 2018). After blooming and seed laying, the nymph and adult phases of the Indian mustard aphid are responsible for the vast majority of yield reductions (Bose et al. 2019). Furthermore, aphid tolerance development attempts have failed due to a shortage of efficient tolerant Brassica cultivars/germplasm lines. Certain wild Brassica species, such as *B. fruticulosa*, were utilized to introduce the aphid tolerance allele into farmed *Brassica juncea* types. Plant lectins are sugar proteins with agglutination capabilities that help them manage sap-sucking insects, aphids inclusive. Aphid resistance was also demonstrated for wheat germ-related proteins, a chitin-binding lectin molecule found in wheat insects. The wheat germ agglutinin (WGA) attaches to a glycoprotein contained in aphids' midgut, preventing them from absorbing nourishment and causing them to starve and die. By activating the RiD gene in Indian mustard, Sarkar et al. (2014a, b) were able to reduce aphid larva abundance by 46.67%. The (E)-farnesene (EF) gene is involved in the manufacture of an essential volatile sesquiterpene molecule that repels mustard aphids. The immediate impact of the EF allele introduced into mustard on aphid colonies was reported by Verma et al. (2015). Aphid colonization was discovered to be directly inhibited by the EF gene isolated from *Mentha arvensis*.

4.7.3 Disease-Resistant Mustard

A variety of pathogenic fungi including *Sclerotinia sclerotiorum* which causes stem rot, *Alternaria brassicae* which causes dark spot infection, *Erysiphe cruciferarum* which causes white rust, *Albugo candida* which causes powdery mildew, and *Hyaloperonospora parasitica* which causes downy mildew are all wreaking havoc on mustard crop yields. Among the key biological stressors limiting mustard yield, according to Meena et al. (2010), is *A. brassicae*. This fungus thrives by creating black patches on leaves and stems that create concentric circles. White rust fungus generates yield reduction of 20–60% in Indian mustard, according to Awasthi et al. (2012). White rust is a pathogen that affects almost all kinds of Indian mustard that are now produced in India (Saharan et al. 2014). During the initial vegetative phase until the flowering, the fungal pathogen induces enlargement of blooming components, resulting in the formation of stagheads that carry no fruits (Meena et al. 2014). In the last 4–5 years, stem rot has been a significant concern to Indian mustard. With the exception of white rust disease, that could be deployed in resistant genetic improvement, no tolerance sources for these biotic stressors have been documented. Furthermore, because of the ever-evolving nascent fungus, chemical fungicides were proven to be useless in controlling it. To tackle such dangerous infections, these circumstances led to the development of transgenic techniques.

4.7.4 Herbicide-Tolerant Mustard

To address the effects of environmental pressures, transgenic plants were established that demonstrate genetic traits for such biosynthetic pathways of different chemical compounds such as metal ion transporters, mannitol, glycine betaine, and ion metal transporters that play particular roles in ion homeostasis, membrane reliability, and osmolyte accumulation (Saha et al. 2016). Owing to the medical implications associated with utilizing antibiotic resistance molecular markers as selective agents in functional genomic investigations, a trend has been documented towards adopting novel genes such as herbicide tolerance genes. Furthermore, weedy crops struggle for nourishment with mustard species in Indian fields, resulting in yield reductions. Herbicides with broad leaves are commonly employed in mustard fields to eliminate weeds while also killing mustard crops. As a result, the introduction of recombinant *B. juncea* with glyphosate-resistant feature provides a glimpse of optimism for the management of weeds, especially against noxious weeds like *Orobanche aegyptiaca*. Because the transgenic mustard crops exhibit a herbicide tolerance feature, an application of the intended herbicide would destroy the weedy vegetation while leaving the mutant mustard plants unharmed. For recombinant seed development, barnase and barstar alleles have been introduced into the transgenic Varuna and EHII strains of Indian mustard, respectively (Ray et al. 2007). As a selective marker, they used a transgenic ALS gene that develops tolerance to the imidazolinone-based herbicide "Pursuit." Imidazolinone was employed as a preference symbol in the in vitro establishment of recombinant plants as well as the creation of hybrid varieties.

4.7.5 Transgenic Mustard for Improved Nutrient-Use Efficiency

Plant species with greater nitrogen uptake and nutritional usage efficiencies are necessary in the current context to cut the expense of applying chemical fertilizers and to eliminate ecological contamination caused by increased fertilizer utilization. By including specific desired traits, genetic modification gives a strategy to improve mineral nutrient utilization. It has been discovered that plants' increased sulfur uptake is complemented by increased nitrogen uptake due to metabolic linkage processes. The impact of increased sulfur absorption caused by activation of the LeST 1.1 gene (from tomato), which codes for an increased sulfur exporter, on nitrogen absorption in mustard was examined (Akmal et al. 2014). In both sulfur-deficient and -sufficient circumstances, the transgenic mustard plants had better nitrogen absorption and dissolved crude protein than non-transformed plants.

4.7.6 In Planta Modification in *B. juncea*

Although *B. juncea* is a very receptive species to plant modification, inadequate conversion recurrence was traditionally a problem in this plant. Furthermore, following the whole genetic analysis of *B. juncea* var. *tumida*, it was necessary to evaluate the roles of a large number of proteins, necessitating the creation of a reliable transformation methodology for this important oilseed crop that did not rely on tissue culture-based approaches. By sprinkling a transformation solution comprising GUS and HPT-II genes onto flowery shoots, a great success was recorded via in planta modification technique (Aminedi et al. 2019) in mustard (*B. juncea*) plant.

4.8 Transgenics for Pest Resistance

Crop bioengineering has emerged as a provider of agricultural technologies, bringing innovative answers to long-standing issues. Among the most prominent theaters of action for plant biotechnology is pest management (Juan et al. 1997). The obvious explanation for this concern is that, amidst the deployment of advanced plant protective methods, such as synthetic insecticides, the global crop loss caused by phytophagous insects is enormous (Juan et al. 1997). The production and marketing/distribution of Bt crops that are resilient to various pests' attack were one of the earliest achievements of plant bioengineering. Crop plants with only one pesticide Bt genotype that were resistant to main corn and cotton insects were included in the first developmental innovative biotechnological products (Ferry et al. 2006). Edible crop plants as tomato, potato, cotton, cucumber, beans, cassava, watermelon, cabbage, pepper, eggplant, and aesthetic plants like hibiscus and chrysanthemum are among the crops extensively attacked by the insect pests. So many insect organisms, both underdeveloped and adult, cause damage to the plant by explicitly eating on phloem sap, resulting in reduced pace of growth, stem diminishing, untimely plant drooping, and foliar discoloration, which can contribute to plant destruction (Zaidi et al. 2017). Honeydew excretion on the surfaces of plants and fruits causes incidental harm to the agricultural crops. Dark sooty molds thrive on honeydew, interfering with photosynthetic activity and texture as well as socioeconomic viability of several agricultural and aesthetic crop varieties. Insect pest damages, which are predicted to be between 35% and 69% for important crops, are a large determinant in reducing food productivity. One of the real accomplishments of transgenic technology was the modification of crop plants for increased insect pest tolerance (Thomas 2014). A particular insect may or may not be designated a nuisance relying on the nature of the ecosystem in a particular location and human's perspective. Nareshkumar et al. (2017) described insect pests as "any insect in the inappropriate area."

Recent advancements in understanding the genetic basis of insect–plant relationships, as well as biotechnology approaches, are offering remedies to the insect-mediated crop damages. Insect resilience is conferred in large part via genetic manipulation practices (Birkett and Pickett 2014). Transgenic technology has been used to develop a variety of crop plants, strengthening their resistance to pest species and increasing agricultural productivity, as shown in Table 4.1. Koul et al. (2014b) developed transgenic tomato crop expressing *Cy1Ab* that is tolerant to insect infestation, demonstrating 100% death rate to the larvae of *Helicoverpa armigera* and *S. litura* with minor damages to the fruits and foliage of the tomato plant. *Tuta absoluta* fatality rates ranged from 38% to 100% when the Cry1Ac protein was expressed in tomato. According to Hatice et al. (2017), gallery configuration was lowered in 57–100% of transgenic crops. Hanur et al. (2015) found that using the Bt

Table 4.1 Crop plants developed through transgenic technology for insect pest tolerance

Crop name	Strategy	Target insects	Transgene	References
Cotton	Cry proteins	*Helicoverpa armigera*	*cry2AX1*	Jadhav et al. (2020)
Cotton	Cry proteins	Lepidopteran pest	*cry2Ab*	Katta et al. (2020)
Sweet Potato	Cry proteins	*Spodoptera litura*	*cry1Aa*	Zhong et al. (2019)
Castor	Cry proteins	*Achaea janata*	*cry1AC*	Muddanuru et al. (2019)
Pigeon pea	Cry proteins	*H. armigera*	*cry2Aa*	Singh et al. (2018)
Tomato	Cry proteins	*Tuta absoluta*	*cry1Ac*	Selale et al. (2017)
Tomato	Cry proteins	*H. armigera*	*cry1Ab*	Koul et al. (2014b)
Rice	Cry proteins	Leaf folder	*cry2A*	Gunasekara et al. (2017)
Chickpea	Cry proteins	Pod borer	*CryIIAa*	Sawardekar et al. (2017)
Cotton	Cry proteins	*H. armigera*	*cry1Ab + ptII*	Khan et al. (2011)
Potato	Cry proteins	*Tuber moth*	*Cry1Ab*	Salehian et al. (2021)
Pigeon pea	Cry proteins	Gram pod borer	*Cry2Aa*	Singh et al. (2018)
Maize	Cry proteins	Lepidoptera	cry19c	Gassmann et al. (2011)
Maize	Cry proteins	Lepidoptera	cry3Bb1	Koziel et al. (1993)
Sugarcane	Chimeric proteins	Shoot borer	cry2Aa + cry1Ca	Koerniati et al. (2020)
Cotton	Chimeric proteins	Lygus bug	cry51Aa2	Gowda et al. (2016)
Rice	Chimeric proteins	Stem borer, leaf folder	cry1Ac + ASAL	Boddupally et al. (2018)
Maize	Chimeric proteins	*Sophora exigua*, *Harmonia axyridis*	cry1Ab/cry2Aj	Chang et al. (2017)
Cotton	Chimeric proteins	Cotton leafworm	cry1Be + ry1Fa	Meade et al. (2017)
Rice	Chimeric proteins	Lepidopteran	cry1Ab + ip3A	Xu et al. (2018)
Rice	Chimeric proteins	*Chilo suppressalis*	cry2Aa + cry1Ca	Qiu et al. (2019)

(continued)

Table 4.1 (continued)

Crop name	Strategy	Target insects	Transgene	References
Cotton	VIP proteins	Cotton bollworm and tobacco budworm	Vip3A + ry1Ab	Bommireddy et al. (2011)
Cotton	VIP proteins	Lepidopteran	Vip3AcAaa (Vip3Aa1 + Vip3Ac1)	Chen et al. (2018b)
Sugarcane	VIP proteins	*Chilo infuscatellus*	*Vip3A*	Riaz et al. (2020)
Cowpea	VIP proteins	Bean pod borer	*Vip3Ba1*	Bett et al. (2017)
Cowpea	Lectin genes	Bruchids	*Arcl1*	Grazziotin et al. (2020)
Rice	Lectin genes	Brown planthopper and hemipteran pest	Mannose-specific GNA	Bharathi et al. (2011)
Mustard	Lectin genes	Mustard aphid	CEA (*Colocasia esculenta* tuber agglutinin)	Das et al. (2018)
Cotton	Lectin genes	Chewing, sucking insect pests	*Sclerotium rolfsii* protein (lectin) that binds to the insect gut	Vanti et al. (2018)
Mustard	Lectin genes	Aphids	Lentil lectin-LL CPPI	Rani et al. (2017b)
Potato	Lectin genes	Aphids	Hv1a/GNA	Nakasu et al. (2014)
Wheat	Lectin genes	Aphids	*Pinellia pedatisecta* agglutinin (PPA)	Duan et al. (2018)
Tomato	Lectin genes	*Meloidogyne incognita*	*Remusatia vivipara* (rvl 1) and *Sclerotium rolfsii* (srl 1)	Bhagat et al. (2019)
Potato	Lectin genes	Aphids	GNA	Mi et al. (2017)
Brassica juncea	Chimeric proteins	Aphids	Enzyme lectin (protease)	Rani et al. (2017b)
Maize	RNA interference	Lepidopteran	Sprinkle of dsRNA	Li et al. (2015a)
Cotton	RNA interference	*H. armigera*	Juvenile hormone	Ni et al. (2017)
Soybean	RNA interference	Pod borer	SpbP0-dsRNA	Meng et al. (2017)
Soybean	RNA interference	Aphids	TREH, ATPD, ATPE, CHSI	Yan et al. (2020)
Rice	RNA interference	Yellow stem borer	AchE-acetylcholine esterase	Kola et al. (2019)
Rice	RNA interference	*C. suppressalis*	APN1, APN2 genes	Qiu et al. (2017)
Maize	RNA interference	Coleopterans	*dvvgr*, *dvbol*	Niu et al. (2017)

(continued)

Table 4.1 (continued)

Crop name	Strategy	Target insects	Transgene	References
Tobacco	RNA interference	*H. armigera*	Chitinase gene-HaCHI	Mamta et al. (2016)
Tobacco	RNA interference	*Bemisia tabaci*	v-ATPase A	Thakur et al. (2014)
Agricultural crops	RNA interference	*B. tabaci*	Acetylcholine receptor subunit α, alpha-glucosidase 1	Vyas et al. (2017)
Agricultural crops	RNA interference	*B. tabaci*	Aquaporin (AQP), calcitonin (CAL), cyclophilin B (CYCP), knottin-1 (k-1), heat-shock proteins (Hsp20, Hsp40, and Hsp70)	Kaur et al. (2020)

gene (Cry2A) in a tomato crop species resulted in resistance to *Helicoverpa armigera* neonate larvae under laboratory settings. These findings indicated that transgenic lines produced a significant amount of Bt Cry2A proteins, which was effective in controlling *Helicoverpa armigera*. For almost 20 years, modified maize which synthesizes pesticidal poisons/toxins obtained first was developed via *Bacillus thuringiensis* (Bt) to control the western corn rootworm. Following multiple years of effective control of many other important maize and cotton insect infestations, the initial transgenic products for corn rootworm control were released (Tabashnik et al. 2000; Siegfried and Hellmich 2012). Majumder et al. (2018) created a recombinant jute species with reduced nutrient intake, body composition, body mass, and dry weight of excrements than non-transgenic controls. The semilooper and hairy caterpillars died at a rate of 66–100%, whereas the indigo caterpillar died at a rate of 87.50% among modified feeders. With respect to yield components and fiber disposition, the GM crops were comparable to the susceptible cultivars. As a result, these Bt jutes in the field would indeed be able to withstand lepidopteran pest infestation, reduce the use of toxic pesticides, and produce high-quality fiber. Cotton lines displaying VIP3A alone as well as a pyramided VIP3A and cry1Ab showed significantly higher tolerance to both *H. zea* and *H. virescens*. The pests, on the other hand, wreaked havoc on cotton fruits that are not genetically transformed (Bommireddy et al. 2011). In another study, insect resistance was increased in cotton lines expressing proteins that have been fused (a combination of Vip3Aa1 and Vip3Ac1), demonstrating that the hybridized VIP3AaAc1 protein is a good choice for insects' control (Chen et al. 2018a, b, c). Transgenic sugarcane containing a unique gene VIP3A neurotoxin in the control of the polyubiquitin activator exhibited 100% mortality to *Chilo infuscatellus* and deposited 5.35–8.89 lg/mL of protein (Riaz et al. 2020). The level of effectiveness of a transgenic hybridized protein (Hv1a/GNA carrying venomous toxins x-ACTXHv1a from spider coupled to snowdrop lectin) expressed in *Arabidopsis* against peach potato and grain aphids was demonstrated (Nakasu et al. 2014). In contrast to non-Bt plants, the sweet potato plants having GNA gene expressed were tolerant to aphids (Mi et al. 2017). At T1 and T2 generations, genetically modified *B. juncea* displaying *Colocasia esculenta* tuber agglutinin (CEA) and a GNA displayed greater activity on the aphid species found on mustard plant. When compared with control groups, mutant mustard producing CEA resulted in a 70–81% increase in insect death, indicating that CEA is an effective pesticidal protein without allergens (Das et al. 2018). Via using embryogenic node explants and Agrobacterium-mediated modification, a combination protein arising from protease and lectin suppressor has been transmitted to mustard plant (Rani et al. 2017a), where plants resulted were resistant to phytophagous insects and had a 40% survival rate when fed up with genetically modifying crops. This fusion protein expression was 2.8-fold higher than that in the non-transgenic control, as reported by Rani et al. (2017b). Chang et al. (2017) reported that the modified Bt maize with a combination of protein (cry1Ab/cry2Aj) in the kernel exhibited tolerance to insect pests such as *S. exigua* and *Harmonia axyridis*. Bt cotton lines harboring the genes cry1Be + cry1Fa showed increased resistance to some insect pests including *S. litura* and *O. nubilalis* (Meade et al. 2017). RNase III enzyme Dicer converts dsRNA that is exceptional to an insect pest's critical gene into siRNAs, which cause RNA-induced silencing complex (RISC) in the Argonaute protein to destroy the mRNA complementary structure as reported by Scott et al. (2013). *Oryza sativa* knockout lines for APN1 and APN2 (aminopeptidase N) alleles reduced the susceptibility to *C. suppressalis*, a significant rice insect, and imparted tolerance to two different Bt cultivars, TT51 (*cry1Ab* and *cry1Ac*) and T1C-19 (*cry1Ca*). This study reveals the crucial role of APN receptors in *C. suppressalis* for *cry1A/C*-mediated toxicity (Qiu et al. 2017). Two genes of barley cultivar (β-1–3 glucanase) were mutated using CRISPR-Cas9 technology. As a result, the aphid *Rhopalosiphum padi* was unable to acquire the phloem sap, which harmed the aphid's development and reduced its predilection for the host (Kim et al. 2020). Insect genetic traits like those of *H. armigera* and *S. exigua* have been knocked out using the CRISPR-Cas9 tool. Wang et al. (2020) demonstrated, for both *H. armigera* and *S. exigua*, CRISPR-Cas9 knockdown genetic changes in the a-6-nicotinic cell receptors (nAchR) conferring tolerance to the insecticide spinosyn (Zuo et al. 2020). In another investigation, CRISPR-Cas9 was used to strike out the ABC transporter HaABCA2 in *H. armigera*, resulting in increased resistance to *cry2Aa* and *cry2Ab*. This study suggests the crucial role of HaABCA2 in mediating toxicity of both the genes against *H. armigera* (Wang et al. 2017). In a similar study, cotton bollworm (*H. armigera*) showed dominant resistance to *cry1Ac* after a point transformation in the "tetraspanin gene." The knockout of tetraspanin gene created using CRISPR technology restored the susceptibility of resistant *H. armigera*. By comparing resistant and susceptible strains, the mutations were further studied using genome-wide association and genetic mapping (Jin et al. 2018). Several pest-tolerant transgenic crops employing various biotechnological techniques are shown in Table 4.1.

4.9 Bt Technology

Bacillus thuringiensis is referred to as a soil-dwelling species of bacteria that releases spores/toxins throughout its stationary developmental phase. Crystals found in the spores exhibit powerful and selective pesticidal action (Ibrahim et al. 2010). Numerous Bt variants produce various sorts of toxins, each of which kills a group of pests. Bt is distinguished by its capacity to create proteinaceous particles referred to as endotoxins upon spore germination (Bravo et al. 2005). Toxins Cry and Cyt make up the majority of such particles (crystals) (Kaur 2000; Bravo et al. 2005). Bt strain genomes are between 2400 and 5700 kb in length. It has chromosomal as well as 1 to >12 extrachromosomal elements in its chromosome (Carlson et al. 1994; Carlson and Kolsto 1994). The sizes of the plasmid range from 4.56 to 228 kb in various species (Lereclus et al. 1982). The bacterium produces globular shaped endotoxins that do not stretch the sporangium and preferentially use enzyme lecithinase for sustenance. A phase-contrast microscope can detect one or even more polycrystalline structures (also known as the crystal) inside the fruiting bodies, indicating Bt (Schnepf et al. 1998). Bt has two phases in its life cycle: vegetative and sporulative stage. The vegetal step phase includes cell division, whereas the sporulation stage involves the generation of spores (Lambert and Peferoen 1992).

Biopesticides centered on *Bacillus thuringiensis* are now an important feature of lepidopteran pest control plans. This bacterium has been effectively used as a repository of cry proteins for crop genetic manipulation to generate GM plants with tolerance to lepidopteran pests' (Tohidfar and Salehi Jouzani 2008; Tohidfar et al. 2013; Melo et al. 2016; Tabashnik and Carrière 2017, 2019). Since this bacterium produces metalloproteinase, thuringiensin, and chitinase, that have been harmful to plant-damaging nematodes, Bt has the potential to be used as a nematicide (Salehi Jouzani et al. 2008; de la Fuente-Salcido et al. 2013; Iatsenko et al. 2014). Bt variants often generate numerous pesticidal enzymes throughout the vegetal growth period as complement to the Cry and Cyt toxins. Vegetative pesticidal proteins (Vip) (Estruch et al. 1996; Warren et al. 1998) and secretory insecticidal protein (Sip) are the two types of toxins that are released into the growth media (Donovan et al. 2006). Cry proteins adhere to specified ligands on the membrane surface of selected insects' epithelial cells, causing them to burst via a cytotoxic action. Many species (such as mankind, other mammals, and nontargeted pests) missing in the necessary receivers in their guts are unharmed by the cry proteins, and thus are unaffected by Bt (Hall 2006).

4.9.1 History of Bt

Bt's pesticidal qualities had been known long before the bacterium was discovered, and there is even evidence that Bt spores have been utilized in ancient Egypt (Sanahuja et al. 2011). During an examination into wilt disorder in bollworms in 1901, Shigetane Ishiwatari identified the bacterium, which he termed *Bacillus sotto* (Khetan 2001). In Thuringia, Ernst Berliner isolated *B. thuringiensis* out of a sick

Mediterranean wheat insect (*Ephestia kuehniella*) around 1911 (Siegel 2000). The maize worm, *Ostrinia nubilalis* Hübner (Lepidoptera: Crambidae), was initially controlled with Bt crystal preparations in 1928 in Europe (Siegel 2000). Sporine, the earliest industrial Bt pesticide, was developed in France in 1938 to combat wheat moths. *B. thuringiensis* toxin was initially developed commercially in the USA around 1958, and by 1961, it had become a household name (Khetan 2001). Robert A. Zakharyan discovered a plasmid in a *B. thuringiensis* strain in 1976 and hypothesized that the plasmid was involved in endospore and crystal production. *B. thuringiensis* is linked to *B. cereus*, a soil-living sp. of bacteria, and *B. anthracis*, the anthrax-causing bacteria; the three species vary fundamentally in their plasmids. All three are anaerobic bacteria proficient of forming endospores, as are many other relatives of the genus (Zakharian et al. 1979). Following that, further Bt toxins were found, resulting in a breakthrough of *B. thuringiensis*-expressing genetically engineered plants (Lambert and Peferoen 1992). A range of crucial plants were raised expressing insect-tolerant cry proteins derived from *B. thuringiensis* and tweaked for plant-preferred codon activation (Letourneau et al. 2003). Arthropod-resistant GM plants, sometimes referred to as Bt crops, have been grown all over the world since 1996 (Kleter et al. 2007; James 2012). Until 2008, South Africa was the only African nation to reap the benefits of Bt agricultural industrialization (James 2012). Burkina Faso, Sudan, and Egypt have recently begun to produce biotech crops like Bt cotton and Bt corn (James 2012).

4.9.2 The Structure, Variety, and Toxicity of Bt Proteins

Numerous scientists have tried systematic categorization of Bt utilizing diverse parameters including serotyping, phage sensitivity, and plasmid patterns, resulting in roughly about 100 subspecies (Sanahuja et al. 2011). Even though there is a strong link among Bt strains and their host species at the family level, the link degrades at both species and genus levels as many Bt variants could produce multiple toxins, yielding in a complex as well as heterogeneous host profiles (Ibrahim et al. 2010). In order to identify the host range, Bt isolates are physiologically categorized at the genus and species levels based on which toxic proteins they release (Sanahuja et al. 2011). The patterns of amino acid, protein sequences, as well as action mechanisms of the biocontrol agents in the insect's midgut have all been characterized (Crickmore et al. 1998). Cry toxins were divided into 51 different classes and subgroups in the past (Crickmore et al. 1998). Depending on the pest host particularities, the poisons were categorized into six groups as follows: group 1: lepidopteran (Cry1, Cry9, and Cry15); group 2: lepidopteran and dipteran (Cry2); group 3: coleopteran (Cry3, Cry7, and Cry8); group 4: dipteran (Cry4, Cry10, Cry11, Cry16, Cry17, Cry19, and Cry20); group 5: lepidopteran and dipteran (Cry1I); and group 6: nematodes (Cry6). Cry1I, Cry2, Cry3, Cry10, and Cry11 toxins (73–82 kDa) appear to be spontaneous shortenings of the larger Cry1 and Cry4 proteins (130–140 kDa) based on the amino acid sequence homology (Crickmore et al. 2013). Cry1Aa, Cry2Aa, Cry3Aa, Cry3Bb, Cry4Aa, and Cry4Ba are all Cry proteins with a broadly analogous three-dimensional configuration, despite their sequence disparity (Crickmore et al. 2013). Because it is cut off in the insect stomach, the C-terminal piece is important in the formation of crystal; however, it does not form a component of the developed toxin. The poison itself has been found at the N-terminus, and it is divided into three regions (Ibrahim et al. 2010). Phase I is a bunch of seven helices, six of which are amphoteric and around the final repellent helix, and is essential for implantation of membrane and creation of pore. The second domain is made up of three antiparallel structures containing visible loop zones, whereas domain III is made up of two alternative splices sandwiched together to produce a "jelly-roll" topology. The host range is defined by both domains, which enable receptor activation selectivity (Boonserm et al. 2006). According to the research reports, both the domains (II, III) attach to predominant receptor proteins (cadherins), which further break the toxin inside of domain I and stimulate oligomerization, which encourages adhesion to supplementary receptors tied along the membrane through glycosylphosphatidylinositol anchor points (Soberon et al. 2009). Cry1Ab-dominant negative mutants (Rodríguez-Almazán et al. 2009) have recently been isolated, confirming the necessity for oligomerization in toxin activity. Furthermore, Zhang et al. (2006) provided an alternate scenario wherein primary midgut binding initiates an Mg^{2+}-dependent receptor complex that results in the buildup of G-protein-dependent cAMP and protein kinase A stimulation. Phylogenetic/molecular study revealed that the Bt family's variety arose through the autonomous development of the three domains and the shifting of domain III within toxic substances (poisons) (Sanahuja et al. 2011). Cyt proteins have a solitary domain that is comprised of two sheets of α-helix loops coiled across a β-sheet (Li et al. 1996). Volvatoxin A2, a pore-producing cardiotoxin synthesized by the hay fungus *Volvariella volvacea*, is similar in structure to cytotoxin (Lin et al. 2004). To trigger the toxic substance, tiny parts of the N-terminus and C-terminus are eliminated (Li et al. 1996). Cyt toxins have distinct detrimental consequences than Cry toxins. Cyt toxins attach to membrane phospholipids and react on them immediately (Bravo et al. 2007). Several Cry and Cyt poisons work together in some cases. Cry and Cyt's complex interactions have been found to increase toxic chemical adherence to targeted cells and lethality to mosquito culex (Pérez et al. 2005). Cyt1Aa protein, via acting as a sensor enzyme, has been shown to enhance Cry11Aa cytotoxicity (Pérez et al. 2005). The hydrophobic responses involving toxins and the primary cell wall of intestinal lumen epithelium may be the basis for these combinatorial interactions, according to certain theories (Berry et al. 2002; Sanahuja et al. 2011). The structures of four Bt cry toxins are presented in Fig. 4.1.

4.9.3 Types of Receptors

When it was discovered that unique strong neurotoxin interaction domains existed in the insect intestinal lumen, researchers focused their studies on identifying and cloning toxin receptors. The aminopeptidase N (APN) receptors (Sangandala et al.

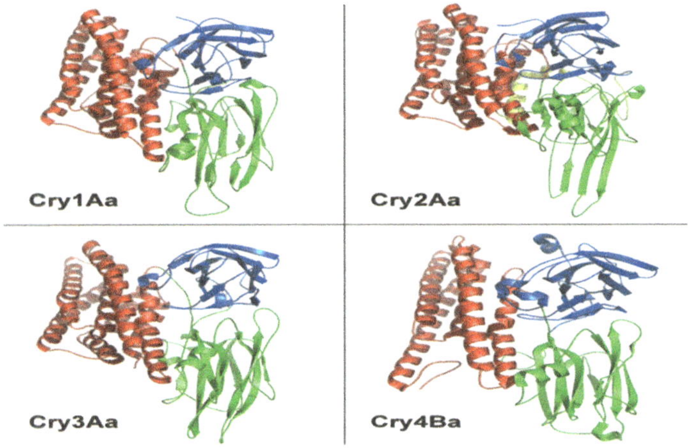

Fig. 4.1 The three-dimensional configurations of some cry proteins produced by *B. thuringiensis* (Talakayala et al. 2020)

1994) and also the cadherin-like receptors (Gahan et al. 2001; Pigott and Ellar 2007) found in lepidopterans seem to be the most documented. Glycolipids are thought to represent a key type of Cry toxin receptor in worms (Griffitts et al. 2005). Alkaline phosphatases (ALPs) (Jurat and Adang 2006), a glycoconjugate of 270 kDa (Valaitis et al. 2001), are all potential receptors (Hossain et al. 2004). Some of the binding site categories are explained in the subsequent paragraphs, with a special emphasis on toxin-receptor ratified connections and receptors' potential to accord toxic chemical vulnerability.

4.9.3.1 Cadherin Receptors

Cellular adherence, mobility, cytoskeletal architecture, as well as morphogenesis are just a few of the roles performed by the cadherin family of enzymes (Grochulski et al. 1995). Cadherin production is heavily controlled both regionally and chronologically and is typically specific to a type of cells. The existence of repetitive attachment domains for calcium, or repeats of cadherin, of roughly 110 amino acids in diameter, characterizes the enzymes. There are five cadherin repetitions in traditional cadherins; however, up to 34 loops have been observed (Dunne et al. 1995). Mucin and epidermal stimulating elements like repeats can also be found in certain cadherins (Nakayama et al. 1998). The proteins are glycosylated and normally include a mono-transmembrane motif that anchors them to the membrane, but there have been reports of seven-transmembrane (Usui et al. 1999) and GPI-anchored versions (Vestal and Ranscht 1992). Considering the virtue of its adherence capacity for Cry1Ab (Vadlamudi et al. 1993), a new cadherin-like protein was identified out from the midgut of *M. sexta* in 1993. A signaling protein, 12 adhesive loops, a membrane proximate external domain, a transmembrane site, and a short cytoplasmic zone were forecasted by genomic evaluation when the protein was first synthesized in 1995 (Vadlamudi et al. 1995). Subsequently, other lepidopteran cadherins were discovered, all of which had been found to possess a domain structure that is comparable (Wang et al. 2005a, b). Other characteristics in the extracellular domain of *M. sexta* cadherin have been found, including the cellular adhesion pattern HAV and the integrin-binding sequences RGD (Ruoslahti 1996) and LDV (Tselepis et al. 1997); nevertheless, the physiological function of such sequences is yet to be determined (Dorsch et al. 2002). An examination of the cytoplasmic domain, on the other hand, revealed no nucleotides recognized to associate with cytoplasmic molecules including catenins (Dorsch et al. 2002). Whereas conventional cadherins are predominantly found in adherens intersections, which are related to cell adhesion (Angst et al. 2001), cadherin in lepidopterans was discovered on the surface of intestinal lumen epithelial cells (Aimanova et al. 2006), which is the potential approach of Cry toxins (Chen et al. 2005a, b, c). Cadherin activity in *M. sexta* larvae varies with transitional level and improves steadily from the initial to the fifth larval stage (Midboe et al. 2003). Cadherin expression has not been found in eggs or adults, however. Even though the precise physiologic or biological role of intestinal lumen cadherins was unknown, it has been suggested that the strict regulation of cadherin stages all through instar improvement indicates their relevance in preserving midgut epidermal entity (Midboe et al. 2003). As Cry1A receptors, lepidopteran cadherin-like proteins have been widely researched, and there is pretty compelling proof that they perform a crucial influence in toxin vulnerability (Pigott and Ellar 2007).

4.9.3.2 APN Receptors

The APN is a group of proteins (enzymes) that breaks down proteins' and peptides' neutral amino acids from the N-terminus. They perform a number of activities in a lot of organisms, although in the lepidopteran larval midgut, they collaborate with endopeptidases and carboxylase to break down polypeptides from the food consumed by the insect (Wang et al. 2005a, b). The proteins are members of the zinc-binding metalloprotease/peptidase complex, specifically the gluczincins subgroup (Hooper 1994). The truncated zincin repeat HEXXH, in which X represents amino acid, is accompanied by a preserved remnant of glutamic acid, 24 amino acids downwards from the initial histidine, which distinguishes representatives of such a group. The amino acids (histidine) and the final glutamic acid residues act as zinc ligands, whereas the initial glutamate acid residue aids in enzyme activity. The effective domain is also thought to contain a GAMEN motif, which is extensively maintained (Laustsen et al. 2001). APNs had been widely investigated as possible Cry toxin sensors in addition to their role in digestion. Numerous distinct kinds of Cry toxins were identified and described as it was originally revealed that they could adhere to APN (Sangandala et al. 1994). The APNs have been split into five major forms, as illustrated by Herrero et al. (2005). Within each class, the mean nucleotide identity ranges from 56% (class 5) to 67% (class 4). Type II is the least equivalent to some others, with an overall sequencing similarity of just 25–26% when compared to the other categories, but classes 1 and 3 seem to be the best identical, having an average pattern similarity of 38%. Most documented APNs within species were discovered to aggregate into multiple categories so far. In fact, certain APNs have more sequence similarity with non-lepidopteran APNs than in other APNs in the same species. Numerous common characteristics have arisen from the numerous distinct APNs that have been investigated (Pigott and Ellar 2007). The proteins encode for 1000-amino-acid peptides, which go through a variety of posttranslational modifications to become finished molecules weighing around 90 and 170 kDa. The molecules contain a noncovalent N-terminal signal sequence that steers emerging protein complexes to the intracellular membrane's external layer. In opposition to primates, in which a polar N-terminal stalk is used for adhesion, they are linked to the membrane via a glycosylphosphatidylinositol (GPI) anchorage (Agrawal et al. 2002; Knight et al. 2005). The anticipated quantity of O-linked glycosylation regions varies significantly across the three forms of APN, although variations in portended N-linked glycosylation regions are less noticeable. Many linkages involving Cry1Ac and APN are thought to be influenced by sugar molecules, such as GalNAc (Knight et al. 1994). Cry1 toxins are poisonous to lepidopterans, and various distinct poisons have been found to attach to APNs, such as Cry1Aa, Cry1Ab, Cry1Ac, Cry1Ba, Cry1Ca, and Cry1Fa (Pigott and Ellar 2007). APNs and toxins from these families have multiple interaction patterns, according to the investigations performed. While several poison-APN adhering complexes had yet to be investigated, exploratory research is shedding light on the factors that influence receptor binding (Pigott and Ellar 2007).

4.9.3.3 ALP Receptors

Cry toxin receptors were also discovered in ALPs. By comparison, investigation on the APN and cadherin-like receptors has become very restricted, and none of the potential receptors has been transcribed or proved to play a central impact on cytotoxicity. However, initial reports reveal that ALP functions as a receptor for Cry1Ac in *M. sexta* (McNall and Adang 2003) and *H. virescens* (Hua et al. 2004), as well as a Cry11Aa receptor for *Aedes aegypti* (McNall and Adang 2003; Fernandez et al. 2006). ALP is a 68 kDa GPI-fastened transmembrane polypeptide in *H. virescens*. The availability of an N-linked oligosaccharide bearing a proximal GalNAc residue seems to be required for attachment to Cry1Ac, as shown by ligand blot investigation of BBMV. ALP development was lowered in a resilient strain of *H. virescens*, implying that it plays a vital component in cytotoxicity. The occurrence of a GPI anchor, as well as the relevance of GalNAc in toxic chemical adherence, bears striking similarities to APN and its engagement with Cry1Ac (Valaitis et al. 1997). Two-dimensional gel electrophoresis accompanied by ligand blot screening revealed a 65 kDa BBMV protein as a Cry1Ac-adhering polypeptide in *M. sexta* (McNall and Adang 2003). Database analyses of mass spectrometric fingerprinting and testing with ALP-specific antibodies were used to confirm its identity as ALP. Although the peptide was anticipated to be GPI fixed, it was not found in the array of molecules liberated by PI-PLC preparation from BBMV. The function of ALP glycosylation in toxin attachment has yet to be determined, but the protein was already found to co-localize with Cry1A toxins to the epithelial cells of *M. sexta* midgut (Chen et al. 2005a, b, c). The characteristics of ALP from *Aedes aegypti* have been compared to those of ALP from lepidopterans. The protein is 65 kDa and has a

GPI anchor that anchors it to the membrane. It is abundant in BBMV, accounting for 15–20% of the entire protein. According to immunofluorescence investigations (Fernandez et al. 2006), ALP is primarily found in the ceca and posterior midgut, with a dispersion arrangement that is comparable to that of coupled Cry11Aa. Cry11Aa binding to ALP was prevented, and toxicity was reduced by phage containing ALP-specific peptides, implying that ALP plays a role in modulating Cry toxin sensitivity (Fernandez et al. 2006). *A. aegypti* ALP and the identical Cry-restraining protein/polypeptide demonstrated by Krieger et al. (1999) and Buzdin et al. (2002) were thought to be analogous because of their comparable features. Both Cry11Aa and Cry4Ba bind to ALP and battle for the same epitope in these investigations. Cry9Aa binds to ALP as well, but it does not interfere with the other toxins and is not poisonous to *Aedes aegypti* (Buzdin et al. 2002). Both N-acetylglucosamine (GlcNAc) and GalNAc were unable to extract the receptors via Cry11Aa- or Cry4Ba-Sepharose after processing with numerous polysaccharides, particularly GalNAc. As a result, unlike the GalNAc-dependent binding identified involving Cry1Ac and *H. virescens* ALP, no involvement for polysaccharides in affinity can be proven (Hua et al. 2004).

4.9.4 Mechanism of Bt Toxin Action

Cry proteins undergo a multistep transition from a fairly innocuous crystallized protoxin precursor to a lethal state. To commence, a vulnerable larva must consume inclusions. The midgut's region enhances crystal solubility and, as a result, protoxin generation. Host proteolytic enzymes detect and remove breakage points on the protoxin, resulting in effective poison which binds to particular receptors on the midgut epithelium. The protein (toxins) subunits oligomerize to generate pore geometries prepared to enter into the membrane, as it is widely believed. Such channels enable ions and fluids to readily enter the cells, causing the victim to enlarge, lyse, and eventually die (Fig. 4.2). An alternate explanation has recently been offered, claiming that Cry toxicity is unrelated to toxic chemical complex formation (Craig and David 2007).

4.9.4.1 Pore Formation Model

Adhesion towards the Bt-R1 (receptor) may be the initial process in the engagement with the midgut membrane, according to the pore creation concept. This first attachment causes a morphological shift in the toxic substance, allowing for the proteolytic degradation of α-1 helix membrane-bound protease and the development of a pre-pore oligomeric structural system. The toxin attaches to the APN, causing the toxin to deform and become a molten gelatinous mass, which is subsequently incorporated into membrane lipids, causing pore development and cell enlargement (Bravo et al. 2007). The toxin blocks amino acid absorption and K^+ transport in the intestinal lumen after becoming inserted into the membrane lipid molecules, resulting in pH, ion, and other macromolecule imbalances, and eventually insect mortality (Ma 2005). The earliest contact of stimulated Cry1A toxic substance is a

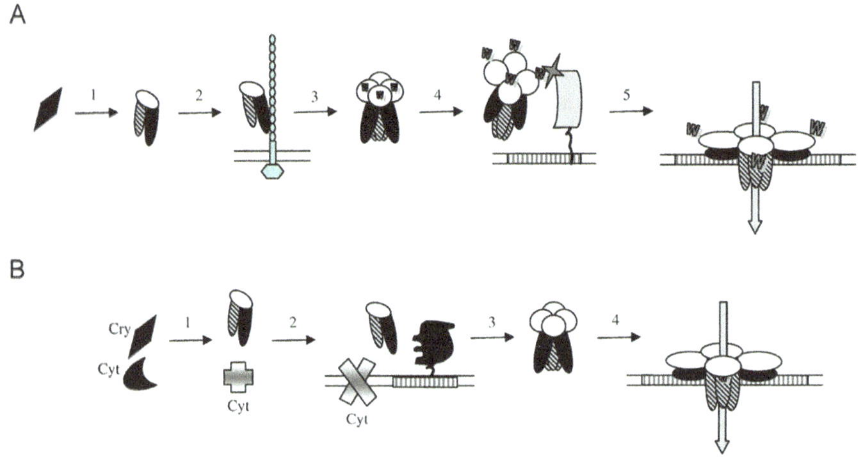

Fig. 4.2 The cry and cyt toxins' mechanism of action is depicted in this diagram. (**a**) Cry toxins bind sequentially with various receptor components in the larvae of lepidopterans. (1) Dissolution and stimulation of the toxic substance; (2) adhesion of Cry poison monomer to the first binding site (CADR or GCR), which causes a crystal structure transformation in the toxin and cleavage of α-helix 1; (3) oligomer establishment; (4) pressing down of oligomeric toxic chemical to the second binding site (GPI–APN or GPI–ALP), which causes a crystal structure alteration in the toxic chemical, thereby inducing the formation of molten globular state; (5) the oligomeric toxin's entry through lipid rafts and the creation of pores; (**b**) toxicity of dipteran larval stage by Cyt and Cry poisons/toxins. (1) Cry and Cyt neurotoxins are hydrolyzed and energized; (2) Cyt toxin penetrates further into membranes, and Cry toxin attaches to receptor inside the membranes; (3) Cry toxic chemical oligomerization is triggered; (4) the oligomer is introduced into the membranes, which results in the creation of an aperture (Bravo et al. 2007)

minimal contact to ALP and APN receptors, as per a previous study by Pardo Lopez et al. (2013), which was an extended version of the pore formation model. Domain II's exposed loop 3 interacts with APN, while domain III's strand 16 interacts with ALP (Pacheco et al. 2009; Arenas et al. 2010). These receptors (ALP, APN) are prevalent polypeptides with a glycosylphosphatidylinositol attachment that keeps them tethered to the membrane (Upadhyay and Singh 2011). The contact with these receptors compresses the active toxic substance in the intestinal lumen cells, allowing the toxin to adhere to the cadherin receptor with greater affinity (Gómez et al. 2006; Pacheco et al. 2009; Arenas et al. 2010).

4.9.4.2 Signal Transduction Model

In another paradigm, referred to as signal conduction/transduction, it was proposed that toxicity of Bt was linked to G-protein-reconciled mortality following sensor engagement (Zhang et al. 2006). Cry toxin coupling to Bt-R1 causes cell apoptosis by initiating a signaling loop that includes excitation of the mitogenic G-protein-subunit (G-s) and adenylyl cyclase (AC), which enhances cyclic adenosine monophosphate (AMP) concentrations and stimulation of protein kinase A (PKA).

When the AC/PKA signaling cascade is activated, a sequence of cytologic processes occur, including membrane blebbing, nuclear ghost appearance, cell enlargement, and ultimately tissue necrosis (Zhang et al. 2006). Figure 4.2 depicts a diagram of the two methods of Cry toxin activity. Broderick et al. (2006) made the important finding that *B. thuringiensis* cytotoxicity is dependent on interactions with typical gut microbes. Pesticidal lethality was eliminated by administering antibodies to the gastrointestinal microbial population, whereas *B. thuringiensis*-mediated mortality was reinstated by reintroducing an *Enterobacter* sp., which commonly dwells in the midgut microbial community. Transgenic plants producing *B. thuringiensis* toxins were utilized to play a protective role against a large number of agricultural insect pests with tremendous results. With the improvement of the early GM tobacco and tomato crops with natural Bt genes (Vaeck et al. 1987; Fischhoff et al. 1987; Barton et al. 1987), great strides were made in the advancement of enticing transgenic plants with greatly altered Bt-cry genes for mRNA reliability and high affirmation (Vaeck et al. 1987; Fischhoff et al. 1987; Barton et al. 1987; Gatehouse 2008). In the laboratory and in the outdoors, a vast amount of viable Bt plants of diverse families have been created that produce several Bt-cry genes and demonstrate considerable pest resistance (Hilder and Boulter 1999; Sharma et al. 2000; Tabashnik et al. 2003). The signal conduction paradigm does not include interactions with some other receptors, oligomerization, or transmembrane incorporation. The toxic effects are generated, according to this hypothesis, through the engagement of the monomeric three-dimensional cry toxin with the CAD receptor, which activates a Mg^{2+}-dependent signal cascade pathway. This connection activates a G-protein, which then activates an adenylyl cyclase, causing cytoplasmic cAMP to be produced. Excessive cAMP levels cause protein kinase A to activate, which initiates an intracellular cascade that leads to apoptosis (Zhang et al. 2006). This procedure relied on studies conducted in the cell lines and immunoblotting solely with the cadherin gene of *Manduca sexta* (Soberon et al. 2010). Figure 4.3 depicts the models of cry toxins' action in lepidopterans at molecular level.

4.9.5 Mode of Action of Vegetative Insecticidal Proteins (VIPs)

VIPs are another important protein naturally produced by *B. thuringiensis*. Many studies have shown that Bt variants secrete VIPs while in the vegetative developmental transition stage, and because the proteins have no systemic or sequence homology with Cry proteins, they can be used as an augment or replacement origin of Cry toxic substances in tolerance management and crop defense. Vip3 and Cry proteins are combined in a variety of commercial Bt crops. Cry1Ab, cry2Ab, cry1Ac, vip3A, cry1F, and cry2Ae are seven cry and vip genes that have been used to boost tolerance to insect pests. To limit the possibility of insect pests developing resistance to Bt toxins generated in genetically engineered crops, gene-pyramiding techniques had been built wherein the Bt crops possess multiple genes (Jouzani et al. 2017). The poisons Vip1 and Vip2, which are specific to coleopterans and homopterans, respectively, function in a combinatorial manner ("A-B type")

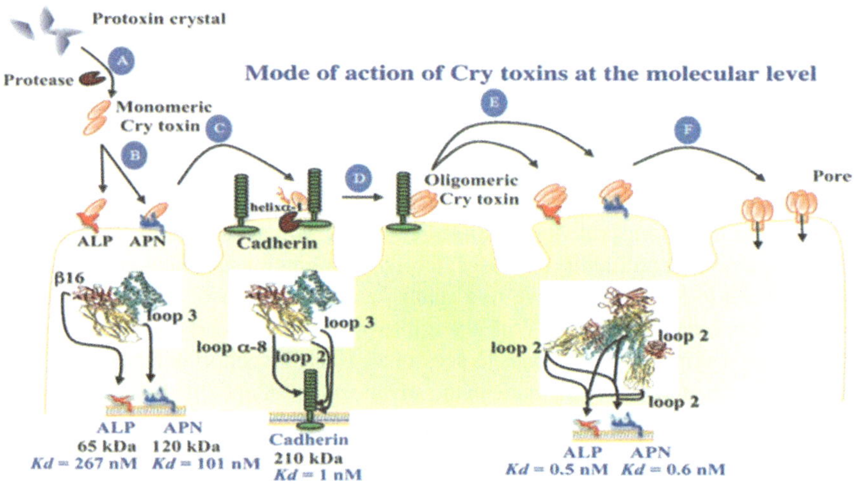

Fig. 4.3 A conceptual illustration of the model of activity of Cry proteins in lepidopterous insects at the molecular level: (A) the larvae consume the three-domain Cry protoxin, which is dissolved in the intestinal lumen owing to excessive pH and lowering circumstances and triggered by gut proteolytic enzymes, resulting in the toxin component; (B) in a low-affinity contact, the Cry toxin (monomeric) engages two receptors (ALP and APN), and the toxin is then positioned near the membranes; (C) the monomeric three-domain Cry toxin attaches the cadherin receptor with strong specificity, causing proteolysis of the toxin's N-terminal end, containing helix-1 of domain I; (D) the shattered Cry toxin can subsequently oligomerize into a toxin that is formed before it enters the pore; (E) the oligomeric three-domain complex interacts to ALP and APN receptors with high affinity; (F) the pre-pore integrates into the membranes, resulting in aperture development (Koul 2020)

toxins (Warren et al. 1998; de Maagd et al. 2003; Yu et al. 2011). Vip1's receptor adherence component attaches to the targeted insect's intestinal lumen membrane receptors as a monomer or oligomer, forming pores, whereby the apoptotic domain of Vip2 protein acts via ADP-ribosyl transferase in the insect's cell. Endocytosis is another way for Vip2 to penetrate the cytoplasm (Leuber et al. 2006; Pardo-Lopez et al. 2013; Chakroun et al. 2016). The availability of glutamic acid that is charged negatively (positions 340 and 345), lysine remnants that are charged positively, and histidine are thought to affect the specificity of channel creation in Vip1 (Leuber et al. 2006). There are two theories about how Vip2 gets into targeted cells. Initially, Vip2's homologous recombination with the dual-structure toxin C2 clostridial (I subunit) facilitates its endocytosis (Barth et al. 2004). The second idea is that the midgut's alkaline liquid exerts a strong proton gradient pressure that immediately penetrates the cytoplasm of the target tissue via a channel produced from Vip1 (Leuber et al. 2006). Vip2 protein enters the cell cytoplasm and operates towards actin in target tissues, limiting the synthesis of microfilaments and cytoskeleton dissolution by ADP-ribose catalytic translocation from NAD to actin, leading to the death of the target organism (Han et al. 1999; Jucovic et al. 2008; Aktories et al. 2011). The proposed mechanism of action of Vip1/Vip2 toxin is shown in Fig. 4.4.

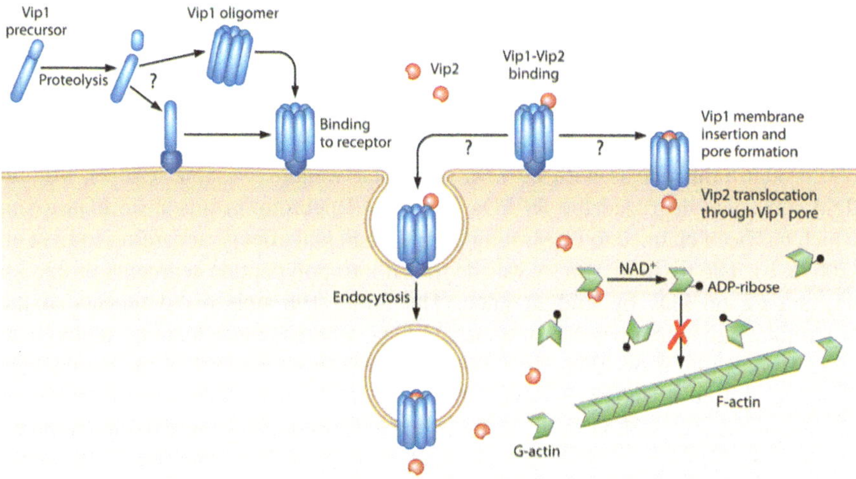

Fig. 4.4 The Vip1/Vip2 toxin's suggested mechanism of action. Midgut proteolytic enzymes break down the Vip1 protoxin. The stimulated toxin connects to targeted monomeric or oligomeric receptors. Vip2 afterward works by binding to Vip1 protein and reaches the cytoplasm either via endocytosis of the entire complex or via the porous structure created by Vip1. Vip2 catalyzes the transfer of the ADP-ribose group from NAD to actin monomers while inside the cell cytoplasm, restricting them from polymerizing (Chakroun et al. 2016)

Numerous investigations have been carried out in order to deduce the processes involved in the model of activity in Vip3 proteins. Vip3A toxins could be dissolved at pH 5–10, allowing them to be dissolved in the intestinal lumen of the insect, which is alkaline in nature. The solubility of these proteins, their enzymatic fragmentation, and adhesion in the insect's midgut were first established and linked to cytotoxic consequences similar to Cry toxins, such as feeding suppression, gut motility cessation, immobility, and mortality of the targeted insect. However, symptoms appear 48–72 h after consumption of Vip3A toxins, whereas they appear within 24 h after taking cry toxins (Yu et al. 1997). Vip3A protein putative glycosylation neurotransmitters on BBMV of the insect intestinal lumen are also distinct from Cry1A receptors. Vip3Aa proteins form complexes of 80 kDa and 110 kDa, while Cry1Ab reacts with 120 kDa and 210 kDa proteins, reflecting APN and cadherin receptors, respectively, according to ligand blot evaluation in the stomach of *Manduca sexta* (Lee et al. 2003). Additionally, in the lepidopteran pest species *H. zea* and *H. virescens*, the distinct binding locations of Vip3A and Cry2Ab or Cry1Ac have been illustrated (Lee et al. 2006). In other research findings, Vip3Aa proteins with a 65 kDa protein and Cry1Ac proteins with a 210 kDa protein were found to attach in the midguts of *Prays oleae* (Mesrati et al. 2009) and *A. segetum* (Mesrati et al. 2009; Hamadou-Charfi et al. 2013). Mesrati et al. (2011) established the interaction of stimulated Vip3A with 55 and 100 kDa ligands in the European flour moth (*Ephestia kuehniella*). Upon consumption of Vip3A-like proteins, substantial destruction was found in the midguts of vulnerable insects, including symptoms such as cell enlargement, cytoplasmic demyelination, ejection of cell substance, rupture in endothelial cells, and cell deconstruction. Such findings revealed that the midgut is the protein's targeted binding region (Chakroun and Ferré 2014; Boukedi et al. 2015). These Vip3A toxins are composed of a protein with an N-terminal and C-terminal (conserved and unconserved region) of around 800 protein unit repeats with 89 kDa molecular weight (variable region). Midgut fluid substance degraded the 89 kDa Vip3A full neurotoxin into a 62 kDa toxin, which has been claimed to be essential for its pesticidal effects, according to Li et al. (2007a, b). The toxin of Vip3A was degraded into polypeptide segments of 22, 33, 45, and 62–66 kDa by midgut juice. The initial 198-amino-acid repeats of the N-terminal generate a peptide of about 22 kDa, whereas the remaining amino acid repeats constitute a 66 kDa core, which when hydrolyzed yields 33 and 45 kDa peptides. Bel et al. (2017) revealed that this 66 kDa toxin binds to brush border membrane vesicles (BBMV) in the lumen of vulnerable insect pests. Vip3Aa toxic effects were lowered in *S. frugiperda* and *S. exigua* by downregulating receptor manifestation. Concurrently, recombinant protein appearance of this phagocytic receptor in the *Drosophila melanogaster* midgut markedly expanded Vip3Aa infectivity against *D. melanogaster* larvae. The scavenger binding site of *D. melanogaster* shared only 27% of its DNA with this neurotransmitter. The formation of motif II of Vip3AaII protein (Jiang et al. 2020) and Vip3B2160 protein (Zheng et al. 2020) revealed the crucial stages required in Vip3 activity as (a) domain I fragmentation and (b) transition states in domain II. This group notably presented the "membrane insertion model" (Jiang et al. 2020). Even though the ion route generation scenario is the extensive plausible explanation for Vip3Aa protein action, the endocytotic function of Sf-SR-C and the apoptotic activity of FGFR and S2 proteins could also be held accountable for its cytotoxicity. Working with targeted insect midgut epithelium might yield additional understanding into endocytosis and apoptosis models, because this paradigm is founded from investigations with Sf9 cells generated from reproductive organs. After stimulation of certain cyclins, the apoptotic activity was triggered in reaction to therapy with Vip3Ca enzymes (Hernández-Martnez et al. 2013a, b). Chakrabarty et al. (2020) and Syed et al. (2020) had also previously demonstrated models pertaining to the pesticidal properties of Vip3A proteins. The 65 kDa and 19 kDa parts of Vip3Aa produce a composite of >240 kDa following the proteolytic procedure, demonstrating a potential insecticidal interactive effect (oligomer formation) for Vip3 proteins (Fig. 4.5). The replacement of alanine or proline for serine at position 164 prevents the development of this complex and results in the deficit of toxic effects against *S. litura*. However, replacing S164 with threonine had no effect on the building process but helped cut insecticidal performance by 35%, indicating that S164 is a crucial protein unit site deployed in the creation of >240 kDa protein complexes with pesticidal action (Shao et al. 2020).

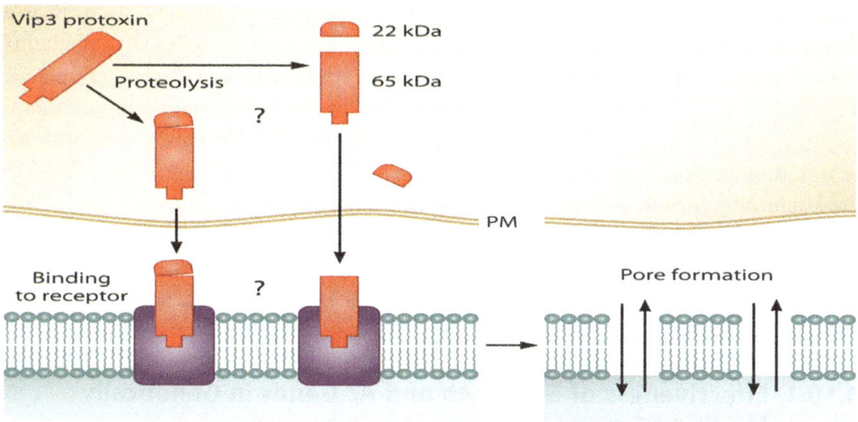

Fig. 4.5 Vip3 proteins' putative mode of action. Midgut proteases break down the overall protoxin into smaller pieces. Specific receptors connect to the 65 kDa segment with segment of kDa (22 kDa) still bound or not. The cell dies as a result of the formation of apertures (Chakroun et al. 2016)

4.10 Transgenic Plants with Bt Crystal Protein Genes

Despite the fact that microbiological Bt formulations were used in agriculture for years and are the most powerful and selective biopesticides available, they have a number of drawbacks. High costs, inconsistency in precipitation and sunshine, and ineffectiveness towards insects eating on crop interior cells are the major limitations (Ely 1993). The benefits of Bt-based transgenic insect tolerance in crops involve defense of target organs sometimes against insect interior penetration, weather-independent protection, reduced usage of synthetic pesticides, and excellent target pest selectivity. The initial efforts at transforming crops with cry genes failed to produce enough Bt proteins. The main reasons for transgenic expression at low levels are life-form changes in codon affinity and other regulatory elements that control transcription. The codon frequencies of a genome can be altered to fit the desired codon utilization frequencies of the host species for improved production of alien proteins in a crop system. Codon standardization is an investigative approach in which in vitro modification is used to shift codons inside a transgene to the chosen codons without modifying the protein units (amino acids) in the synthetic polypeptide. The variation in codon utilization among microbes and crops (Fennoy and Baileyserres 1993) is well established, and genomic uniqueness impacts codon utilization rate (Grantham et al. 1980). The prokaryotic modified sequence patterns that are improper for transcription in the plants are replaced with the crop variety's favored sequence during the transfer of a gene from a bacterium to a plant genome. Destabilizing and deteriorating mRNA sequences are also eliminated (Kang et al. 2004). The production of cry alleles in crops necessitates significant sequence alteration (Kumar et al. 2005). Bt signaling pathway in plants often necessitates enhancement of the cry genes to incorporate crop-favored codons, as well as a decrease in the gene's total AT content (typical of bacteria genomes) by enhancing the GC composition (typical of plant genomes). Also necessary for effective gene expression in a plant system is the elimination of mRNA-destabilizing elements such as ATTTA sequences, polyadenylation signaling sequences, transcriptional terminating regions (Misztal et al. 2004), as well as destabilizing sequences. For the bacterial genes to be expressed in plants, other aspects such as optimizing the bacterial sequence according to the crop codon priority sequences (using codon utilization tables), without changing the amino acid, are needed (Sutton et al. 1992).

4.10.1 Effectiveness of Bt-*cry1Ab* and *Ac* Genes in Genetically Modified Crops

The incorporation of Bt-cry protein into crop plants has resulted in the development of highly effective new tactics for agricultural crop security against prominent insect pests of agricultural importance. Cry1A genes contain a toxin that is extremely effective against the lepidopteran family, which causes significant crop loss. A study of the interactions of Bt proteins (Cry1Aa, Cry1Ab, and Cry1Ac) encrypted by genetic material associated with *Helicoverpa armigera* midgut attachment regions (receptors) revealed that they compete for prevalent adhesion points to distinct epitopes of the receptor proteins arranged as Cry1Ac > 1Ab > 1Aa order, causing cytotoxic activity on the insect pests (Estela et al. 2004; Bravo et al. 2007). Tobacco, cotton, tomatoes, potatoes, soybean, and rice all have cry1Ac genes that were substantially edited and codon optimized, as well as other alterations, enabling amplification (Ferry et al. 2004). The strongest remarkable tale is the introduction of *cry1Ac*-expressing GM *Gossypium* sp. as Bollgard I in 1996 and Bollgard II containing two *cry* alleles (*cry1Ac* and *cry2Ab*) in 2000, which has provided crop producers with substantial improvements over synthetic pesticides and productivity (Perlak et al. 2001). Because of the fragility and untimely cessation of the transcripts, the transcription of the entire *cry1Ac* gene in crops remained reduced (Perlak et al. 1990). Numerous improvements to the cry gene have been made to allow for increased expression, with the most significant discovery being the creation of artificial variants of the allele with codon adjustments to delete purported polyadenylation sequences and favorable codon utilization for top-level representation in crops (Perlak et al. 1990). The foreign gene's transcription and translation initiation is regulated by the 5′ and 3′ UTR leader sequences, which are critical for transgenic expression (Lu et al. 2008). The aggregation of transgenic proteins has been found to be considerably boosted by viral leaders at the 5′ UTR (Dowson Day et al. 1993). The information for transcript polyadenylation is found in the 3′ UTR, and it has a significant effect on mRNA stabilization (Chan and Yu 1998). To solidify the establishment of the transcript, heterologous 3′ UTRs from plants or plant viruses were utilized (Ko et al. 2003). Bt transgenics has been shown to be most efficient against lepidopteran pests when a synthetic truncated BT-toxin gene (1.85 kb) was used (Sardana et al. 1996; Perlak et al. 1990). Nevertheless, comprehensive transformed *cry1Ac* gene is used in the more successful biotech cotton crop that was commercialized (Purcell et al. 2004; Perlak et al. 2001). It is essential to mention that the production of viable modified tomato plants carrying the Cry1Ab poison for insect deterrence has been established (Kumar and Kumar 2004). The removal of the 3′ half of the cry1Ab and cry1Ac toxin-coding regions increased production of the 66 kDa toxins in tobacco and tomato, resulting in complete against tobacco insect pest (Perlak and Fischhoff 1993; Perlak et al. 1991). The expression of modified and truncated *Bt-cry1A* genes has been observed to be weak in resistant varieties of crops such as cotton and cereal legume (Jenkins et al. 1993; Stewart et al. 1996; Sanyal et al. 2005). Furthermore, the growth and survival of transformants expressing Bt toxin encoded by the full-length gene are relatively rare. Monsanto 531, a single event in cotton, was created using full-length altered *cry1Ac* and released as Bollgard cotton. These insect-tolerant mutant cotton cultivars developed by this single event are thriving effectively in the fields all over the world in various different agroclimatic areas (James 2012).

4.11 The Story of Bt Cotton

One of several biotechnology-based plant protection products was Bt cotton. Bt cotton crops are transgenic because they possess one or even more exogenous DNA obtained from *Bacillus thuringiensis*, which is a soil-residing bacterium. Cotton plant cells express crystalline pesticide, commonly known as Cry proteins, and these insecticidal proteins are efficient against the larvae of tobacco budworms and bollworms, which are among the highly destructive insect pests affecting cotton. The US Environmental Protection Agency (EPA) certified this innovative insect pest management method for commercialization in October 1995, and it is presently accessible from different seed firms in the USA and various different cotton-cultivating countries world over. Prior to the advent of Bt cotton, cotton farmers encountered a series of pest stress. Farmers have been losing a lot of their crop to *H. virescens* and the pink bollworm (*Pectinophora gossypiella*), due to development of resistance to synthetic pesticides. According to the USDA, 94% of cotton grown in the USA is GM cultivar (James 2016). According to a study conducted by the California University, the mean expense decreases in insecticides utilized in GM cotton areas through 1996 to 1998 were around \$25 and \$65 per acre, with an output predicted to be 5% higher than the conventional cotton for the same time frame. Furthermore, Bt cotton reduced the frequency of leaf applications of pesticides against other cotton pests and, as a result, the expense of pesticide utilization has been minimized (Anonymous 2000). Bollgard cotton had been the initial Bt cotton that was released in the USA in 1996. It expresses Cry1Ac toxin, which exhibits insecticidal activity against tobacco budworm. Growers in the Western Cotton farmland used Bt cotton to control the pink bollworm, while farm owners in the

Mid-South and South-East used it to control the tobacco budworm and, to a slightly lesser degree, the fall armyworm, *Spodoptera frugiperda*, and *Spodoptera* spp. and *S. exigua* (Stewart 2007). Bollgard II, the next version of Bt cotton, was debuted in 2003. Cry2Ab poison was being produced by it. Cry1Ac and Cry1F were added to WideStrike cotton (a Dow AgroSciences trademark) in 2004. Bollgard II and WideStrike equally outperform the original Bollgard on a wide spectrum of caterpillar insects (Stewart 2007). In poor nations, Bt cotton seems to be the only Bt crop grown (James 2016). During 2006 and 2007, the farmed area of Bt cotton in India and China rose dramatically, reaching 25 million acres (2.5 million ha). Bt cotton was first planted in India in 2002 (James 2016). In 2016, the landmass of cotton in the globe (in 18 countries) was 35 million ha, with 22.3 million ha (64%) being GM cotton. But in the USA, cumulative cotton acreage reached 4 million ha, with 3.2 million ha (80%) being mixed Bt and herbicide-tolerant cotton (James 2016). In the USA, 18 distinct complexes of 11 Bt genes produced by Bt corn and Bt cotton cultivars were established. Each species generates 1–6 Bt poisons, which are lethal to the beetles, larvae, or both (Tabashnik et al. 2009; Mohamed 2018).

4.11.1 Development of Bt Cotton

In some countries like India, serious attempts to exploit genetic engineering techniques for tolerance of bollworm in *Gossypium* species emerged in the 1990s alongside the import of GM cotton and the establishment of national laboratory research program (Karihaloo and Kumar 2009). Throughout 1998 and 2001, significant Bt cotton outdoor experiments were done in India, confirming excellent and efficient management of bollworm and other insects. When contrasted to traditional cotton, the field testing revealed that Bt cotton has the ability to enhance outputs by approximately 40% while lowering the pesticide applications by 50% or higher (Karihaloo and Kumar 2009). As a result, Bt cotton was viewed as possessing the ability to enhance the livelihoods of cotton producers by providing advantageous ecological and financial outcomes. Furthermore, a comparison of the achievement of Bt cotton varieties made by various private firms over various places all over three cotton-producing regions for both 2001 and 2005 demonstrated that Bt hybrids outperformed non-Bt equivalents with respect to crop yield and some other contributory attributes. Such varieties too were better in the quantity of bolls per plant, and seed yield, in addition to increased cotton seed production (Manickam et al. 2008). Table 4.2 summarizes the development of Bt cotton by Mahyco. Indeed, the establishment of Bt cotton in India via Monsanto's modified cotton followed a rigorous legislative system prior to reaching the farmer's grounds. The Indian Council of Agricultural Research (ICAR) gave a satisfactory evaluation on the field testing of Bt cotton to the Ministry of Environment in February 2002. Having followed that, the environmental ministry's GEAC certified Bt cotton for industrial production (Deepak 2012).

Table 4.2 Progress in Bt cotton development in India

Period (years)	Activities
1994	Institutional Biosafety Committees (IBSCs) were formed, and a proposal for modified Bt cotton seed import was submitted.
1995	Permission was granted by the Department of Biotechnology (DBT) to acquire 100 g of modified seed of the Coker 312 strain from Monsanto in the USA. Cry1Ac gene in *Bacillus thuringiensis* was contained in this strain.
1996	Shipped seeds were planted, and a greenhouse trial began. Pollen escape was tested in a small experiment (one site). Backcrossing breeding in the greenhouse for the transmission of the Bt gene into superior progenitor lineages was conducted.
1997–1998	Pollen escape was assessed in a small number of field experiments (five sites). Allergenicity (BNR model) and toxicity using farm animal (goat prototype) research were carried out. In the same period, Mahyco and Monsanto partnered. The Department of Biotechnology (DBT) granted Monsanto the authority to conduct modest experiments of Bt cotton (100 g each trial).
1998/1999	Investigations spanning 15 + 25 sites to examine the effectiveness of the Bt gene in Indian superior genotypes were conducted. To examine the efficiency of the Bt gene in Indian premier cultivars, researchers conducted multicentric research experiments (11 places). The Review Committee on Genetic Manipulation (RCGM) was pleased with the test findings at 40 sites and directed Mahyco to present proposals for testing at 10 sites by MEC on April 12th.
1999/2001	(1) Massive field trials (100 ha) were used to examine the viability of the Bt genome in Indian superior stock and the productivity of Bt hybrids, (2) composite seeds were developed (150 ha), (3) multiple biosafety investigations were performed, and (4) ICAR tests were carried out in six selected sites.
2001–2002	(1) There were vast experiments (100 ha) that tested the efficiency of the Bt genes in Indian prime cultivars and the effectiveness of Bt hybrids, (2) composite seeds of about 300 ha were generated, (3) biosafety evaluation was done on the composite cultivars, as well as (4) ICAR investigated these prime cultivars at 11 localities.

Adapted from Deepak (2012)

Three mutant species (MECH 162 Bt, MECH 184 Bt, and MECH 12 Bt) have been permitted for production under specific circumstances. These hybrids generate moderate to lengthy staple fiber, thereby generating a positive output. Industrial Bt cotton planting began in April 2002 in six Indian states: Gujarat, Andhra Pradesh, Madhya Pradesh, Karnataka, Maharashtra, and Tamil Nadu (Deepak 2012). Bt cotton variants have assisted in increasing productivity from 156 lakh bales in 2001 to an expected 356 lakh bales in 2011. Cotton output was 309 kg/ha in 2001, prior to the introduction of Bt cultivar, and increased to 495 kg/ha in 2010. According to research undertaken by the Central Institute of Cotton Research (CICR), there was widespread farmers' acceptance for Bt-modified cotton, as indicated by the reality that Bt cotton now covers well over 90% of the land throughout all cotton-growing Indian states. Gujarat, Andhra Pradesh, Maharashtra, Haryana, Punjab, and Tamil Nadu have seen the most significant increases in productivity. The CICR has been motivating producers via exhibitions and educational courses about every element of genetically engineered crops, including biosafety and acceptable strategies for realizing long-term advantages using proper crop production practices, in collaboration with the State Agricultural Universities (Deepak 2012).

4.11.2 Bollgard I Cotton (First-Generation Bt Cotton)

In 1996, Australia and the USA certified the earliest biotech cotton (BG I) harboring the lepidopterous insect pest resistance allele *Cry1Ac* for practical exploitation, while on March 26, 2002, the GEAC authorized it for commercial use in India (Singh et al. 2020). This marked the dawn of a revolutionary era in cotton production, with the GEAC annually approving a new Bt cotton cultivar from the corporate companies. Multiple studies have compared Bt vs. non-Bt cotton and shown that Bt cotton is more successful at eliminating cotton bollworm. Genetically modified strains 62 Bt and 65 Bt had reduced 95% rosette flowers when correlated to non-GM plants (Douglas et al. 1994; Wilson et al. 1992). Pink bollworm (PBW) viable larva was obtained from cultured bolls, and seed destruction has been decreased by 97–99% across transformants compared to non-modified lines. In Bt cotton lines harboring either *Cry1Ac* or *Cry1Ab*, Benedict et al. (1996) found that normal bollworm prevalence was lower (198 larvae/60 plants). In this kind of Bt lineages, floral bud and square damage too was minimal (20.60% and 11.77%, respectively). In comparison to non-Bt Coker 312 (1050 kg/ha), they observed an aggregate productivity of 1460 kg/ha across all improved varieties. Throughout the years 1992–1995, Harris et al. (1996) discovered that transgenic cotton provided good management of the bollworm *Helicoverpa zea* in Mississippi. The bugs have been obtained over 44 generations using Bt farmlands (containing *Cry1Ac*), and it was discovered that the larval tolerance of the Bt-selected individuals to *Cry1Ac* toxicity significantly enhanced 106-fold when compared to a susceptible strain (Wu and Guo 2004). In Mexico, two Bt cotton cultivars (DP 33B and DP 35B) outperformed a traditional strain (DP-5690) in terms of BW suppression, and the former required one fewer pesticide application than the latter (Burd et al. 2003). Viable PBW larvae as well as escape pores have only been identified in DP-50 and not in Bt cotton lines (MONS-1 and MONS-2), but there had been considerable disparity in the frequency of PBW impacts on the carpel wall between cultivars in terms of Bt cotton performance. Because lepidopteran stress was modest throughout the season, no substantial yield variations between the kinds were observed (Sieglaff et al. 1999). According to Wu et al. (2000a, b), two Gossypium Bt hybrids (GK-1 and GK12) are extremely tolerant to ABW in the entire period of Bt cotton cultivation. The effectiveness of GK-1 and GK-12 in the control of the second, third, and fourth lines of ABW, respectively, has been 88.71–95.45%, 92.75–97.65%, and 93.33%.

4.11.3 Bollgard II (Second-Generation Bt Cotton)

The earlier BG I was unable to safeguard the plant from BW species developed in the relatively late phases of the crop for the entire period of plant growth. This is due to the fact that it had a solitary gene (*Cry1Ac*), which made it resilient to the bollworms in India, including the American bollworm (*Helicoverpa armigera* Hubner), spotted bollworms (*Earias vittella* Fabricius, *Earias insulana* Boisduval), as well as pink bollworm (*Pectinophora gossypiella* Saunders), and yet highly vulnerable to other insect pests. If the target lepidopterans digest *Cry1Ac* Bt toxin on a routine basis for several years, then resistant strain would develop. With this in mind, a second version of Bt cotton (BG II) with dual alleles (*Cry1Ac* + *Cry2Ab*) had been launched, which further accelerated the death rates of different vital species of insect pests (destroying various crops) including tobacco budworm, fall armyworm, beet armyworm, cabbage looper, as well as soybean looper (Somashekara 2009). A better universal market and a widespread acceptance of insect-tolerant cotton were the key reasons for the 8% rise in overall Bt cotton acreage in 2017 (Anonymous 2002). Gore et al. (2002) found that BW larvae damaged 6.4 fruiting processes/10 plants in BG II, 11.5 blossoming patterns in BG I, and 25.0 flowering patterns in non-Bt cotton when BGs I and II were tested with synthetic BW contamination. Fruiting forms' damage in BG II averaged 0.8 per BW larvae, 3.5 in BG I, and 6.6 in non-Bt cotton. Chitkowski et al. (2003) investigated the viability of BG II containing two *cry* proteins, *Cry1Ac* and *Cry2Ab*, against BW where he discovered that BG II cotton had a considerable proportion of *H. armigera* and soybean looper, Pseudoplusia larvae, although it did not exceed the handling criterion of 3/100 crops on most of the days of sampling. Conversely, it occurred on one or two sampling days in BG I cotton (*Cry1Ac*) and on multiple occasions in traditional genotype. Analogously, the percent decline in BW-infiltrated sections for the BG I and II rows was 10-fold as well as 19-fold, in both, over the traditional cotton cultivars, whereas the % areas sustaining destruction had been lowered by 6-fold by BG I and 16-fold by BG II genetic variants (Jackson et al. 2003). According to Strickland and Annells (2005), BG II has good efficiency against lepidopteran pests, with far fewer caterpillars than ordinary and INGARD cultivars. *S. litura* remained the only spodopteran insect that survived on BG II, which could be attributable to Bt protein inefficacy. Employing analogous pesticides, BG II produced a greater output (5.44 bales/ha) than traditional (1.60 bales/ha) and INGARD (4.83 bales/ha). Udikeri (2006) discovered that the novel breed of GM Bt cotton (MRC-7201BG II) containing two *cry* alleles (*Cry1Ac* + *Cry2Ab*) was the most effective towards cotton bollworms. In MRC-7201 (BG II) with 5.05% injury, the frequency of *H. armigera* proved to be 0.13 larva/plant, which was comparable to MRC-6322 Bt, indicating that BG II had been stronger to all BG I genotypes in vulnerable environments. In the subsequent season, Bheemanna et al. (2008) did not find *H. armigera* juvenile growth on BG II (MRC-7201) compared to BG I (MRC-6322), which exhibited 0.12 and 0.19 larvae per plant.

4.11.4 Bollgard III

Gene pyramiding of biotech crop alleles is another strategy to tackle insect tolerance by halting/delaying the emergence of tolerance. In the case of Monsanto's Bollgard III cotton, *Vip3A* was added to the *Cry1Ac* and *Cry2Ab* proteins present in Bollgard II (Edpuganti 2018). TwinLink cotton from Bayer CropScience comprises two Bt genes, *Cry1Ab* and *Cry2Ae*, and TwinLink Plus involves three Bt toxins (*Cry1Ab*, *Cry2Ae*, and *Vip3Aa19*), while Dow AgroSciences' WideStrike 3 comprises *Cry1Ac*, *Cry1F*, and a *Vip3A* for superior insect pest control (Edpuganti 2018). BG I cotton was unlawfully imported into India in 2000 in Gujarat, while illicit importation and plantation of BG III crops in the Vidarbha area had also been mentioned (Shrivastav 2017). Monsanto had planned to launch BG III (with an extra chromosome vip3) in India, but it backed out due to the anti-GM sentiment. Besides Gujarat, it crossed the border surreptitiously via dealers in Yavatmal, Andhra Pradesh (Shrivastav 2017). Bollgard 3 (*Cry1Ac* + *Cry2Ab* + *Vip3A*), TwinLink Plus (*Cry1Ab* + *Cry2Ac* + *Vip3Aa19*), and WideStrike 3 (*Cry1Ac* + *Cry1F* + *Vip3A*) were the three alleles in the most current third progeny of Bt cotton (Vyavhare 2017).

4.11.5 Impact/Benefits of Bt Cotton

In India, there was a flurry of research on Bt cotton-related issues. The acceptance of Bt cotton has led to substantial increase in women's employability. Subramanian and Quim (2010) found that planting insect-resistant Bt cotton increases overall wage revenue by US$40 per unit of land, showing a greater stage of evolution of cash flow and job opportunities for rural laborers, particularly employed females. Between 1998 and 2010, organizations carried out several other researches on the impacts of modified Bt cotton. All of these investigations have repeatedly shown that Bt cotton yields 50–110% more than traditional cotton. Bt cotton farming generates profits ranging from US$76 to US$250 per hectare (Subramanian and Quim 2010). Profits are being accrued for small and resource-poor cotton growers in India's numerous cotton-growing areas. The increase in output was in the range of 30–60%, while the number of insecticide treatments was reduced by roughly 50% as mean. Farmers' real experiences during the commercialization of Bt cotton are comparable with the benefits observed in pre-commercialization field studies (Choudhary and Gaur 2010). During the pre-commercialization of Bt cotton, research performed by Naik (2001) concluded that the mean economic benefit of Bt cotton was in the cost of approximately US$76 to US$236 per hectare, which is comparable to a 77% rise over ordinary cotton (Deepak 2012).

4.12 The Story of Bt Brinjal

Maharashtra Hybrid Seed Company (Mahyco) in India created a transgenic manipulated, insect-repellent brinjal containing *cry1Ac* gene (Bt brinjal) that offers efficient management of EFSB (eggplant fruit and shoot borer). The United States Agency for International Development-funded Agricultural Bioengineering Support Project II at Cornell University provided a large amount of the Bt brinjal event ("EE-1") to the Bangladesh Agricultural Research Institute (BARI), so this feature was transferred into several locally and industrially prevalent open-pollinated brinjal cultivars (Shelton et al. 2018). BARI tested the effectiveness and pollution control of the nine Bt cultivars that resulted over a 7-year period in greenhouses and restricted outdoor experiments in numerous places across Bangladesh. In October 2013, Bangladesh's National Committee on Biosafety (NCB) authorized four of the nine Bt cultivars for growing. BARI Bt Begun-1, BARI Bt Begun-2, BARI Bt Begun-3, and BARI Bt Begun-4, which are Bt isolines of Uttara, Nayantara, Kazla, and ISD006, respectively, are really the Bt types that have been issued (Shelton et al. 2018). Bt brinjal-1, Bt brinjal-2, Bt brinjal-4, and Bt brinjal-4 are the names given to them in the research. Farm owners can preserve seed for future use because these four Bt variants are open-pollinated. Producers are advised from utilizing conserved seed for several years due to the risk of cross-pollination with other types, particularly non-Bt brinjal utilized in boundary lines as a sanctuary in a resilience control plan (Shelton et al. 2019). Following permission, the authorities distributed Bt brinjal seeds to 20 producers in four provinces for production in 2014, authorizing BARI employees to provide farmers with sustainable farming coaching, counselling, and oversight/supervision. The introduction of Bt brinjal had proven to be fast since 2014. Bt brinjal has been studied extensively as BARI researchers performed research in 35 provinces involving 505 Bt brinjal producers and 350 non-Bt brinjal farmers throughout the 2016/2017 cultivation period, finding that gross returns/ha from Bt brinjal reached US$2151/ha in contrast to US$357/ha for non-Bt brinjal, a sixfold improvement (Rashid et al. 2018). Producers spent 61% lesser on insecticides than non-Bt brinjal growers, and there were negligible yield reductions attributable to the EFSB, according to this report. Prodhan et al. (2018) found that all four Bt brinjal cultivars offered nearly full EFSB suppression without using pesticides and also had greater yields over their non-Bt counterparts in 2-year research. Ahmed et al. (2019) have carried out research to evaluate the implications of Bt brassica technologies on agricultural processes, commercial viability, and public health. Throughout the cold weather of 2017–2018, the researchers examined the findings of 600 Bt brinjal producers with 600 non-Bt brassica producers in 200 communities across four provinces in Bangladesh's northwestern part. Bt brinjal producers saw decreased insecticide utilization, smaller average cost of production, better productivity, and greater profitability as outcomes of the study's findings. Nevertheless, just one of the four commercially available Bt types was included in the research (Shelton et al. 2020). A modified Bt brinjal event "EE-1" has been approved recently by the Department of Agriculture-Bureau of Plant Industry (DA-BPI) in the Philippines after extensive biosafety assessments. The event was approved for immediate utilization as food and feed as well as for processing. It is locally referred to as "Bt talong" with a biosafety permit number 21-078FFP, and it was found to be as safe as the traditionally cultivated eggplant and thus can serve as a substitute to its conventional counterpart (ISAAA 2021).

4.12.1 Development of Bt Brinjal

Bt brinjal is India's recent genetically modified edible produce. It was created by expressing *Cry1Ac* protein from *Bacillus thuringiensis* via the *Agrobacterium tumefaciens*-mediated approach by Maharashtra Hybrid Seed Firm Ltd. (Mahyco), which is a renowned Indian grain corporation (ABSP-II 2014). The appearance of the *cry1Ac* genes in the brassica cultivar offers an appropriate constructed control against FSB, reducing pest-related damages and protecting the ecosystem from pesticide's negative impacts. It is also projected to decrease the price of brinjal production, because the cost of synthetic insecticides for brinjal production is high. The *cry1Ac* protein synthesized in Bt cultivars is structurally and functionally identical to *cry1Ac* proteins discovered in naturally and commercialized B.t.k. microbiological preparations. *Bacillus thuringiensis* and *Bacillus* t.k. microbiological compositions were discovered for being incredibly precise to targeted pest species, with no negative consequences on pollinators, bird, fishes, or primates, particularly humans. Unlike insecticides administered topically, the protein released by Bt brinjal seedlings is hardly wiped away or damaged by sunshine since the crops have an innate system of crop protection against target organisms. The plant is therefore shielded against the FSB 24 h a day, 7 days a week, for the rest of its existence (Ahmed et al. 2019).

4.12.2 Benefits of Bt Brinjal

Utilizing information from massive field experiments undertaken by the Indian Institute of Vegetable Research (IIVR), a prestigious scientific investigation program under the Indian Council of Agricultural Research (ICAR), the success of Bt varieties versus non-Bt and widely known brinjal varieties in terms of output and reduced pesticide adoption had been investigated. During 2007–2008 and 2008–2009, studies have been undertaken at eight localities to evaluate commercial fruit production for seven brinjal hybrid cultivars carrying the Bt gene, non-Bt hybrids, and also most common hybrids. The statistics on pesticide application decrease comes through the All-India Vegetable Improvement Project (AICVIP) experiments carried out in 2004/2005 and 2005/2006.

The introduction of Bt technology has provided a considerable decrease in the adoption of synthetic chemicals. Ultimately, 77.2% lesser pesticides had been applied for the control of fruit and shoot borer (FSB), resulting in a 41.8% reduction in total pesticide usage in brinjal. Similarly, results from field experiments demonstrated that Bt brassica hybrids routinely outperformed non-Bt hybrids in terms of productivity. Bt hybrids increased output by 37.3% beyond non-Bt hybrids and by 54.9% over common cultivars. This big discrepancy in output suggested that using the Bt gene in brinjal to prevent FSB damage was far more efficient than using pesticides. As a result, yield reduction in Bt brassica cultivars was substantially reduced. Generally, Bt hybrids had a very reduced prevalence of stem destruction, with 0.24 percentage points to 4.64% in checks and 4.86% in non-Bt varieties.

Growers have earned profit on several stages: they have been able to reduce the frequency of pesticide application, which has resulted in a reduction in the expenses for buying the pesticides and also the cost of labor. Minimal FSB impairment resulted in increased productivity and greater revenue per unit region, resulting in a significant boost in productivity. Brinjal yield of 30,000 t can be added to overall productivity from the current area under brinjal based on the estimated acceptance levels of Bt varieties. Owing to the decrease in pesticide application to manage FSB, the Bt approach could also result in considerable discounts (47 crores to 187 crores) and, as a result, a large rise in profitability. At a 15% acceptance rate, the yearly incremental expected revenues might be 11,029/ha, and at a 60% deployment level, they might be 44,117/ha.

Reduced immediate pesticide contact would result in fewer medical complications. According to certain research, this could certainly provide farmers with significant ecological and medical advantages (Krishna and Qaim 2007, 2008). According to the study, adopting Bt hybrids might well help customers by lowering the market value of brinjal by 3–15%. Additionally, increased supply of brinjal (30–119 thousand tonnes) could lead to increase in utilization, improving the food security and nourishment of resource-poor customers while also improving the nation's ecological sustainability. The possible financial benefits to suppliers and buyers from Bt brinjal uptake have also been calculated in terms of monetary excess. According to the estimates, the deployment of Bt brinjal might boost customer profit by 381 crores and supplier profit by 196 crores. Total revenue would rise by 1576 crores, while producer gains would rise by 811 crores, if Bt brinjal coverage was extended to 60% of the present area under brinjal cultivation. The economic gains have been divided in a 66:34 proportion among customer and supplier.

4.13 Safety Issues Related to Bt Crops

The majority of the insecticidal molecules released into the soil by Bt transgenic crops are quickly destroyed, so they exhibit little influence on field microorganisms, streptomyces, molds, parasites, cyanobacteria, worms, or earthworms. The colonies of helpful insect pests were discovered to be unaffected by Bt corn or cotton. Furthermore, nontarget species in Bt agricultural regions were unaffected by Bt crop residues or spores (Mendelshon et al. 2003). The pollen grains in Bt corn showed bad impacts on caterpillars of the royal butterfly, *Citheronia regalis*, according to lab research conducted at Cornell University (USA) in 1999 (Losey et al. 1999). The findings of six laboratory and field investigations demonstrated that the concentration of Bt toxic substance in Bt maize pollen grains is not adequate to injure insect caterpillars, according to the Publications of the National Academies of Sciences (Sears et al. 2001). It should be observed that the monarch butterflies possess a lovely color and an airfoil diameter of around 15 cm, and it is a popular household ornament mostly in the USA. According to Lu (2010), yearly Bt cotton production led to a significant invasion rate of draining mirid bugs in China, which is now the primary enemy of Bt cotton. Furthermore, in India, the persistent production of Bt cotton resulted in a visible aphid and moth outbreak (Losey et al. 1999). Relative to *Aphis gossypii*-administered non-Bt cotton, *Aphis gossypii*-consumed Bt cotton exhibited lower proliferative period, ultimate longevity, and an early climax of routine death rates in the first and second phases, according to lab experiments done by Liu et al. (2005). Furthermore, Lu et al. (2012) found a significant drop in aphid densities in Bt cotton plantations in 36 regions across six provinces in northern China within 20 years (1990–2010). Pollen carrying Cry poisons was not really poisonous to lady beetle (Coccinellids), green lacewings (*Chrysoperla* spp.), or bees in experimental investigations, according to Mendelshon et al. (2003). Useful species of insects have been proven to be extremely significantly more prevalent in transgenic plants than it was in synthetic pesticide-treated plants, according to field research. According to Lu et al. (2012), aphid numbers in Bt cornfields in 36 places across six districts in northern China have decreased dramatically.

4.13.1 Disadvantages of Bt Crops

Despite the fact that the industrialization of Bt-modified plants is widely regarded as one of the biggest, significant milestones in the management of farm insect pests ever, a lot of prospective dangers related to the technology's ecological sustainability and prospective applicability were explored. Lethality to nontarget species, transgene leakage into the ecosystem, as well as development of resistance among the target insect communities have been among the main significant hazards discovered. Despite the fact that these challenges were under investigation, measures to mitigate these hazards have been presented and applied in certain instances.

4.13.1.1 Allergenicity of Bt Gene/Unknown Effects on Human Health

Food allergies develop whenever the immune systems perceive anything as strange, unusual, or unpleasant. By definition, all transgenic products contain something alien and unusual. They also seem to elicit reactions, according to various studies. Rats' immune systems responded relatively sluggish after eating transgenic potato (Tudisco et al. 2006). In addition, transgenic soybeans elicited a chronic inflammation in rodents, implying that they could trigger fatal allergic responses in humans. Modified beans entail a distinct, unanticipated protein in addition to the herbicide-tolerant protein, which is presumably the result of alterations made through the genomic modification process. Researchers showed that such a novel protein could link to IgE antibodies, implying that it could cause serious allergies. As a means of pest management, local farmers and others have sprinkled plants with formulations comprising native Bt bacterium. The poison destroys them by causing perforations in their stomachs. Biotechnologies introduce the toxin-producing gene into the genome of plants, allowing the plant to do the task rather than the farmers. The reality that each piece of Bt corn contains that lethal insecticide is scarcely appealing. Natural Bt toxin, on the other hand, is not completely eliminated after ingestion and reacts with primates, according to researches. Aventis Industries in the USA produces StarLink, a Bt corn cultivar for utilization in livestock feed. Because the Bt toxin employed in StarLink is less swiftly absorbed compared to other Bt poisons, there had been a requirement that the product not be utilized as food for humans. Twenty-eight persons developed allergic responses after consuming corn products contaminated with the StarLink allergen. The US Centers for Disease Control and Prevention, on the other hand, examined the plasma of these individuals and found no evidence that the adverse allergic response was linked to the StarLink toxic chemical (Seralini et al. 2007). The microbial (Bt toxins) employed in natural and conventional agriculture and forest products are significantly dissimilar from the Bt toxins produced in GMOs. The model created by plants is intended to be more poisonous than normal variants (Romeis et al. 2004). The Bt protein in improved maize cultivars, like the biotech soybean products, possesses a segment of its peptide sequence that is similar to a renowned allergen (egg yolk), making it too impervious to dissolve during metabolism or due to heat. If Bt poison induces food intolerances, backcrossing could have severe repercussions. If Bt genes are transferred to human gut microbes, our intestinal bacteria could be transformed into life weedkiller production lines, generating Bt toxin on a yearly basis (Verma et al. 2011). Several American children and Europe had experienced life-threatening peanuts as well as other food intolerances. It is possible that inserting a DNA into a plant will lead to formation of innovative allergens or an adverse reaction in vulnerable people. Because of concerns about generating unanticipated severe allergies, a project to introduce a gene from Brazil peanuts into soybeans was postponed (Nordlee et al. 1996). The implications of transgenic potatoes on the gastrointestinal tract in rats were studied in a Lancet article (Hartmann et al. 1999; Mitchell 1999). Furthermore, the trait that was transferred into the potato was such a snowdrop floral glycoprotein, which is poisonous to animals (Verma et al. 2011). The Flavr Savr tomato, the initial plant presented to the FDA's (Food and Drug Administration) voluntary consultation process, tested positive for poisons. Seven of the 20 female rats who were administered the transgenic tomato had gastrointestinal sores. Tomato-related stomach ulcers can cause life-jeopardizing bleeding, particularly in the aged ones that take aspirin to do away with blood clotting (Pusztai 2002). The gastrointestinal tract, which is the first and biggest line of interaction with food stuffs, could indeed expose various responses to toxic materials, according to Dr. Pusztai, and should be the first destination of genetically modified food hazard analysis. Rodents given Bt toxin-producing potatoes evolved anomalous and compromised cells, and also tumorigenic cell expansion, in the bottom region of their gastrointestinal tract (Fares and El-Sayed 1998). Rats given potatoes that produced a diverse sort of pesticide (GNA lectin from the snowdrop plant) had increased cell proliferation in their stomachs and wall of the intestine. A modified corn developed by Monsanto (Mon 863), which was targeted to generate Bt toxin, caused inflammation of liver along with other signs of lethality in rats. Rabbits given engineered soybean exhibited improved metabolism rate and changed enzymatic synthesis in their liver tissues (John 2002). The livers of rats given Roundup Ready soybeans exhibited conformational changes as well (Irina 2007).

4.13.1.2 Horizontal Gene Transfer

Lots of concerns were put forward regarding the potentiality of Bt genes escaping through biotech crops into the ecosystem. Owing to excessive selective pressures, vast production of genetically modified insect-tolerant plants could raise the likelihood of acquired resistance in the targeted insect pest as well as weeds, respectively. The increased evolutionary pressures could result in the growth of a new group of insect germplasms, resistant to genetic modification technique and development of superweeds (Bawa and Anilakumar 2013; Gilbert 2013). Field-evolved insect resilience to GM Bt corn had also been disclosed for three key insect species, including *Busseola fusca* in South Africa to *cry1Ab*-displaying maize (Rensburg 2007), *Spodoptera frugiperda* in Puerto Rico to *cry1F*-overexpressing corn, and *Diabrotica virgifera* in the USA to cry3Bb-expressing corn (Rensburg 2007; Gassmann et al. 2011). Furthermore, a review of 77 investigations found that 5 of the 13 lepidopteran pests had gained tolerance to transgenic plants as a result of field evolution (Tabashnik et al. 2013). In Puerto Rico, fall armyworm resilience to Bt maize was discovered in the least amount of time (3 years), prompting the crop's reversal. Despite these rare occurrences, research using transgenic plants has shown that crop tolerance could be effective against so many insects even beyond 10 years (Tabashnik et al. 2008; Carrière et al. 2010). Furthermore, combination of numerous insect-resistant alleles had been shown to be a viable technique for delaying tolerance degradation. Introgression into wild corn relatives was proposed in certain papers (Quist and Chapela 2001), but this finding was swiftly debunked due to poisoning and analytical difficulties (Christou 2002). To inhibit transgene leakage from transformants, two primary types of techniques have been taken into account: biological and nonbiological. Mechanical control, such as removing pollination blossoms or isolating biotech plants from non-transgenic cultivars, becomes a nonbiological strategy (Rong et al. 2007; Kausch et al. 2010). Absolute transgenic removal from grain and pollen (Luo et al. 2007), seeds (Daniell 2002), as well as male infertility are among the biological ways for bioengineering gene confinement that have been explored (Castagnola and Jurat-Fuentes 2012).

4.13.1.3 Toxicity to Nontarget Insect Pests

One of several obligatory elements of GMO product certification is the impact of transgenic plants on nontarget animals, while there are other investigations assessing the toxicity of Bt crops to nontargets in field trials. Stark (1997) used NewLeaf potatoes as a model experiment and found no negative impacts on nontarget, helpful, or aggressive insect communities, owing to reduced insecticide administrations. Transgenic potatoes yielding Cry3Aa toxin, which is effective at killing highly associated *L. decemlineata* larvae, had no effect on lady beetles (Coccinellidae) and their essential feed intake behavior on aphids. While no pesticides were employed, meta-analysis research suggests that Bt crops sustain fewer pollinators than traditional crops (Marvier et al. 2007), which could be related to a fall in prey populations in Bt flowers. In opposition, if pesticidal treatments were employed, these studies indicated that Bt crops sustained larger numbers of beneficial organisms than non-Bt crops, owing to the greater wide usage required for non-Bt crops. Bt crops have a broadly detrimental effect on predaceous and parasitoid insect pests, according to a compendium of lab research findings on the effects of Bt crops on 48 species of advantageous exploitative pests (Lovei et al. 2009), but it is ambiguous whether it is mainly owing to the existence of the plasmid DNA or a decline in insect prey populaces. As a nontarget experimental species, green lacewings (*Chrysoperla carnea*) were widely employed. The life expectancy, preoviposition duration, proliferation, productivity, and dry weight of mature lacewings supplied with Bt corn pollen displaying Cry1Ab or Cry3Bb1 showed no substantial effect (Li et al. 2008). Lacewing caterpillars were not adversely influenced by Bt toxin (Rodrigo-Simon et al. 2006), even though there have been reports of negative consequences stemming from poor prey composition or accessibility in Bt crop areas (Romeis et al. 2004). Aphids (some other nontarget insect model) do not collect Bt toxin after feeding on Bt plants, which has been proven to protect predators from becoming unintentionally exposed (Lawo et al. 2009). The likelihood for such Bt toxins to seep into neighboring water bodies, as well as the potential effects on aquatic animals, has indeed sparked the debate (Waltz 2009). Although winds and soil overflow could cause Bt poisons to leak into groundwater near rivers (Tank et al. 2010; Viktorov 2011), their pesticidal capabilities on species found in the water body have yet to be proven. Bt crops have been proved to have almost no negative impact on symbiotic mycorrhizae in agricultural soil (Liu 2010), while relatively brief activity adjustments in microbial populations have been revealed due to Bt maize expressing the Cry1Ab toxin (Mulder et al. 2006). In another study, no depletion of soil bacterial community or interaction was observed in a 2-year field experiment using identical Bt corn variants (Oliveira et al. 2008). Generally, the potential advantages of decreased synthetic pesticide application are thought to counterbalance the minor negative effects on useful insect communities (Gatehouse et al. 2011).

4.13.1.4 Bt Resistance in Insects

Tolerance to toxic substances generated by Bt had been discovered in a variety of lab-chosen insect lines, demonstrating the genomic possibility for Bt toxin resilience in the fields. Tolerance in the bulk of lab instances is linked to changes in toxin interaction to intestinal lumen receptors, which is usually passed down as a unique inherited hidden allele (Ferre and Van Rie 2002). The first higher dosage method for delaying the emergence of tolerance to single-Bt crops was coupled with the utilization of mandatory 20% refugia with the production of large Bt cytotoxin doses (25-fold the dose killing 99% of the pathogen populace) within the crop. When such modified crops fail to convene the high-dose threshold for some insects, manufacturer-recommended refuge quantities might be increased up to 50% of the crop. Refugia are frequently located near Bt farmlands, and this closeness considerably improves the chances of resilient insects originating from GM crops breeding with nonselected insects living in the refuge. If tolerance is passed on as a recessive characteristic, pairing events result in heterozygotes, which, if combined with a lethal poison concentration, lowers the fraction of vigorous/robust species (Gould 1998). Persistent Bt toxin generation in the crop throughout the growth period, as well as grower compliance with refuge seeding standards, is required for this hefty dose approach to be successful. Both concepts are challenging to implement, as Bt toxin gene expression fluctuates based on a variety of conditions (Adamczyk et al. 2009), and refuge conformity fluctuates between farmers and seasons (Gray 2010). Furthermore, pest dispersal within a crop is a critical prerequisite for the efficiency of refugia. Modeling investigations imply that as producer adherence to refugia falls, resilience establishment massively enhances in insects with reduced mobility, for instance, larvae of rootworm in western corn (Pan et al. 2011). Previous observations reveal the evolution of modified Bt cotton tolerance in an insect species (*Helicoverpa armigera*) community in a northern Chinese site with a storied record of transgenic cotton deployment (Zhang et al. 2011a, b, c). Concerns in ensuring conformity and a deficiency of supervision over unauthorized modified cultivars have been identified as major challenges in India's Bt crop impedance mitigation efforts (Jayaraman 2002). Despite these challenges, field-evolved Bt crop tolerance is a rare occurrence, given the quantity of uptake after more than a decade of use. Field-evolved Bt crop resilience (Storer et al. 2010; Dhurua and Gujar 2011; Zhang et al. 2011a, b, c) is frequently linked to semi-optimal plant growth settings, and also majority of such were not demonstrated to lead to yield reductions of crop plants.

4.14 Transgenics for Disease Resistance

Plant disease and insect infestations have posed a significant threat to the plant growers from the beginning of agricultural production. Despite all attempts to tackle phytopathogens, even after splitting the atom, traveling to the lunar space, and connecting the universe, plant diseases persist in posing a serious threat to food crop safety. The world average damage due to pathogens and insects is estimated to be 11–30% (Savary et al. 2019). Yield losses are disproportionately high in areas where poor nutrition previously exists (Savary et al. 2019). Disease damages might be substantially severe if farming measures like cultural management, pesticide application, and plant hybridization had not progressed steadily in the past. To address diseases in plants, a coordinated strategy is required, integrating the latest accessible technology and techniques. The control of stem rust in wheat, a condition that produced recurrent expensive outbreaks in the USA around 1918 and 1960, demonstrates the advantages of an incorporated strategy (Pardey et al. 2013). Only a combination of cultural practices (separation of barberry, the pathogen's sexual host), enhanced pesticide influence, as well as a comprehensive breeding program organized by Norman Borlaug have empowered such a disorder of wheat to be contained. Modern farming must supply enough resources to support the world's high rate of growth in terms of population that is anticipated to expand from 7.3 billion in 2015 to approximately 9.8 billion people by 2050. Crop destruction due to diseases rendered this target much more difficult. Viral pathogens lower crop yields by 3%, while fungal and bacterial diseases diminish production by 15% (Oerke and Dehne 2004). Infestation by pathogens is thought to be responsible for up to 30% of crop losses in certain plants, including potatoes (Oerke and Dehne 2004). Scientists are modifying the genetic makeup of crops to increase tolerance to microbial pathogens as an option to using chemical treatments. Traditional breeding is important for crop development, but it requires producing and evaluating massive populations of plants spanning several years, a process which requires longer time and effort. When contrasted to traditional breeding, genetic modification, which pertains to the deliberate manipulation of an individual's genetic makeup via technologies (Christou 2013), has various benefits. For instance, it allows for the insertion, elimination, or alteration of individual relevant genes with little unfavorable effects on the remainder of the crop chromosome. As a result, compared to traditional breeding, plants with specified agricultural qualities can be acquired in few years. Secondly, genetic modification allowed for cross-species genomic exchange. As a result, the fundamental genetic information that might be used in this procedure is not limited to the alleles found among the species. Thirdly, the technique also permits the transfer of foreign genes into crops that can be propagated in a vegetative way like cassava and potatoes. Such characteristics made genetic engineering a potent technique for improving plant disease tolerance. Traditional transgenic techniques or more contemporary genome editing techniques are used in the majority of crop genetic engineering applications. Genes encoding desirable agricultural qualities are injected into the DNA at randomized places by genetic modification in traditional transgenic procedures (Lorence and Verpoorte 2004). Normally, these approaches produce cultivars with alien DNA. Genome editing, on the other hand, enables for alterations to native crop genome, including removals, insertions, and substitutions of DNA of varying sizes at specific sites (Barrangou and Doudna 2016). The outcome may or may not incorporate alien DNA, depending on the sort of alterations used. Regarding microbial diseases, crops have developed complex defense systems (Chisholm et al. 2006). Prefabricated mechanical and biological boundaries, as well as their reinforcements, keep prospective infections out of the cell (Uma et al. 2011). Microbes are perceived by plasma membrane-bound and cytoplasmic immunological sensors, which trigger systemic resistance either immediately by mechanically engaging with microbe immunogens or implicitly by tracking pathogen-induced changes to host targets (Kourelis and van der Hoorn 2018). Antibiotic proteins and some other chemicals generated from plants can reduce toxicity either directly or by inhibiting the function of virulent determinants (Ahuja et al. 2012). RNA interference (RNAi) is also used by plants to identify invasive viral infections as well as intended RNA of the virus for breakage (Rosa et al. 2018). Pathogens that are robust have developed techniques to defeat their crop hosts' defenses. Several fungal and bacterial diseases, for example, generate and secrete proteases that degrade cell walls (Kubicek et al. 2014). Mostly all plant viruses develop viral RNAi silencers to overcome plant

RNAi-based defenses (Zamore 2004). Numerous viruses employ the host's RNAi machinery to suppress host genomes in order to increase viral lethality (Wang et al. 2012). Greater understanding of the molecular mechanisms underpinning plant-pathogen relationships, as well as advances in biotechnology, has opened up new avenues for designing plant tolerance to microbial pathogens. Using genetic engineering and genome editing to broaden the genetic resources accessible to farmers is among the most successful and long-term approaches to regulate plant diseases (Xiaoou and Ronald 2019).

4.14.1 Transgenic Crops Developed for Disease Resistance

Transgenic technology has been deployed in the development of a number of crop plants that are tolerant to diverse pathogenic organisms that destroy them by causing different abnormalities that lead to reduction in yield and quality of the farm produce. When both the R gene in the plant host and the cognate avr in the pathogen are present, the plant-pathogen interaction is incompatible and the host exhibits full resistance to the pathogen (Flor 1971). The effectiveness of gene-mediated resistance was first demonstrated by British scientist Rowland Biffen in wheat (*Triticum* sp.) breeding in the early twentieth century (Biffen 1905). Since then, other R alleles have been transferred into separate species of plants, between taxa, and also between distinct genera/families as reported by De Wit et al. (1985), Song et al. (1995), and Tai et al. (1999). According to the decade-long field testing performed under industrial marketing conditions, pepper Bs2 R-encoding vegetables protect *Xanthomonas* sp., a causative agent of microbial (bacterial) spot disorder/syndrome as demonstrated by Horvath et al. (2015) and Kunwar et al. (2018).

Productive pathogenic species frequently elude recognition by the R genomes of their hosts (Jones and Dangl 2006). Because pathogens could mutate to elude identification by changing the associated avr gene, resistance to diseases acquired by a solitary R gene frequently lacks longevity in the outdoors setting. Numerous R alleles are frequently inserted at the same time to boost robustness and widen the resilience range, a process termed as stacking (Mundt 2018). The emergence of a strain of pathogenic species that can overcome tolerance given by several R genes at the same time is a rare event; therefore, resistance acquired by stacked R genes is expected to endure a lengthy time. Crossbreeding preexisting R loci is one way to stack R genes. The offspring with the required R gene component can subsequently be identified via marker-assisted screening (Das and Rao 2015). Three R genes, Xa21, Xa5, and Xa13, that impart protective immunity to blight in rice, got transferred to rice cultivar by interbreeding and marker-assisted selection (Pradhan et al. 2015). For eight Xoo strains examined, the resultant variant having tightly packed R genes showed a greater potential of field tolerance (Pradhan et al. 2015). Despite the fact that marker-assisted selection had greatly enhanced the effectiveness of the selection process, merging numerous loci using this method might still take a long time. As an option to gene clustering using marker-aided choice, researchers could construct many transgene units on a single vector and then introduce the entire R gene assembly into a specific genomic region using crop modification (Dafny-Yelin and Tzfira 2007). Such a technique, termed as stacking, renders filtering simpler since all of the gene products are transferred as a single chromosomal cluster. To promote molecular stacking of the three diverse potatoes, late blight R proteins including Rpisto1, Rpi-vnt1.1, and Rpi-blb3 utilized *Agrobacterium*-based alteration of a vulnerable potato variety (Zhu et al. 2012). Under warm conditions, the triple-genome transgenic plants showed wider endurance equivalent to the combined amount from each of the three Rpi genes' strain-specific susceptibility (Zhu et al. 2012). Agrobacterium-mediated transformation of three potato hybrids introduced a unique DNA strand including Rpi-vnt1.1 and Rpi-sto1 (Jo et al. 2014). With the insertion of both R genes, the R biomarker alleles demonstrated broad late blight resistance (Jo et al. 2014). Tolerance to fungal disease in potato was proven in the doubled gene-stacked and three gene-stacked potatoes (Haverkort et al. 2016). The right regional and chronological placement of such R gene stacks in potatoes had minimized fungicide usage by over 80% (Haverkort et al. 2016). Ghislain et al. (2018) revealed that molecular mounting of three R genes led to considerable field tolerance to the late blight infection. The R gene stacking potato varieties double the average income. These findings imply that this technology's resistivity has no effect on yield (Ghislain et al. 2018). The aforementioned discovery shows that molecular stacking might be used to build expansive pathogen immunity in vegetatively generated crops where breeding layers/stacks are deemed impracticable. Tailored gene introduction may allow for the piling of massive quantities of R genes with specified virus sensitivities at a specific locus, allowing for long infection tolerance levels while easing replication.

Transgenic crops expressing genetic traits deduced from viral pathogenic organisms have long been discerned to possess resistance to the microbe and associated species (Lomonossoff 1995). These findings led researchers to believe that appearance of genetic makeup coding wild-type or genetically mutated virus particles might disrupt the life span of the virus (Sanford and Johnston 1985). New research has shown that this protection is influenced by RNAi, which plays a key function in plant antimicrobial protection. Considering the fact that viral pathogens rely on physiological function of the host cell to complete their life span, activating RNAi has been demonstrated to be an efficient strategy for engineering virus resistance against such infections (Lindbo and Dougherty 2005; Lindbo and Falk 2017). The genetic material of most plant viruses is ssRNA. Double-stranded RNA (dsRNA) replicative intermediates typically develop upon viral genome duplication controlled by enzyme polymerase, eliciting RNAi inside the host species. Genetically modified viral RNA overexpression frequently results in the creation of dsRNA, which activates RNA interference as described (Galvez et al. 2014). Co-suppression is a term used to describe this procedure. This approach was used to create the vast majority of established transgenic crops with virus-resistant traits that were accepted for production and utilization. Genetically modified squash and papaya cultivars developed with this method were industrially produced in the USA for over two decades.

The knowledge that dsRNA effectively induces RNAi inspired the design of transgenes encoding inverted repeat sequences, the transcripts of which form dsRNA (Waterhouse and Helliwell 2003). This strategy was used to develop a transgenic common bean variety exhibiting resistance to the DNA virus bean golden mosaic virus (BGMV) (Bonfim et al. 2007). The genetically engineered bean exhibited strong and robust resistance in greenhouse conditions (Bonfim et al. 2007) as well as field conditions (Aragao and Faria 2009). The discovery of microRNAs (miRNAs), a class of endogenous noncoding regulatory RNAs (Reinhart et al. 2002), led to further refinements of genetic engineering for viral resistance. The miRNA machinery (Xie et al. 2015) has been exploited in engineering resistance against RNA viruses by replacing specific nucleotides in the miRNA-encoding genes to alter targeting specificity. Such artificial miRNAs (amiRNAs) have been used in many lab studies in engineering resistance to a wide range of plant viral pathogens (Niu et al. 2006; Qu et al. 2007; Zhang et al. 2011a, b, c). Antiviral approaches based on artificial miRNA appear to be quite promising, according to these studies. The pathogen resistance of this approach will be field-tested in the future. Table 4.3 shows the reports on crop plants that have been genetically altered to resist infection by a variety of microbial diseases for increased agricultural productivity.

4.15 Pathogenesis-Related Proteins (PR Proteins)

Plants being sessile are continually attacked by harmful microbes like pathogenic fungi, bacteria, nematodes, as well as viral species that threaten plant existence and viability (Cramer et al. 2011). These diseases cause considerable yearly agricultural output reductions and constitute a substantial danger to future food availability. Plants protect themselves against these predators by employing a variety of protection methods so as to remain living and retain their fitness as reported by Roux et al. (2014). Phyto-immunotherapy uses microbiological genomic pattern-triggered immunology (PTI) and effector-triggered immunology (ETI). PAMPs are microbiological or pathogenic elements that are recognized by patterns recognizing receptors (PRRs), culminating to PTI (Zipfel and Felix 2005). To counteract this, bacterial pathogens produce signaling molecules that are recognized by a subgroup of resistivity (R) proteins. Such activator peptides are critical for the aggressiveness of a fungi to crops, notably throughout the phytopathogenic period (Sonah et al. 2016). Nonetheless, PR peptides were recognized as vital in plant-fungal pathogen interactions, and an increasing number of harmful activator peptides that effectively communicate with such proteins upon infestation have been found (Breen et al. 2017). The plant immune system's complexity and efficacy in fending off pathogen incursion vary from species to species (Jones and Dangl 2006).

PRs are a type of polypeptide that builds up in plants in reaction to pathogenic attack to defend them from harm (Li et al. 2021). They encompass a broad range of phytopathogen-induced compounds along with defense-associated signaling proteins. They constitute important components of the plant innate immunity,

Table 4.3 Reports on transgenic crops developed for disease resistance

Crop name	Targeted microbes	Gene expressed	References
Papaya	Ringspot virus	Coat protein	Gonsalves (2006)
Papaya	Ringspot disease of pawpaw	Replicase proteins	Ye and Li (2010)
Bean	Viruses of the golden pea	RNA (+/−) sequences encoding viral replicating enzymes	Tollefson (2011)
Potato	Potato leafroll virus	Helicase and replicase	Thomas et al. (1997)
Potato	A Y virus that infects potatoes	Proteins with a protective coating	Newell et al. (1991)
Squash	Mosaic viruses infecting cucumbers and melon	Proteins with a protective coating	Tricoli et al. (1995)
Sweet pepper	Virus that causes cucumber malformation/discoloration	Proteins with a protective coating	Zhu et al. (1996)
Tomato	Cucumber mosaics/ discoloration virus	Proteins with a protective coating	Yang et al. (1995)
Plum	The virus that causes the plum pox	Proteins with a protective coating	Illardi and Tavazza (2015)
Potato	*Phytophthora infestans*	Protcin that confers tolerance	Foster et al. (2009)
Papaya	Papaya ringspot	Proteins with protective coating	Davis and Ying (2004)
Squash	Mosaic virus 2 of melon as well as yellow mosaic virus of zucchini	Proteins with protective coating	Tricoli et al. 1995
Tobacco	Cucumber mosaics/ discoloration virus	Replicase protein	Singh et al. (1998)
Apple	*E. amylovora*	Attacin E	Norelli et al. (1994)
Rice	*Xanthomonas oryzae*	Chimeric protein	Sharma et al. (2000)
Pear	*E. amylovora*	Lactoferrin	Malnoy et al. (2003)
Tobacco	*Pseudomonas syringae* pv. *tabaci*	Lysozyme	Nakajima et al. (1997)
Tomato	*Ralstonia solanacearum and Xanthomonas campestris pv. vesicatoria*	Cecropin B	Jan et al. (2010)
Potato	*Erwinia carotovora*	Msr A1	Osusky et al. (2000)
Wheat	*Pseudomonas syringae* pv. *oryzae*	AtEFR	Schoonbeek et al. (2015)
Wheat	*Blumeria graminis* f. sp. *tritici*	TaRLK1, TaRLK2	Cheng et al. (2021)
Barley, maize, dream wheat	Multiple biotrophic pathogens	TaLr34	Risk et al. (2013), Krattinger et al. (2016)

(continued)

Table 4.3 (continued)

Crop name	Targeted microbes	Gene expressed	References
Barley	*Fusarium graminearum*	FgCYP51A, FgCYP51B and FgCYP51C	Koch et al. (2013)
Rice	*Xanthomonas oryzae* pv. *oryzae*	OsSWEET13	Zhou et al. (2015)
Wheat	Wheat dwarf virus	Virus gene of dwarf wheat	Kis et al. (2015)
Wheat	*Fusarium graminearum*	TaHRC	Su et al. (2019a)

especially in the case of developed resistivity that has spread throughout the body, and are frequently deployed as diagnostics genomic indicators of defensive systems of signals. Despite the advent of improved scientific methods and the isolation of PR proteins and peptides, their physiological function remains mainly unclear. Prior study has demonstrated that PR genes provide increased resistance to pathogen invasion, making them among the most viable choices for generating multiple stress-tolerant crop types. Plant genetic modification is widely regarded as among the most exciting approaches of developing disease-resistant GM crops employing various antibacterial genes such as PR genes in this regard (Ali et al. 2018). Upregulation of PR peptides alone or in combinations significantly increased the amount of immune reaction in crops against a host of health conditions. Nonetheless, a thorough understanding of the communication pathways that drive the synthesis of such adaptable enzymes is essential for agricultural plant tolerance to a wide range of challenges, which will be the focus of future plant stress genetic investigations (Ali et al. 2018).

4.15.1 Discovery and Categorization of PR Proteins

PR proteins were first reported in tobacco plants affected with the tobacco mosaic virus (TMV) (Bol et al. 1990). On the basis of molecular and biochemical techniques, five main categories of PR proteins, PR1, PR2, PR3, PR4, and PR5, have been reported from tobacco. Yet, several additional PR proteins were extracted and discovered in diverse plants in subsequent investigations. PR proteins have been divided into distinct groups using an appropriate naming approach on the basis of physiological, genomic, biochemical, immunological, and other functional parameters. Subsequently, PR proteins in tobacco and tomato crops have been categorized into 11 groups, which served as a framework for extracting PR homologous proteins in other species of plants, particularly monocotyledonous and dicotyledonous species (Van Baarlen et al. 2007). For a freshly isolated protein to be added to the PR protein family, it should initially exhibit basal level expression in cells and yet substantially enhanced expression upon pathogen attack, and this overexpression must be affirmed in numerous plant pathological laboratories or take place in a comparable pattern during various plant-pathogen relationships. Depending on the protein sequence homology, catalytic activity, and other biochemical traits, PR proteins are currently categorized into 17 different groups (Sels et al. 2008).

4.15.2 PR Proteins and Pathogenic Resistance

Fungal species were among the most damaging phytopathogenic organisms, producing considerable yield reduction in a large number of agronomically vital crop varieties around the globe (Dean et al. 2012). Phytopathogenic fungi are classified as biotrophs, hemibiotrophs, or necrotrophs, depending on their lifestyle. Cutinases, pectinases, cellulases, and proteolytic enzymes are among the hydrolytic proteins produced by fungi to get ingress into the plant cell walls. Plants employ a variety of immunological responses to combat fungal diseases, including pathogen detection, stimulation of defensive signal networks, and synthesis of antimicrobial chemicals such as PR proteins, all of which help to limit pathogen penetration and multiplication (Sels et al. 2008). Additional methods of avoiding fungal infections include genetic modification of important defense components. PR peptides were among the such, and they are great candidates for developing lengthy, expansive fungal pathogen resistance in agricultural crops (Stuiver 2011; Ali et al. 2018). Several transcriptomic investigations have revealed that upon fungus infestations, PR proteins are upregulated in numerous crops, revealing their function in disease tolerance. Under normal settings, PR gene expression is low, but it rises rapidly upon fungal attack, in both infected area and noninfected area of the hosts, triggering the SAR system (Návarová et al. 2012; Ali et al. 2017). In addition, numerous in vitro investigations indicated that PR peptides hit or hydrolyze fungal cell walls, causing apoptosis. PR2, PR3, PR4, PR5, and PR12 are the most effective antimicrobial polypeptides in plants. Furthermore, overexpression of PR genes, either singly or in mixture, results in increased disease tolerance against different fungal attacks in various crops (Jiang et al. 2015; Dai et al. 2016).

One of the really intriguing relationships is that between plants and bacteria, which can be advantageous or destructive to the plants. Plant bacterial infections have been recognized for a decade, but fire blight of apples and pears has been the earliest proven bacterial infection in 1878 (Burrill 1878). Following that, a variety of bacterial pathogens have been discovered and characterized from a variety of major agricultural crops, resulting in enormous crop failure. Bacterial pathogens enter the hosts by a variety of mechanisms, including stomata, lenticles, external injury, insect grazing on foliage, and chemoattraction. Plants use a variety of innate immunities to combat bacterial infections. The detection of bacterial pathogens by host PRRs is a fundamental element of the first line of defense against them. This plant-bacterium struggle then triggers two important immunological reactions in the host, PTI and ETI. Surprisingly, the immune reaction generated by PAMP provides a quick and easy way to prevent disease from spreading to the host (Zeidler 2004; Ausubel 2005). The PR compounds are well-known substances that were used in the plants to produce microbial tolerance. Rice plants overexpressing the lipid transfer protein

(PR14) demonstrated improved tolerance to fungal and bacterial diseases (Patkar and Chattoo 2006). Antimicrobial effect of other PR polypeptides and AMPs against a broad array of bacterial diseases in commercially significant crops will require more research.

Viruses infect plants and damage the host tissue, impairing plant metabolism and inhibiting immune responses. Plants release antiviral chemicals or enzymes like RNA-binding proteins (RBPs), ribosome-inactivating proteins (RIPs), and PR proteins to react to viral attack (Park et al. 2004a, b; Musidlak et al. 2017). AMPs build in noninfected cells during viral invasion, preventing virus propagation. Antiviral proteins PR2a and PR3 exhibit anti-TMV action (Sindelarova and Sindelar 2005). Also, a new PR family member, PR9 (peroxidase), is antiviral (Nawrot et al. 2014). CaPR10, a protein from *Capsicum annuum*, inhibits TMV. CaPR10's antiviral efficacy against TMV was previously shown to increase when phosphorylated (Park et al. 2004a, b; Nawrot et al. 2014). Upregulation of the PR1b protein in tobacco plants has previously been demonstrated to result in increased TMV tolerance (Cutt et al. 2005). Apart from their antifungal and antibacterial capabilities, PR proteins and AMPs appear to be potential candidate genes for producing virus-tolerant mutant plants.

4.16 Antimicrobial Peptides and Disease Tolerance

Antibacterial proteins made up a substantial portion of AMPs, as they exert a wide suppressive activity on harmful bacteria like *Acinetobacter baumannii* in clinical practice, *S. aureus*, *Listeria monocytogenes*, *E. coli* in foods, and *Salmonella* and *Vibrio parahaemolyticus* in marine commodities. Numerous natural and synthesized AMPs, such as nisin, cecropins, and defensins, had exhibited excellent Gram-positive and Gram-negative bacterial inhibitory action. According to the current study, the AMPs P5 and P9, that are built on *Aristicluthys nobilia* interferon I, inhibit MRSA while causing minimal mortality (Su et al. 2019b). Because of their all-around effectiveness against many challenges including antifungal, antibacterial, and antiviral and their function in nonliving stress endurance, AMPs were garnering increased emphasis for enhancing pathogen resilience. For example, upon bacterial and fungal attacks, AMP gene concentrations in tomato seedlings rise considerably, showing that they play a function in immune function. PR6 proteins were expressed to have significant antibacterial action against a wide range of fungal infections in in vitro tests (Terras et al. 1993). Plant defensins (PR12) are the most significant antifungal polypeptides in antimicrobial compounds. Plant defensins have antifungal action against several fungal infections, according to in vitro investigations (Ali et al. 2018). Furthermore, in both experimental and agricultural plants, upregulation of plant defensin molecules has resulted in improved and lengthy disease tolerance (Kaur et al. 2017) (Table 4.4). PR13 and PR14 peptides are another key category of AMPs that play a function in microbial defense. Increased expression of PR13 proteins, for example, has been demonstrated to improve disease resilience in tomato and potato plants towards fungal infections (Muramoto et al. 2012). In vitro

Table 4.4 Reports on crops expressing PR proteins for enhancement of disease tolerance

PR category	Gene involved	Host organism	Function	References
PR1	Antifungal	Tobacco	Pathogenic tolerance	Sarowar et al. (2005)
PR1	Antifungal	Rice	Resistance to *Magnaporthe grisea race 003*	Mitsuhara et al. (2008)
PR1	Antifungal	Rice	Resistance to *Alternaria alternata*	Mitsuhara et al. (2008)
PR2	β-1,3-Glucanase	Wheat	*Fusarium graminearum* tolerance	Mackintosh et al. (2007)
PR2	β-1,3-Glucanase	Groundnut	*Cercospora arachidicola* and *Aspergillus flavus* resistance	Sundaresha et al. (2010)
PR2	Endo-β-1,3(4)-glucanase	*P. pastoris GS115*	Barley β-glucan and CMC-Na, birchwood xylan	Jinyang et al. (2017)
PR3	CHIT33, CHIT42	*T. harzianum*	Resilience to pathogenic and environmental pressure	Cruz et al. (1992)
PR3	Chitinase	*Daucus carota*	*Alternaria radicola*, *Botrytis cinerea*	Jayaraj and Punja (2007)
PR3	Chitinase	Tobacco	*Ralstonia solanacearum* resistance	Tang et al. (2017)
PR3	Chitinase	Rice	Tolerance to fungal attack	Li et al. (2009)
PR4	Chitinase II	Rice	Drought stress and pathogen response	Wang et al. (2011)
PR5	Thaumatin-like TaLr19TLP1	Wheat	*Puccinia triticina* tolerance	Yanjun et al. (2017)
PR6	Proteinase inhibitor	*P. ginseng* Meyer	Environmental stress, hormonal imbalances, and toxic substances	Myagmarjav et al. (2017)
PR10	Ribonuclease-like	Rice	Biotic and ecological pressure	Wu et al. (2016)
PR10	Ribonuclease-like	*Capsicum annuum*	Ribonucleolytic activity against TMV	Park et al. (2004a, b)
PR12	Defensins	Rice	*Magnaporthe grisea*	Kanzaki et al. (2002)
PR12	Defensins	Peanut	Fungal and bacterial pathogens	Anuradha et al. (2008)
PR12	VrPDF1	Mung bean	Weevils	Thao et al. (2017)
PR13	Thionin	*Solanum tuberosum*	*B. cinerea*	Hoshikawa et al. (2012)
PR13	Thionin	*S. tuberosum*	*Fusarium* spp. tolerance	Hammad et al. (2017)

Adapted from Ali et al. (2018)

experiments, on the other side, had indicated that a number of PR14 peptides or lipid transfer proteins exhibit antimicrobial properties (Wang et al. 2004a, b). Antifungal proteins are a type of AMP which is used to treat fungal diseases, which are resistant to antibiotics. Several AFPs had exhibited strong antimicrobial activity against prevalent fungal strains in medical field, such as *Aspergillus* and *Candida albicans*, yeast, filamentous fungus (e.g., *Aspergillus flavus*), as well as mold in food production. Numerous synthesized proteins, with the exception of brevinin, ranatuerin, and cecropins, have antifungal action. For example, AurH1, which is produced from aurein 1.2, is beneficial in treating *C. albicans* infection, which has a 40% mortality rate (Madanchi et al. 2020).

4.17 Ribosomal Inactivating Proteins

The presence of ribosome-inactivating peptides has been discovered in plants, fungus, algae, and bacteria. RIPs have rRNA N-glycosylase functionality and thus cause an adenine nucleotide to be cleaved at a preserved location on the 28S rRNA. The irrevocable breakage of this solitary N-glycosidic bridge disrupts the connection involving the elongation elements and the ribosome, resulting in protein biosynthesis suppression (Pizzo and Di Maro 2016). The type 1 and type 2 RIPs (monomeric and dimeric) are the two types of RIPs that can be distinguished. The increased lethality of type 2 RIPs compared to type 1 RIPs is attributable to the identification of mammalian cell surface galactose monomers, which is facilitated by the A-chain. Because the lectin chain is missing, type 1 RIPs have a much harder time getting inside cells, resulting in reduced mortality. Other noncanonical RIPs have been discovered, including tetrameric ebulin (Jiménez et al. 2015) and proteolytic activated corn b-32 (Hey et al. 1995).

Biochemically, phyto-RIPs are of three types: types 1, 2, and 3 (De Virgilio et al. 2010). The most common is type I, which contains a single polynucleotide pattern of around 30 kDa with N-glycosidase function (Stirpe 2004). Because type I RIPs maintain a few enzymatic function cleavage sites, their entire genomes and post-translational alterations vary dramatically (Husain et al. 1994). In contrast to type I, type II RIPs featured two domains: a catalytically active A domain of around 30 kDa and a bonding B domain of around 35 kDa with lectin characteristics. The disulfide bridge links A and B domains, while the B-lectin motif binds sugar molecules. The A domain could then travel back into the cytoplasm, reducing protein production (Steeves et al. 1999). Without a B domain, type I RIPs have limited side effects. Nontoxic type II RIPs are classified depending on their toxicity. Indiscriminate poisoning is done by modeccin, viscumin, volkensin, abrin, and ricin. The reasons for cytotoxicity variations are unknown. Given that type II RIPs have only been discovered in plants, the association among RIP and lectin sites is presumed to be recent.

4.17.1 Role of RIPs in Plant Pathogen Resistance

The potential of using ribosome-inactivating peptides (RIPs) as host defense against viruses and fungi has sparked interest. Antifungal activities were discovered against two pathogens (*Coprinus comatus* and *Physalospora piricola*) (Ng et al. 2003). Variable susceptibility was observed in various fungal species belonging to the same genus, which was intriguing. *Pythium irregulare*, for example, had previously been found to be susceptible, whereas *Pythium ultimum* had previously been shown to be resistant. Roberts and Stewart (1979) stated that on a solid media, the modified barley type I (RIP) was found to suppress the proliferation of a fungal strain, *Trichoderma reesei*. Notwithstanding, barley RIP had only a small effect on *T. reesei* growth in aqueous media; however, suppression was elevated by the addition of an enzyme chitinase (Leah et al. 1991). Antifungal activities are found in both hairy melon RIP and small RIP luffacylin (Wong et al. 2010). According to another study, MMC derived from *Momordica charantia* seed exhibits antifungal effects (Zhu et al. 2013). MbRIP-1's antifungal activities were evaluated using a radial proliferation suppression experiment (Kushwaha et al. 2012). According to the findings, MbRIP-1 exhibited a high antifungal effectiveness against *Aspergillus niger* (Kushwaha et al. 2012). *R. solani* was shown to be resistant to curcin 2 expression in mutant cannabis plants (Huang et al. 2008). Chopra and Saini (2014) mentioned that the GM crops had a higher resistance to the fungal disorder, Corynespora leaf spot. GM rice with the RIP gene has recently been revealed to have developed tolerance to the fungus infection (Qian et al. 2014).

RIPs were known to have antiviral properties for almost 75 years. Antiviral weapons known as RIPs were proven to be efficient for many animal, plant, and human being viruses. According to previous studies, exogenous PAP treatment improves *N. benthamiana's* systemic tolerance to TMV infections (Zhu et al. 2016). When dianthins derived from *Dianthus caryophyllus* foliage were coupled with TMV, the number of localized tumors on *Nicotiana glutinosa* leaf tissue was dramatically decreased (Thorpe et al. 1981). When acquired from *M. jalapa*, cucumber mosaic virus (CMV), tomato mosaic virus (TMV), potato Y virus, cucumber green mottle mosaic virus, and turnip mosaic potyvirus (TuMV) all showed excellent resistance to mechanical virus transmission (Kubo et al. 1997). CMV and TMV disease tolerance enhanced after the creation of a type I RIP TCS (Krishna et al. 2002). Generalized infectious symptoms seemed to be decreased and impeded in recombinant plants overexpressing TCS genes. Upregulation of type I RIPs in transgenic crops appears to increase plant sensitivity to a variety of plant viruses, according to the majority of evidence. The antimicrobial property of RIPs towards phytopathogenic viruses as well as their possible involvement in broad immunity have all been investigated in detail. The *Clerodendrum aculeatum*-generalized resilience-initiating protein, which is RIP active, is essential for establishing substantial protective immunity in vulnerable plants against a variety of plant virus infections, including viruses that cause blight (Kumar et al. 1997). The transgenic plant species that produce RIPs have shown increased tolerance to several plant pathogens, as shown in Table 4.5.

Table 4.5 Reports on transgenic crops expressing RIPs

RIP	Targeted pathogen	Host plant	References
Pokeweed antiviral protein	Cucumber mosaic virus, potato virus Y, potato virus X	Tobacco	Lodge et al. (1993)
Trichosanthin	Cucumber mosaic virus, tobacco mosaic virus	Tobacco	Lam et al. (1996)
Type II RIP SNA-I′	Tobacco mosaic virus	Tobacco	Chen et al. (2002)
Phytolacca insularis antiviral protein	Potato virus Y, potato virus X, potato leafroll virus	Potato	Moon et al. (1997)
Curcin 2	Tobacco mosaic virus	Tobacco	Huang et al. (2008)
Alpha-momorcharin	Chili veinal mottle virus, cucumber mosaic virus, tobacco mosaic virus, turnip mosaic potyvirus	*Momordica charantia*	Zhu et al. (2013), Yang et al. (2016)
PAP	Tobacco mosaic virus	*Phytolacca americana*	Dallal and Irvin (1978)
BDP-30	Tobacco mosaic virus	*Boerhavia diffusa*	Srivastava et al. (2015)
Clerodendrum aculeatum systemic resistance-inducing (CA-SRI) protein	Tobacco mosaic virus, Sunn-hemp rosette virus	*Clerodendrum aculeatum*	Verma et al. (1996), Kumar et al. (1997)
New single-chain RIPs	Artichoke mottled crinkle virus	*Basella rubra*	Bolognesi et al. (1997)
Barley RIP	*Rhizoctonia solani*	*Nicotiana tabacum*	Logemann et al. (1992)
Curcin 2	*Rhizoctonia solani*	*Nicotiana tabacum*	Huang et al. (2008)
Modified maize RIP (MOD1)	*Rhizoctonia solani*	*Oryza sativa*	Kim et al. (2003)
PhRIP I	*Botrytis cinerea*, *Rhizoctonia solani*	*Solanum tuberosum*	Gonzales-Salazar et al. (2017)
Barley RIP	Corynespora leaf spot fungal disease	*Vigna mungo*	Chopra and Saini (2014)
α-MMC	Rice blast fungus	*Oryza sativa*	Qian et al. (2014)
Alpha-momorcharin	*Bipolaris maydis*, *Fusarium graminearum*, *Aspergillus oryzae*, *Aspergillus niger*	*Momordica charantia*	Zhu et al. (2013)
Luffacylin	*Fusarium oxysporum*, *Mycosphaerella arachidicola*	*Luffa cylindrica*	Parkash et al. (2002)
Tobacco RIP (TRIP)	*Pseudomonas solanacearum*, *Erwinia amylovora*, *Shigella sonnei*, *Salmonella typhimurium*, *Rhizobium leguminosarum*	*Nicotiana tabacum*	Sharma et al. (2004)

(continued)

Table 4.5 (continued)

RIP	Targeted pathogen	Host plant	References
Balsamin	*Staphylococcus aureus*, *Salmonella enterica*, *Staphylococcus epidermidis*, *Escherichia coli*	*Momordica balsamina*	Ajji et al. (2016)
ME2	*Agrobacterium tumefaciens*, *Agrobacterium radiobacter*	*Mirabilis expansa*	Vivanco et al. (1999)

Source: Zhu et al. (2018)

Insects are less likely to attack plants with RIPs (Stirpe 2013). Pesticidal effects of RIPs have been demonstrated against Gram-negative insects such as Lepidoptera (Wei et al. 2004; Dowd et al. 2006), Coleoptera (Kumar et al. 1993), and Diptera (Kumar et al. 1993; Shahidi-Noghabi et al. 2008). The fertility and longevity of *Anticarsia gemmatalis* and *Spodoptera frugiperda* were both decreased when they were fed a fictitious meal enriched with many type I RIPs (Bertholdo-Vargas et al. 2009). Previously, it was discovered that apple (*Malus domestica* Borkh) RIP type I and type II had strong aphicidal properties (Hamshou et al. 2016). Type I and type II RIPs from apples (*Malus domestica* Borkh) have been shown to exhibit considerable aphicidal action in previous studies (Hamshou et al. 2016). Findings demonstrated that transgenic rice plant lines expressing version I or type II RIP decreased the green peach aphid (*Myzus persicae* Sulzer) larva viability significantly. Overexpression of a corn RIP in tobacco plants improved tolerance to *Helicoverpa zea* (Dowd et al. 2003). This maize foliage was robust to *Spodoptera frugiperda* and corn earworms because of the presence of high amounts of MRIP and wheat germ agglutinin (WGA) (Dowd et al. 2012). The specific mode of action by which RIPs perform their pesticidal activity is unclear. RIPs have been proven to increase mortality in several studies (Das et al. 2012). When *A. pisum* was administered a synthetic meal enriched with SNA-I, caspase-3 was triggered, resulting in damages in the intestinal lumen (Shahidi-Noghabi et al. 2010).

4.18 Use of Antimicrobial Proteins

Antimicrobial peptides (AMPs) constitute a type of polynucleotide found in people, animals, and plants, which are small and generally positively charged. Given the rise of drug-resistant infections, the antibiotic action of AMPs has gotten a lot of interest (Li et al. 2021). Several species rely on AMPs with wide-ranging effectiveness towards bacteria and fungus as a defense against infections. Numerous other biological effects of plant AMPs, in complement to their antibacterial activities, were discovered via extensive study, including regulating development and proliferation of plants as well as managing a variety of disorders with great efficacy. The possibility for plant AMPs to be used in agricultural productivity, as food supplements, and as illness therapies has piqued people's curiosity (Li et al. 2021). Plants had evolved complex defensive mechanisms to safeguard themselves from infections and insects. These systems enable crops to efficiently fight against pathogenic microbes and insects (Iqbal et al. 2019; Yang et al. 2007a, b). Physical criteria to pathogen intrusion and dispersal, like waxy cuticular parts and trichomes, and chemical barriers to microbiota advancement and improvement, such as advanced and powerful cell-recognizing systems, complex plant hormone channels, countless transcriptional processes, diversified enzymes, and secondary toxic metabolites, are among all such pathogen defense measures (Campos et al. 2018). AMPs are a very frequent kind of polypeptide and well-known pharmacologic defenses developed by plants to shield themselves from plant pathogens (Kulaeva et al. 2020). AMPs, which are found in virtually all living things, including bacteria, arthropods, mammals, and plants, play a crucial role in pathogen defense (Zaslof 2002). AMPs are short molecular peptides synthesized by ribosomes, whereas developed peptides are synthesized through disintegration of larger protein precursors and other posttranslational alterations. Nonribosomal peptide synthetases can also synthesize various AMPs (Tyagi et al. 2019). Numerous AMPs have antibacterial properties against a wide range of microbes and parasitic species (Marcocci et al. 2020). Plant AMPs contain features that are similar to those seen in bacteria, insects, and humans. Plant AMPs operate on microbes in a distinct way than animal AMPs. Hevein-like peptides, for instance, hold chitins, knottin-type proteins hinder proteolytic enzymes, and lipid transfer proteins attach lipids to prevent microbe's infiltration into cellular membrane (James et al. 2015).

4.18.1 Classification and Functions of AMPs

Plant AMPs are grouped considering their sequence similarities, motifs, and disulfide bond configurations that determine their folding three-dimensional structure. According to these criteria, plant AMPs are categorized as snakins, defensins, hevein-like peptides, lipid transfer proteins, and unclassified CRPAMPs. With their diverse actions, architectures, and expression patterns, and unique targets, plant AMPs have a complicated and time-consuming categorization process. Plant AMPs are classified as cationic or anionic peptides depending on the type of charge in them (Prabhu et al. 2014). Plant AMPs are categorized based on sequence homology, cysteine motifs, and tertiary structures (Hammami et al. 2009).

Thionins are hydrophobic and are thought to cause cytotoxicity in bacteria, fungi, animals, and plants by interacting with membranes via hydrophobic residues or a positive surface charge. The postulated hazardous pathway is cellular membrane disruption, but this is currently being investigated. To explain why cell membranes are solubilized and lysed, Steer provided a framework of interaction between thionin and phospholipids (Steer and Carapetis 2004). Apart from protein receptors, thionins have been found to react immediately with membrane lipids (Richard et al. 2002). *Pyrularia thionin*, derived from *Pyrularia pubera* nuts, mediates Ca^{2+} inflow during some cell signaling, whereas Tyr iodination decreases activated alkaline phosphatase A2, neurotoxicity, and hemolysis. Studies on the relationship between structure and functionality have found that Lys1 and Tyr13 are mostly retained in thionins, apart from nontoxic, nonlytic crambin, and are considered to be crucial for their lethality. Conversely, crambin has Thr1 and Phe13 residues (Steer and Carapetis 2004).

There are several different antimicrobial molecules found in plants, including plant defensins. There are several membrane-soluble crop AMPs, but plant defensins are the greatest and largest group of these AMPs (Graham et al. 2008). Defense proteins are an important AMP class found in almost all plants (Taylor et al. 2008). Apart from molecules that make up strong disulfide bridges, their protein compositions are rather diverse with their very stable architectures (Parisi et al. 2019). Much other metabolic pathways are also affected by them, such as altering self-incompatibility, acting as epigenetics, and changing vitamin oxidative redox status (Sitaram 2006; Carvalho and Gomes 2011).

There are around 30 amino acids in plant knottins, which are a family of trypsin, amylase, and carboxypeptidase antagonists, as well as cyclic proteins. They were identified about 20 years ago (Le Nguyen et al. 1990). Generally, plant AMPs of the knottin family are the smallest. Because of their chemical specificity, they may bind to many different types of chemicals and accomplish a vast array of physiological tasks, including improving environmental tolerances, stimulating root development, and serving as stimulating factors. They are also antibacterial, antifungal, and antiviral, as well as cytotoxic, insecticidal, and HIV fighting (Hwang et al. 2010; Aboye et al. 2015). A cystine wound created by maintained disulfide linkages between many cysteine partners is a distinguishing property of knottins. Cysteine motifs differ among subfamilies of plant knottins. In various species, knottin structures from functionally unrelated protein families seem similar. Defensins and protease inhibitors are plant proteins having cystine patterns that are similar to knottin-type proteins. Despite the fact that plant defensins have a cystine-knotting motif, their cysteine positioning is different from that of knottin-like peptides (Tam et al. 2015).

The snakin class of the plant AMP family has 12 cysteine regions and contains the greatest number of disulfide bonds. Snakin-1, the first identified snake peptide, was seen in the tubers *Solanum tuberosum* and had pattern/sequence similarities to snake venom. These proteins include a high concentration of cysteine (19% Cys), comprise 63 amino acids, and contain six disulfide connections, with other forms of AMPs possessing 2–4 disulfide linkages (Tavares et al. 2008). In reaction to microbial or environmental stress, snakins can be produced inherently or inducibly in a number of cells, including the rhizome, branches, foliage, flowers, and seeds. Repressing snakin-1 in potatoes changes the height, leaf number, and leaf structure of the potato through altering cell division, primary metabolism, and content of the cell wall (Nahirñak et al. 2012). Both snakin-1 and snakin-2 could decrease fungal and bacterial growth (Berrocal-Lobo et al. 2002).

The α-hairpinin family is a subfamily of Lys/Arg-rich plant resistance proteins with unique Cys motifs that create a helix-loop-helix secondary structure (Slavokhotova and Rogozhin 2020). This helix is made up of two helices that are alternately connected. Two disulfide connections maintain the tertiary configuration (Sousa et al. 2016). Antibacterial, trypsin-inactivating, and ribosome-inactivating characteristics were among the metabolic actions of this AMP group (Tam et al.

Table 4.6 Bioactivities and classification of several prominent plant-based antimicrobial compounds

AMPs	Class	Function	References
alfAFP	Plant defensins	Antifungal	Gao et al. (2000)
Rs-AFP4	Plant defensins	Antifungal	Terras et al. (1995)
ZmD32	Plant defensins	Antibacterial, antifungal	Kerenga et al. (2019)
Fa-AMP1	Plant defensins	Antibacterial, antifungal	Fujimura et al. (2003)
Rs-AFP3	Plant defensins	Antifungal	Terras et al. (1995)
NaD1	Plant defensins	Antibacterial, antifungal	Kerenga et al. (2019)
C. fistula PI	Plant defensins	Trypsin inhibitor	Wijaya et al. (2000)
Vv-AMP1	Plant defensins	Antifungal	de Beer and Vivier (2008)
Rs-AFP2	Plant defensins	Antifungal	Terras et al. (1995)
ZmD32	Plant defensins	Antibacterial, antifungal	Kerenga et al. (2019)
Purothionins	Thionin	Antibacterial	Fernandez et al. (1972)
Tk-AMP-X1	α-Hairpinin	Antifungal	Utkina et al. (2013)
MBP-1	α-Hairpinin	Antibacterial, antifungal	Duvick et al. (1992)
Tk-AMP-X2	α-Hairpinin	Antifungal	Utkina et al. (2013)
Cp-thionin II	Thionin	Antifungal	Schmidt et al. (2019)
CaThi	Thionin	Antibacterial, antifungal	Taveira et al. (2014)
kalata B6	Cyclotide family	Insecticidal	Colgrave et al. (2008)
cyO2	Cyclotide family	Antibacterial, anticancer	Pränting et al. (2010)
kalata B1	Cyclotide family	Insecticidal	Colgrave et al. (2008)
Ps-LTP1	Lipid transfer protein	Antibacterial, antifungal	Bogdanov et al. (2016)
Lc-LTP3	Lipid transfer protein	Antibacterial, antifungal	Bogdanov et al. (2015)
Lc-LTP1	Lipid transfer protein	Antibacterial, antifungal	Bogdanov et al. (2015)
McLTP1	Lipid transfer protein	Antibacterial	Souza et al. (2018)
bevuTI-I	Hevein-like peptides	Trypsin inhibitory	Retzl et al. (2020)
NTC	Hevein-like peptides	Antifungal	Van Parijs et al. (1991)
Psacotheasin	Hevein-like peptides	Antibacterial, antifungal	Hwang et al. (2010)
Ep-AMP1	Hevein-like peptides	Antibacterial, antifungal	Aboye et al. (2015)
pnAMP-h2	Hevein-like peptides	Antifungal	Koo et al. (2002)
Snakin-1	Snakins	Antifungal, antibacterial	Segura et al. (1999)

Adapted from Li et al. (2021)

2015; Slavokhotova and Rogozhin 2020). In a variety of plant-damaging microorganisms, MBP-1, a 33-amino-acid AMP isolated from corn, has been used to suppress sporulation and mycelial expansion (Duvick et al. 1992). Cyclotides were lengthy nonlinear polypeptides produced by plant species that comprise around 28 and 37 amino acids; the cyclotide family constitutes the majority of plant anionic AMPs (Harris et al. 2009; Prabhu et al. 2014). They have a cyclic backbone with six coils (referred to as cyclic cystine knots) produced by six distinct residues organized and cross-linked in a convoluted manner, and three disulfide connections that keep them together (Craik 2009; Pränting et al. 2010). Table 4.6 summarizes some plant antimicrobial peptides as well as their physiological uses/bioactivities.

4.19 Pathogen-Derived Resistance for Viral Diseases

Sanford and Johnston established the simple but fundamental idea of parasite- or pathogen-derived tolerance (Sanford and Johnston 1985). Having followed this, several attempts have been undertaken to confer viral tolerance on transgenic plants by displaying gene fragments derived from the virus (Lomonossoff 1995). Numerous successful initiatives have culminated in the introduction of virus-resistant potato and squash cultivars (Baulcombe 1996). PDR was first employed to limit viral incursion and disease 11 years back; since that period, there have been an escalating frequency of examples of robustness and the advent of a variety of tolerance techniques. In 1995, Asgrow Co (Kalamazoo, MI, USA) made the earlier commercialization of virus-resistant transgenic crops in the USA, including virus-resistant squash; additional instances are anticipated to hit the market in the coming years. The molecular and cellular processes at work in the different forms of PDR are still a mystery, but they are not completely unknown. The upcoming task for researchers in this discipline will be to devise techniques to widen the scope and heighten the level of tolerance. This will be accomplished by basic research that elucidates the molecular underpinnings of resilience and then applies that information to the development of second- and third-generation resistance genes that are more effective (Beachy 1997).

4.19.1 Strategies of PDR

One could think that knowing about the pathogen would lead to the invention of PDR techniques to inhibit virus infection and replication. In reality, PDR has a better grasp of virus replication and pathogenesis than previously thought. There are two types of PDR techniques: ones that need the creation of proteins and ones that just need the buildup of viral nucleotide sequences. The former confers tolerance to a wider spectrum of virus strains, while the latter confers exceptionally high degrees of tolerance to a single virus strain. The following subheadings summarize the results of protein-mediated and RNA-mediated tolerance.

4.19.2 Protection Conferred by Nucleic Acids

The creation of alleles that generate nucleic acids incapable of encoding proteins is employed in a variety of PDR strategies. The production of antisense RNA sequences to suppress RNA virus multiplication was among the first strategies; in certain situations, virus invasion was barely impacted (Powell et al. 1989), while in others, infection was severely hindered (Hammond and Kamo 1995). Even though RNA-mediated inhibition may have been liable for some of the (−) sense-induced rebellion revealed to date, other pathways such as the disruptions of replicase blueprint selection or the establishment and eventual deterioration of double-stranded RNA could also be held responsible. Antisense RNA-mediated tolerance is likely to be somewhat restricted, protecting the virus, wherein the patterns are generated but not isolates with major differences from the transgene's genomes. In geminivirus infections (Bendahmane and Groenborn 1997), where multiplication and transcription occur in the nucleus, antisense RNA techniques have also been moderately successful. The reality that certain transgenic strains have a high level of susceptibility suggests that the transgene can produce enough antisense RNA to lower virus multiplication and/or gene expression frequencies. The greatest instance of nucleic acid-mediated tolerance is RNA inhibition, a process that governs the controlled degradation of nucleotide sequence after transcription. While there is a strong connection between RNA inhibition and more widespread criteria of transgene suppression, the similarities and differences between these two processes are reportedly being studied (Lindbo et al. 1993).

4.19.3 Protection Through Movement Proteins (MP)

The majority of CPMP cases rely on transgenic development of wild-type CP genes. Pathogen-derived resistance can also be engineered using dominant-negative mutant versions of viral proteins. Transgenic production of viral movement proteins (MP) demonstrated the efficiency of this method by conferring protection once the transgene designated a defective MP. The expression of a modified functioning MP showed neither effect nor enhanced vulnerability to virus infection (Ziegler-Graff et al. 1991). Transgenic production of a malfunctioning TMV MP confers resistance owing to struggle for plasmodesmata-binding sites between the mutated MP and the injected virus's wild-type MP (Lapidot et al. 1993). The tolerance mechanism's wide array effectiveness is an intriguing and highly useful feature of MP-mediated protection (MPMP). In addition to tobamoviruses, the protection afforded by TMV's genetically mutated MP facilitates tolerance to viruses from the potex-, cucumo-, and tobraviral families (Cooper et al. 1995). TMV tolerance was achieved in a nonhost plant by genetic overexpression of brome mosaic virus (BMV) MP (Malyshenko et al. 1993). These instances of extensive array tolerance suggest that MPs from a wide range of viruses might associate with plasmodesmata constituents in a comparable manner. Three MPs were expressed by overlapped encoding areas known as the triple gene block (TGB) in the potex, carla, hordei, and furovirus families (Beck et al. 1991). The TMV MP gives protection against a wider array of viruses than transgenic production of a mutated TGB protein (Beck et al. 1994).

4.19.4 RNA (or DNA)-Mediated Resistance

In the aforementioned situations, tolerance was conferred via transgene-encoded proteins. Additionally, pathogen-derived tolerance was induced when the transgenic or its RNA transcript inhibited the viral assault cycle directly. If the transformed nucleic acid behaves as a deception particle, this RNA-mediated resistance (or DNA-mediated resistance) may be successful. Such camouflage would compete with the viral chromosome for the ability to direct host- or viral protein interactions in ways that would be detrimental to the virus's reproduction and spread in the plant affected. This form of suppression of rivalry may emerge whenever the modified gene is classified as a defective interfering (DI) RNA/DNA, encircles a *cis*-acting region in the genetic material of the virus, or is generated via RNA satellite. There are several examples of pathogen-derived resistance that are likely due to the deception action of a genetically engineered RNA. These include resilience imparted by satellite RNA (Gerlach et al. 1987; Harrison et al. 1987), impedance to geminiviruses (Stanley et al. 1990), and resistance to the cymbidium ringspot virus (Stanley et al. 1990). Transgenically produced 3′ DI RNA competes effectively with the virus DNA to promote turnip yellow mosaic virus resilience (Zaccomer et al. 1993).

4.19.5 Tolerance Conferred by a Coat Protein

A number of field crops have benefitted from coat protein-mediated resistance (CPMR) since it was originally reported in 1986 in the TMV tobacco model system. For instance, TMV's CP offered excellent security against highly associated TMV isolates, while reducing the vulnerability to tobamoviruses with reduced CP sequence homology, CPMR may give wide or specific coverage for CP. PVR strain N60.5's CP allele conferred tolerance on GM potato plants treated with PVY variant N605 plus 0803 variant, whereas the CP allele of papaya ringspot virus (PRV) variant HA offered immunity exclusively to PRV strain HA. To put that into perspective, the noninfectious soybean mosaic virus (SMV) CP provided tolerance in tobacco to PVY and the tobacco etch virus (TEV) in contrast (Stark and Beachy 1989). It is perplexing how certain CPs have a wide or robust CPMR while some have a minimal or medium amount of robustness. Three different tomato spotted wilt virus strains were given massive protection by combining the nucleoprotein alleles from each of their nucleoprotein genes. The transgene-derived CP has to be functional to pair monomers yet not necessarily create viral particles in order to perform CPMR on TMV (Clark et al. 1995). According to Register and Beachy (1988), CP limits infection progression by interacting with ThlV degradation. Furthermore, there is a significant connection involving CP concentration and disease tolerances (Powell et al. 1990).

4.19.6 Resistance Modulated by a Replicase

Near-immunity to infection can be imparted by genetic composition that encodes whole or partial replicase polypeptides, which is typically but not always confined to the viral strain that provided the nucleotide sequence. TMV replication-mediated resistance (Rep-MR) has been initially discovered in transgenic plants harboring a gene coding for a 54 kDa replicase fraction. Even though certain types of Rep-MR have been proposed to be aided by RNA rather than protein (Baulcombe 1996), others require an accessible reading frame and protein synthesis. A shorter replicase mutant isolated from a subgroup I cucumber mosaic virus (ChlV) conferred increased resistance on all subgroup I ChlV strains in tobacco plants, but not on group 2 species strains or other viruses (Zaitfin et al. 1994). While it has been established that plants with Rep-MR may significantly inhibit multiplication and are resistant to higher stages of challenge inoculum, the mechanisms behind Rep-MR remain unclear. It is believed that the transgene's protein begins interfering with the replicase function of the virus in some way, potentially by attaching to host or viral peptides, which control the replication and expression of viral genome. Rep-MR against CMV has been shown to limit both viral accumulation and systemic illness (Helwald and Palukaitis 1995); this might be owing to decrease in movement protein produced by virus replication suppression.

4.20 Non-pathogen-Derived Resistance for Viral Diseases

Agricultural production faces a big difficulty in protecting food crops from viral diseases. Aside from pathogen-derived resistance, plant viral disease resistance can also be achieved through RNAi, CRISPR-Cas9 targeting DNA, CRISPR-Cas9 targeting RNA, and host factor editing.

4.20.1 Protection Against Plant Viruses Through RNA Silencing

The RNA silencing in plants was first observed in 1990 (Napoli et al. 1990), and since then it has been intensively investigated in a number of eukaryotes, including fungi, mammals, and plants (Fig. 4.2) (Hannon 2002; Baulcombe 2004). RNA silencing, also referred to as RNA interference (RNAi), is triggered when double-stranded RNA molecules (dsRNAs) are present. This restricts or suppresses the expression of genes in a genomic sequence-specific way (Hannon 2002; Voinnet 2005). In crops, many polypeptide groups are involved in RNA silencing regulation, notably Dicer-like (DCL), Argonaute (AGO), RNA polymerase (RDR), and silencing of genes (SGS). dsRNA and miRNA progenitors are converted to siRNA and miRNA of 20–24 nucleotides with a two overhang at the 3′ end by DCL enzyme, which are type III RNases. The RNA-generated silencing complexes are created when siRNAs or miRNAs are integrated into restriction enzyme AGO complexes. These may connect with the target mRNA or noncoding RNA and subsequently mute the targeted gene's activation by chelating and destroying the destination RNA or by engaging DNA and histone regulators and blocking the targeted gene's transcription, as instructed by the siRNA/miRNA contained within RISC. RDR proteins may be able to detect the damaged RNA and increase the amount of dsRNA that is produced, thus increasing the impact of the silencing. Due to the fact that they protect the additional siRNA produced by DCLs, SGS enzymes

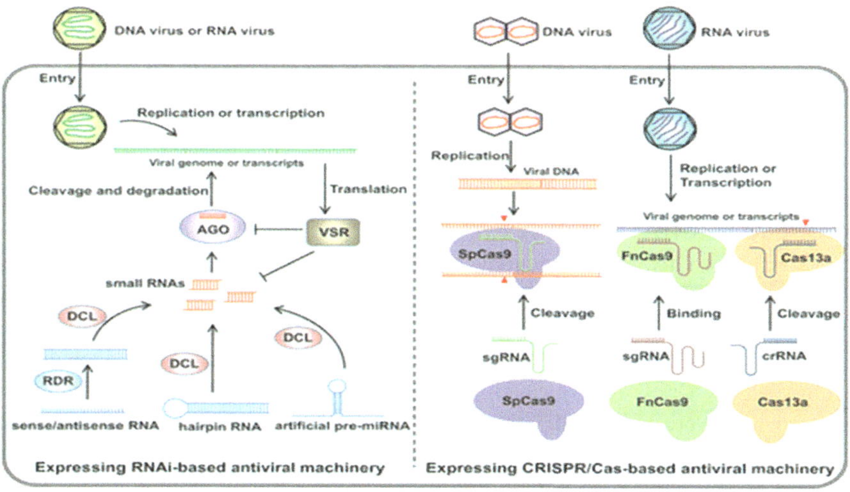

Fig. 4.6 The processes involved in RNA silencing/CRISPR-Cas method for eliminating plant pathogenic viruses. On the left, the technique of antiviral bioengineering via RNA silencing is illustrated. Crop tissues are modified or exogenously given virus-developed sense/antisense RNA, hairpin RNA, or synthetic pre-miRNA that are used to generate short RNAs that target genomes of the virus and/or the transcripts. The short RNAs are deposited in the AGO enzyme to drive the breakage of viral RNA, resulting in the breakdown RNA genome or viral messenger RNA. To fight RNA silencing-based tolerance, plant viruses that target the AGO protein or short RNAs can express VSR. The picture on the other sight (right) illustrates the CRISPR-Cas-dependent method for antiviral protection against viruses. This system is composed of two components: single-stranded RNA (sgRNA) and Cas protein. The virus can be effectively suppressed by recombinant/transit expression of the virus that specifically targets sgRNA as well as its related Cas protein. When a DNA virus, such as the geminivirus, infects a plant cell, it changes the plant's genome to a double-stranded DNA precursor that the *Streptococcus pyogenes* Cas9 polypeptide can target and break (SpCas9). It has been demonstrated that Cas9 from *Francisella novicida* (FnCas9) and Cas13a confer viral tolerance on RNA viruses. FnCas9 and Cas13a are capable of binding to or cleaving viral genomes or transcripts, respectively, when directed by their matching sgRNA or crRNA. The viral genome's cleavage sites are indicated by red arrowheads

improve the RNA silencing process (Fig. 4.6; Ding 2010; Ipsaro and Joshua-Tor 2015; Voinnet 2005). Numerous techniques for constructing virus-resistant mutant plants have been identified, the bulk of which rely on numerous precursor RNAs for siRNA synthesis, include sense/antisense RNA, helical/hairpin RNA (hpRNA), and artificial miRNA progenitors (Duan et al. 2012).

4.20.2 CRISPR-Cas-Based Plant Viral Disease Resistance

Using an innate immune response, the CRISPR-Cas system stops alien genomes or viral DNA from invading bacteria and archaea via cleavage. Genomic engineering techniques such as CRISPR-Cas employ the Cas enzyme, an endonucleolytic enzyme, and single-guide (sg) RNA, which direct the Cas protein to the DNA or RNA target. Additionally, sgRNA has a scaffold for Cas protein interaction and a 20-nt-long spacer sequence chosen by the user for genome targeting. Numerous laboratories in a variety of areas have chosen this technique over TALEN because of its simplicity, great efficiency, and cost. Numerous studies are being conducted to determine how CRISPR-Cas9 may be utilized to manage viral infections that affect humans such as HIV as well as viruses that affect plant. *Streptococcus pyogenes* created the first CRISPR-Cas genome editing mechanism, which targets DNA. As a result, the CRISPR-Cas9 system was initially employed to fight geminiviruses by specifically targeting their viral genomic DNA during replication. Three groups reported successful generation of geminivirus resistance in tobacco and Arabidopsis using CRISPR-Cas9. They created sgRNAs that targeted just certain IR, Rep, and CP loci and were able to drastically decrease or eradicate illness symptoms in various geminiviruses. Along with its usage in model species, the same technique was previously applied to barley, establishing very effective immunity to wheat dwarf virus (WDV) (Kis et al. 2019). Cauliflower mosaic virus (CaMV) lethality in Arabidopsis was shown to be effectively eradicated using CRISPR-Cas9 (Liu et al. 2018).

4.20.3 TALEN/ZFN-Based Resistance Against Viral Diseases

Some years back, a new technique known as genome editing developed, allowing scientists to modify genomic information in many cell types and creatures. TALENs and zinc finger nucleases were the first generation of gene editing technology's tools. ZFNs were used to alter DNA (Moscou and Bogdanove 2009). Zinc finger proteins (ZFPs) and transcription activator-like effectors (TALENs) are two types of recombinant peptides that were produced by linking the nonspecific proteolytic region of the protein FokI with a DNA-binding domain (DBD). As a result, the DBD can both recognize a specific nucleotide inside a given DNA strand and generate suitable double-strand breaks (DSBs) (Boch et al. 2009; Moscou and Bogdanove 2009; Urnov et al. 2010). In eukaryotes, homologous recombination and NHEJ are two methods for mending DSBs, and both have the potential to result in abnormalities in the damaged genome sequence (Wyman and Kanaar 2006). Using gene editing platforms, plant viruses gain a new weapon as well as the ability to merge, delete, and/or modify genes of concern. AZP, a synthetic zinc finger protein created by Sera in 2005, specifically targets the beet severe curly top virus's internal transcribed region (IR) in Arabidopsis, although it lacks ZFN's lytic domain (Sera 2005).

When geminiviruses replicate, they must bind to the viral replication initiator polypeptide (Rep), which has a stem-loop pattern in its infrared spectrum (Hanley-Bowdoin et al. 2013). By blocking Rep binding to BSCTV's IR, the transgenically generated AZP reduces viral transmission (Sera 2005). Rice tungro bacilliform virus (RTBV) proliferation has been limited by developing an AZP that can identify and destroy viral promoter sequences in Arabidopsis (Ordiz et al. 2010). Tobacco curly

Table 4.7 CRISPR-Cas technology to help plants resist viral pathogens

Target factor	CRISPR type	Host plant	Targeted virus	References
Viral genome	SpCas9	Tobacco	Yellow leaf curl virus in tomatoes	Ali et al. (2015, 2016)
			Leaf curl virus of cotton	
		Tobacco and Arabidopsis	Beet severe curly top virus, beet curly top virus	Ali et al. (2015), Ji et al. (2015)
		Tobacco	Bean yellow dwarf virus	Baltes et al. (2015)
		Barley	Wheat dwarf virus	Kis et al. (2019)
		Arabidopsis	Cauliflower mosaic virus	Liu et al. (2018)
	FnCas9	Tobacco	Tobacco mosaic virus	Zhang et al. (2018a, b)
		Tobacco and Arabidopsis	Cucumber mosaic virus	Zhang et al. (2018a, b)
	LshCas13a	Tobacco	Turnip mosaic virus	Aman et al. (2018)
		Tobacco	Tobacco mosaic virus	Zhang et al. (2019)
		Rice	Southern rice black-streaked dwarf virus	Zhang et al. (2019)
		Rice	Stripe mosaic virus of rice	Zhang et al. (2019)
		Potato	Y virus of potato	Zhang et al. (2019)
Host factor	SpCas9	Cucumber	Cucumber vein yellowing virus	Chandrasekaran et al. (2016)
		Cassava	Cassava brown streak virus	Gomez et al. (2019)
		Cucumber and Arabidopsis	Zucchini yellow mosaic virus	Chandrasekaran et al. (2016), Pyott et al. (2016)
			Papaya ringspot virus	Pyott et al. (2016)
			Turnip mosaic virus	Chandrasekaran et al. (2016)

shoot virus (TbCSV) has been utilized to assault the Rep gene in tobacco plants, with substantial viral replication suppression, whereas the ZFN method exploits both DBD and DNA breakage domains, unlike AZP (Chen et al. 2014). To combat these begomoviruses, researchers developed TALEs without the nuclease domain seen in TALEN. TALEs were developed to deal with the conserved motifs seen in begomoviruses. As for resistance to tomato leaf curl Yunnan virus (TLCYnV), only modest tolerance was found in tobacco plants that expressed a TALE gene (Cheng et al. 2015). Table 4.7 shows the reports on CRISPR-Cas technologies to help plants resist viral pathogens.

4.21 Transgenic for Stress Tolerance

Abiotic stress refers to a variety of pressures caused by abnormal environmental conditions like extreme sunlight, ultraviolet rays, cold, drought, and salt. Crop damages owing to adverse climatic conditions have been more common in recent years (Boyer et al. 2013), and it is predicted that there will be a rise in the incidence of water scarcity (droughts), extensive heat, and excess water (floods) in different parts of the world (Gourdji et al. 2013). As a result, the productivity and outcome of a number of food crops like maize, rice, and wheat will be declining due to adverse environmental conditions, posing a serious threat to world food supply (Rosenzweig et al. 2014). Abiotic challenges have a deleterious impact on crop survivability, biomass development, and productivity. Crops' resistance to distinct abiotic issue is governed by many genes that respond differently to different stress situations (Sarkar et al. 2019). A decrease in land for crop production of up to 22% was related to high salt concentration and scarcity of water. This has led to reduction of 50% in agricultural production in the concerned places (Agarwal et al. 2017; Bhatnagar-Mathur et al. 2008; Sarkar et al. 2016). Nonliving factors may cause leaf withering, dieback, a drop in absolute moisture concentration, a decline in chlorophyll level and membrane durability, enhanced osmotic potential, and formation of ROS (reactive oxygen species) (Patel et al. 2016). As a result, plant cultivars that can tolerate detrimental effects of the environment must be developed to increase crop yield under numerous abiotic pressures (Sarkar et al. 2014a, b; Bosamia et al. 2015). Advances in biotechnology have aided in the development of genetically engineered crops with increased abiotic stress resistance, particularly in crop productivity (Sarkar et al. 2019). The use of traditional and marker-aided breeding strategies was reported to increase plant resistance to such stresses, although with little impact (Sarkar et al. 2014a, b).

Abiotic stresses are detrimental to crop development because they induce alterations in structural, physiological, biochemical, and genomic activity of the crops as reported by Rai et al. (2013) and Kumar and Verma (2018). Depending on the phase of plant life cycle, even minimal pressure from certain climatic conditions may have a major impact on crop output. Transgenic techniques allow us to discover potential genes, and transcription factors (TFs) used in certain activities in plants, allowing us to get a unified understanding of the plant functional system/mechanisms of action so as to determine plant tolerance and production. The reliability and efficiency of such pattern ensure considerable achievement in subsequent plant modifications. As a result, genetic modification has shown to be a potential strategy for agricultural abiotic stress mitigation (Anwar and Kim 2020).

To fulfil expanding nutritional requirements and mitigate the detrimental consequences of the adverse environmental pressure on plant productivity, there is need to develop GM crops that can resist a broad variety of environmental challenges (Talaat and Shawky 2016; Pandey et al. 2017). Traditional breeding procedures were employed to strengthen the plant's tolerance to environmental challenges using tissue culture via hybridization techniques. However, these strategies have a number of drawbacks (Dita et al. 2006). It takes a lot of time to launch novel crops, multiple unwanted genes could be transported alongside preferred alleles, yield improved performance under environmental stresses is limited due to the ambiguity of the response to stress and its pathways, and there is no assurance of acquiring a specific genomic pairing over large numbers of crosses. Recent technical advancements have significantly increased the capacity of gene detection and functional genomics in crops to control a certain characteristic (Choudhary et al. 2012; Noman et al. 2017). Biotechnology methods facilitate the analysis of proteins and metabolites, allowing us to get a better understanding of physiologically complicated phenomena and cell activities (Dita et al. 2006). Current attempts to enhance plant resistance to environmental stress have shown substantial results (Chen et al. 2018a, b, c). Plant bioengineering techniques to abiotic stress resistance (Mushtaq et al. 2019) are centered on the expression of alleles encoding abiotic-tolerant enzymes for the manufacture of bioactive molecules (Hong et al. 2002a, b).

4.21.1 Transgenic Crops Developed for Resistance to Abiotic Stresses

Transgenic techniques have been deployed by different researchers in an attempt to develop plants that can resist the challenges of such stresses. Many alleles expressing stress-resistant proteins, enzymes, and antioxidants have been studied in allied and unrelated species. The outcomes of some of such studies are discussed below.

4.21.1.1 Drought-Resistant Transgenic Crops

Numerous transcription factors (TFs), including Apetala 2/ethylene-responsive factor (AP2/ERF), MYB, and NAC TF families, have been identified for deployment in producing new crop plants (Marco et al. 2015). Tomato plants transformed with the Arabidopsis DNA construct (CBF/DREB) were more tolerant to water scarcity than the non-modified crops (Kidokoro et al. 2015). Moreover, transgenic apple expressing the OsMYB4 allele from rice showed enhanced less water resistance compared to control group (Buti et al. 2018). Genetically modified grape crops altered using DREB1b gene complex outperformed non-transformed plants in terms of water stress. Another investigation used the WRKY71 genome to modify banana, and the resultant GM crops displayed resilience to a variety of abiotic stimuli (Zhou et al. 2015). GM rice tolerant to drought was developed using SNAC1 and OsNAC6 genome as reported by Todaka et al. (2015) under field conditions. Additionally, researchers have shown that converting transgenic plants with the TaNAC67 gene complex improves drought resistance by boosting physiological features, strengthening membrane integrity, and activating many genes involved in stress defenses (Sharma et al. 2017). Various researches have demonstrated that overexpression of late embryogenesis abundant (LEA) polypeptides improves resistance to dehydration and water loss conditions by reducing membrane degradation and enhancing the manufacture and storage of osmolytes proline and glycine betaine

(Artur et al. 2018). Additionally, overexpression of *Arthrobacter globiformis* choline oxidase gene has been demonstrated to provide drought resistance on transgenic potato plants by increasing the aggregation of osmolytes proline and glycine betaine. Not only did the transgenic potato plants accumulate more osmolytes, but they also had a greater chlorophyll concentration, less lipid peroxidation, and increased antioxidant enzyme activity as compared with non-modified lines (Chen et al. 2016).

4.21.1.2 Metal Stress Tolerance Through Transgenic Crops

Metals may cause oxidative stress in plants via a variety of activities such as oxidation/reduction reactions that generate ROS, interfering with different plant functions including photosynthesis activities. Furthermore, metal buildup (e.g., Hg^{+} and Cu^{+}) exhibits acute sensitivity against the thiol group, which has a significant impact on the morphology and function of protein (Berni et al. 2018). Soil mining is the traditional technique of metal removal; cleaning and reburial have not yet helped significantly in the cleanup of metal pollution of agricultural lands. As a result, using a transgenic method to drive metal cleanup by crops is a viable approach. Many genes had been identified as being implicated in metal ion absorption, transportation, and sequestration, and increased exogenous application of such metal-responsive genetic makeup is an efficient technique for designing agricultural crops exhibiting greater metal stress tolerance (Marco et al. 2015). According to a study, upregulation of metal chelating agent, metallothioneins (MTs), in tobacco and oilseed crop increased the tolerance of mutant crops to environmental stresses (Peng et al. 2017). Furthermore, transformed cauliflower species with exogenous production of the metal resistance-associated metallothionein (CUP1) genotype showed 15-fold superior resilience to Cd and Cu destruction compared to non-transgenic counterparts (Ruta et al. 2018). Genetically modified mustard expressing the glutamyl cysteine synthase gene showed improved resilience to Cd stress via increasing ion accumulation and absorption (Yuan et al. 2015). Another significant gene (NtCBP4) capable of transporting metal from tobacco plants was discovered to be implicated in the adsorption of protein calmodulin. This gene's expression, on the other hand, led to increased Ni and Pb accumulation in modified plants (Mosa et al. 2016). Resilience to iron and cadmium exposure was enhanced in modified tobacco expressing *A. thaliana's* natural resistance-associated macrophage protein 1 (NRAMP1). Likewise, upregulation of a number of other iron carrier genes, such as IRTI, has been reported to enhance Fe, Zn, Cd, and Mn metal transportation (Agorio et al. 2017).

4.21.1.3 Transgenics for Heat Stress Tolerance

Extreme heat, also referred to as heat stress, is among the main damaging factors of crop production (Wahid et al. 2012). Plants suffer from oxidative pressure due to high temperatures, which damages membranes, reduces plant photosynthetic capacity, increases the generation of reactive oxygen species, and decreases their recycling (Dwivedi et al. 2016). Because ROS take part in numerous signaling routes at the baseline level, and once their level surpasses a definite limit, antioxidative defense pathway peptides like superoxide dismutase (SOD), glutathione reductase (GR), peroxidase (POX), and catalase (CAT) are efficient at scavenging them; hybrid crops with enhanced ROS scavenging potential are needed for genetic manipulation for thermal tolerance in plants (Fancy et al. 2017). Genetically modified tomato overexpressing the cAPX gene also demonstrated higher thermal stability of up to 40 °C in open-field settings, reduced heat stress-induced oxidation injury, cholesterol degradation, and increased antioxidant enzymatic scavenging (Sadiq and Akram 2018). Cu/Zn-SOD production in okra improved heat resistance in potato species through activation of several ROS-scavenging enzymes, allowing effective suppression of ROSs in GM potato cultivars (Sadiq and Akram 2018). In transgenic tomato plants, overexpression of *S*-adenosyl-I-methionine decarboxylase (SAMDC) from *Saccharomyces cerevisiae* boosted polyamine synthesis, which improved heat tolerance (Parmar et al. 2017). Heat-shock proteins (HSPs) act as molecular chaperones, maintaining protein homeostasis, thus inhibiting protein denaturation (Jacob et al. 2017). Studies have demonstrated that overexpressing heat-shock factor (HSF) from *Glycine max* improves resilience soybean to high temperature by triggering numerous genes related to stress defense process (Zhang et al. 2015). Similarly, overexpressing VpHSF1 from *Vitis pseudoreticulata* increases thermal dissipation in hybrid tobacco crops (Hu et al. 2006).

4.21.1.4 Transgenics for Salt Stress Tolerance

Salinization of cropland is a major threat to the ecosystem and agricultural sustainability. Abiotic variables influencing plant growth and productivity globally include salt concentration of the soil (Bless et al. 2018). Presently, there are over 1.5 billion ha of farmland, and about 25% of that land is unsuitable for agriculture due to severe soil salinity (Bless et al. 2018). Like other factors, salinity destruction is controlled by a sophisticated regulatory structure (Munns and Gilliham 2015). Many studies have demonstrated that crops that can tolerate environmental stresses, like salt stress, may produce pathogenesis-related proteins to limit the impact of pressure/stress (Negrao et al. 2017). When a plant is subjected to such stress, osmotin peptides are the fastest to express their presence, stimulating the plant's intrinsic defense to battle detrimental effects of numerous abiotic stresses (Wan et al. 2017). In order to test this idea, strawberry crops expressing the osmotin gene demonstrated increased resistance to salt stress relative to non-modified cultivar (Sripriya et al. 2017). Transgenic chili plants modified using tobacco osmotin genome in the T2 progeny showed higher resilience to salt stress as a consequence of enhanced productivity, reduced lipid peroxidation, and greater oxidative enzyme activities (Ullah et al. 2018). Reactive oxygen species sequestration in modified tomato enhanced salt stress resistance by increasing cytoplasmic ascorbate peroxidase (cAPX) production (Jiang et al. 2016a, b). Transgenic cherry tomato lines overexpressing GalUR gene from strawberry demonstrated improved salt tolerance, ascorbic acid accumulation, and protection system for other stresses (Lim et al. 2016). Table 4.8 shows numerous transgenic agricultural plants resistant to abiotic stresses.

Table 4.8 Transgenic crops for resistance to biotic and abiotic stresses

Common name	Scientific name/ variety	Gene	Gene source	Useful trait introduced	Expression	Reference(s)
Biotechnological strategies to combat biotic stresses						
Alfalfa	*Medicago sativa*	*cryIC*	*Bacillus thuringiensis*	Resistance to *Spodoptera littoralis* and *Spodoptera exigua*	0.01–2%	Strizhov et al. (1996)
Arabidopsis thaliana	*Arabidopsis thaliana*	Vq STS36	Grape (*Vitis vinifera*)	Resistance to powdery mildew (*G. cichoracearum*)	–	Wang et al. (2005a, b)
Arabidopsis thaliana	*Arabidopsis thaliana*	TdLTP4	*Triticum turgidum*	Resistance to fungus such as *Alternaria solani* and *Botrytis cinerea*	–	Safi et al. (2015)
Arabidopsis thaliana	*Arabidopsis thaliana*	ViwRKY3	*Vitis labrusca* × *Vitis vinifera*	Resistance to *Golovinomyces cichoracearum* and *P. syringae*, *B. cinerea*	12–48 hpi	Xu et al. (2006)
Arabidopsis thaliana	*Arabidopsis thaliana*	GsMYB15	Soybean (*Glycine max*)	Resistance to *Helicoverpa armigera* larva	–	Shen et al. (2018)
Arabidopsis thaliana	*Arabidopsis thaliana*	CBf2	*Muscadinia rotundifolia*	Resistance to downy mildew	22 ng/g fresh wt.	Merdinoglu et al. (2003)
Chickpea	*Cicer arietinum*	*cry1Ac*	*B. thuringiensis*	Resistance against pod borer *Heliothis armigera*	0.003%	Kar et al. (1996, 1997)
Chickpea	*Cicer arietinum*	*cry1Ac*	*B. thuringiensis*	Pod borer insect *Helicoverpa armigera*	14.5–23.5 ng/ mg	Sanyal et al. (2005)
Chickpea	*Cicer arietinum*	*cry1Ac*	*B. thuringiensis*	Protection from *H. armigera* and *S. litura*	6–20 ng/mg	Indurker et al. (2007)

Corn	*Zea mays*	*cry1Ab*	*B. thuringiensis*	Resistance to European corn borer (*Ostrinia nubilalis* Hubner)	0.4	Koziel et al. (1993)
Corn	*Zea mays*	*cry1Ab*	*B. thuringiensis*	Resistance to *Ostrinia nubilalis*	14–213 ng/g FW	Fearing et al. (1997)
Corn	*Zea mays*	*cry1Ab*	*B. thuringiensis*	Protection against *Spodoptera littoralis*	46.8–85.3 ng/cm^2	Dutton et al. (2005)
Eggplant	*Solanum melongena*	*cryIIIb*	*B. thuringiensis*	Resistance to Colorado potato beetle (*L. decemlineata* Say)	–	Arencibia et al. (1997)
Eggplant	*Solanum melongena*	*cry1Ab*	*B. thuringiensis*	Significant insecticidal activity against *Leucinodes orbonalis*	0.02%	Kumar et al. (1998)
Eggplant	*Solanum melongena*	*cry1Ab*	*B. thuringiensis*	Resistance to tuber moth (*Phthorimaea operculella* Zeller)	–	Peferoen et al. (1990)
Eggplant	*Solanum melongena*	*cryIIIA*	*B. thuringiensis*	Tolerance to Colorado beetle (*Leptinotarsa decemlineata* Say)	–	Perlak et al. (1993)
Eggplant	*Solanum melongena*	*cry*V Bt *cry1Ab*	*B. thuringiensis*	Resistance to potato tuber moth (*P. operculella* Zeller)	–	Douches et al. (1998)
Eggplant	*Solanum melongena*	*cry9Aa2*	*B. thuringiensis*	Resistance to potato tuber moth	–	Gleave et al. (1998)
Eggplant	*Solanum melongena*	*cry1Ab*	*B. thuringiensis*	Resistance to *Helicoverpa armigera* (Hubner)	0.005–0.04	Chakrabarti et al. (2000)
Eggplant	*Solanum melongena*	*cry1Ac*	*B. thuringiensis*	Resistance to *Tecia solanivora*	0.02–17μg/g FW	Milena Valderrama et al. (2007)

(continued)

Table 4.8 (continued)

Common name	Scientific name/ variety	Gene	Gene source	Useful trait introduced	Expression	Reference(s)
Mexican cotton	*Gossypium hirsutum*	*cry1Ab* *cry1Ac*	*B. thuringiensis*	Resistance to cotton bollworm (*H. armigera* Hubner)	0.05–0.1%	Perlak et al. (1990)
Mexican cotton	*Gossypium hirsutum*	*cry2Ab*	*B. thuringiensis*	Resistance to pink bollworm (*Pectinophora gossypiella*)	–	Tabashnik et al. (2002)
Mexican cotton	*Gossypium hirsutum*	*Bean chitinase*	–	Resistance against *diverticulosis* in cotton	–	Tohidfar et al. (2005)
Moth bean	*Vigna aconitifolia*	*cry1Ac*	*B. thuringiensis*	Protection from *H. armigera*	–	Kamble et al. (2003)
Peanut	*Arachis hypogea*	*cry1Ac*	*B. thuringiensis*	Resistance against lesser cornstalk borer	0.18%	Singsit et al. (1997)
Pigeon pea	*Cajanus cajan*	*cry1EC*	*B. thuringiensis*	Resistance to *Spodoptera litura*	–	Surekha et al. (2005)
Pigeon pea	*Cajanus cajan*	*cry1Ab*	*B. thuringiensis*	Protection from *Helicoverpa armigera*	–	Verma and Chand (2005)
Pigeon pea	*Cajanus cajan*	*cry1Ab*	*B. thuringiensis*	Protection from *Helicoverpa armigera*	–	Sharma et al. (2006)
Papaya	*Carica papaya*	*Viral coat protein*	–	Increased resistance to papaya ringspot virus	–	Ferreira et al. (2002)
Rapeseed	*Brassica napus* L.	*cry1Ac*	*B. thuringiensis*	Resistance to *H. zea* Boddie and *S. exigua* Hubner	0.4%	Stewart et al. (1996)
Rapeseed	*Brassica napus*	hrf2 gene encoding harpinXooc protein	*Xanthomonas oryzae*	*Resistance to Sclerotinia sclerotinorium*	–	Ma et al. (2008)

Rice (Asian rice)	*Oryza sativa*	*cry1Ab*	*B. thuringiensis*	Resistance to striped stem borer (*Chilo suppressalis* Walker), and leaf folder (*Cnaphalocrocis medinalis* Guenee)	0.05%	Fujimoto et al. (1993)
Rice (Asian rice)	*Oryza sativa*	*cry1Ac*	*B. thuringiensis*	Resistance to yellow stem borer (*S. incertulas* Walker)	–	Nayak et al. (1997)
Rice (Asian rice)	*Oryza sativa*	*cry1Ab*	*B. thuringiensis*	Resistance to yellow stem borer (*Scirpophaga incertulas*)	–	Wu et al. (1997)
Rice (Asian rice)	*Oryza sativa*	*cry1Ab/Ac*	*B. thuringiensis*	Resistance to striped stem borer and yellow stem borer	3%	Cheng et al. (1998)
Rice (Asian rice)	*Oryza sativa*	*cry1Ab*	*B. thuringiensis*	Resistance to yellow stem borer (*S. incertulas* Walker)	–	Datta et al. (1998)
Rice (Asian rice)	*Oryza sativa*	*cry2A*	*B. thuringiensis*	Effective control of yellow stem borer and rice leaf folder	5%	Maqbool et al. (1998)
Rice (Asian rice)	*Oryza sativa*	*cry1B*	*B. thuringiensis*	Resistance to striped stem borer	0.01–0.4%	Breitler et al. (2000)
Rice (Asian rice)	*Oryza sativa*	*cry1Ab*	*B. thuringiensis*	Resistance to eight lepidopteran rice pests	1%	Shu et al. (2000)
Rice (Asian rice)	*Oryza sativa*	*cry1Ab* *cry1Ac* Hybrid	*B. thuringiensis*	Resistance to leaf folder (*C. medinalis* Guenee) and yellow stem borer (*S. incertulas* Walker)	0.01–0.2%	Tu et al. (2000)

(continued)

Table 4.8 (continued)

Common name	Scientific name/ variety	Gene	Gene source	Useful trait introduced	Expression	Reference(s)
Rice (Asian rice)	*Oryza sativa*	*cry1Ab*	*B. thuringiensis*	Resistance to eight lepidopteran rice pests	–	Shu et al. (2000)
Rice (Asian rice)	*Oryza sativa*	*cry1Ac*	*B. thuringiensis*	Resistance to yellow stem borer (*Scirpophaga incertulas*)	0.1%	Khanna and Raina (2002)
Rice (Asian rice)	*Oryza sativa*	*cry1Ab*	*B. thuringiensis*	Resistance to rice leaf folder, *Cnaphalocrocis medinalis*	–	Ye et al. (2003)
Rice (Asian rice)	*Oryza sativa*	*cry1Ab/Ac*	*B. thuringiensis*	Resistance to stem borer	–	Ramesh et al. (2004)
Rice (Asian rice)	*Oryza sativa*	*cry2A*	*B. thuringiensis*	Resistance to lepidopteran rice pest	9.65–12.11μg/g FW	Chen et al. (2005a, b, c)
Rice (Asian rice)	*Oryza sativa*	*Chitinase*		Fungal disease resistance of rice	0.20%	Itoh et al. (2003)
Rice	*Oryza sativa*	*Xa21*		Resistance to bacterial blight of rice	20.02%	Song et al. (1995)
Rice	*Oryza sativa*	Ace-Amp1	*Allium cepa*	Resistance to *Magnaporthe grisea* and *Rhizoctonia solani*, *Xanthomonas oryzae*	–	Patkar and Chattoo (2006)
Rice	*Oryza sativa*	Ace-AMP1	*Allium cepa*	Resistance to *Magnaporthe grisea*, *Rhizoctonia solani*, and *Xanthomonas oryzae*	–	Patkar and Chattoo (2006)

Rice	*Oryza sativa*	CcHyPRP	Pigeon pea (*Cajanus cajan*)	Resistance to fungal pathogen *Magnaporthe grisea*		Mellacheruvu et al. (2016)
Rice	*Oryza sativa*	Cry1Ab	*B. thuringiensis*	Resistance to lepidopteran insects		Qi et al. (2009)
Rice (Indica, Basmati)	*Oryza sativa*	Cry1Ac, Cry2A	*B. thuringiensis*	Resistance to YSB		Bashir et al. (2005)
Rice (Indica, Minghui-63)	*Oryza sativa*	Cry2A	*B. thuringiensis*	Resistance to YSB	10μg/g leaf fresh weight	Chen et al. (2005a, b, c)
Rice (Indica, Minhuli 63)	*Oryza sativa*	Cry1Ac, Cry2A, Cry9c	*B. thuringiensis*	Resistance to YSB and Asiatic rice borer		Chen et al. (2008)
Rice (Elite Vietnamese)	*Oryza sativa*	Fused gene, Cry1Ab-1B, and hybrid Bt gene, Cry1A/Cry1Ac	*B. thuringiensis*	Resistance to YSB	Bt protein (within the range of 0.8–1.3% of the total soluble protein)	Ho et al. (2006)
Indica Pusa Basmati 1, Japonica, Tainung 67	*Oryza sativa*	Potato proteinase inhibitor 2 (Pin 2)	*B. thuringiensis*	Resistance to YSB		Bhutani et al. (2006)
Indica Basmati 370	*Oryza sativa*	Cry1Ac, Cry2A	*B. thuringiensis*	Resistance to YSB		Riaz et al. (2006)
Rice (Korean varieties: PI, P-II and P-III)	*Oryza sativa*	Cry1Ab	*B. thuringiensis*	Resistance to YSB		Kim et al. (2008)
Rice (Zhuxian B)	*Oryza sativa*	Sbti + GNA	*Galanthus* and *Glycine max*	Resistance to leaf folder + BPH		Li et al. (2005a, b)

(continued)

Table 4.8 (continued)

Common name	Scientific name/ variety	Gene	Gene source	Useful trait introduced	Expression	Reference(s)
Indica rice	*Oryza sativa*	Cry1Ab, Cry1Ac, GNA	*B. thuringiensis*	Resistance to YSB		Ramesh et al. (2004)
Indica rice	*Oryza sativa*	Cry1Ab, Cry1Ac	*B. thuringiensis*	Resistance to YSB	0.26–160 ng/ mL	Alcantara et al. (2004)
Indica rice	*Oryza sativa*	Cry1Ac, Cry2A, GNA	*B. thuringiensis*	Resistance to lepidopteron insects		Rahman et al. (2007)
Indica rice	*Oryza sativa*	Chitinase + β-1,3-glucanase genes	*Trichoderma viride*	*Rhizoctonia solani*		Sridevi et al. (2008)
Japonica rice	*Oryza sativa*	Pi-d2	*Variety (Digu)*	Rice leaf blast and neck blast		Chen et al. (2010a, b)
Coffee	*Coffea canephora/ Coffea arabica*	*cry1Ac*	*B. thuringiensis*	Resistance to leaf miner	>0.1%	Leroy et al. (2000)
Soybean	*Glycine max*	*cry1Ac*	*B. thuringiensis*	Resistance to bollworm (*Helicoverpa zea* Boddie) and budworm (*H. virescens* F.)	0.02%	Stewart et al. (1996)
Soybean	*Glycine max*	*Viral coat protein*	–	Resistance to soybean dwarf virus	–	Tougou et al. (2006)
Soybean	*Glycine max*	*Rps1-k*	–	Enhanced resistance to root and stem disease caused by *Phytophthora sojae* in soybean	2.12%	Gao et al. (2005)
Soybean	*Glycine max*	Rhg-1	Soybean (*Glycine max*)	Resistance to *Heterodera glycines*		Kandoth et al. (2011)

Soybean	*Glycine max*	Rhg-4	Soybean (*Glycine max*)	Resistance to *Heterodera glycines*	–	Liu et al. (2012), Matthews and Youssef (2016)
Sugarcane	*Saccharum officinarum*	*cry1Ab*	*B. thuringiensis*	Resistance to stem borer (*Diatraea saccharalis* F.)	–	Arencibia et al. (1997)
Sugarcane	*Saccharum officinarum*	*cry1Ac*	*B. thuringiensis*	Control against stem borer in field trials	50 ng/mg TSP	Weng et al. (2011)
Sweet potato	*Ipomoea batatas*	*cryIIIA δ-endotoxin*	*B. thuringiensis*	Resistance against sweet potato weevil (*Cylas formicarius*)	–	Moran et al. (1998)
Tobacco	*Nicotiana tabacum*	*cry1Aa*	*B. thuringiensis*	Resistance to tobacco hornworm (*M. sexta* L.)	–	Barton et al. (1987)
Tobacco	*Nicotiana tabacum*	*δ-Endotoxin var. kurstaki HD1*	–	Resistance to lepidopteran insects	–	Barton et al. (1987)
Tobacco	*Nicotiana tabacum*	*cry1Ab*	*B. thuringiensis*	Resistance to tobacco hornworm (*M. sexta* L.) and budworm (*H. virescens* Fabricius)	0.001%	Vaeck et al. (1987)
Tobacco	*Nicotiana tabacum*	*cry1Ac*	*B. thuringiensis*	Resistance to tobacco hornworm (*M. sexta* L.)	0.03%	Perlak et al. (1991)
Tobacco	*Nicotiana tabacum*	*Cry1Ab*	*B. thuringiensis*	Resistance to lepidopteran pests	400 ng/μg/g FW	Carozzi et al. (1992)
Tobacco	*Nicotiana tabacum*	*cry1Ac*	*B. thuringiensis*	Resistance to tobacco budworm (*H. virescens* Fabricius)	3–5%	McBride et al. (1995)

(continued)

Table 4.8 (continued)

Common name	Scientific name/variety	Gene	Gene source	Useful trait introduced	Expression	Reference(s)
Tobacco	*Nicotiana tabacum*	*cry1C*	*B. thuringiensis*	Resistance to *Spodoptera litura* and *Spodoptera exigua*	0.01–0.2%	Strizhov et al. (1996)
Tobacco	*Nicotiana tabacum*	*cry1Ia5*	*B. thuringiensis*	Protection against *Heliothis armigera*	0.06%	Selvapandiyan et al. (1998)
Tobacco	*Nicotiana tabacum*	*cry1Aa2*	*B. thuringiensis*	Resistance to *Heliothis virescens*, *Helicoverpa zea*, *Spodoptera exigua*	2–35	Kota et al. (1999)
Tobacco	*Nicotiana tabacum*	*cry2Aa2*	*B. thuringiensis*	Resistance to cotton bollworm (*H. zea* Boddie)	35.55	De Cosa et al. (2001)
Tobacco	*Nicotiana tabacum*	*δ-Endotoxin*	*B. thuringiensis*	Control of polyphagous pest *Spodoptera litura*	–	Singh et al. (2004)
Tobacco	*Nicotiana tabacum*	*cry2Aa2*	*B. thuringiensis*	Effective control of *Heliothis virescens*	0.215	Zaidi et al. (2005)
Tobacco	*Nicotiana tabacum*	*cry1Ac*	*B. thuringiensis*	Control of *Heliothis virescens* and *Manduca sexta*	0.0835	Gulbitti-Onarici et al. (2009)
Tobacco	*Nicotiana tabacum*	VvWRKY2	*Vitis vinifera*	Resistance to necrotrophic fungal pathogen		Marchive et al. (2013)
Tobacco	*Nicotiana tabacum*	AnnBj1	*Brassica juncea*	Resistance to oomycete pathogen *Phytophthora parasitica var nicotianae*		Jami et al. (2008)

Tobacco	*Nicotiana tabacum*	SpWRKY1	Tomato (*Solanum pimpinellifolium*)	Resistance to *Phytophthora nicotianae*	1.43 times	Li et al. (2015b)
Tobacco	*Nicotiana tabacum*	LeCDJ2	Tomato (*L. esculentum*)	Resistance to pathogen *Pseudomonas solanacearum*	–	Wang et al. (2014a, b)
Tobacco	*Nicotiana tabacum*	GbTLP1 gene	*Sea Island cotton* (*Gossypium barbadense* L.)	*Resistance to Verticillium dahliae*	1.2 mg/g FW	Munis et al. (2010)
Tobacco	*Nicotiana tabacum*	Magi6 peptide	*Macrothele gigas*	*Resistance to Spodoptera frugiperda*		Hernández-Campuzano et al. (2009)
Tomato	*Solanum lycopersicum*	*cry1Ac*	*B. thuringiensis*	Resistance to tobacco hornworm (*Manduca sexta* L.)	0.0015	Fischhoff et al. (1987)
Tomato	*Solanum lycopersicum*	Bt(k)	*B. thuringiensis*	Resistance to tobacco hornworm (*Manduca sexta* L.), tomato pinworm (*Keiferia lycopersicella*), and tomato fruitworm (*Heliothis zea*)	1 ng/mg TSP	Delannay et al. (1989)
Tomato	*Solanum lycopersicum*	*cry1Ac*	*B. thuringiensis*	Resistance to fruitworm (*Helicoverpa armigera* Hubner)	0.06–0.425	Mandaokar et al. (2000)
Tomato	*Solanum lycopersicum*	*cry1Ab*	*B. thuringiensis*	Protection against fruit borer (*Helicoverpa armigera* Hubner)	–	Kumar and Kumar (2004)

(continued)

Table 4.8 (continued)

Common name	Scientific name/ variety	Gene	Gene source	Useful trait introduced	Expression	Reference(s)
Tomato	*Solanum lycopersicum*	*Bt*	*B. thuringiensis* (*kurstaki*) HD-1	Resistance to tobacco hornworm (*Manduca sexta* L.), tomato pinworm (*Keiferia lycopersicella*), and tomato fruitworm (*Heliothis zea*)	–	Delannay et al. (1989)
Tomato	Vollendung	*Vst1* and *vst2*	*Vitis vinifera*	*Phytophthora infestans* resistance	–	Thomzik et al. (1997)
Tomato	Pusa Ruby	*Cry1Ac*	*B. thuringiensis*	Resistant to fruit borer	–	Mandaokar et al. (2000)
Tomato	Pusa Early Dwarf (PED)	*taf.4b*	*A. thaliana*	Pathogen-tolerant transgenic plant	–	Sawant et al. (2010)
Tomato	Pusa Early Dwarf (PED)	Chitinase gene	Fern (*Tectaria*)	Resistance against whiteflies	–	Singh et al. (2011); Koul et al. (2014a)
Tomato	Pusa Early Dwarf (PED)	*cry1Ac*	*B. thuringiensis*	Resistance against *Helicoverpa armigera*	–	Koul et al. (2012)
Tomato	Asc/Asc, VFNT cherry	*p35*	*Baculovirus*	Broad-spectrum disease resistance	–	Lincoln et al. (2002)
Tomato	Pusa Ruby	*CP*	*Tomato leaf curl virus (TLCV)*	Resistance against TLCV	–	Raj et al. (2005)
Tomato	Moneymaker	*PI-II* and *PCI*	*S. tuberosum*	Multiple insect resistance	–	Abdeen et al. (2005)
Tomato	Moneymaker	*Ep5c* (antisense)	*Pseudomonas syringae*	Resistant to *P. syringae*	–	Alberto et al. (2005)

Tomato	L. 276-76, L.149-88, INB777	Nucleoprotein gene	Tomato spotted wilt virus	Resistant to TSWV	–	Nervo et al. (2003)
Tomato	LEPA	*cry1Ab*	*B. thuringiensis*	Resistant to fruit borer	–	Kumar and Kumar (2004)
Tomato	*Solanum lycopersicum*	Mi-1.2	Wild tomato (*S. peruvianum*)	Resistance to *Meloidogyne incognita*	–	Milligan et al. (1998)
Tomato	*Solanum lycopersicum*	CeCPI	Taro (*Colocasia esculenta*)	Resistance to *Meloidogyne incognita*	–	Chan et al. (2010)
Tomato	*Solanum lycopersicum*	VqSTS36	Grape (*Vitis vinifera*)	Resistance to *Botrytis cinerea*	–	Wang et al. (2005a, b)
Wild cabbage	*Brassica oleracea* L.	*cry1Ac or cry1C*	*B. thuringiensis*	Resistance to diamondback moth larvae	0.5 ng/g FW	Cao et al. (2005)
Sugar beet	*Beta vulgaris*	Hs1pro-1	Wild beet (*B. procumbens*)	Resistance to *Meloidogyne schachtii*	–	Cai et al. (1997)
Sugar beet	*Beta vulgaris*	SpTI-1	Sweet potato (*Ipomoea batatas*)	Resistance to *Heterodera schachtii*	–	Cai et al. (2003)
Potato	*Solanum tuberosum*	Gpa-2	Potato (*S. tuberosum*)	Resistance to *Globodera pallida*	–	Van der vossen et al. (2000)
Potato	*Solanum tuberosum*	Hero A	Tomato (*S. lycopersicum*)	Resistance to *Globodera pallida* and *Globodera rostochiensis*		Sobczak et al. (2005)
Potato	*Solanum tuberosum*	Gro 1-4	Potato (*S. tuberosum*)	Resistance to *Globodera rostochiensis*	–	Paal et al. (2004)

(continued)

Table 4.8 (continued)

Common name	Scientific name/ variety	Gene	Gene source	Useful trait introduced	Expression	Reference(s)
Potato	*Solanum tuberosum*	CpTI	Cowpea (*Vigna unguiculata*)	Resistance to *Globodera pallida* and *Meloidogyne incognita*		Hepher and Atkinson (1992)
Potato	*Solanum tuberosum*	Oc-1D86	Rice (*Oryza sativa*)	Resistance to *Globodera pallida* and *Meloidogyne incognita*	–	Urwin et al. (2003)
Potato	*Solanum tuberosum*	RB resistance gene	*Solanum bulbocastanum*	Potato late blight resistance		Liu et al. (2009)
Potato	*Solanum tuberosum*	Cry1Ab	*B. thuringiensis*	Potato tuber moth resistance		Kumar et al. (2010)
Potato	*Solanum tuberosum*	StPUB17 (UND/ PUB/ARM) repeat-type gene	*Solanum tuberosum*	*Phytophthora infestans resistance*		Ni et al. (2010)
Wheat	*Triticum aestivum*	Cre loci	*Aegilops* spp.	Resistance to *Heterodera avenae*	–	Safari et al. (2005)
Wheat	*Triticum aestivum*	Ace-AMP1	*Allium cepa*	Resistance to some fungus spore *Blumeria graminis*	–	Roy-Barman et al. (2006)
Wheat	*Triticum aestivum*	PIN 2	Potato (*S. tuberosum*)	Resistance to *Heterodera avenae*	–	Vishnudasan et al. (2005)
Wheat	*Triticum aestivum*	Ta-Tlp (thaumatin-like protein gene)	*Haynaldia villosa*	Powdery mildew and *Fusarium* head blight resistance	100.0 ± 0.0	Xing et al. (2008)

Plantain	*Musa* spp.	CCII	Maize (*Zea mays*)	Resistance to *Radopholus similis*, *Helicotylenchus multicinctus*, and *Meloidogyne*		Roderick et al. (2012)
Groundnut	*Arachis hypogaea*	TPWNV	TSWV-L	Resistance to tomato spotted wilt virus	–	Li et al. (1997)
Groundnut	*Arachis hypogaea*	*Cry*IA(c)	*Bacillus thuringiensis*	Resistance to lesser cornstalk borer	–	Singsit et al. (1997)
Groundnut	*Arachis hypogaea*	AGLUI	*Medicago sativa*	Resistance to *Sclerotinia minor*	–	Chenault et al. (2003)
Groundnut	*Arachis hypogea*	Cry1EC	*Bacillus thuringiensis*	Resistance to *Spodoptera litura*	–	Kumar et al. (2009)
Groundnut	*Arachis hypogea*	BjNPR1 and Tfgd	*Brassica juncea* and *Trigonella foenum-graecum*	Resistance to *Aspergillus flavus* and *Cercospora arachidicola*		Sundaresha et al. (2016)
Biotechnological strategies to combat abiotic stresses						
Alfalfa	*Medicago sativa*	TPS1 (trehalose synthesis)	*Yeast*	Drought, salt, temperature tolerance	55.5–83.5μg/g fresh weight	Suarez et al. (2009)
Alfalfa	*Medicago sativa*	MsALR (aldose/aldehyde reductase)	*Medicago sativa*	Drought and heavy metal tolerance	180μg/g fresh weight	Oberschall et al. (2000)
Alfalfa	*Medicago sativa*	A1fin1 (transcription factor)	*Arabidopsis thaliana*	Salt tolerance		Winicov (2000)
Alfalfa	*Medicago sativa*	*MsSPSA* and *MsSPSB*	*Zea mays*	Glyphosate tolerance	–	Sayed Gebril et al. (2015)
Arabidopsis	*Arabidopsis thaliana*	AtHsp 17.6A (small heat-shock protein)	*Arabidopsis thaliana*	Drought and salt tolerance		Sun et al. (2001)

(continued)

Table 4.8 (continued)

Common name	Scientific name/ variety	Gene	Gene source	Useful trait introduced	Expression	Reference(s)
Arabidopsis	*Arabidopsis thaliana*	TPS and TPP (trehalose synthesis)	*Saccharomyces cerevisiae*	Drought, salt, temperature tolerance	29.9μg/g fresh weight	Miranda et al. (2007)
Arabidopsis	*Arabidopsis thaliana*	pdc1 and pdc2 (pyruvate decarboxylase overexpression)	*Arabidopsis thaliana*	Hypoxic stress survival		Ismond et al. (2003)
Arabidopsis	*Arabidopsis thaliana*	SPDS (spermidine synthase)	–	Salinity tolerance	20 pmol [14CO_2] mg/ prot/h	Bagni et al. (2006)
Arabidopsis	*Arabidopsis thaliana*	ProDH (proline dehydrogenase)	*Arabidopsis thaliana*	Salt stress tolerance		Nanjo et al. (1999)
Arabidopsis Rice Brassica	*Arabidopsis thaliana*	COD1; COX (choline oxidase)	*Arthrobacter globiformis*	Salt, cold, light stress tolerance	960 nmol/g fresh weight	Hayashi et al. (1997), Sakamoto et al. (1998), Huang et al. (2000), Prasad et al. (2000)
Arabidopsis	*Arabidopsis thaliana*	HAL3 (FMN-binding protein)	*Saccharomyces cerevisiae*	Salt and osmotic tolerance		Espinosa-Ruiz et al. (1999)
Arabidopsis	*Arabidopsis thaliana*	HAL1	*Saccharomyces cerevisiae*	Salt stress tolerance		Ellul et al. (2003)
Arabidopsis	*Arabidopsis thaliana*	DREB1A	*Lilium longiflorum*	Drought, salt, and cold tolerance		Kasuga et al. (1999), Liu et al. (1998)
Arabidopsis	*Arabidopsis thaliana*	OsDREB1A	*Oryza sativa*	Drought, salt, and cold tolerance		Dubouzet et al. (2003)

Arabidopsis	*Arabidopsis thaliana*	DREB1A, DREB2A (transcription factor)	*Arabidopsis thaliana*	Drought and cold tolerance		Maruyama et al. (2009)
Arabidopsis	*Arabidopsis thaliana*	OsSMCP1 (transcription factor)	*Oryza sativa*	Salt stress tolerance		Yokotani et al. (2009)
Arabidopsis	*Arabidopsis thaliana*	OrbHLH2 (transcription factor)	*Zizania*	Salt and osmotic stress tolerance		Zhou et al. (2009)
Arabidopsis	*Arabidopsis thaliana*	OsWRKY45 (transcription factor)	*Oryza sativa*	Drought tolerance		Qiu and Yu (2009)
Arabidopsis	*Arabidopsis thaliana*	CBF4	*Arabidopsis thaliana*	Drought tolerance		Haake et al. (2002)
Arabidopsis	*Arabidopsis thaliana*	ABF3/ABF4	*Arabidopsis thaliana*	Drought tolerance		Kang et al. (2002)
Arabidopsis	*Arabidopsis thaliana*	AtMYC2/AtMYB2	*Arabidopsis thaliana*	Drought tolerance		Abe et al. (2003)
Arabidopsis	*Arabidopsis thaliana*	CpMYB10	*Craterostigma plantagineum*	Drought and salt tolerance		Villalobos et al. (2004)
Arabidopsis	*Arabidopsis thaliana*	Mn-SOD	*Saccharomyces cerevisiae*	Oxidative stress tolerance		Wang et al. (2004a, b)
Arabidopsis	*Arabidopsis thaliana*	DHAR1 (dehydroascorbate reductase)	*Oryza sativa*	Salt tolerance		Ushimaru et al. (2006)
Arabidopsis	*Arabidopsis thaliana*	AtALDH3 (aldehyde dehydrogenase)	*Craterostigma plantagineum*	Drought, salt, and oxidative stress		Sunkar et al. (2003)
Arabidopsis	*Arabidopsis thaliana*	AtGSK1	*Arabidopsis thaliana*	Drought and salt stress tolerance		Piao et al. (2001)
Arabidopsis	*Arabidopsis thaliana*	AtNDPK2 (nucleotide diphosphate kinase)	*Saccharomyces cerevisiae*	Salt, cold, methyl viologen tolerance		Moon et al. (2002)
Arabidopsis	*Arabidopsis thaliana*	SOS1 (plasma membrane Na^+/H^+ antiporter)	*Arabidopsis thaliana*	Salt stress tolerance	18 mol g FW	Shi et al. (2003)

(continued)

Table 4.8 (continued)

Common name	Scientific name/ variety	Gene	Gene source	Useful trait introduced	Expression	Reference(s)
Arabidopsis	*Arabidopsis thaliana*	AVP1 (K^+/Na^+ transport regulation)	*Arabidopsis thaliana*	Drought and salt tolerance		Gaxiola et al. (2001)
Arabidopsis	*Arabidopsis thaliana*	CaXTH3 (xyloglucan endotransglucosylase)	*Capsicum frutescens*	Drought and salt stress tolerance		Cho et al. (2006)
Arabidopsis	*Arabidopsis thaliana*	ZmOPR1 (12-oxo-phytodienoic acid reductases)	*Arabidopsis thaliana*	Osmotic and salt stress tolerance		Gu et al. (2008)
Arabidopsis	*Arabidopsis thaliana*	SPCP2 (papain-like cysteine protease)	*Ipomoea batatas*	Salt and drought stress tolerance		Chen et al. (2010a, b)
Arabidopsis	*Arabidopsis thaliana*	CodA	*Arthrobacter globiformis*	Salt and cold tolerance	–	Hayashi et al. (1998)
Arabidopsis	*Arabidopsis thaliana*	*PgCYP736A12*	Ginseng	Herbicide resistance	–	Khanom et al. (2019)
Arabidopsis	*Arabidopsis thaliana*	*CYP1A2*	Human	Herbicide resistance	–	Azab et al. (2018)
Arabidopsis	*Arabidopsis thaliana*	CMO + BADH	*Spinacia oleracea*	Salt tolerance	–	Hibino et al. (2002)
Arabidopsis	*Triticum aestivum*, *Arabidopsis thaliana*	TaPEPKR2	*Triticum aestivum*	Heat and dehydration tolerance	–	Zang et al. (2018)
Arabidopsis	*Arabidopsis thaliana*	TaHSP23.9	*Triticum aestivum*	Heat and salt stress tolerance	–	Wang et al. (2020)
Arabidopsis	*Arabidopsis thaliana*	CodA	*A. globiformis*	Salt tolerance	–	Sulpice et al. (2003)
Apple	*Malus pumila*	Osmyb4 (cold-induced transcription factor)	*Oryza sativa*	Drought and cold tolerance	44% and 9.5 nmol O_2/g/h	Pasquali et al. (2008)

Bean	–	P5CS (Δ1-pyrroline-5-carboxylate synthase)	*Vigna aconitifolia*	Drought, salt, and cold tolerance		Chen et al. (2009)
Bottle gourd	*Lagenaria siceraria*	AVP1 gene	*A. thaliana*	Resistance to salinity	–	Han et al. (2015)
Cauliflower	*Brassica oleracea* var. *botrytis*	*DREB1A*	*Arabidopsis thaliana*	Increased tolerance to drought and salt		Pasapula et al. (2011)
Corn	*Zea mays*	*aroA*	*Zea mays*, *E. coli*	Glyphosate herbicide resistance	0.01–0.2%	Howe et al. (2002)
Cotton	*Gossypium*	*AnnBj1*	Mustard	Improved salt tolerance	0.16%	Divya et al. (2010)
Cotton	*Gossypium*	*ahCMO*	Garden orache	Enhanced salt tolerance	2.12%	Zhang et al. (2009)
Cotton	*Gossypium*	*AVP1*	*Arabidopsis thaliana*	Increased tolerance to salinity, cold, and drought	4.25%	Kasuga et al. (1999)
Citrus	*Citrus aurantium*	P5CSF129A	*V. aconitifolia*	Drought tolerance and enhanced oxidative capacity	–	Molinari et al. (2004)
Citrus	*Citrus aurantium*	P5CSF129A	*V. aconitifolia*	Drought tolerance and enhanced oxidative capacity	–	Molinari et al. (2004)
Chickpea	*Cicer arietinum*	P5CS	*V. aconitifolia*	Salt tolerance with increased proline	–	Ghanti et al. (2011)
Chickpea	*Cicer arietinum*	P5CS	*V. aconitifolia*	Salt tolerance with increased proline	–	Ghanti et al. (2011)
Chickpea, pigeon pea	*Cicer arietinum*, *Cajanus cajan*	RuvB and p68	–	Salt stress tolerance	–	Kharb et al. (2021)
Chili pepper	*Capsicum frutescens*	Osmotin gene	*Triticum aestivum*	Drought and salinity resistance	–	Bulle et al. (2016)

(continued)

Table 4.8 (continued)

Common name	Scientific name/ variety	Gene	Gene source	Useful trait introduced	Expression	Reference(s)
Daisy	*Bellis perennis*	CmWRKY1	*Chrysanthemum morifolium*	Tolerance to drought stress	–	Fan et al. (2016)
Finger millet and soybean	*Eleusine coracana* and *Glycine max*	*TUAm*	*Eleusine indica*	Trifluralin herbicide resistance	–	Yemets et al. (2008)
Groundsel	–	*PSBA gene*	Groundsel (mutation)	Triazine resistance	–	Ryan (1970)
Groundsel	*Bidens subalternans*	*TIPT*	*Bidens subalternans* (mutation)	Herbicide resistant	–	AlcaNtara-De La Cruz et al. (2016)
Groundsel	*Amaranthus hybridus*	*TAP-IVS*	*Amaranthus hybridus* (mutation)	Herbicide resistant	–	Gaines et al. (2010)
Grass pea	*Lathyrus sativus*	GpZF	*Lathyrus sativus*	Drought tolerance	–	Parsa et al. (2021)
Lettuce	*Lactuca sativa*	ME-leaN4 (LEA protein)	*Brassica napus*	Salt tolerance	2.8 cm and 2.5 g/plant	Park et al. (2005a, b, c)
Mulberry	*Morus indica*	HVA1 (group 3 LEA protein gene)	*Hordeum vulgare*	Salt and drought		Lal et al. (2008)
Mustard	–	AtNHX1 (vacuolar Na^+/H^+ antiporter)	*Arabidopsis thaliana*	Salt tolerance		Zhang et al. (2001)
Maize	*Zea mays*	MAPKKK (NPK1)	*Tobacco*	Drought tolerant	–	Shou et al. (2004)
Mustard	*Brassica juncea*	CodA	*A. globiformis*	Salt tolerance	–	Prasad et al. (2000)
Maize	*Zea mays*	*Dicamba monooxygenase*	*Pseudomonas maltophilia*	Dicamba tolerance		Cao et al. (2011)
Maize	*Zea mays*	BetA	*Escherichia coli*	Drought tolerance	–	Quan et al. (2004)
Maize	*Zea mays*	ARGOS8	*Zea mays*	Drought stress tolerance	–	Shi et al. (2017)

Orchids		*Pat, bar*	*Streptomyces hygroscopicus*	Resistance to methionine sulfoximine in orchids	>0.1%	Chai et al. (2007)
Potato	*Solanum tuberosum*	P5CS (Δ1-pyrroline-5-carboxylate synthase)	*Arabidopsis thaliana*	Salt tolerance	900μm/g FW	Hmida-Sayari et al. (2005)
Paspalum grass	*Paspalum dilatatum*	DREB1A (transcription factor)	*Hordeum spontaneum*	Drought tolerance		James et al. (2008)
Petunia	*Petunia × atkinsiana*	ZPT2-3 (Cys2/His2-type zinc finger protein)	*Petunia*	Drought tolerance		Sugano et al. (2003)
Potato	*Solanum tuberosum*	BADH	*S. oleracea*	Salinity and drought tolerance	–	Zhang et al. (2011a, b, c)
Potato	*Solanum tuberosum*	stnsLTP1	*Saccharomyces cerevisiae*	Heat, drought, and salinity tolerance	–	Gangadhar et al. (2016)
Potato	*Solanum tuberosum*	StPUB17 (UND/PUB/ARM) repeat-type gene	*Solanum tuberosum*	Salt tolerance		Ni et al. (2010)
Rice	*Oryza sativa*	AtOAT (ornithine amino transferase)	*Escherichia coli*	Drought and salt stress tolerance	0.31–1.076 mg/g fresh weight	Jang et al. (2003)
Rice	*Oryza sativa*	pdc1 (pyruvate decarboxylase overexpression)	*Oryza sativa*	Submergence tolerance		Minhas and Grover (1999)
Rice	*Oryza sativa*	P5CS (Δ1-pyrroline-5-carboxylate synthase)	*Vigna aconitifolia*	Salt and drought stress tolerance	68μg^{-1} FW	Zhu et al. (1998)

(continued)

Table 4.8 (continued)

Common name	Scientific name/ variety	Gene	Gene source	Useful trait introduced	Expression	Reference(s)
Rice	*Oryza sativa*	TPP1 (trehalose synthesis)	*Arabidopsis thaliana*	Salt and cold stress tolerance	115μg/g FW	Ge et al. (2008)
Rice	*Oryza sativa*	ADC (polyamine synthesis)	*Datura stramonium*	Drought stress tolerance		Capell et al. (2004)
Rice	*Oryza sativa*	Os LEA3-1 (LEA protein)	*Hordeum vulgare*	Drought stress tolerance		Xiao et al. (2007)
Rice	*Oryza sativa*	OsNAC10 (transcription factor)	*Oryza sativa*	Drought stress tolerance	Increasing grain yield by 25–42%	Jeong et al. (2010)
Rice	*Oryza sativa*	OsCDPK (calcium-dependent protein kinase)	*Arabidopsis thaliana*	Drought and salt stress tolerance	250μg/g fresh weight	Saijo et al. (2000)
Rice	*Oryza sativa*	Mn-SOD	*Saccharomyces cerevisiae*	Oxidative stress tolerance		Tanaka et al. (1999)
Rice	*Oryza sativa*	AtNHX1 (vacuolar Na^+/H^+ antiporter)	*Atriplex gmelinii*	Salt stress tolerance		Ohta et al. (2002)
Rice	*Oryza sativa*	TSRF1 (ethylene-responsive factor)	*Oryza sativa*	Drought stress tolerance		Quan et al. (2010)
Rice	*Oryza sativa*	*Mx PPO*	*Myxococcus xanthus*	Butafenacil herbicide resistance	–	Lee et al. (2007)
Rice (Asian rice)	*Oryza sativa*	*ostA* and *ostB*	*E. coli*	Enhanced tolerance to salinity, drought, and cold	0.20%	Gao et al. (2007)
Rice	*Oryza sativa*	*nahA1*	*Escherichia coli*	Soil salinity tolerance	–	Wu et al. (2005)
Rice	*Oryza sativa*	HVA1	*Hordeum vulgare*	Drought resistance	0.3–2.5% in leaf 0.3–1.0% in root	Xu et al. (1996)

Rice	*Helianthus tuberosus*	CYP76B1	*Helianthus tuberosus*	Phenylurea herbicide	–	Robineau et al. (1998)
Rice	*Amaranthus palmeri*	EPSPS	*Amaranthus palmeri*	Glyphosate tolerance		
Rice	*Oryza sativa*	P5cs	*Oryza sativa*	Drought and salt tolerance	–	Su and Wu (2004)
Rice	*Oryza sativa*	OAT	*A. thaliana*	Drought and salt stress tolerance	–	Wu et al. (2003)
Rice	*Oryza sativa*	OAT	*O. sativa*	Drought and oxidative stress tolerance	–	You et al. (2012)
Rice	*Oryza sativa*	CodA	*A. globiformis*	Resistance to cold/ freezing	–	Konstantinova et al. (2002)
Rice	*Oryza sativa*	GSMT and DMT^{av}	*Aphanothece halophytica*	Cold and salt tolerance	–	Niu et al. (2014)
Rice	*Oryza sativa*	CYP81A6	Cytochrome P450	Bentazon and metsulfuron-methyl tolerance	–	Lu et al. (2015)
Rice	*Oryza sativa*	BADH	*Hordeum vulgare*	Cold, heat, and salt tolerance	–	Kishitani et al. (2000)
Rice	*Oryza sativa*	Cox		Salt tolerance	–	Su et al. (2006)
Rice	*Oryza sativa*	*AtPLC9*	*Arabidopsis thaliana*	Heat tolerance	–	Liu et al. (2020)
Rice	*Oryza sativa*	OsRab7	–	Tolerance to heat stress	–	El-Esawi and Alayafi (2019)
Rice	*Oryza sativa*	AtDREB1A	*Arabidopsis thaliana*	Drought and salinity tolerance	–	Muthurajan et al. (2021)
Rapeseed	*Brassica napus*	*DtNHX*	*Arabidopsis thaliana*	Salt tolerance	–	Zhang et al. (2001)

(continued)

Table 4.8 (continued)

Common name	Scientific name/ variety	Gene	Gene source	Useful trait introduced	Expression	Reference(s)
Strawberry	Fragaria	Osm1 to Osm4 (osmotin protein accumulation)	*Agrobacterium tumefaciens*	Salt and drought stress tolerance	350μg/g fresh weight	Husaini et al. (2008)
Sugarcane	*Saccharum officinarum*	*TSase*	*Grifola frondosa*	Osmotic stress tolerance	–	Wang et al. (2003)
Soyabean	*Glycine max*	*CP4 EPSPS*	*Agrobacterium tumefaciens*	Glyphosate resistance	–	Chinnadurai et al. (2018)
Soybeans	*Glycine max*	*Glufosinate N-acetyltransferase*	*Streptomyces*	Enhanced resistance to herbicide (dicamba) in soybean	–	Behrens et al. (2007)
Tobacco	*Nicotiana tabacum*	W6 (ethylene-responsive factor gene)	*Boea hygrometrica*	Salt tolerance		Lu et al. (2008)
Tobacco	*Nicotiana tabacum*	betA (choline dehydrogenase)	*Escherichia coli*	Salinity and low temperature	35 ± 15 nmol/g FW	Holmstrom et al. (2000)
Tobacco	*Nicotiana tabacum*	EctA, ectB, ectC	*Halomonas elongata*	Salinity tolerance		Nakayama et al. (2000)
Tobacco	*Nicotiana tabacum*	OstA, OstB (trehalose-6-P synthase, trehalose-6-P phosphatase)	*Escherichia coli*	Salt, drought tolerance	17 ± 5μg trehalose per g of fresh weight	Garg et al. (2002)
Tobacco	*Nicotiana tabacum*	TPS and TPP (trehalose synthesis)	*Arabidopsis thaliana*	Salt and mannitol tolerance	2.3μmol/g FW plantlets	Roosens et al. (2002)
Tobacco	*Nicotiana tabacum*	WCOR15 (cold-induced gene)	*Triticum aestivum*	Freezing tolerance		Shimamura et al. (2006)
Tobacco	*Nicotiana tabacum*	SAMDC (polyamine synthesis)	*Homo sapiens*	Drought, salinity tolerance		Waie and Rajam (2003)

Tobacco	*Nicotiana tabacum*	P5CS (Δ1-pyrroline-5-carboxylate synthase)	*Vigna aconitifolia*	Salt and drought tolerance	1120μg/g fresh weight	Kishor et al. (1995)
Tobacco	*Nicotiana tabacum*	IMT1 (myo-inositol-*O*-methyl transferase)	*Mesembryanthemum crystallinum*	Salt and drought tolerance	35μmol/g fresh weight	Sheveleva et al. (1997)
Tobacco	*Nicotiana tabacum*	BhLEA1, LEA2 (LEA protein)	*Boea hygrometrica*	Drought tolerance		Liu et al. (2009)
Tobacco	*Nicotiana tabacum*	DREB1A (transcription factor)	*Arabidopsis thaliana*	Salt tolerance	160 soluble sugar mg/g DW	Cong et al. (2008)
Tobacco	*Nicotiana tabacum*	Tsi1 (EREBP/AP2 DNA-binding motif)	*Brassica napus*	Salt and pathogen tolerance		Park et al. (2001)
Tobacco	*Nicotiana tabacum*	Fe-SOD (superoxide dismutase)	*Arabidopsis thaliana and Pisum sativum*	Salt and oxidative stress tolerance	50μg/g fresh weight	Van Camp et al. (1996)
Tobacco	*Nicotiana tabacum*	Glutathione-*S*-transferase/ glutathione peroxidase	*Nicotiana tabacum*	Salt and cold tolerance		Roxas et al. (2000)
Tobacco	*Nicotiana tabacum*	KatE (catalase)	*Cyanobacteria*	Salt and oxidative stress tolerance		Al-Taweel et al. (2007)
Tobacco	*Nicotiana tabacum*	Ascorbate peroxidase	*Nicotiana tabacum*	Drought and salt stress tolerance		Badawi et al. (2004)
Tobacco	*Nicotiana tabacum*	GlyI and GlyII (glyoxalase)	*Arabidopsis thaliana*	Salt tolerance		Yadav et al. (2005)
Tobacco	*Nicotiana tabacum*	Cnb1 (calcineurin)	*Saccharomyces cerevisiae*	Salt tolerance		Pardo et al. (1998)
Tobacco	*Nicotiana tabacum*	DnaK (heat-shock proteins)	*Aphanothece halophytica*	Salt tolerance	180μg/g fresh weight	Sugino et al. (1999)
Tobacco	*Nicotiana tabacum*	*betB*	*Escherichia coli*	Salt stress tolerance	–	Holmstrom (1998)

(continued)

Table 4.8 (continued)

Common name	Scientific name/ variety	Gene	Gene source	Useful trait introduced	Expression	Reference(s)
Tobacco	*Nicotiana tabacum*	*TPS1*	*Yeast*	Improved drought tolerance	–	Romero et al. (1997)
Tobacco	*Nicotiana tabacum*	TaALMT1	*Triticum aestivum*	Aluminum tolerance	–	Sasaki et al. (2004)
Tobacco	*Nicotiana tabacum*	*Pat, bar*	*Streptomyces hygroscopicus*	Resistance to herbicide phosphinothricin	0.24%	Greef et al. (1989)
Tobacco	*Nicotiana tabacum*	*GmGSTU4*	Soybean	Herbicide resistance	–	Benekos et al. (2010)
Tobacco	*Nicotiana tabacum*	*CsGSTUs*	*Citrus sinensis*	Fluorodifen herbicide resistance	–	Cicero et al. (2015)
Tobacco	*Nicotiana tabacum*	GmGSTU4	Soyabean	Chloroacetanilide tolerance	–	Benekos et al. (2010)
Tobacco	*Nicotiana tabacum*	PPOX	*Arabidopsis thaliana*	Acifluorfen tolerance	–	Inna Lermontova, Bernhard Grimm (2000)
Tobacco	*Nicotiana tabacum*	*Aro M-1 gene*	*E. coli*	Glyphosate-resistance	–	He et al. (2001)
Tobacco	*Nicotiana tabacum*	*tfdA gene*	*Ralstonia eutropha*	2,4-D resistance	–	Streber et al. (1987)
Tobacco	*Nicotiana tabacum*	*bar*	*Streptomyces hygroscopicus*	Bialaphos resistance	–	Thompson et al. (1987), Lutz et al. (2001)
Tobacco	*Nicotiana tabacum*	P5cs	*Vigna aconitifolia*	Salt tolerance	10–18-fold more Pro	Kavi Kishor et al. (1995)
Tobacco	*Nicotiana tabacum*	P5CSF129A	*Vigna aconitifolia*	Tolerance to osmotic stress	–	Hong et al. (2000)

Tobacco	*Nicotiana tabacum*	ProDH	*A. thaliana*	Increased proline level with tolerance to various abiotic stresses	–	Ibragimova et al. (2012)
Tobacco	*Nicotiana tabacum*	BADH	*Spinacia oleracea*	Heat tolerance	–	Yang et al. (2007a, b)
Tobacco	*Nicotiana tabacum*	BADH1	*O. sativa*	Tolerance to salinity stress	–	Hasthanasombut et al. (2010)
Tobacco	*Nicotiana tabacum*	AhCytb6	*Stenotrophomonas maltophilia*	Nitrogen-deficit tolerance	–	Alexander et al. (2021)
Tomato	*Solanum lycopersicum*	*AtNHX1*	*A. thaliana*	Improved salt tolerance	0.2%	Zhang et al. (2001)
Tomato	*Solanum lycopersicum*	*HAL2*	*Saccharomyces cerevisiae*	Salt tolerance	–	Arillaga et al. (1998)
Tomato	P73	*HAL 1*	*Saccharomyces cerevisiae*	Salt tolerance	–	Gisbert et al. (2000)
Tomato	Heinz 902, Heinz 1439	*ACC deaminase*	*Enterobacter cloacae UW4*	Flood tolerance	–	Grichko and Glick (2001)
Tomato	CL5915-93 D4-1-0-3	*CBF1*	*A. thaliana*	Tolerance to water-deficit stress	–	Hsieh et al. (2002)
Tomato	Bailichun	*BADH*	*Atriplex hortensis*	Salt tolerance	–	Jia et al. (2002)
Tomato	UC82B	*TPS1*	*S. cerevisiae*	Increased tolerance to abiotic stress	–	Cortina and Culianez-Macia (2005)
Tomato	*Solanum lycopersicum*	*APX*	*Pisum sativum*	Tolerance to chilling and salt stress	–	Wang et al. (2005a, b)
Tomato	*Solanum lycopersicum*	*aro A*	*Salmonella typhimurium*	Glyphosate herbicide tolerance	–	Fillatti et al. (1987)

(continued)

Table 4.8 (continued)

Common name	Scientific name/ variety	Gene	Gene source	Useful trait introduced	Expression	Reference(s)
Tomato	*Solanum lycopersicum*	BADH1 (betaine aldehyde dehydrogenase)	*Sorghum bicolor* *Atriplex hortensis*	Salinity tolerance	4.9 nmol/min/ mg protein	Jia et al. (2002)
Tomato	*Solanum lycopersicum*	PPO (polyphenol oxidases suppression)	*Solanum tuberosum*	Drought tolerance	13μmol quinone formed/min/ mg protein	Thipyapong et al. (2004)
Tomato	*Solanum lycopersicum*	CBF1 (DREB1B)	*Arabidopsis thaliana*	Drought tolerance		Hsieh et al. (2002)
Tomato	*Solanum lycopersicum*	AtNHX1 (vacuolar Na^+/H^+ antiporter)	*Arabidopsis thaliana*	Salt tolerance	80μg/g fresh weight	Zhang and Blumwald (2001)
Tomato, tobacco	*Solanum lycopersicum* *Nicotiana tabacum*	TERF2/LeERF2 (ethylene-responsive factor)	*Solanum lycopersicum*	Freezing tolerance		Zhang and Huang (2010)
Tomato	*Solanum lycopersicum*	D-Galacturonic acid reductase gene (GalUR)	*Strawberry*	Heat and drought tolerance		Lim et al. (2016)
Tomato	*Solanum lycopersicum*	BADH	*Atriplex hortensis*	Salt tolerance	–	Zhou et al. (2008)
Tomato	*Solanum lycopersicum*	BADH	*S. oleracea*	Heat and light inhibition tolerance	–	Li et al. (2014)
Tomato	*Solanum lycopersicum*	Coda	*A. globiformis*	Tolerance to oxidative, chilling, and salt stresses	–	Park et al. (2007)
Tomato	*Solanum lycopersicum*	DHAR	*Pseudomonas*-resistant tomato	Oxidative stress tolerance with high photosynthesis	–	Li et al. (2010)

Tomato	*Solanum lycopersicum*	f SlGRAS10	*Arabidopsis thaliana*	Multiple abiotic stress tolerance	–	Habib et al. (2021)
Watermelon	*Citrullus lanatus*	HAL1	*Arabidopsis thaliana*	Salt tolerance		Yang et al. (2001)
Wheat	*Triticum aestivum*	P5CS (Δ1-pyrroline-5-carboxylate synthase)	*Vigna aconitifolia*	Drought tolerance	22 mmol proline/g of leaf dry weight	Vendruscolo et al. (2007)
Wheat	*Triticum aestivum*	mt1D (mannitol-1-phosphate dehydrogenase)	*Escherichia coli*	Salt and osmotic stress tolerance	1.7–3.7 mol/g fresh weight	Abede et al. (2003)
Wheat	*Triticum aestivum*	*bar*	*Streptomyces hygroscopicus*	Basta tolerance	–	Vasil et al. (1992)
Wheat	*Triticum aestivum*	HvSUT1	–	Heat stress tolerance	–	Weichert et al. (2017)
Wheat	*Triticum aestivum*	TaFER-5B	*Triticum aestivum*	Thermo tolerance	–	Zang et al. (2017)
Wheat	*Triticum aestivum*	TaDREB2, TaERF3	*Triticum aestivum*	Abiotic stress tolerance	–	Kim et al. (2018)
Wheat	*Triticum aestivum*	TaASR1-D	*Triticum aestivum*	Multiple abiotic stress tolerance	–	Qiu et al. (2021)
Wheat	*Triticum aestivum*	*EPSP synthase*	*Petunia hybrid*	Glyphosate herbicide resistance		Zhou et al. (1995)
Wheat	*Triticum aestivum*	*AtNHX1*	*Arabidopsis thaliana*	Salt tolerance	–	Xue et al. (2004)

4.21.2 Production of Osmoprotectants in Plants

Adaptation techniques are used by plants to cope with a wide range of abiotic challenges. Several families of low-molecular-weight chemicals, generally referred to as osmoprotectants, are used to counteract the detrimental effects of abiotic stressors. There are different types of chemicals that may act as osmoprotectants. In addition to acting as metabolic signals, these benign chemicals also regulate the structure cells and enzymes as well as scavenge reactive oxygen species developed during stressful situations. Due to the recent dramatic fluctuations in planetary climate, plants that are more suited to abiotic challenges must be developed (Zulfiqar et al. 2020). Different abiotic stresses or specific stress combinations alter plant cellular functioning in different ways (Wang et al. 2018). One of the most significant of these alterations is the creation and buildup of a wide range of organic molecules that are extremely soluble in water, characterized with a lower molecular weight and neutral charges, and are hence harmless (Slama et al. 2015; Per et al. 2017; Riaz et al. 2019). To some extent, this is true; however, the tissue/cellular levels of these compounds are dependent on a variety of parameters, including the tissue or cell's growth level as well as the kind and duration of an adverse condition of the environment (Joshi et al. 2010).

In plants subjected to abiotic challenges, osmoprotectants aggregate and accumulate in high concentrations, and their structure and compartmentalization are determined by a variety of parameters, including plant growth stage, type and intensity of stress, and plant species (Ashraf and Foolad 2007; Kumar 2009; Evers et al. 2010). The primary function of these substances' (osmoprotectants) accumulation in plants under adverse circumstances is to regulate the equilibrium in osmotic pressure in the plant. During extreme circumstances, they maintain cell turgor pressure by osmoregulation, replenishing ions, safeguarding cell structures, and reducing the harmful effects of pollutants on cellular constituents. In a nutshell, osmoprotectants have a great effect on stress tolerance, including the protection of biological membranes, the stabilization of protein and other cellular structures, the elimination of ROS, and the preservation of plant physiological oxidative balance (Suprasanna et al. 2016). Moreover, when paired with various antioxidants, osmoprotectants co-control polypeptide folding that assists in the regulation of stress responses (Rosgen 2007). These organic chemicals also aid in the stabilization of thylakoid membranes, which results in greater photosynthetic rate (Alam et al. 2014). These chemicals help plants to strengthen their antioxidant defense mechanism by actively eliminating damaging reactive oxygen species (ROS) and safeguarding critical antioxidant enzymes (Hasanuzzaman et al. 2014). More importantly, under a variety of stressors, osmoprotectants play an important role in the stimulation of genomes associated with plant defenses (Wani et al. 2018).

4.21.3 Na^+/H^+ Antiporters for Improved Salt Tolerance

Salt stress affects roughly 20% of agricultural land and 50% of irrigated area globally (Rhoades and Loveday 1990), culminating to a 0.5–1% degradation of irrigated field annually (Munns and Tester 2008). Crops in a highly saline land must withstand water shortages, ion-specific toxic effect, and other challenges that have a significant impact on plant physiology (Blumwald et al. 2000). Plants may adapt to high-salt environments via a variety of mechanisms, including osmotic tolerance, Na^+ subcellular localization, and Na^+ elimination, all of which have developed through natural selection (Zhu 2001). The salt overly sensitive 1 (SOS1) antiporter that eradicates sodium ions (Na^+) from the cytoplasm, and the vacuole membrane Na^+/H^+ exchangers (NHXs), which compartmentalize Na^+ into the vacuole to ameliorate its harmful effects are two Na^+/H^+ antiporters that play critical functions in sustaining homeostatic ion balance, consequently imparting salinity resistance on crops. Vacuolar Na^+/H^+ antiporters are found in a variety of species, ranging from unicellular algae to higher plants (Chanroj et al. 2012; Sami and Alemzadeh 2016). Plant NHXs participate in a number of activities, notably vacuolar pH control (Bassil et al. 2011), flower growth and coloration (Ohnishi et al. 2005), nutrient delivery (Bowers et al. 2000), as well as cellular progression (Brett et al. 2005).

Furthermore, many studies have shown that NHXs may impart salt tolerance on plants, most likely through preserving ion homeostasis and improving the ability to alter osmotic and antioxidant levels (Bassil et al. 2011; Wu et al. 2015; Wang et al. 2016). In modified tobacco and tomato leaves with greater amount of water and catalase function, exogenous production of *Triticum aestivum* TNHX1 boosted K^+ accumulation and lowered Na^+ concentration (Gouiaa et al. 2012; Gouiaa and Khoudi 2015). In engineered potato plants, upregulation of IbNHX2 gene from *Ipomoea batatas* boosted proline level, SOD function, as well as rate of photosynthesis while decreasing malondialdehyde (MDA) and H_2O_2 levels (Wang et al. 2016; Li et al. 2017).

Duan et al. (2007) found that the use of H^+-PPase (transgenic crops) led to excessive proton electrical potential all across vacuolar membrane, which hastened the absorption of Na^+ into the vacuole and boosted cellular salt resilience. As a result, it was hypothesized that transgenic plants with Na^+/H^+ antiporter and H^+-PPase might have superior salinity tolerance than those with either of the two genes alone. Similarly, Zhao et al. (2006) found that the combined effect of two genes (SsNHX1 and AtAVP1) in modified rice has led to a better resistance to salt than the use of SsNHX1 alone. These examples of pyramiding Na^+/H^+ antiporters with a vacuolar transporting peptide (H^+-PPase or V–H^+-ATPase) show the success of such strategy. It was also used to stack genes involved in drought and salinity tolerance (Singla-Pareek et al. 2003; Zhou et al. 2009; Wei et al. 2011).

4.21.4 COR and Heat-Shock Regulons

Plants can reprogram their transcriptome, proteomics, and metabolome to adapt to a diverse array of temperatures. The CBF–COR route in cold resistance and the HSF–HSP system in extreme heat tolerance were discovered to be part of signaling pathways in earlier studies (Hua 2009). One of the major critical ecological elements that control plant physiology is temperature (Penfield 2008). Temperature is perceived differently by plants depending on its intensity, variations, and overall thermal readings at a given period. The plant response to the impact of temperature might show up right away or take a long time to show up. The previous two decades have seen incredible growth in identifying key factors in cold and heat acclimatization, and also vernalization. Recent research is revealing the molecular processes that occur in plants' reactions to non-extreme temperatures, as well as the cross talk involving temperature and other environmental stimuli. Cold acclimation (Chinnusamy et al. 2007), vernalization (Schmitz and Amasino 2007), thermotolerance (Guy et al. 2008), and temperature association with sunlight and diurnal clock have all been the subject of outstanding studies (Hotta et al. 2007).

4.21.5 CBF Route/Pathway Regulation

Cold acclimatization is an adaptation reaction in which plants get a higher freezing sensitivity after being exposed to a mild nonfreezing temperature (Thomashow 1999). Upon cold exposure, a large amount of transcriptional regulation ensues, resulting in the induction of cold-regulated (COR) genes that are important for the production of cryoprotective chemicals. The dehydration response element (DRE) genes, which encode the AP2/ERF group signaling enzymes, are crucial to such a transcriptional regulation (Thomashow 1999). When three CBF alleles (CBF1–3) became overexpressed in Arabidopsis, they increased chilling resilience. These genes have distinct expression trends and do not control the same collection of genes, implying a perfect regulatory mechanism for effective cold adaptation (Hua 2009). The CBF peptides connect to the (CRT/DRE) motif (CCGAC) found in the promoters of the CBF regulon, a group of COR genes. Both monocots and dicots have shown the relevance of the CBF network in thermotolerance (Stockinger et al. 2007).

4.21.6 Acclimatization to Cold Temperatures Without Activation of CBF Transcripts

In addition to the activation of CBF genes, freezing tolerance is controlled at numerous levels. HOS9 and HOS10 deficiency causes a higher or faster stimulation of the CBF regulon without changing CBF production, indicating that they may have a broader role in suppressing COR genes. SFR6 (SENSITIVE TO FREEZING 6) is an additional promoter of COR genes, but not CBF activation (Knight et al. 2008).

Due to a deficiency in upregulating the CBF regulon, the sfr6 mutant is unable to acclimate to cold temperatures. SFR6 is necessary for the activation of COR enzymes by CBF1 and CBF2 overexpression; hence, it works downward of the CBFs. Delayed flowering is among the additional abnormalities of the sfr6 mutation (Knight et al. 2008). The SFR6 gene provides instructions for making a new protein, which is found in the nucleus, implying that it could directly impact gene expression to affect a variety of responses. In a group of 50 procurements, the use of CBF and COR genes does not completely correspond with cold resistance (McKhann et al. 2008), demonstrating that various factors may be under adaptive selection. Variables apart from cold may activate the CBF pathways. A temperature reduction from 28 to 22 °C generated two COR genes via the CBF alleles and the CRT components, according to another research (Wang et al. 2009). As a result, the CBF route is activated in response to minor temperature drops that could perhaps be an adaptive strategy for crops to prepare for hot pressures.

4.21.7 Thermotolerance via HSF and HSP

A prior high-temperature treatment may aid in the improvement of plants' thermal tolerance. Due to this adaptation, heat-shock proteins (HSPs), which are biological chaperones, aggregate (Sun et al. 2002). Heat-stress transcription factors (HSFs) bind to the heat-shock element (HSE) "GAANNTTC" in the regulators of heat-shock proteins (HSPs) and induce their production. There are 44 HSP genes classified into four groups in Arabidopsis, and also 21 HSF genomes (Swindell et al. 2007). These genes may be triggered by a variety of abiotic and biotic stimuli in addition to temperature (Swindell et al. 2007), indicating that they might be involved in several stress response networks. Despite the fact that taking out a particular HSP, HSF, or heat-induced allele had minimal effect on chilling resistance in most instances, mutant analysis revealed a function for HSP101I, Hsa32, and HSFA2 in Arabidopsis and HSFA1a in tomatoes through thermotolerance (Larkindale and Vierling 2008). Hsp110, HSFA7a, and HSFA3 have all been implicated in this process, according to previous research (Schramm et al. 2008). Thermosensitive male sterile 1 (TMS1), which is critical for pollen tube cold tolerance, was discovered to be a Hsp40 homologue (Yang et al. 2009). Posttranscriptional modifications may affect HSF performance. The AtCBK3 deletion lines had lower baseline thermotolerance and could not adequately transcribe HSP genes. AtCBK3 regulates HSF binding activity to HSEs by phosphorylating AtHSFA1a. Thermotolerance involves many HSP-independent pathways including ABA, hydrogen peroxide, and ethylene (Larkindale et al. 2005). This list includes beta-aminobutyric acid (BABA), a nonprotein amino acid. It enhances heat resilience but not basal cold tolerance by upregulating HSP101 and via the ABA pathway (Zimmerli et al. 2008).

4.21.8 Thermotolerance Mediated by Temperature-Sensitive Transcription Factors

Cold resistance is aided by the presence of extra heat-inducible signaling pathways. The thermal stimulation rhythm of the nuclear transcription factor X-box binding 1 (NF-X1) genotype is similar to that of a cluster of genes with a DRE domain in the promoters. Improved thermotolerance and salt resistance are both aided by it. Multiprotein bridging factor 1c (MBF1c), a transcriptional coactivator implicated in a variety of stress reactions, aggregates quickly and is found in the nucleus after extreme heat. It seems to have a role in thermotolerance in the presence of trehalose and ethylene, although it is not essential for the expression of HSFA2 and other HSPs (Suzuki et al. 2011). The dual/multifunctional nature of DREB2A, MBF1c, and NF-X1 adds to the link between heat acclimatization and other stress responses (Hua 2009).

4.21.9 Expression of Enzymes Involved in Scavenging of ROS

ROS are a class of oxygen-derived compounds and free radicals that may be innate (xenobiotics, radiation, pollutants) or external in origin. Because of their high catalytic activity, ROS have a variety of hazardous and damaging consequences. This antioxidant complex is composed of enzyme-mediated components such as superoxide dismutases and catalase as well as nonenzymatic components such as vitamins C and E, glutathione, and thioredoxins. Additionally, it has enzyme activity that inhibits the formation of reactive oxygen species, such as heme-oxygenase. Due to the fact that the intracellular compartment is the primary origin of internal reactive oxygen species (ROS), cells contain a larger amount of antioxidant material than the exterior segment. They have a role in a variety of diseases; thus, cells have evolved a comprehensive antioxidant system for scavenging ROS. There are two sides to ROS: positive and negative. The initial one is characterized by reduced ROS quantities, while the latter is characterized by substantial ROS levels (Larkindale and Vierling 2008). As they cause problems to macromolecules (lipids, proteins, and DNA) and subcellular organelles, ROS are widely recognized to be linked to a variety of diseases; yet, they are also engaged in regulating different physiological functions (Yoshida et al. 2008).

4.21.10 The SOD Enzyme Family

The enzyme dismutase is used to catalyze the process of dismutation of superoxide to hydrogen peroxide and oxygen molecule. SOD is found in most subcellular components of plant cells that produce reactive oxygen and is thought to contribute a major role in resistance to oxidative stress. SOD isozymes are divided into three categories based on the metallic cofactor: copper/zinc (Cu/Zn-SOD), manganese (Mn-SOD), and iron (Fe-SOD). The Cu/Zn-SOD isozymes are found in the cytoplasm or plastids, whereas mitochondria constitute the Mn-SOD. Although not usually detectable in crops, the Fe-SOD isozymes are generally located in the chloroplasts. The nuclear genome contains all of the plant's SOD genetic makeup, which is directed to their appropriate cellular components via an amino-terminal targeting domain. They are, however, not controlled in concert, but separately, depending on the level of oxidative stress in the various cellular components. Each of the enzymes has been developed and evaluated for its capacity to tolerate the oxidative stress in transformed tobacco plants. There has been a demonstrable improvement in the capacity to endure oxidative stressors such as ozone exposure in each instance. Mn-SOD has also been translated to alfalfa and directed to the chloroplast or mitochondria. In this example, the findings of a 3-year field study revealed considerable gains in transgenic plant productivity and longevity. It is unclear whether the defensive benefits are attributable to the enzyme's elimination of superoxide or if the hydrogen peroxide by-product boosts stress resistance by triggering other stress-related proteins. Furthermore, combining Mn-SOD, Cu/Zn-SOD, and Fe-SOD into alfalfa has been found to improve the crop's winter robustness (Slater et al. 2003).

Plant stress responses rely heavily on ROS scavenging mechanisms. The nucleotide diphosphate kinase from *A. thaliana* is overexpressed in *Solanum tuberosum*, which results in improved resilience to heat adverse effect (Grover et al. 2013). When the GDP-mannose pyrophosphorylase (GMPase, ROS-scavenging enzyme) allele from *S. lycopersicum* was overexpressed in *N. tabacum*, it led to improved ascorbic acid buildup, which worked as a potent reducing agent in transgenic lines, resulting in great heat resistance (Grover et al. 2013). Antisense transgenic crops possessed reduced foliage, while flowering started sooner. When the chloroplast protein-enhancing stress tolerance allele from *O. sativa* was highly expressed in *A. thaliana*, it led to a great thermotolerance (Grover et al. 2013). Furthermore, ectopically expressing *A. thaliana* gene for cytokinin oxidase/dehydrogenase (cytokinin-deactivating enzyme) in *N. tabacum* led to great heat tolerance (Macková et al. 2013). The stress-associated protein 1 (SAP1, a membrane-bound protein) gene from *Xerophyta viscosa* has been upregulated in *A. thaliana*, resulting in remarkable thermal resilience (Grover et al. 2013). When the stable protein gene from *Populus tremula* got upregulated in *A. thaliana*, it resulted in a significant increase in heat resistance also (Grover et al. 2013).

4.21.11 Production of Antioxidants

Plants are regularly subjected to a multitude of unfavorable or even harmful climate factors known as abiotic stresses, which pose significant dangers to crop sustainable cultivation (Bhatnagar-Mathur et al. 2008). Abiotic stressors continue to be the most significant limitation to agricultural production across the globe. They are thought to be responsible for more than half of all yield reductions (Rodríguez et al. 2005; Acquaah 2007). They also cause structural, physiological, biochemical, and genomic alterations in plants that have a negative impact on their development and yield

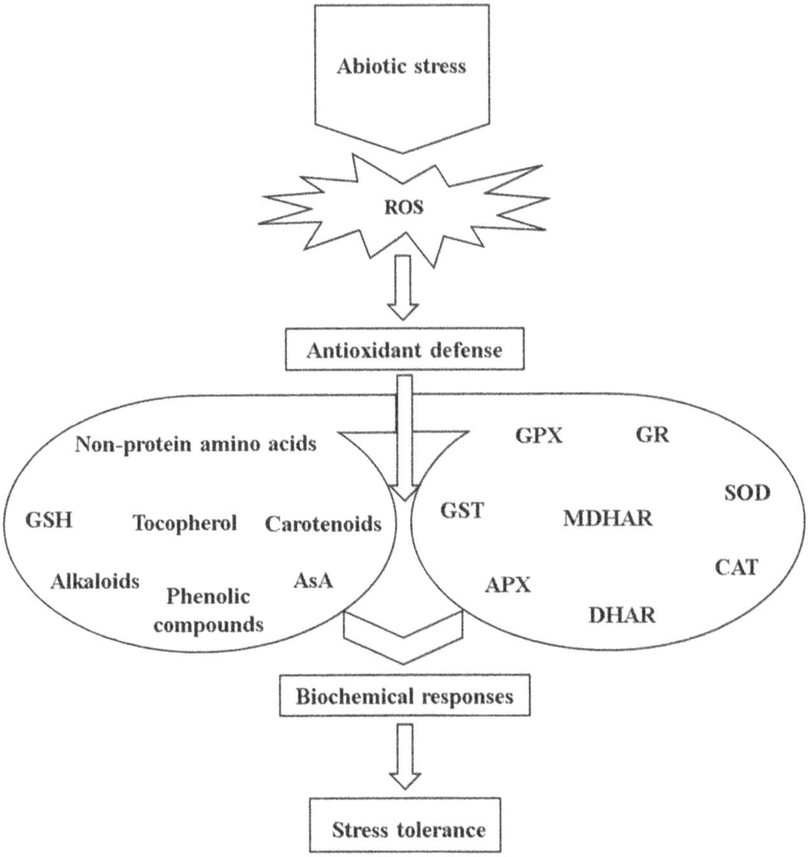

Fig. 4.7 Major antioxidants, both nonenzymatic and enzymatic, significantly participate in the antioxidants' defensive strategy

(Wang et al. 2000). Plant cells are well equipped to keep ROS concentrations within the limits generated by normal cellular physiological functions. Yet, at varied stress conditions, ROS generation often exceeds the overall antioxidative capability of the cells, culminating to detrimental stress-induced effects on plant growth and metabolism. A steady balance is required to protect plant organs from oxidative damage. Plants include both nonenzymatic and enzymatic antioxidants (Gill and Tuteja 2010) (Fig. 4.7). The relevance of ROS detoxification for cellular viability is shown by the presence of antioxidant defense mechanisms in practically every cellular compartment (Mittler et al. 2004). These defenses are present not just in the inner membrane, but also in the extracellular matrix, although to a limited amount (Mittler 2002; Gill and Tuteja 2010).

4.21.12 Nonenzymatic Antioxidant Components

4.21.12.1 Ascorbate (AsA)

Ascorbate (AsA) is a potent antioxidant that occurs naturally in crop cells. It is primarily synthesized in the cytoplasm of advanced species of plants by the conversion of D-glucose to AsA. It mixes with a different ROS, such as H_2O_2, $O_2^{\bullet-}$, and $^1O_2$, establishing the basis for its antioxidant effect. AsA, the ultimate electron donor in such processes, acts as a scavenger of free radical within the aqueous/hydrophilic environment of cell membranes. Further, it is a diffusion-dependent scavenger of $OH^{\bullet}$ (Smirnoff 2005). In the AsA-GSH cycle, ascorbate peroxidase reduces H_2O_2 to liquid while concurrently producing MDHA. MDHA is a short-lived antioxidant that breaks to DHA and AsA. Normally, NADPH serves as the electron donor, and the activity is catalyzed by MDHAR or ferredoxin in the chloroplasts' water–water cycles (Gapper and Dolan 2006). AsA is the principal reducing reagent for H_2O_2 removal in plant cells (del Rio et al. 2006; Wu et al. 2007). Moreover, it is hypothesized that AsA retains the chloroplast antioxidant α-tocopherol in its reduced form. Additionally, AsA is required for the stabilization of the reduced condition of therapeutic metallic ions, thereby sustaining the antioxidant activity of numerous enzymes (De Tullio 2004). Stress tolerance in plants is dependent on AsA content (Sharma and Dubey 2005; Hossain et al. 2010; Hasanuzzaman et al. 2011; Hossain et al. 2011). External delivery of AsA synergistically acts with different antioxidants to control the performance of numerous enzymes and lowers the damage caused by oxidative systems (Shalata and Neumann 2001).

4.21.12.2 Glutathione (GSH)

Glutathione is an antioxidant that contributes significantly to the removal of the bulk of ROS (Noctor and Foyer 1998). It is crucial for antioxidative defense since it replenishes certain putative water-soluble antioxidants like AsA via the AsA-GSH cycle (Foyer and Halliwell 1976). Additionally, it helps partly in membrane preservation by maintaining low levels of α-tocopherol and zeaxanthin. During extreme conditions, GSH safeguards proteins from thermal degradation caused by oxidation of thiol monomers. Also, GSH is a precursor for the catalysts GPX and GST, which are implicated in the decontamination of ROS (Noctor et al. 2002). Among GSH's additional roles is the synthesis of phytochelatins (PCs), which possess a sensitivity for HM and are transported into the cell as compounds, imparting resilience to crops (Sharma and Dietz 2006). Additionally, GSH aids in the purification of toxins and acts as a kind of sulfur-reduced retention and translocation (Srivalli and Khanna-Chopra 2008). Previous research indicates that raising GSH levels enhances tolerance to a number of abiotic stresses (Hossain and Fujita 2010; Hasanuzzaman and Fujita 2011).

4.21.12.3 Tocopherols

Tocopherols are abundant in thylakoid membranes, which also comprise polyunsaturated fatty acids (PUFA) and are located in immediate proximity to reactive oxygen species (ROS) produced during photosynthesis (Fryer 1992). Indeed, contextual and corroborated evidence strongly suggests that tocopherol possesses antioxidant properties (Munne-Bosch and Alegre 2003). Tocopherol and tocotrienol are isomers that fall into four different categories (a, b, g, and d). The antioxidant capacity of tocopherol isomers in vivo is as follows: a > b > g > d, with α-tocopherol showing the highest antioxidant capacity (Garg and Manchanda 2009). Through transforming lipid peroxyl radicals ($LOO^{\bullet -}$) to their corresponding peroxides, tocopherols contribute to the decrease of reactive oxygen species (ROS) in plastid walls as well as the limiting of lipid oxidation (Maeda et al. 2005). A single-tocopherol molecule may block about 120 $^1O_2$ radicals before decomposition using resonance energy transmission (Munné-Bosch 2007). Further, since tocopherols are such a portion of a complex signaling network controlled by reactive oxygen species (ROS), antioxidants, and plant hormones, they represent interesting candidates for influencing cellular signaling in plants (Munné-Bosch 2007).

4.21.13 Enzymatic Components

Antioxidant enzymes are found in many locations throughout plant cells and function cooperatively to eliminate reactive oxygen species (ROS). The enzymes operate in concert to detoxify H_2O_2 via a series of cyclic processes and to renew AsA and GSH.

4.21.13.1 Superoxide Dismutases (SODs)

SODs act as the primary barrier of defense against reactive oxygen species (ROS) in plant cells. It removes $O_2^{\bullet -}$ by facilitating its dismutation, leading to reduction of one $O_2^{\bullet -}$ to H_2O_2 and the oxidation of the other to O_2. SODs are classified by the metal ion present in their active region, which can be copper and zinc (Cu/Zn-SOD), manganese (Mn-SOD), or iron (Ir-SOD) (Fe-SOD). Cu/Zn-SOD is found in the cytosolic region and plastids of flowering plants, Mn-SOD is found in the matrix of mitochondria and peroxisomes, and Fe-SOD is found in the chloroplasts of some higher plant species, as well as in prokaryotes (Scandalios 1993). Increased SOD activity mitigates inorganic oxidative damage and is critical for plant tolerance to abiotic stress (Mobin and Khan 2007; Singh and Dwivedi 2008).

4.21.13.2 Glutathione Reductase (GR)

Glutathione reductase (GR) is a AsA-GSH cycle protein that performs a viral function in the defense mechanism against ROS. Enhanced GR expression imparts stress resistance and has the potential to change the redox state of critical ETC subunits. This protein promotes the reduction of GSH, which is relevant in numerous physiological regulations and antiapoptotic plant functions. In particular, GR facilitates the NADPH-dependent reduction of the disulfide bond of GSSG, which is critical for sustaining the GSH pool (Chalapathi Rao and Reddy 2008). So, GR promotes a high GSH/GSSG proportion in plant tissue, which is also required for speeding the H_2O_2 detoxification process, especially under stress circumstances (Pang and Wang 2010). GR is important in defining a plant's resistance to diverse stimuli by regulating the antioxidant mechanism of the cell and conferring stress resistance (Sumithra et al. 2006; Hossain et al. 2011).

4.21.13.3 Catalases (CATs)

Catalases (CATs) are multimeric enzymes that oxidize H_2O_2 to water (H_2O) and oxygen (O_2), substances that protect plant structures from oxidative damage (Sanchez-Casas and Klessig 1994). CATs are present in peroxisomes, glyoxysomes, as well as other cell compartments that generate hydrogen peroxide (Agarwal et al. 2008). Each molecule of CAT may transform roughly six million molecules of hydrogen peroxide (H_2O_2) to H_2O and O_2 each minute, making it among the largest turnaround rates of all proteins. Therefore, CAT is vital in eliminating H_2O_2 from peroxisomes that is produced by oxidases involved in lipid beta-oxidation, aerobic respiration, and purine degradation (Gill and Tuteja 2010). CAT has been shown to interact with certain hydroperoxides, in addition to H_2O_2. During diverse abiotic pressures, CAT function displays varying potentials (Singh and Dwivedi 2008; Hasanuzzaman et al. 2011; Hasanuzzaman and Fujita 2011).

4.21.13.4 Glutathione *S*-Transferases (GSTs)

Plant GSTs constitute a versatile enzyme family that catalyzes the GSH linkage of electrophilic xenobiotic substrates (Dixon et al. 2010). GST isoenzymes constitute about 1% of a plant's overall soluble polypeptide (Marrs 1996) within the proteins involved in GSH metabolism. GSTs catalyze the attachment of different xenobiotics and their electrophilic intermediates to GSH, resulting in reduced cytotoxic and greater water-soluble conjugates (Edwards et al. 2000). GST isoenzymes have POX action in addition to catalyzing the mixing of electron-loving substances to GSH (Gullner and Kömives 2001). GST activity in plants is induced by a variety of abiotic stressors (Dixon et al. 2010). Crop glutathione *S*-transferases (GSTs) are a versatile enzyme family that catalyzes the attachment of electrophilic xenobiotic substrates to GSH (Dixon et al. 2010).

4.21.13.5 Glutathione Peroxidases (GPXs)

Glutathione peroxidases (GPXs) comprise a broad group of different isozymes, which utilize glutathione (GSH) to decrease hydrogen peroxide (H_2O_2) and organic and lipid hydroperoxides (LOOHs), thus protecting crop cells from oxidative stress (Noctor et al. 2002). Additionally, GPX is a major cellular enzyme effective for reversing membrane lipid peroxidation and serves as a significant antioxidant against oxidative membrane degradation (Kühn and Borchert 2002). Several GPX alleles were isolated from species of plants in recent decades (reviewed by Kumar et al. 2010). Apart from detoxicating H_2O_2, GPX acts as an oxidative signaling transmitter (Miao et al. 2006).

4.22 Transgenics for Nutrient Biofortification and Yield

Biofortification of crops refers to the process of producing crops with increased nutritional content. This may be accomplished using either traditional selective breeding or genetic engineering (Malik and Maqbool 2020). Biofortification is distinct from fortification in that it tries to organically increase the nutritional value of plant products instead of incorporating nutritional additives via food preparation (Malik and Maqbool 2020). Nearly 800 million people worldwide are estimated to be malnourished, with nearly 98% living in poor nations (Sinha et al. 2019). Also, around two billion individuals worldwide suffer from a category of starvation described as concealed starvation that is driven by insufficient consumption of key micronutrients in the daily diet (Muthayya et al. 2013; Gillespie et al. 2016). Biofortified crops are a practical and cheap technique for supplying nutrients to the inhabitants of underdeveloped nations. Micronutrient enrichment is the practice of supplementing foods regularly eaten by the general public with one or more nutrients, such as milk, lipids, grains, and condiments. Additionally, basic foodstuffs like wheat flour or cooking oil may be fortified (mass fortification) (Dary and Hainsworth 2008), as well as food eaten by a particular demographic, such as supplemental meals for homeless persons and small children (Targeted fortification). Additionally, food producers may enrich foods sold in the open market. Wheat foods were fortified with a variety of micronutrients, including riboflavin, niacin, zinc, and iron. Nutritional additives are supplied to it in the best possible form to keep its appearance, texture, and taste. Nevertheless, fortification of each micronutrient is unique. Compounds with a moderate oxidation reactivity are easier to work with in food fortification. Generally, the government designs, mandates, initiates, and regulates this sort of fortification. When a considerable proportion of the population is lacking in a particular micronutrient or is on the verge of being insufficient, massive enrichment of staple crops is the right strategy (Bernier et al. 2008).

4.22.1 Crop Plants Biofortified for Nutrients

Many modified crops have been generated by inserting new genes, overexpressing preexisting genes, suppressing the production of certain genes, or deleting genes that hinder the production process (Malik and Maqbool 2020; Hefferon 2020). A number of crops were modified transgenically to increase their nutrient contents as presented in Table 4.9.

4.23 Engineering Plant Protein Composition

One of the targets of plant biotechnology was to establish food crops that are optimized for the benefits of man and his domestic animals. A primary objective was to enhance the amino acid profile of protein content—specifically the lysine level of grain crops and the methionine concentration of leguminous crops (Tabe and

Table 4.9 Transgenic plants with improved nutritional value

Crop name	Gene	Gene source	Useful trait introduced	Reference(s)
Carrot (*Daucus carota*)	H^+/Ca^{2+} transporter Cation Exchanger 1 gene (*cax*)	*Arabidopsis thaliana*	Increase in calcium (Ca) content	Park et al. (2004a, b)
	Beta-carotene ketolase gene	*Haematococcus pluvialis*	Increased expression of beta-carotene hydroxylase	Jayaraj et al. (2008)
	Bacterial phytoene synthase (*psy*) gene (*crt*B)	*Bacteria*	Increased carotenoid biosynthesis	Maass et al. (2009)
Cassava (*Manihot esculenta*)	*Pantoea agglomerans* (*crtB*)	*Pantoea agglomerans*	Increase in vitamin A content	Welsch et al. (2010)
	Iron-specific assimilatory protein (*fea1*) gene	*Chlamydomonas reinhardtii* (codon optimized)	Increase in iron (Fe) content	Sayre et al. (2011)
	ZAT *transporter gene* (*zat*)	*Arabidopsis thaliana*	Increase in zinc (Zn) content	Sayre et al. (2011)
	ZIP plasma membrane zinc transporter (*zip*) gene	*Arabidopsis thaliana*	Increase in zinc (Zn) content	Sayre et al. (2011)
Corn (*Zea mays*)	*Rice dehydroascorbate reductase* (*dhar*) gene	*Oryza sativa*	Increase in vitamin C content	Naqvi et al. (2009)
	Gm ferritin and *Af phytase* gene	*Glycine max*, *Aspergillus*	Increase in iron (Fe) content	Drakakaki et al. (2005)
	Phytoene synthase (*psy*)	*Zea mays*	Increase in vitamin A content	Zhu et al. (2008)
	Beta-carotene hydroxylase (*β-CHX*)	*Gentiana lutea*	Increase in vitamin A content	Zhu et al. (2008)
	Lycopene ε-cyclase (*lcyE*)	*Gentiana lutea*	Increase in vitamin A content	Zhu et al. (2008)
	Beta-carotene ketolase (*crtW*)	*Paracoccus*	Increased levels of beta-carotene and other carotenoids, including complex mixtures of hydroxycarotenoids and ketocarotenoids	Zhu et al. (2008)
	Phytoene desaturase (*crtI*)	*Pantoea ananatis*	Increase in vitamin A content	Zhu et al. (2008)
	Phytoene synthase (*crtB*)	*Pantoea ananatis*	Increase in vitamin A content	Aluru et al. (2008)
	Phytoene desaturase (*crtI*)	*Pantoea ananatis*	Increase in vitamin A content	Aluru et al. (2008)
	ζ-carotene desaturase (*zds*)	*Pantoea ananatis*	Increase in beta-carotene content	Aluru et al. (2008)
	Phytoene synthase (*psy*)	*Zea mays*	Increase in vitamin A content	Naqvi et al. (2009)
	Carotene desaturase (*crtI*)	*Pantoea ananatis*	Increase in beta-carotene content	Naqvi et al. (2009)
Indian mustard (*Brassica juncea*)	Plastidic ATP sulfurylase (*aps*) gene	*Arabidopsis thaliana*	Increase in selenium (Se) content	Pilion-Smits et al. (1999)
	Selenocysteine methyltransferase (*smt1*) gene	*Astragalus bisulcatus*	Increase in selenium (Se) content	LeDuc et al. (2004)
Lettuce (*Lactuca sativa*)	H^+/Ca^{2+} transporter sCAX1 (cation exchanger 1)	*Arabidopsis thaliana*	Increase in calcium (Ca) content	Park et al. (2009)

(continued)

Table 4.9 (continued)

Crop name	Gene	Gene source	Useful trait introduced	Reference(s)
Potato (*Solanum tuberosum*)	Phytoene synthase (*crt*B), phytoene desaturase (*crt*I), and lycopene β-cyclase (*crt*Y)	*Erwinia uredovora*	Increase in vitamin A content	Diretto et al. (2007)
	Or gene	*Brassica oleracea*	Increase in vitamin A content	Lopez et al. (2008)
	Zeaxanthin epoxidase gene	*Arabidopsis thaliana*	Increase in vitamin A content	Romer et al. (2002)
	Phytoene synthase (*psy*) *crtB* gene	*Erwinia uredovora*	Increase in vitamin A content	Ducreux et al. (2005)
	GDP-l-galactose phosphorylase	*Arabidopsis thaliana*	Increase in vitamin C content	Bulley et al. (2011)
	H^+/Ca^{2+} transporter, cation exchanger 1 (*cax1*) gene	*Arabidopsis thaliana*	Increase in calcium (Ca) content	Park et al. (2005a, b, c)
	Lycopene epsilon cyclase (*lcy-e*) (*crtE*)	*Pantoea ananatis*	Increase in beta-carotene levels	Diretto et al. (2006)
	Beta-carotene hydroxylase gene (*bch*)	*Solanum tuberosum*	Increase in carotenoids: beta-carotene and lutein content	Van Eck et al. (2007)
	Orange cauliflower (*Or*)	*Brassica oleracea*	Increase in carotenoid accumulation	Lopez et al. (2008)
Rice (*Oryza sativa*)	Mammalian GTP cyclohydrolase I (*gtpchi*) gene, aminodeoxychorismate synthase (*adcs*) gene	*Mammal*, *Arabidopsis thaliana*	Increase in folic acid content	Storozhenko et al. (2007)
	Nicotianamine synthase genes (*nas*)	*Oryza sativa*	Increase in iron (Fe) content	Johnson et al. (2011)
	Nicotianamine synthase (*nas*) gene	*Oryza sativa*	Increase in iron (Fe) content	Lee et al. (2009)
	Nicotianamine synthase 2 (*nas*2) gene	*Oryza sativa*	Increase in zinc (Zn) content	Lee et al. (2011)
	Ferritin genes, nicotianamine synthase (*nas2*) gene	*Oryza sativa*	Increase in iron (Fe) and zinc (Zn) content	Johnson et al. (2011)
	*nas*1, *ferritin*, *phytase* gene	*Arabidopsis thaliana*: *nas*, *Phaseolus vulgaris*: *ferritin*, *Aspergillus fumigates*: *phytase*	Increase in iron (Fe) content Increase in zinc (Zn) content	Wirth et al. (2009)
	Phytoene synthase (*psy*)	*Zea mays*	Increase in total carotenoids	Paine et al. (2005)
	Phytoene desaturase (*crtI*)	*Erwinia uredovora*	Increase in total carotenoids	Paine et al. (2005)
	Phytoene synthase (*crtB*)	*Narcissus pseudonarcissus*	Increase in total carotenoids	Ye et al. (2000)
	Phytoene desaturase (*crtI*)	*Erwinia uredovora*	Increase in total carotenoids	Ye et al. (2000)
Tomato (*Solanum lycopersicum*)	GDP-1-galactose phosphorylase (GGP) gene	*Actinidia chinensis*	Increase in vitamin C content	Bulley et al. (2011)

(continued)

Table 4.9 (continued)

Crop name	Gene	Gene source	Useful trait introduced	Reference(s)
	GTP-cyclohydrolase I, aminodeoxychorismate synthase	*Arabidopsis thaliana*	Increase in folic acid content	Storozhenko et al. (2007)
	NADP-dependent glutamate dehydrogenase (*gdh* A) gene	*Aspergillus nidulans*	Improved fruit taste (increased glutamate levels)	Kisaka and Kida (2003)
	Phytoene synthase (*crt B*) gene	*Erwinia uredovora*	Increased carotenoid content	Fraser et al. (2002)
	Yeast *S*-adenosylmethionine decarboxylase (*Samdc*) gene	*Saccharomyces cerevisiae*	Increased phytonutrient, carotene in fruit	Mehta et al. (2002)
	Phytoene desaturase (crtI)	*Erwinia uredovora*	Increase in β-carotene content	Römer et al. (2000)
	Lycopene β-cyclase (lcyB)	*Arabidopsis thaliana*	An increase in total carotenoid content	Rosati et al. (2000)
	Beta-carotene hydroxylase (*β*-Chy) genes	*Capsicum annuum*	Increase of beta-carotene, beta-cryptoxanthin, and zeaxanthin content	Dharmapuri et al. (2002)
	Phytoene synthase (crtB)	*Erwinia uredovora*	An increase in total carotenoid content	Fraser et al. (2002)
	Lycopene β-cyclase (lcyB)	*Solanum lycopersicum*	Increased beta-carotene content	D'Ambrosio et al. (2004)
	3-Hydroxymethylglutaryl CoA (hmgr) gene	*Arabidopsis thaliana*	Elevated levels of phytosterols	Enfissi et al. (2005)
	1-Deoxy-D-xylulose-5-phosphate synthase (dxs) gene	*Escherichia coli*	Increased carotenoid content	Enfissi et al. (2005)
	Cryptochrome (cry1)	*Solanum lycopersicum*	Increased accumulation of lycopene in fruits	Giliberto et al. (2005)
	Fibrillin	*Capsicum annuum*		Simkin et al. (2007)
	Lycopene β-cyclase (crtY)	*Erwinia herbicola, Narcissus pseudonarcissus*	Increase in provitamin A and total carotenoid accumulation	Apel and Bock (2009)
	SlNCED1	*Solanum lycopersicum*	Increase in carotenoid content and associated carotenoid-derived flavor volatiles	Sun et al. (2012)
Wheat (*Triticum aestivum*)	*Phytoene desaturase (crtI)*	*Erwinia uredovora*	Increase in vitamin A content	Cong et al. (2009)
	Phytoene synthase (psy)	*Zea mays*	Increase in vitamin A content	Cong et al. (2009)
Canola (*Brassica napus*)	*Phytoene synthase (crtB)*	*Erwinia uredovora*	Increase in carotenoids	Shewmaker et al. (1999)
	Phytoene synthase (*crt*B) + geranylgeranyl diphosphate synthase (*crtE*)	*Erwinia uredovora*	Increase in the levels of total carotenoids	Ravanello et al. (2003)
	Phytoene synthase (*crt*B) + *lycopene β-cyclase (crtY)*	*Erwinia uredovora*	Increase in the level of total carotenoids	Ravanello et al. (2003)

(continued)

Table 4.9 (continued)

Crop name	Gene	Gene source	Useful trait introduced	Reference(s)
	Lycopene ε-cyclase (lcyE)	*Arabidopsis thaliana*	Increased levels of beta-carotene, zeaxanthin, violaxanthin, and lutein	Yu et al. (2008)
	Isopentenyl pyrophosphate isomerase (idi)	*Paracoccus* spp.	Increased levels of ketocarotenoids	Fujisawa et al. (2009)
	Beta-carotene ketolase (crtW)	*Brevundimonas* spp.	Increased levels of ketocarotenoids	Fujisawa et al. (2009)
	Beta-carotene hydroxylase (crtZ)	*Pantoea ananatis*	Increased levels of ketocarotenoids	Fujisawa et al. (2009)
	Phytoene desaturase (crtI)	*Pantoea ananatis*	Increased levels of ketocarotenoids	Fujisawa et al. (2009)
	Lycopene β-cyclase (crtY)	*Pantoea ananatis*	Increased levels of ketocarotenoids	Fujisawa et al. (2009)
	Lycopene ε-cyclase (crtE)	*Pantoea ananatis*	Increased levels of ketocarotenoids	Fujisawa et al. (2009)
	Phytoene synthase (crtB)	*Pantoea ananatis*	Increased levels of ketocarotenoids	Fujisawa et al. (2009)
	microRNA miR156b	*Arabidopsis thaliana*	Enhanced carotenoid level	Wei et al. (2010)
Citrus	Beta-carotene hydroxylase gene (Csβ-CHX)	*Citrus sinensis*	Increase in beta-carotene content	Pons et al. (2014)

Higgins 1998). Essential amino acids cannot be synthesized by humans or animals; therefore, they should be acquired from foreign sources, most commonly crops. Such amino acids include lysine (Lys), methionine (Met), threonine (Thr), phenylalanine (Phe), tryptophan (Trp), valine (Val), isoleucine (Ile), and leucine (Leu) (Galili and Hofgen 2002). Four of these amino acids (Lys, Met, Thr, and Trp) have been the most limited among the amino acids in plant crops, signifying that they are supplied in little amounts in comparison to the amounts necessary for proper development (Galili and Amir 2013). Huge numbers in poor nations, whose diets are mostly plant based, are deficient in such important amino acids, which may result in serious, debilitating disorders (Wang et al. 2017). Livestock feed compositions often contain proteins from a range of origins/sources to complement the amino acid content of protein in the diet—but even still, they are typically augmented with unadulterated amino acids. In cereal crops, Lys is the first limiting amino acid that is often employed as the primary caloric provider in man and animal nutrition. On the other hand, leguminous seeds, which comprise a significant source of protein, often possess an acceptable amount of lysine and yet are lacking in the S-amino acids methionine and cysteine. Approximately, 0.8% of pea protein content is methionine and 1.0% is cysteine (Evans and Boulter 1980). This goes below the animal development needs for S-amino acids, which vary between 3% and 5% of food protein by volume. Because animals are capable of converting methionine to cysteine and not the other way around, methionine may provide the whole demand for S-amino acids while cysteine is not capable to do that (Tabe and Higgins 1998). Progress in plant tissue culture and genetic manipulation methods has provided novel potentials for altering the amino acid composition of crops. Efforts to alter the proportion of S-amino acids come in two broad types. One strategy is to modify amino acid biosynthesis regulations in order to boost methionine availability. Another strategy has attempted to incorporate and activate alleles expressing sulfur-containing peptides into modified lines/crops (Tabe and Higgins 1998).

Methionine belongs to the aspartate group of amino acids, which contains the physiologically crucial amino acids which include lysine, threonine, and isoleucine. The aspartate kinase and dihydrodipicolinate synthase (DHPS) catalyze key regulatory mechanisms in the aspartate pathway. In plants, there were often two isozymes of aspartate kinase, each of which is hindered by threonine or lysine. Lysine inhibits DHPS via a feedback mechanism. With the influence of continuous or organ-specific regulators, alleles expressing feedback-unresponsive versions of aspartate kinase and DHPS were synthesized and introduced to crops. In genetically engineered crops like soybean as well as canola seeds, overexpression of such genes via a seed-specific regulator led to more than 100-fold improvements in free lysine, equal to nutritionally relevant improvement in the overall seed lysine of 25% and 100%, respectively (Falco et al. 1995). The production of a feedback-insensitive aspartate kinase in the seeds of modified tobacco culminated in a 17-fold rise in the threonine and a 3-fold rise in free methionine (Karchi et al. 1993).

It has been stated that leguminous crops provide 10–20% of the overall nutritional caloric source in several of the world's disadvantaged nations (Akibode and Maredia 2011). Furthermore, cereals account for 68% of all nutritious calories consumed globally. For instance, pulse reserve proteins are abundant in Lys and are deficient in sulfur-containing amino acids, especially Met. In comparison, cereal crops are deficient in Lys and Trp (Galili et al. 2005; Galvez et al. 2008; Wenefrida et al. 2009).

4.23.1 Plant Proteins Improved Through Genetic Engineering Techniques

Conventional breeding techniques and mutagenesis were used to raise the essential amino acid level of different crops in the past few decades, so we have acquired a thorough insight of the enzymes associated with essential amino acid biogenesis, degeneration, and regulatory frameworks in *A. thaliana* as well as other model plants in the last few years. These findings have enabled the use of genetic engineering techniques to increase the quantities of vital amino acids in horticulture species. Notwithstanding, boosting the concentrations of such amino acids in plants, particularly agricultural crops, remains challenging because of the following reasons: (a) the pathways for synthesizing some of the essential amino acids, like Lys, Leu, Ile, Val, Phe, and Trp, are heavily controlled by a negative feedback system and (b) the specifically aimed amino acids are effectively deteriorated via catabolism, for example, Lys, that is diminished in the tricarboxylic (TCA) cycle (Wang and Galili 2016). Lys concentrations in plants are relatively inadequate, and major attempts have been undertaken to enhance Lys composition in Arabidopsis and numerous plant cultivars, notably maize and rice, resulting in much greater Lys quantities (Zhu and Galili 2003, 2004; Angelovici et al. 2011). Because Lys production and catabolism in plants are well understood, it was easy to get favorable outcomes in horticulture plants. Several early researches targeted at increasing lysine levels in horticultural plants focused on expressing the lysine feedback-intensive DHPS enzymes, such as expression in potato, soybean, and canola, which resulted in a significant increase in free lysine (Perl et al. 1992; Falco et al. 1995). Crops overexpressing microbial DHPS, on the other hand, commonly displayed the usual aberrant morphology, including a gradual decrease of apical dominance, late blooming, and irregular foliar structure (Shaul and Galili 1993). Similarly, when Hacham et al. (2007) crossed homozygous tobacco plants overexpressing both feedback-insensitive DHPS and AtCGS, the amount of lysine was found to be comparable to those expressing just DHPS. Surprisingly, the amount of methionine appeared considerably higher in crops exhibiting co-expression of both transgenes relative to AtCGS expression only. De novo production of α-helical coiled-coil protein increased lysine buildup in tobacco kernels as well (Karchi et al. 1993). Lys is effectively destroyed by catabolism; hence, preventing its breakdown through the TCA cycle can be another productive method for raising Lys concentrations. But, with the exception of the findings achieved in Arabidopsis, maize, and rice (Zhu and Galili 2004; Angelovici et al. 2011; Wang and Galili 2016), neither of these approaches were effective in horticultural plants. Having followed the model plant investigations, the microbial DHDPS has been produced in soybean, rapeseed, and maize embryos in a seed-specific manner. Contrasting to tobacco and Arabidopsis, modified soybean and rapeseed exhibited a large increase in unbound Lys in mature seeds, which almost quadrupled entire seed Lys quantity in certain instances (Falco et al. 1995; Mazur et al. 1999; Frizzi et al. 2008). The discrepancies in Lys content across species of plants might be attributed to the utilization of distinct DHDPS enzymes derived through various bacterial origins. Lys excess supply was also linked to higher levels of different Lys catabolites within those species (Falco et al. 1995; Mazur et al. 1999; Huang et al. 2005; Frizzi et al. 2008). A variety of methods (conventional breeding, mutagenesis, and genetic engineering) have been used to boost Met content. Conventional breeding techniques had extremely little impact (Galili and Amir 2013); hence, most recent attempts are centered on employing genetic modification techniques. Similar to lysine, most initiatives to increase content have focused on increasing Met synthesis or minimizing Met catabolism. Upregulation of cystathionine-synthase (CGS), the first enzyme in the Met biosynthetic pathway, resulted in 6.5-, 12.8-, and 32.7-fold increases in dissolved Met composition in genetically modified potato, tobacco, and alfalfa leaves, disclosing the key role in the regulation of CGS in Met buildup in cultivated crops. A further successful method for boosting Met level is to express sulfur-rich peptides like 2S albumin from Brazil nut or sunflower. Transgenic expression of the Brazil nut 2S albumin gene increased Met concentrations in crops like canola, tobacco, and Vicia (Molvig et al. 1997; Tiger et al. 2003; Wang et al. 2017). Thr, like Lys, is generated by a subset of the Asp subfamily route, and Thr inhibits the first (Asp kinase, AK) and third (homoserine dehydrogenase, HSD) enzymes in the process. A growing body of data suggests that AK is the primary rate-limiting protein in Thr biosynthesis in crops. Earlier research found that genetically mutated tobacco and alfalfa plants containing or overexpressing AK accumulated a significant amount of Thr (Frankard et al. 1991; Karchi et al. 1993; Jiang et al. 2016a, b). Furthermore, transgenic tobacco plants expressing Thr synthase (TS), the final enzyme in Thr production, demonstrated a fivefold increase in Thr buildup (Hacham et al. 2008). Furthermore, since Thr and Met split from the homologous branches of the Asp biosynthesis route, their biochemical processes struggle for the similar carbon source to a certain point. As a result, CGS, the initial distinctive protein in Met production, is anticipated to serve a significant impact in Thr buildup in horticulture species (Zeh et al. 2001). Overexpressing feedback-insensitive AS (anthranilate synthase) has been shown to increase Trp content in Arabidopsis and rice (Wakasa et al. 2006; Ishahara et al. 2006). Genetically modified tobacco and *Astragalus sinicus* had a 10-fold rise in free Trp concentration, whereas potato had a 431-fold increase (Yamada et al. 2004). Recent research shows that sunlight, moisture, and dark-induced senescence all increase Trp levels (Arajo et al. 2010; Galili et al. 2016).

4.24 Engineering Plants for Vit A Composition

Providing nutritional sustainability to all people is regarded as a major goal for worldwide societal progression. Malnutrition has declined dramatically in recent years, falling from nearly 20% of the global populace in 1990 to slightly more than 10% in 2016 (FAOSTAT 2017). It is undeniable that additional attempts need to be made to decrease the globe's malnourished population, which is already close to 800 million. Supplementation, commercial enrichment, biofortification, and educative programs promoting nutritional variety may all help to prevent vitamin deficits. It should be highlighted that the interventional technique to be used is determined by geographical dietary and sociocultural variables (Bailey et al. 2015). Yet, certain generally applicable observations may be made. Supplementation, either via the delivery of (multi)vitamin tablets or by the enrichment of grain foodstuffs (which is required in many countries), has proven to be a quick and potent way to eliminate vitamin deficits (Sandjaja et al. 2015; Atta et al. 2016; Wang et al. 2016). Regrettably, this strategy is difficult to apply to disadvantaged local people in needs (Blancquaert et al. 2014). Additionally, supplementation may have negative consequences, as seen by higher death rate and a greater risk of colon cancer in men after vitamin A and B9 intake (Benn et al. 2015; Cho et al. 2015). Provitamin A may be found in both animal and plant diets (Bai et al. 2011; Mody 2017). Meat and milk products are often high in retinoic acid derivatives that could be converted to retinol in the human system (Bai et al. 2011). Carotenoids, the most common of which is beta-carotene, characterize the provitamin A composition of plant-derived foodstuffs (Grune et al. 2010). Beta-carotene may be transformed to retinal by the human beta-carotene 15,15′–monooxygenase (Lindqvist and Andersson 2002), which is lacking in mainly carnivores. Provitamin A is abundant in brightly colored fruits and vegetables. Carrots, sweet potatoes, pumpkin, kale, and spinach are examples of high provitamin A carotenoid-containing plants (Harrison 2005).

Biofortification must be done with adequate attention for its impacts on plant metabolism, not only for the customers' vitamin demands. Because of genomic, ecological, and nutritional variables, the health benefit of a biofortified crop may vary by area. Excessive intake of basic foods deficient in one or more micronutrients seems to be a primary contributor worsening the shortage. As a result, biofortification of these plants is recommended. Biofortification efforts, meanwhile, should not concentrate just on vitamin concentration rather, they must take into account all elements impacting vitamin-specific nutritious value of the given crop, like preservation and processing sustainability, as well as accessibility (Blancquaert et al. 2015; Diaz-Gomez et al. 2017).

Vitamin A is a catchall name for various fat-soluble retinoid compounds (Bai et al. 2011), considered as any chemical structure capable of fulfilling the biochemical function of all-*trans*-retinol after human ingestion (Eitenmiller et al. 2016). Carotenoids are made up of approximately 600 distinct molecules, only 3 of which may be biologically transformed into active vitamin A molecules including retinol and its oxidized counterparts retinal and retinoic acid (Asson-Batres and Rochette-Egly 2016). Carotenoids are the primary source of provitamin A in the food and are found across the kingdom Plantae. The overall backbone is generated by head-to-tail joining of eight isoprene monomers, leading to a C40 unsaturated chain, lycopene, a carotenoid precursor (Eitenmiller et al. 2016). The most significant carotenoid, beta-carotene, has cyclized-ionone rings on both ends of the C40 chain. As these substances are made up of long-chain linked polyene components, they are vulnerable to oxidative effect, radiation, temperature, and acidity (Asson-Batres and Rochette-Egly 2016). Their oxidation susceptibility, on the other hand, allows them to function as antioxidants in living organisms, since the radical formed by association with reactive oxygen species (ROS) is rendered far less dangerous by the stabilizing of the polyene rings. Vitamin A's usefulness, on the other hand, much outweighs its antioxidant qualities, since it plays several functions in plant and animal metabolism.

4.24.1 Biosynthesis of Vitamin A (Retinol)

The most important provitamin A for humans as mentioned previously is beta-carotene, which is made up of two identical retinyl units. One such retinyl group comprises a retinyl isoprenoid circuit and a β-ionone loop, both of which are required for vitamin A activity (Send and Sundholm 2007). As a result, since α-carotene, γ-carotene, and β-cryptoxanthin all have one β-ionone chain, they have 50% vitamin A potential. Provitamin A is generated in plastids of all autotrophs using proteins related with the thylakoids, specifically phytoene desaturase (PDS), ζ-carotene desaturase (ZDS), lycopene-β-cyclase (β-LCY), and lycopene-ε-cyclase (ε-LCY), or related in protein complexes (Cunningham and Gantt 1998). Geranylgeranyl diphosphate (GGPP) is the primary progenitor for provitamin A. It is generated by the convergence of the basic components isopentenyl diphosphate (IPP) and three dimethylallyl diphosphate (DMAPP) molecules by GGPP synthase (GGPPS) (Ruiz-Sola et al. 2016). IPP is formed through the plastid-localized 2-C-methyl-D-erythritol 4-phosphate (MEP) route, while DMAPP is the isomerization output catalyzed by isopentenyl diphosphate isomerase (IDI). GGPP is also a progenitor for chlorophylls, ubiquinones, tocopherols, gibberellic acid, and terpenes (Saini et al. 2015). The consolidation of two GGPP compounds by phytoene synthase (PSY) to create 15-*cis*-phytoene is thought to be a rate-limiting phase in the real provitamin A biosynthesis system (Fray and Grierson 1993; Cazzonelli and Pogson 2010). Furthermore, ethylene is recognized to possess a beneficial effect on carotenoid development via increasing PSY activation (Zhang et al. 2018a, b). This factor is specifically vital in the ripening process of fruits and has thus been explored in mango (Ma et al. 2018), durian, (Wisutiamonkul et al. 2017), and tomato plants (Su et al. 2015; Cruz et al. 2018). Previous research revealed the tomato transcription factor (SlCMB1) as a modulator of both ethylene biosynthesis and carotenoid deposition (through PSY and PDS) (Zhang et al. 2018a, b). PSY may thus be regarded as a central controller of carotenoid buildup in most species, considering that it is also induced by sunlight and is explicitly regulated by the transcription factors PHYTOCHROME INTERACTING FACTOR 1 (PIF1) and LONG HYPOCOTYL 5 (HY5) in Arabidopsis photomorphogenesis (Toledo-Ortiz et al. 2010; Llorente et al. 2017). In the next biosynthetic stage, which occurs immediately downstream of PSY, 15-*cis*-phytoene is converted into 9,15,90-tri-*cis*-carotene through a 15,9-di-*cis*-phytofluene intermediates using two sequential desaturation processes mediated by phytoene desaturase (PDS) (Pecker et al. 1992; Li et al. 1996; Qin et al. 2007). Following that, either photoisomerization or isomerization by beta-carotene isomerase (ZISO) (Pecker et al. 1992; Li et al. 2007a, b) results in 9,90-di-*cis*-carotene. Two desaturation processes are done recursively by beta-carotene isomerase (ZDS) yielding neurosporene accompanied by 7,9,70,90-tetra-*cis*-lycopene (prolycopene) (Wong et al. 2004; Dong et al. 2007). Eventually, light or carotene isomerase (CRTISO) isomerizes the *cis* linkages to form all-*trans*-lycopene. This enzyme is a secondary regulator since it is epigenetically controlled by methylation (Cazzonelli et al. 2009).

4.24.2 Biofortification of Vitamin A in Crops

Throughout the past several years, enormous attempt has been made towards increasing provitamin A contents in various plant crops (Giuliano 2017). PSY, which is crucial for the earliest dedicated stage of carotenoid production, has been identified as the rate-limiting step, making it a suitable target gene in genetic improvement efforts (Fitzpatrick et al. 2012). A well-clear example is bioengineered golden rice (*Oryza sativa*) (Beyer et al. 2002; Paine et al. 2005), which has a yellow color attributable to its elevated carotenoid content. The golden rice development approach was further enhanced by substituting the daffodil-derived PSY with a maize ortholog with better catalytic performance in rice than the initial daffodil protein, resulting in higher beta-carotene contents in the golden rice 2 (GR2) (Paine et al. 2005). The endosperm of this rice variety contains up to 3.7 mg/100 g dry weight carotenoids. In 72 g of dry rice, GR2 provides 50% of a child's RDA of provitamin A. Aside from its capacity to be used to reduce VAD, golden rice may be regarded as a robust proof of concept, allowing the deployment of this bioprocess method in a variety of crops. Employing this method in maize resulted in maize grains with 6 mg/100 g DW beta-carotene (Naqvi et al. 2009), amounting to a 112-fold increment in average carotenoid level over the WT corn cultivar employed in this investigation. This technique also had resulted in a tenfold rise in endosperm carotenoid concentrations in wheat crop, attaining about 500 g/100 g dry weight (Cong et al. 2009). Surprisingly, a one-gene molecular genetic method that overexpressed just PSY resulted in the effective biofortification of various crops. PSY introduction resulted in a 50-fold rise in seed carotenoid concentration in canola (Shewmaker et al. 1999). Additionally, the carotenoid content of potatoes was increased to 3.5 mg/100 g DW, owing mainly to significantly increased beta-carotene levels (up to 1.1 mg/100 g DW) (Ducreux et al. 2005). Carotenoid quantities in cassava were increased 20-fold by roots' localized exogenous overexpression of PSY, hitting 2.5 mg/100 g DW (Sayre et al. 2011). Furthermore, a new *cis*-genic PSY overexpression bioengineering method led to banana cultivars with α-carotene comparable concentration of up to 5.5 mg/100 g DW (Paul et al. 2017).

A separate one-gene method was used in tomato plant development (Rosati et al. 2000; Ralley et al. 2016), since this tissue has great expression of genomes regulating lycopene production, like the abovementioned PSY. As a result, a carotenoid production allele downstream of lycopene represented a better alternative for carotenoid biofortification in tomato fruit (Rosati et al. 2000). The lycopene β-cyclase gene (β-LCY), which catalyzes the cyclization of the lycopene monomer by the insertion of β-ionone chains generating beta-carotene (Cunningham et al. 1996), has been generated in tomatoes, leading to enhanced beta-carotene tomato fruits (Rosati et al. 2000; D'Ambrosio et al. 2004; Ralley et al. 2016).

The gene encoding 1-deoxyxylulose-5-phosphate synthase (DXS) is another intriguing factor for carotenoid genetic improvement. The enzyme functions upstream of IPP creation in the chloroplast isoprenoid route (Sayre et al. 2011; Ruiz-Sola and Rodríguez-Concepción 2012), hence operating up front of the production of a variety of compounds dependent on this route, notably tocochromanols. Cassava, tomato, and Arabidopsis have all prospered via this strategy (Enfissi et al. 2005; Sayre et al. 2011). The concept of modifying carotenoid content by enabling a further upstream portion is appropriate, as demonstrated in tomato, where fruit-specific downregulation of Deetiolated1 (DET1) [a light signaling pathway-regulating gene (Schäfer and Bowler 2002)] results in an increase in both carotene and glycoside threshold. Inhibition of an epoxycarotenoid dioxygenase (NCED), a critical enzyme in abscisic acid (ABA) manufacture, by fruit-specific RNAi has led to increased lycopene and beta-carotene quantities (Sun et al. 2012).

Surprisingly, the metabolisms of various vitamins may interact, possibly impacting their deposition and stabilization, as it was the scenario with the joint biofortification of vit E and carotene in "Golden Sorghum" (Che et al. 2016). Molecular approaches have permitted the development of crops with a substantial vitamin content. Biofortified maize with 1.5 mg/100 g DW of carotenoids (Muzhingi et al. 2011; Zunjare et al. 2018), cassava with 800 g/100 g DW of carotenoids (Welsch et al. 2010; Ilona et al. 2017), and sweet potato having 400 g/100 g (Low et al. 2017) are examples of biofortified crops. The latter is now available to over 3 million families in sub-Saharan Africa, an effort of the Sweet Potato for Profit and Health Initiative (SPHI), which intends to serve ten million households with this orange-fleshed sweet potato (OFSP) (Laurie et al. 2018). Regrettably, sufficient variety in rice germplasm was not discovered to permit appropriate breeding for increased provitamin A content of the endosperm (De Moura et al. 2016). Provitamin A is an instance of a micronutrient for which significant biofortification development has been made during the previous years (Giuliano 2017). A significant portion of these gains were made via breeding efforts (Bouis and Saltzman 2017; Ilona et al. 2017), which did not need the use of genetic engineering and were thus relatively easily acceptable for commercial distribution (Potrykus 2017).

4.24.3 Functions of Vitamin A

Vitamin A provides retinaldehyde, which when combined with protein forms rhodopsin, which aids in night vision. Vitamin A may also help maintain normal vision by promoting tissue and structural differentiation. Retinol, retinoic acid, and other vitamin A metabolites are essential in corneal protein biosynthesis and glucose utilization. Vitamin A enhances corneal wound regeneration, proper eyesight, and immune system function by increasing the activation of epidermal growth factor receptor in endothelial cells (Rühl et al. 2004). Deficiency in vitamin A leads to night blindness due to delayed rhodopsin production. Vitamin A is important in nonspecific defense in animals (Jason et al. 2002). It also protects the pulmonary, digestive, and urinary systems against invading pathogens by maintaining epidermal thickness. Immune organs such as the thymus and spleen shrink due to vitamin A deficits, affecting the animal's cellular immunity. Its deficiency may also impact animal humoral immunity by lowering antigen and antibody response. Vitamin A is required for the correct function of epithelial tissue. It contributes to the production of mucosal cells by forming retinol-phosphomannose, and it subsequently influences the proper differentiation of the respiratory, digestive, as well as reproductive and urinary systems (Yuan et al. 2020).

4.25 Biofortified Rice

Considering the significance of rice in feeding half of the worldwide people, genetically modified (GM) rice was designed to explicitly combat "hidden hunger" as it seems to be the most viable among all staple crops (Majumder et al. 2019). Rice, however, is deficient in several important minerals and vitamins. Vitamin A shortage affects 70% of younger children in Southeast Asia (Zimmermann and Hurrel 2002). Enhanced vitamin A supplementation might save one to two million children's lives annually (Zimmermann and Hurrel 2002). To minimize food spoilage while in storage, rice is routinely pulverized in tropical environments to eliminate the oil-rich aleurone layer. The endosperm, like the bran, is deficient in provitamin A. Ye et al. (2000) employed genetic engineering methods to create rice grains rich in beta-carotene, the primary precursor of vitamin A. Brown rice, the richest nutritive type of processed rice, is produced by hulling farm-grown paddy rice. Unrefined ones possess vital elements such as iron, zinc, copper, calcium, and phosphorus, as well as vitamins like B1, B2, B3, B5, and B9. It also contains α-tocopherol E, but vitamins A, D, and C are lacking. Yet, majority of consumers choose white rice grains because of their brightness, smoothness, easier digestion, superior ingesting features, and faster boiling duration. White refined rice lacks the bran cover, as well as the sub-aleurone, embryo, and a tiny portion of the endosperm (Champagne et al. 2004). It also has a relatively low nutritive value than brown type, since its iron level is lessened 2.14 times to 4.75, zinc is reduced 1.83 times, and some other nutrients, vitamins, lipids, polypeptides, and fibers are lessened 1.83 times (Luh 1991; Dexter 1998; Masuda et al. 2009). However, the quantities of mineral loss could differ across rice varieties or grain-processing techniques. Biofortification is thought to be an efficient method for increasing micronutrient levels in agricultural crops like rice. It is also a long-term and viable method for alleviating micronutrient shortages in persons who mostly eat rice and have low accessibility to diverse foodstuffs and appropriate medical care (Datta and Bouis 2000). Various ways were schemed globally over rice biofortification research projects for sustaining, improving, and bringing innovative elements in grains. Iron is a mineral that is necessary for the health of man. Rice iron content is lowered more than any other mineral as a result of postharvest treatment. Iron content in paddy (raw rice) is 38 ppm, which is decreased to 8.8 ppm in brown rice following milling and 4.1 ppm in milled rice (Dexter 1998). In another study, iron content in brown rice was lowered from 19 ppm to roughly 4 ppm in refined grains (a 4.75-fold drop) (Masuda et al. 2009). This obvious loss in iron in edible part of rice is what prompted the development of iron biofortification in parboiled rice. Sufficient nutrient (iron) availability in rice could aid in the well-being of adolescents and pregnant mothers in underdeveloped nations. Iron deficiency anemia (IDA) is caused by a lack of iron and has major effects on human health, particularly in women and children. IDA impacts 32.9%, with the risk being greater in sub-Saharan Africa and South Asia (Masuda et al. 2009).

In comparison to traditional breeding, genetical breeding is increasingly regarded as a more effective and consistent approach of researching the genotype–phenotype link. Ali et al. (2013) created a Pusa Sugandh II (PSII) indica rice hybrid through suppressing the IPK1 genome utilizing an Ole18 seed-specific regulator as well as inhibiting the production of the last-stage critical enzyme inositol-1,3,4,5,6-pentakisphosphate 2-kinase (IPK1) of phytic acid biosynthesis. The genetically engineered seeds had a 3.85-fold downregulation in IPK1 transcripts that also directly linked to a massive decrease in phytate thresholds and an increasing amount of inorganic phosphate (Pi) and stored up 1.8-fold much iron in the seed coat, not impairing growth and development of the modified rice plant. Karmakar et al. confirmed 46.2% phytic acid knockdown by seed-specific RNAi-mediated gene silencing of an inositol triphosphate kinase (ITPK) homologue (OsITP/6K-1) in the Khitish indica rice variety, as well as a 1.3-fold increase in iron aggregation in seed, 1.6-fold zinc buildup, and 3.2-fold Pi bioavailability (Karmakar et al. 2019).

4.25.1 Development of "Golden Rice" (GR)

GR is evidence of a functional genetic manipulation of a specific carotenoid production system in the rice seed. The first step in this technique was the insertion of a daffodil phytoene synthase (PSY) genome in an endosperm-specific regulator to generate modified japonica rice (Taipei-309). Established GM rice contains phytoene in seeds, which is a crucial intermediary of provitamin A (Burkhardt et al. 1997). This PSY gene, along with a microbial phytoene desaturase (CRTI) gene, was inserted under the influence of the endosperm-specific glutelin activator in the same rice variety (Taipei-309) using Agrobacterium-mediated transformation (Ye et al. 2000). This genomic pairing was effective to generate 1.6 g/g of total carotene in rice endosperm (grains). This genetic integration has been shown to develop complete carotenoids in rice grain (1.05 g/g) in indica rice (Datta et al. 2003). Thereafter, additional common indica cultivars like IR64 and BR29 have been modified to increase seed carotenoid content. Overall carotenoids rose to 9.34 g/g in these transgenic varieties, and beta-carotene (provitamin A) in processed IR64 grain has been recorded as 2.32 g/g, while for BR29 it was 3.92 g/g (Paine et al. 2005). Paine et al. (2005) observed the greatest total carotene synthesis (37 g/g) through inserting maize (ZmPSY) and *Erwinia uredovora* (CRTI) genes within endosperm-specific promoters. This GR has been given the designation GR2 (golden rice 2). Bai et al. (2016) created GR by combining another gene from maize (ZmPSY1), bacterial (PaCRT1), and AtDXS for a constant flow of metabolic precursors and AtOR (the ORANGE gene for the establishment of a physiologic sink), which also generated up to 31.78 μg/g total carotenoids in rice grain (Bai et al. 2016).

Parkhi et al. and Baisakh et al. proved what to do to effectively eliminate a marker gene in GR, resulting in "marker-free" GR (Parkhi et al. 2005; Baisakh et al. 2006). Antimicrobial resistant alleles, including *hpt*, *nptII*, *bla*, and *aad*, are commonly deployed in molecular breeding methods. To delete these marker genes off GM crops, unique procedures and improved plant modification methodologies were devised. Transposon-mediated marker gene deletion (Cotsaftis et al. 2002), intrachromosomal homologous recombination (Zubko et al. 2000), site-specific recombination Cre/LoxP (Odell et al. 1990), and FLP/FRT (Lloyd and Davis 1994) approaches were all created. Several ways for marker-free biotech rice production include improved conversion procedures such as co-transformation of genomic regions and target genes and removal of the gene sequence in later generations through genomic partition (Parkhi et al. 2005). Datta et al. (2006) observed an enhancement in beta-carotene in T_2 seedlings relative to T_1 during the growth of indica GR cultivars (IR64 and BR29). Such increase in beta-carotene production was thought to be a favorable post-transgenerational result of carotenoid production in GR. In several laboratories throughout the globe, extensive molecular profiling of GR (Parkhi et al. 2005), field productivity evaluation (Datta et al. 2007), and biochemical and proteomic studies have been undertaken, which have proven GR to be safe for public ingestion (Swamy et al. 2019). In 2018, three main worldwide food safety regulatory agencies, the Food Standards Australia New Zealand, Health Canada, and the United States Food and Drug Administration, endorsed and offered good comments on golden rice (Majumder et al. 2019).

4.25.2 Storing of "Golden Rice"

The appropriate preservation and conservation of the nutritive value of GR are a difficult task. In seeds, there is an enzyme called lipoxygenase (LOX), which catalyzes the incorporation of oxygen molecules into polyunsaturated fatty acids (PUFA), resulting in linked hydroperoxide by-products, which in response oxidize carotenoids and result in a reduction in the nutritious integrity of GR (Carrera et al. 2007). When it comes to LOX protein-encoding proteins, the rice gene comprises 14 different varieties; among such alleles, it is the r9-LOX1 gene which is accountable for grain quality decline (Gayen et al. 2014). When the r9-LOX1 gene in GR was downregulated by RNA interference (RNAi) under the influence of the oleosin-18 promoter, the oxidative durability and viability of GR seeds were significantly enhanced (Gayen et al. 2015). In the future, this method may prove to be beneficial for the long-term preservation of rice seeds.

4.26 Biofortified Maize and Cassava

Cereal crops are generally high in calories but low in nutritional quality (Nuss and Tanumihardjo 2010). The method of genetically enhancing micronutrients in cereals via plant breeding is known as "biofortification," and it is an inexpensive and durable approach wherein micronutrients hit the targeted population in their organic state (Gupta et al. 2015; Neeraja et al. 2017). Maize has a crucial role in the global economic development. Together with rice and wheat, it supplies an nearly 30% of dietary calories to more than 4.5 billion individuals in 94 developing nations, in addition to providing a substantial proportion of livestock nutrition (Shiferaw et al. 2011). Maize is the third most frequent grain in India, following rice and wheat, and is utilized as a diet and fodder source (Yadav et al. 2015). Typical maize protein has less lysine (0.16–0.26%) and tryptophan (0.02–0.06%), which is less than half of the required dosage in human diets (Vivek et al. 2008). Furthermore, in contrast to other crops, conventional yellow maize offers a sufficient amount of kernel carotenes. Yet, it is characterized by non-proA components and provides just 0.25–2.50 g/g of proA carotenoids, which is much less than the dietary need (15 g/g) for men (Pixley et al. 2013). Preferable genotypes of the genes lycopene β-cyclase (lcyE) and beta-carotene hydroxylase-1 (crtRB1) produce an increase in proA in corn (Yan et al. 2010a, b; Babu et al. 2013). The recessive opaque-2 (o2) allele nearly doubles endosperm lysine and tryptophan levels (Mertz et al. 1964). Marker-assisted selection (MAS) utilizing extremely low-cost DNA markers aids in the mounting of numerous target genomes into a genetic makeup without the need for offspring screening (Das et al. 2017). Also, it considerably decreases the number of breeding cycles necessary to reconstruct the recurring parental genotype (RPG) (Gupta et al. 2013). Furthermore, the high expense of high-performance liquid chromatography (HPLC) tests for estimating micronutrients across individuals from segregated groups might be addressed by using molecular markers. In India, notable models of MAS used in the production of nutritive maize lines include the commercial introduction of "Vivek QPM9" (Gupta et al. 2013), "Pusa Vivek QPM9 Improved" (Muthusamy et al. 2013), "Pusa HM4 Improved," "Pusa HM8 Improved," and "Pusa HM9 Improved" (Hossain et al. 2017). Mehta et al. (2020a, b) reported that marker-assisted stacking of crtRB1, lcyE, and o2 in the genetic makeup of four maize cultivars (HQPM1, HQPM4, HQPM5, and HQPM7) largely cultivated in India resulted in a 7.7 fold increase in proA with a mean concentration of 18.98 μg/g, in comparison with the original varieties (3.12 μg/g).

Marker-assisted hybridization has the possibility to improve zinc biofortification (Collard and Mackill 2008). High-throughput single nucleotide polymorphism (SNP) genotyping technologies are being established in the corn (McMullen et al. 2009), providing up a new avenue for Zn molecular biofortification. Transgenic transformation also has the ability to enhance agricultural plants' genetics for several purposes. Each of these solutions are then explored in order to comprehend the function and possibilities for zinc biofortification of maize. Kernel Zn deposition genetic variation has been investigated, and considerable variations in Zn contents have been observed. QTL mapping investigations in maize have been carried out to determine the chromosomal areas linked with Zn buildup. Qin et al. (2012) created an F2:3 maize mapping population by pairing two differing progenitors for Fe and Zn contents in kernels and cobs. In such mapping populations for kernel zinc content with strong inheritance, there were a lot of genetic variability and transgressive partitioning. QTLs were found using genetic studies between single and mixed habitats, and 15 and 16 QTLs were detected in both settings, respectively, with several being identical under consolidated assessment. The majority of the discovered QTLs were found on chromosomes 2, 7, and 9 in both mapping groups. This research also discovered that QTLs for kernel Zn and Fe levels appeared co-localized on chromosomes 2, 7, and 9. The co-localization of QTLs for Fe and Zn contents demonstrated that both minerals' levels might be increased concurrently by attacking the same genomic areas using marker-assisted selections (Qin et al. 2012). Zhang et al. (2017a, b) additionally investigated QTLs for mineral element (Fe, Zn, Cu, and Mn) buildup in maize in single and mixed settings. For simple and multiple environmental studies, this research discovered 64 and 67 QTLs, respectively. Genomic and morphological linkages demonstrated that mineral levels might be improved genetically and phenotypically at the same time. Simic et al. (2012) demonstrated QTL analysis utilizing 121 polymorphic markers as well. The research discovered 32 substantial QTLs for mineral deposition (Fe, Zn, P, and Mg) in maize, with some of these QTLs being co-localized. The discovery of QTLs regulating mineral buildup in maize is a valuable resource that may be used in marker-assisted breeding to generate Zn-enriched maize cultivars (Maqbool and Beshir 2019).

4.26.1 Biofortified Cassava

Cassava (*Manihot esculenta*), a tuber crop, is among the principal sources of carbohydrates for over 250 million Africans. Cassava roots, on the other hand, possess the least protein:energy proportion of any important food crop on the planet. Moreover, a normal cassava-based food supplies just 10–20% of the needed iron, zinc, vitamin A, and vitamin E. The BioCassava Plus initiative used current techniques of genetic engineering to enhance Africans' health by developing and delivering new cassava cultivars with higher nutritional content (Leyva-Guerrero et al. 2012).

Cassava storage tubers have one of the poorest protein contents of any major crop (FAO 2008). It generally contains 0.7–3% protein by dry weight (Ceballos et al. 2006; USDA 2010). The food treatment procedures employed to eliminate cyanogens could worsen this decreased protein content (Montagne et al. 2009). Increasing cassava root protein levels by selective breeding or engineering techniques is one option for reducing protein-energy malnutrition (PEM) caused by dependence on a cassava-based food. One of the earliest efforts to boost cassava protein quantities was the continuous amplification of an artificial storage protein called ASP1 (Kim et al. 1992). GM crops harboring ASP1 across all tissues had a 30% rise in foliage protein content, demonstrating the leaves' significant nitrogen sink efficiency. The protein composition of roots, on the other hand, has not been documented (Zhang et al. 2006). Following that, zeolin, a transgenic storage protein, was produced solely in storage roots using the root-specific patatin promoter (Abhary et al. 2011). Zeolin is a union of phaseolin, a storage protein found in ordinary beans, and a portion of the maize γ-zein protein that stimulates protein structure synthesis in the endoplasmic reticulum (Kim et al. 1994). The inclusion of the γ-zein protein-stabilizing motif, followed by the formation of protein storage structures, was anticipated to promote protein buildup in cassava root system. Genetically modified cassava cultivars that produce zeolin under the influence of the patatin promoter amass close to quadruple more protein compared to wild-type root system (Abhary et al. 2011). Surprisingly, in zeolin-expressing plants, root steady-state linamarin contents have been lowered by 50%, agreeing with linamarin metabolism to give less nitrogen for protein biosynthesis. Due to the fact that some people are intolerant to phaseolin, zeolin-expressing cassava genotypes should not be established for human use. Following that, the Fauquet group created a new chimeric protein composed of sporamin and a shortened maize γ-zein sequence (Kim et al. 1994). The unique protein, sporazein, was discovered to form protein complexes in tobacco and cassava (Mainieri et al. 2004).

Overexpression of metabolic enzymes in cassava storage roots may promote protein storage while also modifying root physiology in a targeted manner. One such illustration is the amplification of the cassava enzyme HNL in roots. HNL catalyzes the conversion of cyanide from acetone cyanohydrin produced by linamarase. HNL is prevalent in the apical membrane of leaflets but not in the storage roots of cassava (White et al. 1998). As a consequence, the primary remaining cyanogen in inadequately treated cassava tubers is acetone cyanohydrin (Tylleskar and Banea 1992). Intake of cyanogens in improperly treated cassava meals may result in serious health issues such as paralysis and even mortality (Tylleskar and Banea 1992; Abhary 2010). Remarkably, overexpression of HNL in root system reduced acetone cyanohydrin concentration in cassava by up to 80% in as little as 15 min following processing, rendering it a safer and healthier commodity (Narayanan et al. 2011a, b). Furthermore, soluble protein quantities rose twice to thrice in early roots producing HNL under the influence of the patatin promoter compared to wild-type plants (Narayanan et al. 2011a, b). Contrary to zeolin, HNL does not seem to be allergic (Tylleskär et al. 1991). A thorough molecular insight of iron equilibrium, encompassing information on the fundamental metabolic mechanisms of iron intake, transport, and preservation across a range of species, would be perfect for biofortifying plants with increased iron levels, but is typically absent for numerous plants. New synthetic molecular techniques that use focused translation of unique iron absorption, transport, and storage proteins have the ability to improve iron intake in crops (Colangelo and Guerinot 2006). Because cassava is clonally reproduced, it is especially suitable for genetic techniques to increase its nutritive value (Sayre et al. 2011). The activation of the FEA1 protein in cassava roots boosted the iron levels from 10 to approximately 40 ppm. It is worth noting that FEA1 protein expression had no effect on iron or zinc deposition in non-root regions. The iron and zinc concentrations in the foliage of FEA1 transgenic lines were adequate, compatible with the selective absorption and deposition of iron facilitated by the FEA1 enzyme in cassava roots (Narayanan et al. 2011a, b; Ihemere et al. 2011; Leyva-Guerrero et al. 2012).

4.27 Engineering Plant Mineral Composition

Adopting molecular genetics and cell biology technologies may be necessary for generating crops with high nutrient density and increased bioavailability. Presently, contemporary agriculture has exceptional exposure to a vast information on plant genes, such as the Arabidopsis and rice genomic sequences. This will offer a broader perspective of the fundamental processes behind vitamin and mineral production and accumulation in the plant cells (Holm et al. 2002). To overcome the disadvantages of mineral element enhancement, biofortification of staple foods is a potential, practical, and successful approach for supplying nutrient-rich foodstuff to battle latent starvation, especially in rural populations. This technique may augment existing initiatives by offering a more consistent and cost-effective way of eliminating undernourished populations that rely on supplements and commercialized crop enrichment for nourishment. The biofortification strategy entails one-time fixed expenses for creating breeding procedures, engineering improved nutritional features into existing cultivars. Once the nutritious varieties are distributed, this technique will need little recurring expenditures. Furthermore, the prices do not rise in proportion to the number of people, and the advantages may be made accessible internationally, particularly to all developing nations. Moreover, breeding for increased trace mineral content in routine crops do not lead to yield penalty (Graham et al. 2012; Kumar et al. 2015).

A number of studies have addressed the possibilities for improving iron and zinc uptake in roots and transport and deposition in the vegetative parts of the plants. However, in wheat and rice, the most widely eaten food by the poor in developing countries, only a small fraction (wheat 20% and rice 5%) (Carvalho and Vasconcelos 2013; Hoekenga 2014) of the iron is transported from the senescing leaves to the grain. In contrast, more than 70% of the zinc is mobilized (Hoekenga 2014). Second, in cereals, the two minerals are almost exclusively stored in the husk, the aleurone, and the embryo, and large proportions, therefore, are lost during milling and polishing. This implies that the full potential of the genotype-determined enhancement in iron and zinc content is not realized for improving human nutrition (Holm et al. 2002). The two strategies to biofortifying crops with minerals such as iron and zinc are conventional and molecular breeding, as well as genetic modification techniques (Pfeiffer and McClafferty 2007; Kumar et al. 2018). Because the absorption and deposition of micronutrients in palatable portions of plants are regulated by polygenes with small effects, traditional breeding-based biofortification techniques have had very limited efficiency (Kumar et al. 2018). Furthermore, the effectiveness of this technique is mostly determined by natural diversity in the gene pool. In the absence of sufficient genetic variability genetic engineering would be a reliable option for enhancing micronutrient levels in crop plants (Bhullar and Gruissem 2013; Dunwell 2014).

4.27.1 Genetically Engineered Crops with Improved Mineral Contents

The major cause of mineral deficiency in humans has been identified as a lack of divalent cation elements in plant-based meals. As a result, it is critical to constantly develop and increase the nutritious ingredients of main food crops (Gitlin 2006). Owing to that, several researchers have attempted to develop crops with improved mineral composition to meet the demand of mankind all over the world. Ramesh et al. (2004) upregulated the AtZIP1 gene (zinc transporter from *Arabidopsis*) in *Hordeum vulgare* to produce Fe-fortified crops, which led to increased seed Zn and Fe concentrations. Kanobe et al. (2013) investigated the impact of a soybean's ferritin gene on the transcript and protein quantities of native proteins in maize plant in their research. The findings indicated that the modified seed had considerably greater amounts of calcium, magnesium, and iron, despite the fact that the seed storage zein as well as the overall protein level were less in the transgenic specimens.

Qu et al. (2005) integrated and overexpressed the ferritin gene in rice under the influence of a rice endosperm-specific glutelin promoter GluB-1, yielding a seed with iron concentrations of up to three times higher compared to wild species. Similarly, modified rice harboring ferritin gene under the regulation of the maize ubiqutin-1 promoter had a twofold increase in iron level of foliage but without substantial rise in grains (Drakakaki et al. 2000). Despite this, when OsZIP4 is upregulated in rice, the Zn contents in the seeds are four times less than in the modified seeds compared to the controls (Ishimaru et al. 2007). AtIRT1, a ZIP member, was initially transcribed from Arabidopsis using the saccharomyces double-mutant gene, fet3fet4 (Eide et al. 1996). Numerous relatives of the IRT1 group were discovered in crops and exhibit substantial similarities and related roles. Along with iron, AtIRT1 appears to convey zinc, manganese, cadmium, and most likely cobalt (Eide et al. 1996; Korshunova et al. 1999; Rogers et al. 2000; Vert et al. 2002). OsIRT1 was effectively extracted from rice employing a PCR-based method, while its cDNA restored a saccharomyces iron-uptake mutant's developmental abnormalities (Bughio et al. 2002). Lactoferrin is abundant in human milk (1–2 g/L) (LF). OsIRT1 has the ability to transfer iron, zinc, copper, and cadmium (Nakanishi et al. 2006; Lee and An 2009). To develop Fe-transformed rice that grows better in limited-iron environments, entire OsIRT1 cDNA has been generated under the influence of ubiquitin promoter from maize. As contrasted to wild relatives' grains, the plants overexpressing OsIRT1 possessed significantly higher Fe (up to 113%) and Zn (up to 112%) (Lee and An 2009).

To improve the iron level of baby formulas, Nandi et al. (2002) generated engineered rice seeds containing the LF gene of human using rice glutelin-1 activator. The recombinant protein was expressed at a substantially greater concentration than the reference lines, attaining about 0.5% of grain mass, while its availability was established using the human Caco-2 bioassay. Lee et al. (2010) further demonstrated that the hLF gene was expressed in GM japonica rice and that it contributed roughly 1.5% of overall protein content. Goto et al. (1999) produced rice transgenic plants expressing SoyferH1 under the endosperm-specific GluB1 rice promoter and observed a threefold improvement in grain Fe concentration relative to non-transformed strains. Furthermore, soybean ferritin cDNA was introduced into wheat and rice (Drakakaki et al. 2000), but the resulting transgenic plants had a higher Fe concentration in foliar regions than in seeds, due to the leaves' significant function as a sink. This also implies that an oversupply of ferritin accumulates Fe in the leaflets, hence decreasing Fe transfer to the grains. Increases in grain Zn, Fe, Mn, and protein abundances were noted in grown wheat following the introduction of the Gpc-B1 locus from exotic polyploid wheat (*Triticum turgidum*) into various transgene chromosome modification lines, implying that Gpc-B1 is involved in the movement of protein and other nutrients like, Zn, Fe, and Mn from the foliar to the grain regions (Distelfeld et al. 2007). Goto et al. (2000) generated modified lettuce crops that overexpressed the iron accumulation protein, the ferritin. The iron contents in GM lettuce were found to be between 1.2 and 1.7 times higher than those in the untreated plants; nevertheless, manganese values in GM cultivars were comparable to those in reference crops. Improvements in Zn concentration in rice were seen when NA synthase (NAS) was overexpressed by the introduction of 35S promoter factors (Lee et al. 2009). Additionally, modified rice harboring HvNAS1 gene from barley accumulated 2–3 times the amount of zinc in processed rice grains (Masuda et al. 2009). As a result of the upregulation of NAS genetic makeup, nicotianamine presents an intriguing opportunity for Zn biofortification. Additionally, biofortifying cereal crops employing NAS individually or in conjunction with ferritin has the possibility to significantly contribute to the worldwide fight against mineral deficiencies (Lee et al. 2009; Zheng et al. 2010).

4.27.2 Biofortified Wheat

Similar to various other main crop grains, wheat is deficient in the critical nutrient components. Iron and zinc deficiency for instance affects up to two billion people globally, having the majority of them living in cereal-based countries. While wheat flour is typically enriched upon preparation, biofortification is a more appealing and efficient approach that needs the development of novel wheat cultivars with innately increased nutritional value in their seeds (Borrill et al. 2014). Currently, wheat research is limited to endosperm-specific production of wheat or soybean ferritin, resulting in 1.5–1.9-fold and 1.1–1.6-fold improvements in grain iron concentration, respectively (Borg et al. 2012; Sui et al. 2012), as well as increased phytase function (Holm et al. 2002). These experiments provide evidence of principle suggesting that grain micronutrients in wheat could be changed using agronomic and genetic techniques. The agronomic biofortification approach includes treating crops using zinc fertilizers that might improve grain Zn composition. For example, Zhang et al. (2012) found that a foliar treatment of 0.4% $ZnSO_4{\cdot}7H_2O$ increased total grain Zn by 58% and wheat flour Zn by 76%. In another work, Zou et al. (2012) used Zn as a foliar application to enhance grain Zn by 84% and 90%, respectively. Conversely, in the case of iron, similar agricultural measures have been shown to be less efficient (Zhang et al. 2010), unless supplemented with additional nitrogen fertilizers (Aciksoz et al. 2011) that might not be financially viable. Backcross breeding involving participative cultivar selections conducted by CIMMYT (El Batán, Mexico), the Indian Institute of Wheat and Barley Research (Karnal, India), and the Punjab Agricultural University (Ludhiana, India) led to the emergence of zinc-biofortified wheat cultivars. Presently, four Zn-biofortified cultivars have been published which include "Zinc Shakti" (formed through transmitting the genetic traits from *Ae. squarrosa* into the Indian cultivar, PBW343), "Zincol 20160" (created by conveying the genetic traits from *T. spelta* into the Pakistani variety, NARC2011), "WB020," and "HPBW-010" (established by transferring the genetic material from Ae). All four varieties are now growing in India and Pakistan. Moreover, human interference studies to investigate the efficacy of ingesting flour manufactured from Zn-biofortified wheat are now underway in Pakistan (Lowe et al. 2018); whereas Zn genetic improvement has been successful, no Fe-biofortified variety has been created employing traditional farming procedures yet (Saini et al. 2020).

Transgenic techniques might not only raise Fe and Zn levels in whole grains, but also boost iron and zinc accumulation in the endosperm. Because milling removes much of the outer tissues of wheat grain, boosting micronutrient concentration in the embryo may be more favorable for boosting human intake of iron and zinc. There were less investigations in wheat than in the model crop rice; nevertheless, greater modification efficiencies (Hayta et al. 2019) and innovative ways to change superior lines (Borrill 2020) will allow for direct description of genome performance in wheat. Despite scientific advances, regulatory obstacles will most certainly postpone the introduction of modified wheat varieties in agricultural production. Nonetheless, transgenic investigations will aid in elucidating the function of gene products, whether they are discovered using mapping methodologies or by orthology to genes discovered in other plant species. The first GM strategy to increasing iron concentration in wheat intended to improve endosperm iron accumulation by activating ferritin alleles (TaFer1/TaFer2) that express iron-binding peptides under an endosperm-specific promoter. This led to an elevated amount of iron in wheat, as well as rice, that was 50–85% greater than in that native species (Borg et al. 2012). A comparable logic underpins a more recent work in which TaVIT2, a vacuolar iron transporter, was produced under an endosperm-specific promoter, resulting in a twofold rise of iron in wheat flour when contrasted to control groups (Connorton et al. 2017). Other investigations have used OsNAS2, which encodes a nicotianamine synthase enzyme, to regulate iron and zinc absorption and transportation.

The benefits on iron and zinc levels have been likewise maintained under field settings without affecting productivity (Beasley et al. 2019), demonstrating the agricultural viability of this strategy. Genetic modification experiments have been hampered by a relative paucity of genome sequencing for wheat, but now that high-quality nucleotide sequences are obtainable and conversion effectiveness is rising, they are probably to play a considerable role in recognizing gene functions in iron and zinc transport in wheat. While transgenic techniques to boosting iron and zinc content in wheat might be helpful, no genetically altered (GM) wheat varieties have been introduced so far (Ali and Borrill 2020). Genome editing techniques such as CRISPR-Cas9 may avoid some of these obstacles since most nations will not classify these techniques as GM, despite the fact that they are classified GM in Europe. Conversely, mutagenesis may result in changes to gene function that are not deemed GM and hence can be easily employed in breeding programs (Ali and Borrill 2020). In wheat, concurrent editing of the three homeologs has been observed (Cui et al. 2019; Zhang et al. 2016); however, only a small fraction of transgenic lines had all three homeologs altered. More complicated editing, such as tailored gene introduction, single-base editing, and epigenetic mark alterations, is likewise now achievable (Zhang et al. 2019). The use of CRISPR-Cas9 may allow for the quick alteration of genes for biofortification effectively in superior wheat varieties, bypassing several generations of backcrossing to eliminate harmful alleles acquired by traditional crossing schemes (Ali and Borrill 2020).

4.28 Enhancement of Photosynthesis for Improved Yield

Photosynthesis is the fundamental factor of crop production, and the effectiveness with which a crop absorbs sunlight and turns it into biomass during the growing period is an important component of ultimate yield (Long et al. 2006). Production capacity is the highest yield obtainable from a plant and may be described as the greatest output achievable when the best suited crop variety is cultivated under ideal circumstances without biotic or abiotic disturbance (Evans and Fischer 1999). Sunlight accessibility, light collection, energy conversion, and plant architecture are all factors that influence yield possibilities (Long et al. 2006; Zhu et al. 2010). Yet, considering that up to 50% of fixed carbon is wasted to aerobic respiration under certain circumstances, the effectiveness of this transformation of energy to useable bioenergy has yet to be properly investigated. CO_2 enhancement experiments, which have repeatedly demonstrated substantial proof that yields may be boosted via enhanced CO_2 absorption, provide corroborating facts that enhanced yields can be attained by strengthening photosynthetic carbon absorption (Ainsworth and Long 2005; Leakey et al. 2009; Weigel and Manderscheid 2012). Even though some studies found a negative association between leaf area photosynthesis and yields (Evans 1993, 1998), a positive link among photosynthetic rates and biomass (Kruger and Volin 2006) and productivity (Fischer et al. 1999) has been shown in wheat.

Direct engineering of the CB cycle in a range of species between 1992 and 2015 indicated that even slight reductions in a restricted number of CB cycle enzymes might have a deleterious influence on carbon absorption and development. Declines in proteins including sedoheptulose1,7-bisphosphatase (SBPase; EC 3.1.3.37) (Harrison et al. 1998, 2001; Lawson et al. 2006), chloroplastic fructose-1,6-bisphosphatases (FBPase; EC 3.1.3.11) (Koßmann et al. 1992; Sahrawy et al. 2004; Rojas-González et al. 2015), or fructose1,6-bisphosphate aldolase (FBPA; EC 4.1.2.13) (Haake et al. 1998) slowed development and reduced ultimate biomass output. Moreover, a 20–40% decline in chloroplast transketolase (TK; EC 2.2.1.1) action in antisense tobacco crops was demonstrated to suppress ribulose-1,5-bisphosphate (RuBP) restoration and photosynthetic activity (Henkes et al. 2001); as light concentrations increased, the downregulation of photosynthesis would become more prominent, with the highest rate of photosynthetic process constrained under both saturating light and saturating CO_2. The theoretical maximum energy conversion efficiency possible in plants that fix ambient CO_2 via the CB cycle enzyme Rubisco is 4.6% for C3 plants (Zhu et al. 2010), although in the field, efficiencies of less than 50% are attained. The result of Poolman et al. (2000)'s model demonstrated that flux control in the CB cycle is mostly shared by SBPase and Rubisco, depending on the environmental circumstances in which the plants are produced. More subsequent research has added sucrose/starch biosynthesis and photorespiration, leading to the construction of a more dynamic model of carbon metabolism based on these early models (Zhu et al. 2007). Zhu et al. (2007) employed an evolutionary method in conjunction with a model that utilized existing kinetic data while restricting the quantity of nitrogen. According on this, it was postulated that boosting SBPase, FBPA, and ADPglucose pyrophosphorylase in the same plant, in conjunction with a minor decrease in photorespiration, might lead to an improvement in photosynthetic carbon absorption efficiency. This model is significant because it emphasized the fact that more than one target is likely to be required and that model has the ability to identify the most effective mixture on targets. This model has yet to be verified empirically (Simkin et al. 2019).

Earlier efforts to increase crop photosynthetic performance by genetic modification concentrated on the amplification of a specific enzyme in the Calvin–Benson (CB) cycle. Overexpression of SBPase in Arabidopsis (Simkin et al. 2017a, b), tobacco (Lennartz et al. 2001; Simkin et al. 2015), and tomato (Ding et al. 2016) has demonstrated that increased SBPase enzymatic action led to increased photosynthetic carbon incorporation and biomass production. Lennartz et al. (2001) shown that in tobacco, photosynthetic CO_2 uptake rates increased, and sucrose and glucose accrued, culminating in a 30% rise in biomass. Nonetheless, no substantial improvement in photosynthetic rates was seen in the same plants' fully extended leaves (Lennartz et al. 2001). The findings of Lennartz et al. (2001) were verified in follow-up tests, which demonstrated that these improvements were retained between generations, 10 years apart, and when cultivated in either high- or low-light conditions (Simkin et al. 2015). Following that, enhanced SBPase activity in tobacco was demonstrated to significantly boost biomass output in field-grown tobacco in an open-air CO_2 elevation (Rosenthal et al. 2011). Studies on Arabidopsis, in which organic matter was enhanced by 42% and CO_2 integration was risen by 20%, as well as in tomato, in which biomass, disaccharide, and starch all piled up, had supported these findings. The studies show that SBPase is among the enzymes that could indeed exhibit control over the flow of carbon in the CB cycle in numerous different species (Simkin et al. 2017a, b). Substantial evidence suggests that increased SBPase activity may have an impact on plant growth and development by altering the metabolites that are produced as a result of increased CO_2 absorption. However, an intriguing alternate theory suggests the opposite.

The physiological adjustments that arise in reaction to altering SBPase activity have not been clarified; however, it is noteworthy to notice the developmental alterations that happened in SBPase antisense plants as well as in reaction to modification of other biochemical activities (Lawson et al. 2006; Raines and Paul 2006). Interestingly, it was demonstrated that boosting SBPase activity in wheat may result in large improvements in photosynthetic rates (Driever et al. 2017). Importantly, observed improvements in SBPase activity led to higher grain production (+30–40%) as well as biomass output. Driever et al. (2017) cultivated these plants using two distinct growth schedules to corroborate their findings. Plants were cultivated at large density in one research in which tillering was inhibited but at minimal density in another study where tillering was promoted. Plants with a greater developing density exhibited less tillers with an improvement in seed number per ear, while plants having a reduced growing density generated more ears with no huge rise in seed number per ear. Under increased CO_2 (700 ppm) conditions, overexpression of the CB cycle enzyme FBPA in tobacco led to enhancement in photosynthetic activities (Uematsu et al. 2012). They observed a 1.4- to 1.9-fold boost in aldolase activity, a 1.5-fold increase in photosynthetic CO_2 fixation, and a 70–120% increase in biomass under such circumstances. However, these benefits were significantly less pronounced if plants were cultivated in ambient CO_2, in which biomass increases varied from 10% to 30% when contrasted to the wild type (Uematsu et al. 2012). Individual gene alterations have shown that increasing the activities of CB cycle enzymes may boost photosynthetic carbon absorption, accelerate growth, and result in large increments in vegetal biomass under controlled circumstances.

In addition to the Rubisco carboxylation process, whereby a CO_2 is linked to RuBP, leading to the transport of carbon via the CB cycle, a competitive reaction of the Rubisco enzyme occurs in fixing O_2. The activity of Rubisco oxygenase competes with carbon fixation at the active center, and oxygen combines with RuBP than CO_2, resulting in the synthesis of a molecule of 3PGA plus 2-phosphoglycolate (2PG) at the expense of one ATP and one NAD(P)H. The metabolite 2PG is not utilized in the CB cycle and must be recycled at a considerable energy cost, lowering the efficiency of CO_2 absorption with major influence on yield (Zhu et al. 2010; Busch 2013; Walker et al. 2016). Modeling investigations revealed that in situations with minor stressors, a slight drop in the amounts of photorespiratory proteins might result in improved nitrogen dispersion, which may contribute to increased CO_2 uptake (Zhu et al. 2007). Previous research has demonstrated that decreasing the flow via the photorespiratory system under extreme photorespiratory circumstances leads to a drop in the efficacy of photosynthetic activities. Knocking down the GCS P-protein in potato and the H-protein in rice, for example, has been demonstrated to reduce flow through the photorespiratory feed-back loop, the rate at which mitochondria oxidize glycine (−70%), photosynthesis, and growth rates (Zhou et al. 2013; Lin et al. 2016). Antisense P-protein potato plants accumulated >100-fold more glycine and demonstrated highly substantial decline in the rate of glycine oxidation (Heineke et al. 2001; Bykova et al. 2005); and deactivation of the GCS P-protein in Arabidopsis has been exhibited to be detrimental under non-photorespiratory situations (Engel et al. 2007). Additionally, in rice grown in ambient CO_2, knocking down the H-protein led to chlorophyll depletion, protein deterioration, lipid peroxidation, and a buildup of reactive oxygen species (ROS), leading to ROS-induced senescence (Zhou et al. 2013).

Modification of the photosynthetic ETC is another possibility for increasing photosynthetic carbon absorption and production. Chida et al. (2007) were the first to show that enhancements in electron transport may promote gains in plant development. These researchers demonstrated that expressing the algal (*Porphyra yezoensis*) cytochrome (Cyt) c6 in Arabidopsis chloroplasts increases chlorophyll and starch content, as well as ATP and NADPH levels. These alterations were followed by an improvement in CO_2 assimilation, photosynthetic electron transport efficiencies, and plant biomass (Chida et al. 2007). Cyt c6 has been demonstrated in cyanobacteria and green algae to substitute plastocyanin as an electron transporter when reacting to copper deprivation (Merchant and Bogorad 1987). Chida et al. (2007) have established that in vivo, algal Cyt c6 may transport electrons from the Cyt b6f complex to Arabidopsis PSI at a quicker pace than Arabidopsis' natural plastocyanin. When the Cyt c6 from *Ulva fasciata* was overexpressed in tobacco, comparable outcomes were obtained (Yadav et al. 2018). The Cyt b6f combination is a critical part of photosynthetic electron transport. It is found in the thylakoid membrane and mediates electron movement across PSII and PSI, as well as produces ATP and NADPH for photosynthetic carbon acquisition by oxidizing PQH2 and reducing plastocyanin (Kurisu et al. 2003; Cramer et al. 2006, 2011; Tikhonov 2014). The structure is made up of eight different subunits, two of which are encoded in the nucleus and the remaining six in the plastid's genetic material (Cramer and Zhang 2006; Cramer et al. 2006; Baniulis et al. 2009; Schöttler et al. 2015).

4.29 Conclusions

There is need to improve agricultural productivity by 70% so as to satisfy the food demands of the projected 9.3 billion human population by 2050 (Tester and Langridge 2010). Considering the increase in the global human population, diminishing agricultural land, and accelerated rate of global warming, improved crop cultivars with enhanced nutrient content, and with enhanced resistance to biotic and abiotic stresses, need to be grown. Crop varieties with higher yields, resilience to biotic and abiotic stressors, and improved nutrient contents have been raised via transgenic technology. Undoubtedly, transgenic technology has aided in the reduction of usage (application) and ecological impacts of herbicide and insecticide, with simultaneous increase in farmer's revenue. But the transgenic plants are given FDA approval only after undergoing stringent nutrient safety checks, including allergenicity, toxic impacts, and compositional assessment and addressing issues related to the safety of human health and environment (due to horizontal gene transfer and development of superweeds) and negative impacts on nontarget life-forms.

Currently, transgenic technology aims at generation of several high-yielding as well as biotic/abiotic stress-resistant crops. However, in a situation where the source of germplasm is not available, specifically for traits such as stem rot, Alternaria blight, and powdery mildew resistance, there is a need to involve genetic engineering for raising transgenic crops. Genetic manipulation regimes must be combined with traditional crop breeding to generate high-yielding and climatic stress-resistant crops (Thakur et al. 2020).

ROS has a dual purpose: it is an unavoidable end product of aerobic metabolism and also serves as a signal under stressful situations on the other. They not only act as damage agents in plants, but they also activate stress-signaling components to avoid additional destruction. ROS biosynthesis is extensive, having manufacturing regions found both intra- and extracellularly. ROS causes severe damages, and its priorities include all macromolecules such as lipids, proteins, and DNA, compromising the cell's structure and eventually resulting in apoptosis. Plants, on the other hand, have evolved a broader variety of defensive mechanisms, including morphological, metabolic, and genetic alterations to adapt to harsh environmental circumstances. Despite tremendous advances in recent years, there are still uncertainties and inadequacies in our knowledge of ROS generation and how they influence plants, owing to their short half-life and highly reactive properties.

Phytophagous insects, such as lepidopterans, coleopterans, and hemipterans, reduce crop output by consuming or spreading pathogens. Bioengineering techniques, such as the production of Cry toxins, complex proteins, RNAi, as well as some genes related to pathogens, were used to introduce novel traits in various crops. For instance, in cotton and maize, transgenic technology via insertion of pesticidal proteins provided excellent resistance against aggressive and damaging pests/insects. As a result, between 1997 and 2017, the global acreage under GM crops expanded to 190 million ha from 1.7 million ha. Despite the use of innovative technology in crops to acquire resistance, agricultural pests normally acquire resistance to insecticidal chemicals, wreaking havoc on crop productivity.

In India, Bt cotton served as a "pioneer transgenic crop." The widespread acceptance and cultivation of Bt cotton have drastically improved India's cotton production efficiency. The widespread acceptance of Bt cotton is seen as a great leap towards ecological and agricultural sustainability. Farmers' earnings have increased dramatically due to adoption of Bt cotton, and it also brought great profits to Bt seed production companies. The frequent use and related detrimental effects of synthetic chemicals on the environment have been reduced through the adoption of pathogen/insect-resistant GM crops.

As aforementioned, biotic stresses have become a serious concern in agricultural crop production, and multiple research teams are working to develop robust cultivars resistant to a multitude of pathogens, insects, etc. using RNAi, CRISPR-Cas, PR proteins, antimicrobial peptides, ribosome-inactivating proteins, and pathogen-derived resistance. Genetically modified plants have already proven incredible in plant disease resistance initiatives and will remain promising in the future. Since 1986, when the first genetically modified crop with disease tolerance was discovered, antiviral tobacco expressing the viral coat protein gene, tremendous advances have been made both in labs and in the field (Abel et al. 1986). R gene enrichment and sequencing (RenSeq) and molecular transformation are the two recent approaches that have greatly increased the accessibility of genetic materials to be used in upgrading crop tolerances to pathogenic attacks. Molecular stacking and specific gene introduction by genome editing are believed to play a key role in the development of crops with broad resistance to viral and nonviral infections. Similarly, CRISPR-Cas genome editing techniques help in the modifications of innate plant genes involved in disease tolerance, thereby increasing the plant resistance to microbial attack. Abiotic stresses also cause significant hazard to agricultural production worldwide, and their impacts are expected to worsen the crop growth and yields in the future. Plants respond to abiotic stressors via extremely complicated signaling networks, and identifying the genes and their mechanism involved in conferring resistance to these stressors requires gigantic efforts.

In a nutshell, plants that are tolerant to insect infestations, pathogenic (bacterial, viral, fungal, etc.) infections, high salinity, drought, waterlogging, and excessive heat have been developed through transgenic approaches. Besides developing crops that are resistant to biotic and abiotic stresses, these biotechnological methods have also been used to enhance plant nutrient compositions. Biofortification of crops is widely recognized as a promising, cost-effective, and humanitarian approach for boosting the nutrient status of the nutrient-deprived communities worldwide and ensuring food and health security of the future generation. The development of biofortified crops with improved nutritional compositions, such as increased iron, zinc, selenium, amino acids, and provitamin A content, shall cater to the needs of the humans for proper growth and development. Biotechnological techniques were similarly involved in improving photosynthetic performance and crop yields, and a positive link was established between the rate of photosynthesis, biomass, and improved crop productivity in genetically modified crops. Thus, transgenic approaches can be deployed for the development and improvement of agricultural crops with novel and useful traits for the benefit of humans and environment.

References

Abdeen A, Virgos A, Olivella E, Villanueva X, Gabarra AR, Prat S (2005) Multiple insect resistance in transgenic tomato plants over-expressing two families of plant proteinase inhibitors. Plant Mol Biol 57:189–202

Abe H, Urao T, Ito T, Seki M, Shinozaki K, Yamaguchi-Shinozaki K (2003) *Arabidopsis* AtMYC2 (b-HLH) and AtMYB2 (MYB) function as transcriptional activators in abscisic acid signalling. Plant Cell 15:63–78

Abede T, Guenzi AC, Martin B, Cushman JC (2003) Tolerance of mannitol-accumulating transgenic wheat to water stress and salinity. Plant Physiol 131:1748–1755

Abel P, Nelson R, De B, Hoffmann N, Rogers S, Fraley R, Beachy R (1986) Delay of disease development in transgenic plants that express the tobacco mosaic virus coat protein gene. Science 232:738–743

Abhary M (2010) Development of a methodology to attain high levels of storage proteins in cassava tuberized roots and its implications. PhD thesis, University of Missouri, St Louis

Abhary M, Siritunga D, Stevens G, Taylor NJ, Fauquet CM (2011) Transgenic biofortification of the starchy staple cassava (*Manihot esculenta*) generates a novel sink for protein. PLoS One 6: e16256

Aboye TL, Strömstedt AA, Gunasekera S, Bruhn JG, El-Seedi H, Rosengren KJ, Göransson U (2015) A cactus-derived toxin-like cystine knot peptide with selective antimicrobial activity. Eur J Chem Biol 16:1068–1077

ABSP II (2014) Fruit and shoot borer resistant (FSBR)/Bt eggplant, Agricultural Biotechnology Support Project II. Cornell University. http://absp2.edu/project/interects.cfm?productid=2&countryid=4

Aciksoz SB, Yazici A, Ozturk L, Cakmak I (2011) Biofortification of wheat with iron through soil and foliar application of nitrogen and iron fertilizers. Plant Soil 349:215–225

Acquaah G (2007) Principles of plant genetics and breeding. Blackwell, Oxford, p 385

Adamczyk JJ Jr, Perera O, Meredith WR (2009) Production of mRNA from the *cry1Ac* transgene differs among Bollgard lines which correlates to the level of subsequent protein. Transgenic Res 18:143–149

Agarwal S, Singh R, Sanyal I, Amla DV (2008) Expression of modified gene encoding functional human alpha-1-antitrypsin protein in transgenic tomato plants. Transgenic Res 17:881–896

Agarwal PK, Gupta K, Lopato S, Agarwal P (2017) Dehydration responsive element binding transcription factors and their applications for the engineering of stress tolerance. J Exp Bot 68(9):2135–2148

Agorio A, Giraudat J, Bianchi MW, Marion J, Espagne C, Castaings L, Lelièvre F, Curie C, Thomine S, Merlot S (2017) Phosphatidylinositol 3-phosphate-binding protein AtPH1 controls the localization of the metal transporter NRAMP1 in Arabidopsis. Proc Natl Acad Sci U S A 114:3354–3363

Agrawal N, Malhotra P, Bhatnagar RK (2002) Interaction of gene-cloned and insect cell-expressed aminopeptidase N of *Spodoptera litura* with insecticidal crystal protein Cry1C. Appl Environ Microbiol 68:4583–4592

Ahmed I, Yadav D, Shukla P, Vineeth TV, Sharma PC, Kirti PB (2017) Constitutive expression of *Brassica juncea* annexin, AnnBj2 confers salt tolerance and glucose and ABA insensitivity in mustard transgenic plants. Plant Sci 265:12–28

Ahmed AU, Hoddinott JF, Islam KM, Khan AS, Abedin N, Hossain NZ, Ghostlaw J, Parvin A, Quabili W, Rahaman ST, Rahman W (2019) Impacts of Bt brinjal (Eggplant) technology in Bangladesh. Project Report prepared for the US Agency for International Development (USAID)

Ahuja I, Kissen R, Bones AM (2012) Phytoalexins in defense against pathogens. Trends Plant Sci 17:73–90

Aimanova KG, Zhuang M, Gill SS (2006) Expression of Cry1Ac cadherin receptors in insect midgut and cell lines. J Invertebr Pathol 92:178–187

Ainsworth EA, Long SP (2005) What have we learned from 15 years of free-air CO2 enrichment (FACE)? A meta-analytic review of the responses of photosynthesis, canopy properties and plant production to rising CO2. New Phytol 165:351–337

Ajji PK, Walder K, Puri M (2016) Functional analysis of a type-I ribosome inactivating protein balsamin from Momordica balsamina with anti-microbial and DNase activity. Plant Foods Hum Nutr 71:265–271

Akibode S, Maredia M (2011) Global and regional trends in production, trade and consumption of food legume crops. Report submitted to SPIA. p 83

Akmal A, Kiran U, Ali A, Abdin MZ (2014) Enhanced nitrogen assimilation in transgenic mustard (*Brassica juncea* L.) overexpressing high affinity sulfate transporter gene. Indian J Biotechnol 13:381–387

Akter S, Mollika SR, Sarker RH, Hoque MI (2016) *Agrobacterium*-mediated genetic transformation of two varieties of *Brassica juncea* (L.) using marker genes. Plant Tiss Cult Biotechnol 26(1):55–65

Aktories K, Lang AE, Schwan C, Mannherz HG (2011) Actin as target for modification by bacterial protein toxins. FEBS J 278:4526–4543

Alam MM, Nahar K, Hasanuzzaman M, Fujita M (2014) Trehalose induced drought stress tolerance: a comparative study among different *Brassica* species. Plant Omics 7:271–283

Alberto C, Vicente R, Phillipe E, Esther M, Pablo V (2005) The H_2O_2-regulated *Ep5C* gene encodes a peroxidase required for bacterial speck susceptibility in tomato. Plant J 42:283–293

Alcantara EP, Aguda RM, Curtiss A, Dean DH, Cohen MB (2004) *Bacillus thuringiensis* δ-endotoxin binding to brush border membrane vesicles of rice stem borers. Arch Insect Biochem Physiol 55:169–177

AlcaNtara-De La Cruz R, Fernández-Moreno PT, Ozuna CV, Rojano-Delgado AM, Cruz-Hipolito HE, Domínguez-Valenzuela JA, Barro F, De Prado R (2016) Target and non-target site mechanisms developed by glyphosate-resistant hairy beggarticks (Bidens pilosa L.) populations from Mexico. Front Plant Sci 7:1492

Alexander A, Singh VK, Mishra A (2021) Overexpression of differentially expressed AhCytb6 gene during plant-microbe interaction improves tolerance to N2 deficit and salt stress in transgenic tobacco. Sci Rep 11(1):1–20

Ali MW, Borrill P (2020) Applying genomic resources to accelerate wheat biofortification. Heredity 125(6):386–395

Ali N, Paul S, Gayen D, Sarkar SN, Datta K, Datta SK (2013) Development of low phytate rice by RNAi mediated seed-specific silencing of inositol 1,3,4,5,6-pentakisphosphate 2-kinase gene (IPK1). PLoS One 8(7):e68161

Ali Z, Abulfaraj A, Idris A, Ali S, Tashkandi M, Mahfouz MM (2015) CRISPR/Cas9-mediated viral interference in plants. Genome Biol 16:238

Ali Z, Ali S, Tashkandi M, Zaidi SS, Mahfouz MM (2016) CRISPR/Cas9-mediated immunity to geminiviruses: differential interference and evasion. Sci Rep 6:26912

Ali S, Mir ZA, Tyagi A, Bhat JA, Chandrashekar N, Papolu PK (2017) Identification and comparative analysis of *Brassica juncea* pathogenesis related genes in response to hormonal, biotic and abiotic stresses. Acta Physiol Plant 39:268. https://doi.org/10.1007/s11738-017-2565-8

Ali S, Ganai BA, Kamili AN, Bhat AA, Mir ZA, Bhat JA, Tyagi A, Islam ST, Mushtaq M, Yadav P et al (2018) Pathogenesis-related proteins and peptides as promising tools for engineering plants with multiple stress tolerance. Microbiol Res 212–213:29–37

Al-Taweel K, Iwaki T, Yabuta Y, Shigeoka S, Murata N, Wadan AA (2007) Bacterial transgene for catalase protects translation of D protein during exposure of salt-stressed tobacco leaves to strong light. Plant Physiol 145:258–265

Aluru M, Xu Y, Guo R, Wang Z, Li S, White W, Wang K, Rodermel S (2008) Generation of transgenic maize with enhanced provitamin A content. J Exp Bot 59(13):3551–3562

Aman R, Ali Z, Butt H, Mahas A, Aljedaani F, Khan MZ, Ding S, Mahfouz M (2018) RNA virus interference via CRISPR/Cas13a system in plants. Genome Biol 19:1

Aminedi R, Dhatwalia D, Jain V, Bhattacharya RC (2019) High efficiency in planta transformation of Indian mustard (*Brassica juncea*) based on spraying of floral buds. Plant Cell Tiss Org Cult 138:22–237

Anderson K (2010) Economic impacts of policies affecting crop biotechnology and trade. N Biotechnol 2:631–637

Angelovici R, Fait A, Fernie AR, Galili G (2011) A seed high-lysine trait is negatively associated with the TCA cycle and slows down Arabidopsis seed germination. New Phytol 189:148–159

Angst BD, Marcozzi C, Magee AI (2001) The cadherin superfamily: diversity in form and function. J Cell Sci 114:629–641

Anonymous (2000) Bt cotton. www.bt.ucsd.edu/bt_cotton.html. pp 1–3

Anonymous (2002) Global transgenic crop area continues to grow; exceeds 50 million hectares. Biotech Glob Update 4:1–5

Anwar A, Kim JK (2020) Transgenic breeding approaches for improving abiotic stress tolerance: recent progress and future perspectives. Int J Mol Sci 21(8):2695

Apel W, Bock R (2009) Enhancement of carotenoid bio-synthesis in transplastomic tomatoes by induced lycopene-to-provitamin A conversion. Plant Physiol 151:59–66

Aragão FJL, Faria JC (2009) First transgenic geminivirus-resistant plant in the field. Nat Biotechnol 27(12):1086–1088

Araujo WL, Ishizaki K, Nunes-Nesi A, Larson TR, Tohge T et al (2010) Identification of the 2-hydroxyglutarate and isovaleryl-CoA dehydrogenases as alternative electron donors linking lysine catabolism to the electron transport chain of Arabidopsis mitochondria. Plant Cell 22:1549–1563

Areal FJ, Riesgo L, Rodriguez-Cerezo E (2013) Economic and agronomic impact of commercialized GM crops: a meta-analysis. J Agric Sci 151(1):7–33

Arenas I, Bravo A, Soberon M, Gomez I (2010) Role of alkaline phosphate from *Manduca sexta* in the mechanism of action of *Bacillus thuringiensis* Cry1Ab toxin. J Biol Chem 258:12497–12503

Arencibia A, Vazquez RI, Prieto D, Tellez P, Carmona ER, Coego A, Hernandez L, de la Riva GA, Selman-Housein G (1997) Transgenic sugarcane plants resistant to stem borer attack. Mol Breed 3:247–255

Arillaga I, Gil-Mascarell R, Gisbert C, Sales E, Montesinos C, Serrano R, Moreno V (1998) Expression of the yeast *HAL2* gene in tomato increases the *in vitro* salt tolerance of transgenic progenies. Plant Sci 136:219–226

Arora H, Lakshmi Padmaja K, Paritosh K, Mukhi N, Tewari AK, Mukhopadhyay A, Gupta V, Pradhan AK, Pental D (2019a) BjuWRR1, a CC-NB-LRR gene identified in *Brassica juncea*, confers resistance to white rust caused by *Albugo candida*. Theor Appl Genet 132:2223–2236

Arora S, Steuernagel B, Gaurav K, Chandramohan S, Long Y, Matny O, Johnson R, Enk J, Periyannan S, Singh N et al (2019b) Resistance gene cloning from a wild crop relative by sequence capture and association genetics. Nat Biotechnol 37:139–143

Artur MAS, Zhao T, Ligterink W, Schranz E, Hilhorst HW (2018) Dissecting the genomic diversification of late embryogenesis abundant (LEA) protein gene families in plants. Genome Biol Evol 11(2):459–471

Ashraf M, Foolad MR (2007) Roles of glycine betaine and proline in improving plant abiotic stress resistance. Environ Exp Bot 59:206–216

Asson-Batres MA, Rochette-Egly C (2016) The biochemistry of retinoid signaling II: the physiology of vitamin A-uptake, transport, metabolism and signaling. Springer, Berlin. https://doi.org/10.1007/978-94-024-0945-1

Atherton KT (2002) Safety assessment of genetically modified crops. Toxicology 181:421–426

Atkinson RG, Sutherland PW, Johnston SL, Gunaseelan K, Hallett IC, Mitra D, Brummell DA, Schroder R, Johnston JW, Schaffer RJ (2012) Down-regulation of polygalacturonase 1 alters firmness, tensile strength and water loss in apple (*Malus domestica*) fruit. BMC Plant Biol 12: 129

Atta CA, Fiest KM, Frolkis AD, Jette N, Pringsheim T, St Germaine-Smith C, Rajapakse T, Kaplan GG, Metcalfe A (2016) Global birth prevalence of *Spina bifida* by folic acid fortification status: a systematic review and meta-analysis. Am J Public Health 106(1):24–34

Ausubel FM (2005) Are innate immune signaling pathways in plants and animals conserved? Nat Immunol 6:973–979

Awasthi RP, Nashaat NI, Kolte SJ, Tewari AK, Meena PD, Bhatt R (2012) Screening of putative resistant sources against Indian and exotic isolates of *Albugo candida* inciting white rust in rapeseed mustard. J Oilseeds Brassica 3:27–37

Azab E, Kebeish R, Hegazy AK (2018) Expression of the human gene CYP1A2 enhances tolerance and detoxification of the phenylurea herbicide linuron in *Arabidopsis thaliana* plants and *Escherichia coli*. Environ Pollut 238:281–290

Babu R, Rojas NP, Gao S, Yan J, Pixley K (2013) Validation of the effects of molecular marker polymorphisms in LcyE and CrtRB1 on provitamin A concentrations for 26 tropical maize populations. Theor Appl Genet 126:389–399

Badawi GH, Kawano N, Yamauchi Y, Shimada E, Sasaki R, Kubo A, Tanaka K (2004) Overexpression of ascorbate peroxidase in tobacco chloroplasts enhances the tolerance to salt stress and water deficit. Physiol Plant 121:231–237

Bagni N, Ruiz-Carrasco K, Franceschetti M, Fornalè S, Fornasiero RB, Tassoni A (2006) Polyamine metabolism and biosynthetic gene expression in *Arabidopsis thaliana* under salt stress. Plant Physiol Biochem 44:776–786

Bai C, Twyman RM, Farré G, Sanahuja G, Christou P, Capell T, Zhu C (2011) A golden era—provitamin A enhancement in diverse crops. In Vitro Cell Dev Plant 47(2):205–221

Bai C, Capell T, Berman J, Medina V, Sandmann G, Christou P, Zhu C (2016) Bottlenecks in carotenoid biosynthesis and accumulation in rice endosperm are influenced by the precursor-product balance. Plant Biotechnol J 14(1):195–205

Bailey RL, West KP, Black RE (2015) The epidemiology of global micronutrient deficiencies. Ann Nutr Metab 66:22–33

Baisakh N, Rehana S, Rai M, Oliva N, Tan J, Mackill DJ, Khush GS, Datta K, Datta SK (2006) Marker-free transgenic (MFT) near-isogenic introgression lines (NIILs) of 'golden' indica rice (cv. IR64) with accumulation of provitamin A in the endosperm tissue. Plant Biotechnol J 4(4): 467–475

Baltes NJ, Hummel AW, Konecna E, Cegan R, Bruns AN, Bisaro DM, Voytas DF (2015) Conferring resistance to geminiviruses with the CRISPR–Cas prokaryotic immune system. Nat Plants 1:15145

Baniulis D, Yamashita E, Whitelegge JP, Zatsman AI, Hendrich MP, Hasan SS, Ryan CM, Cramer WA (2009) Structure-function, stability, and chemical modification of the cyanobacterial cytochrome b6f complex from Nostoc sp. PCC 7120. J Biol Chem 284(15):9861–9869

Bapat VA, Trivedi PK, Ghosh A, Sane VA, Ganapathi TR, Nath P (2010) Ripening of fleshy fruit: molecular insight and the role of ethylene. Biotechnol Adv 28:94–107

Barfield DG, Pua EC (1991) Gene transfer in plants of *Brassica juncea* using *Agrobacterium tumefaciens*-mediated genetic transformation. Plant Cell Rep 10:308–314

Barrangou R, Doudna JA (2016) Applications of CRISPR technologies in research and beyond. Nat Biotechnol 34:933–941

Barth H, Aktories K, Popoff MR, Stiles BG (2004) Binary bacterial toxins: biochemistry, biology, and applications of common *Clostridium* and *Bacillus* proteins. Microbiol Mol Biol Rev 68: 373–402

Barton KA, Whiteley HR, Yang NS (1987) *Bacillus thuringiensis* δ-endotoxin expressed in transgenic *Nicotiana tabacum* provides resistance to lepidopteran insects. Plant Physiol 85: 1103–1109

Bashir K, Husnain T, Fatima T, Riaz N, Makhdoom R, Riazuddin S (2005) Novel indica basmati line (B-370) expressing two unrelated genes of *Bacillus thuringiensis* is highly resistant to two lepidopteran insects in the field. Crop Prot 24:870–879

Bassil E, Ohto MA, Esumi T, Tajima H, Zhu Z, Cagnac O (2011) The Arabidopsis intracellular Na^+/H^+ antiporters NHX5 and NHX6 are endosome associated and necessary for plant growth and development. Plant Cell 23:224–239

Baulcombe DC (1996) Mechanisms of pathogen-derived resistance to viruses in transgenic plants. Plant Cell 8(10):1833–1844

Baulcombe D (2004) RNA silencing in plants. Nature 431:356–363

Bawa AS, Anilakumar KR (2013) Genetically modified foods: safety, risks and public concerns-a review. J Food Sci Technol 50:1035–1046

Beachy RN (1997) Mechanisms and applications of pathogen-derived resistance in transgenic plants. Curr Opin Biotechnol 8(2):215–220

Beasley JT, Bonneau JP, Sánchez-Palacios JT, Moreno-Moyano LT, Callahan DL, Tako E et al (2019) Metabolic engineering of bread wheat improves grain iron concentration and bioavailability. Plant Biotechnol J 17:1514–1526

Beck DL, Guilford PJ, Voot DM, Andersen MT, Forster RLS (1991) Triple gene block proteins of white clover mosaic potex virus are required for transport. Virology 183:695–702

Beck DL, Van Dolleweerd CJ, Lough TJ, Balmori E, Voot DM, Andersen MT, O'Brien IEW, Forster RLS (1994) Disruption of virus movement confers broad-spectrum resistance against systemic infection by plant viruses with a triple gene block. Proc Natl Acad Sci USA 91:10310–10314

Bednarek J, Boulaflous A, Girousse C, Ravel C, Tassy C, Barret P, Bouzidi MF, Mouzeyar S (2012) Down-regulation of the TaGW2 gene by RNA interference results in decreased grain size and weight in wheat. J Exp Bot 63(16):5945–5955

Behrens MR, Mutlu N, Chakraborty S, Dumitru R, Jiang WZ, LaVallee BJ, Herman PL, Clemente TE, Weeks DP (2007) Dicamba resistance: enlarging and preserving biotechnology-based weed management strategies. Science 316(5828):1185–1188

Bel Y, Banyuls N, Chakroun M, Escriche B, Ferré J (2017) Insights into the structure of the Vip3Aa insecticidal protein by protease digestion analysis. Toxins 9:131

Bendahmane M, Groenborn B (1997) Engineering resistance against tomato yellow leaf curl virus (TYLCV) using antisense RNA. Plant Mol Biol 33:351–357

Benedict JH, Sachs ES, Altman DW, Deaton WR, Kohel RJ, Ring DR (1996) Field performance of cotton expressing transgenic Cry1Ac insecticidal proteins for resistance to *Heliothis virescence* and *Helicoverpa zea* (Lepidoptera: Noctuidae). J Econ Entomol 89:230–238

Benekos K, Kissoudis C, Nianiou-Obeidat I, Labrou N, Madesis P, Kalamaki M, Makris A, Tsaftaris A (2010) Overexpression of a specific soybean GmGSTU4 isoenzyme improves diphenyl ether and chloroacetanilide herbicide tolerance of transgenic tobacco plants. J Biotechnol 150(1):195–201

Benn CS, Aaby P, Arts RJ, Jensen KJ, Netea MG, Fisker AB (2015) An enigma: why vitamin A supplementation does not always reduce mortality even though vitamin A deficiency is associated with increased mortality. Int J Epidemiol 44:906–918

Berni R, Luyckx M, Xu X, Legay S, Sergeant K, Hausman JF, Lutts S, Cai G, Guerriero G (2018) Reactive oxygen species and heavy metal stress in plants: impact on the cell wall and secondary metabolism. Environ Exp Bot 161:98–106

Bernier J, Atlin GN, Serraj R, Kumar A, Spaner D (2008) Breeding upland rice for drought resistance. J Sci Food Agric 88:927–939

Berrocal-Lobo M, Segura A, Moreno M, López G, García-Olmedo F, Molina A (2002) Snakin-2 an antimicrobial peptide from potato whose gene is locally induced by wounding and responds to pathogen infection. Plant Physiol 128:951–961

Berry C, O'Neil S, Ben-Dov E, Jones AF, Murphy L, Quail MA, Holden MTG, Harris D, Zaritsky A, Parkhill J (2002) Complete sequence and organization of pBtoxis, the toxin-coding plasmid of *Bacillus thuringiensis* subsp. israelensis. Appl Environ Microbiol 68:5082–5095

Bertholdo-Vargas LR, Martins JN, Bordin D, Salvador M, Schafer AE, Barros NM (2009) Type 1 ribosome inactivating proteins entomotoxic, oxidative and genotoxic action on Anticarsia gemmatalis (Hübner) and Spodoptera frugiperda (J.E. Smith) (Lepidoptera: Noctuidae). J Insect Physiol 55:51–58

Bett B, Gollasch S, Moore A, James W, Armstrong J, Walsh T, Harding R, Higgins TJ (2017) Transgenic cowpeas (*Vigna unguiculata* L. Walp.) expressing *Bacillus thuringiensis* Vip3Ba protein are protected against the Maruca pod borer (*Maruca vitrata*). Plant Cell Tiss Org Cult 131:335–345

Beyer P, Al-Babili S, Ye XD, Lucca P, Schaub P, Welsch R (2002) Golden rice: introducing the beta-carotene biosynthesis pathway into rice endosperm by genetic engineering to defeat vitamin A deficiency. J Nutr 132:506–510

Bhagat YS, Bhat RS, Kolekar RM, Patil AC, Lingaraju S, Patil RV, Udikeri SS (2019) *Remusatia vivipara* lectin and *Sclerotium rolfsii* lectin interfere with the development and gall formation activity of Meloidogyne incognita in transgenic tomato. Transgenic Res 28:299–315

Bharathi Y, Kumar SV, Pasalu IC, Balachandran SM, Reddy VD, Rao KV (2011) Pyramided rice lines harbouring *Allium sativum* (ASAL) and *Galanthus nivalis* (GNA) lectin genes impart enhanced resistance against major sapsucking pests. J Biotechnol 152:63–71

Bhatnagar-Mathur P, Vadez V, Sharma KK (2008) Transgenic approaches for abiotic stress tolerance in plants: retrospect and prospects. Plant Cell Rep 27:411–424

Bheemanna M, Patil BV, Hanchinal SG, Hosamani AC, Bansi AB (2008) Comparative performance and economics of Bollgard-II cotton under irrigated condition. J Cotton Res Dev 22(1): 118–121

Bhuiyan MSU, Min SR, Jeong WJ, Sultana S, Choi KS, Lee Y, Liu JR (2011) Overexpression of AtATM3 in *Brassica juncea* confers enhanced heavy metal tolerance and accumulation. Plant Cell Tiss Org Cult 107:69–77

Bhullar NK, Gruissem W (2013) Nutritional enhancement of rice for human health: the contribution of biotechnology. Biotechnol Adv 31:50–57

Bhutani S, Kumar R, Chauhan R, Singh R, Choudhury VK, Choudhury JB (2006) Development of transgenic indica rice plants containing potato proteinase inhibitor 2 gene with improved defense against yellow stem borer. Physiol Mol Biol Plants 12:43–52

Biffen RH (1905) Mendel's laws of inheritance and wheat breeding. J Agric Sci 1:4–48

Birch RG (1997) Plant transformation: problems and strategies for practical application. Annu Rev Plant Biol 48(1):297–326

Birkett MA, Pickett JA (2014) Prospects of genetic engineering for robust insect resistance. Curr Opin Plant Biol 19:59–67

Bisht NC, Burma PK, Pental D (2004a) Development of 2,4-Dresistant transgenics in Indian oilseed mustard (Brassica juncea). Curr Sci 87(3):367–370

Bisht NC, Jagannath A, Gupta V, Burma PK, Pental D (2004b) A two gene – two promoter system for enhanced expression of a restorer gene (barstar) and development of improved fertility restorer lines for hybrid seed production in crop plants. Mol Breed 14:129–144

Bisht NC, Jagannath A, Burma PK, Pradhan AK, Pental D (2007) Retransformation of a male sterile barnase line with the barstar gene as an efficient alternative method to identify male sterile-restorer combinations for heterosis breeding. Plant Cell Rep 26:727–733

Blancquaert D, Van Daele J, Strobbe S, Kiekens F, Storozhenko S, De Steur H, Gellynck X, Lambert W, Stove C, Van Der Straeten D (2014) Improving folate (vitamin B 9) stability in biofortified rice through metabolic engineering. Nat Biotechnol 10:1076–1078

Blancquaert D, Van Daele J, Strobbe S, Kiekens F, Storozhenko S, De Steur H, Gellynck X, Lambert W, Stove C, Van Der Straeten D (2015) Improving folate (vitamin B9) stability in biofortified rice through metabolic engineering. Nat Biotechnol 33(10):1076–1078

Bless AE, Colin F, Crabit A, Devaux N, Philippon O, Follain S (2018) Landscape evolution and agricultural land salinization in coastal area: a conceptual model. Sci Total Environ 625:647–656

Blumwald E, Aharon GS, Apse MP (2000) Sodium transport in plant cells. Biochim Biophys Acta Biomembr 1465(1–2):140–151

Boch J, Scholze H, Schornack S, Landgraf A, Hahn S, Kay S, Lahaye T (2009) Breaking the code of DNA binding specificity of TAL-type III effectors. Science 326:1509–1512

Boddupally D, Tamirisa S, Gundra SR, Vudem DR, Khareedu VR (2018) Expression of hybrid fusion protein (Cry1Ac: ASAL) in transgenic rice plants imparts resistance against multiple insect pests. Sci Rep 8:1–10

Bogdanov IV, Finkina EI, Balandin SV, Melnikova DN, Stukacheva EA, Ovchinnikova TV (2015) Structural and functional characterization of recombinant isoforms of the lentil lipid transfer protein. Acta Naturae 7:65–73

Bogdanov IV, Shenkarev ZO, Finkina EI, Melnikova DN, Rumynskiy EI, Arseniev AS, Ovchinnikova TV (2016) A novel lipid transfer protein from the pea *Pisum sativum*: isolation, recombinant expression, solution structure, antifungal activity, lipid binding, and allergenic properties. BMC Plant Biol 16:107

Bol BJC, Linthorst JF, Cornelissen HJM (1990) Plant pathogenesis-related proteins induced by virus infection. Annu Rev Phytopathol 28:113–138

Bolognesi A, Polito L, Olivieri F, Valbonesi P, Barbieri L, Battelli MG, Carusi MV, Benvenuto E, Del Vecchio Blanco F, Di Maro A, Parente A (1997) New ribosome-inactivating proteins with polynucleotide: adenosine glycosidase and antiviral activities from Basella rubra L. and bougainvillea spectabilis Willd. Planta 203(4):422–429

Bommireddy PL, Leonard BR, Temple J, Price P, Emfinger K, Cook D, Hardke JT (2011) Field performance and seasonal efficacy profiles of transgenic cotton lines expressing Vip3A and Vip Cot against *Helicoverpa zea* (Boddie) and *Heliothis virescens* (F.). J Cotton Sci 15:251–259

Bonfim K, Faria JC, Nogueira EOPL, Mendes EA, Aragão FJL (2007) RNAi-mediated resistance to Bean golden mosaic virus in genetically engineered common bean (*Phaseolus vulgaris*). Mol Plant Microbe Interact 20:717–726

Boonserm P, Mo M, Angsuthanasombat C, Lescar J (2006) Structure of the functional form of the mosquito larvicidal Cry4Aa toxin from *Bacillus thuringiensis* at a 2.8-angstrom resolution. J Bacteriol 188:3391–3401

Borg S, Brinch-Pedersen H, Tauris B, Madsen LH, Darbani B, Noeparvar S (2012) Wheat ferritins: improving the iron content of the wheat grain. J Cereal Sci 56:204–213

Borrill P (2020) Blurring the boundaries between cereal crops and model plants. New Phytol. https://doi.org/10.1111/nph.16229

Borrill P, Connorton JM, Balk J, Miller AJ, Sanders D, Uauy C (2014) Biofortification of wheat grain with iron and zinc. Front Plant Sci 53:1–8

Bosamia TC, Mishra GP, Radhakrishnan T, Dobaria JR (2015) Novel and stress relevant EST derived SSR markers developed and validated in peanut. PLoS One 10(6):e0129127

Bose S, Gangopadhyay G, Sikdar SR (2019) Rorippa indica HSPRO2 expression in transgenic *Brassica juncea* induces tolerance against mustard aphid *Lipaphis erysimi*. Plant Cell Tiss Org Cult 136:431–443

Bouis HE, Saltzman A (2017) Improving nutrition through biofortification: a review of evidence from HarvestPlus, 2003 through 2016. Glob Food Sec 12:49–58

Boukedi H, Khedher SB, Triki N, Kamoun F, Saadaoui I, Chakroun M (2015) Overproduction of the Bacillus thuringiensis Vip3Aa16 toxin and study of its insecticidal activity against the carob moth *Ectomyelois ceratoniae*. J Invertebr Pathol 127:127–129

Bowers K, Levi BP, Patel FI, Stevens TH (2000) The sodium/proton exchanger Nhx1p is required for endosomal protein trafficking in the yeast *Saccharomyces cerevisiae*. Mol Biol Cell 11: 4277–4294

Boyer JS, Byrne P, Cassman KG et al (2013) The US drought of 2012 in perspective: a call to action. Glob Food Secur 2:139–143

Bradley KW, Sweet LE (2008) Influence of glyphosate and fungicide on weed control spray penetration, soybean response and yield in glyphosate resistant soybean. Agron J 100:1360–1313

Bravo A, Soberón M, Gill SS (2005) *Bacillus thuringiensis* mechanisms and use. In: Gilbert LI, Iatrou K, Gill SS (eds) Comprehensive molecular insect science. Elsevier, New York

Bravo A, Gill SS, Soberon M (2007) Mode of action of *Bacillus thuringiensis* cry and Cyt toxins and their potential for insect control. Toxicon 49:422–435

Breen S, Williams SJ, Outram M, Kobe B, Solomon PS (2017) Emerging insights into the functions of pathogenesis-related protein 1. Trends Plant Sci 22:871–879

Breitler JC, Marfa V, Royer M, Meynard D, Vassal JM, Vercambre B, Frutos R, Messeguer J, Gabarra R, Guiderdoni E (2000) Expression of a *Bacillus thuringiensis cry1B* synthetic gene protects Mediterranean rice against the striped stem borer. Plant Cell Rep 19(12):1195–1202

Brett CL, Tukaye DN, Mukherjee S, Rao RJ (2005) The yeast endosomal $Na^+(K^+)/H^+$ exchanger Nhx1 regulates cellular pH to control vesicle trafficking. Mol Biol Cell 16:1396–1405

Brini F, Yamamoto A, Jlaiel L et al (2011) Pleiotropic effects of the wheat dehydrin DHN-5 on stress responses in Arabidopsis. Plant Cell Physiol 52:676–688

Broderick NA, Raffa KF, Handelsman J (2006) Midgut bacteria required for *Bacillus thuringiensis* insecticidal activity. Proc Natl Acad Sci U S A 103:15196–15199

Brookes G, Barfoot P (2008) GM crops: global socio-economic and environmental impacts 1996–2006. PG Economics Ltd, Dorchester

Brookes G, Barfoot P (2009) Global impact of biotech crops: income and production effects 1996–2007. AgBioForum 12(2):184–208. www.agbioforum.org

Brookes G, Barfoot P (2013) The global income and production effects of genetically modified (GM) crops 1996–2011. GM Crops & Food 4(1):74–83

Brookes G, Barfoot P (2017) Environmental impacts of genetically modified (GM) crop use 1996–2015: impacts on pesticide use and carbon emissions. GM Crops & Food 8(2):117–147

Brookes G, Barfoot P (2018) Farm income and production impacts of using GM crop technology 1996–2016. GM Crops Food 9:59–89

Brower LP, Taylor OR, Williams EH, Slayback DA, Zubieta RR, Ramirez MI (2012) Decline of monarch butterflies overwintering in Mexico: is the migratory phenomenon at risk? Insect Conserv Divers 5(2):95–100

Bucchini L, Goldman LR (2002) Starlink corn: a risk analysis. Environ Health Perspect 110(1):5–13

Bughio N, Yamaguchi H, Nishizawa NK, Nakanishi H, Mori S (2002) Cloning an iron-regulated metal transporter from rice. J Exp Bot 53(374):1677–1682

Bulle M, Yarra R, Abbagani S (2016) Enhanced salinity stress tolerance in transgenic chilli pepper (Capsicum annuum L.) plants overexpressing the wheat antiporter (TaNHX2) gene. Mol Breed 36:36

Bulley S, Wright M, Rommens C, Yan H, Rassam M, Lin-Wang K, Andre C, Brewster D, Karunairetnam S, Allan AC, Laing WA (2011) Enhancing ascorbate in fruits and tubers through over-expression of the L-galactose pathway gene GDP-L-galactose phosphorylase. Plant Biotechnol J 10:390–397

Burd AD, Gould F, Bradley JR, Van Duyn JW, Moar WJ (2003) Estimated frequency of nonrecessive Bt resistance genes in bollworm, Helicoverpa zea (Boddie) (Lepidoptera: Noctuidae) in eastern North Carolina. J Econ Entomol 96(1):137–142

Burkhardt PK, Beyer P, Wünn J, Klöti A, Armstrong GA, Schledz M, von Lintig J, Potrykus I (1997) Transgenic rice (*Oryza sativa*) endosperm expressing daffodil (Narcissus pseudonarcissus) phytoene synthase accumulates phytoene, a key intermediate of provitamin A biosynthesis. Plant J 11(5):1071–1078

Burrill TJ (1878) Pear blight. Trans Ill State Hortic Sci 11:114–116

Busch FA (2013) Current methods for estimating the rate of photorespiration in leaves. Plant Biol 15:648–655

Buti M, Pasquariello M, Ronga D, Milc JA, Pecchioni N, Pucciariello C, Perata P, Francia E (2018) Transcriptome profiling of short-term response to chilling stress in tolerant and sensitive *Oryza sativa* ssp. Japonica seedlings. Funct Integr Genomics 18(6):627–644

Buzdin AA, Revina LP, Kostina LI, Zalunin IA, Chestukhina GG (2002) Interaction of 65- and 62-kD proteins from the apical membranes of the *Aedes aegypti* larvae midgut epithelium with Cry4B and Cry11A endotoxins of *Bacillus thuringiensis*. Biochemistry (Moscow) 67:540–546

Bykova NV, Keerberg O, Pärnik T, Bauwe H, Gardeström P (2005) Interaction between photorespiration and respiration in transgenic potato plants with antisense reduction in glycine decarboxylase. Planta 222(1):130–140

Cai D, Kleine M, Kifle S, Harloff HJ, Sandal NN, Marcker KA, Klein-Lankhorst RM, Salentijn EM, Lange W, Stiekema WJ, Wyss U (1997) Positional cloning of a gene for nematode resistance in sugar beet. Science 275(5301):832–834

Cai D, Thurau T, Tian Y, Lange T, Yeh KW, Jung C (2003) Sporamin-mediated resistance to beet cyst nematodes (Heterodera schachtii Schm.) is dependent on trypsin inhibitory activity in sugar beet (Beta vulgaris L.) hairy roots. Plant Mol Biol 51(6):839–849

Campbell EI, Unkles SE, Macro JA, Van den Hondel C, Contreras R, Kinghorn JR (1989) Improved transformation efficiency of *Aspergillus niger* based on the pyrG Gene. Mol Gen Genet 206:71–75

Campos ML, de Souza CM, de Oliveira KBS, Dias SC, Franco OL (2018) The role of antimicrobial peptides in plant immunity. J Exp Bot 69:4997–5011

Cao J, Shelton AM, Earle ED (2005) Development of transgenic collards (Brassica oleracea L., var. acephala) expressing a cry1Ac or cry1C Bt gene for control of the diamondback moth. Crop Prot 24(9):804–813

Cao M, Sato SJ, Behrens M, Jiang WZ, Clemente TE, Weeks DP (2011) Genetic engineering of maize (*Zea mays*) for high-level tolerance to treatment with the herbicide dicamba. J Agric Food Chem 59(11):5830–5834

Capell T, Bassie L, Christou P (2004) Modulation of the polyamine biosynthetic pathway in transgenic rice confers tolerance to drought stress. Proc Natl Acad Sci U S A 101:9909–9914

Cardoso FM (2014) Single domain antibodies targeting neuraminidase protect against an H5N1 influenza virus challenge. J Virol 88:82788296

Carlson CR, Callgant DA, Kolsto AB (1994) Genotypic diversity among Bacillus cereus and B. thuringiensis strains. Appl Environ Microbiol 60:487–497

Carlson CR, Kolsto AB (1994) A small (2.4 Mb) Bacillus cereus chromosome corresponds to a conserved region of a larger (5.3 Mb) Bacillus cereus chromosome. Mol Microbiol 13:161–169

Carozzi NB, Warren GW, Desai N, Jayne SM, Lotstein R, Rice DA, Evola S, Koziel MG (1992) Expression of a chimeric CaMV35S *Bacillus thuringiensis* insecticidal protein gene in transgenic tobacco. Plant Mol Biol 20:539–548

Carpenter JE (2010) Peer-reviewed surveys indicate positive impact of commercialized GM crops. Nat Biotechnol 28:319–321

Carrera A, Echenique V, Zhang W, Helguera M, Manthey F, Schrager A, Picca A, Cervigni G, Dubcovsky J (2007) A deletion at the Lpx-B1 locus is associated with low lipoxygenase activity

and improved pasta color in durum wheat (*Triticum turgidum* ssp. durum). J Cereal Sci 45(1): 67–77
Carrière Y, Crowder DW, Tabashnik BE (2010) Evolutionary ecology of insect adaptation to Bt crops. Evol Appl 3:561–573
Carvalho AO, Gomes VM (2011) Plant defensins and defensin-like peptides—biological activities and biotechnological applications. Curr Pharm Design 17:4270–4293
Carvalho SM, Vasconcelos MW (2013) Producing more with less: strategies and novel technologies for plant-based food biofortification. Food Res Int 54(1):961–971
Carzoli AK, Aboobucker SI, Sandall LL, Lübberstedt TT, Suza WP (2018) Risks and opportunities of GM crops: Bt maize example. Glob Food Sec 19:84–91
Castagnola AS, Jurat-Fuentes JL (2012) Bt crops: past and future. In: *Bacillus thuringiensis* biotechnology. Springer, Dordrecht, pp 283–304
Castiglioni P, Warner D, Bensen RJ, Anstrom DC, Harrison J et al (2008) Bacterial RNA chaperones confer abiotic stress tolerance in plants and improved grain yield in Maize under water-limited conditions. Plant Physiol 147:446–455
Castle LA, Siehl DL, Gorton R, Patten PA, Chen YH, Bertain S, Cho HJ, Duck N, Wong J, Liu D, Lassner MW (2004) Discovery and directed evolution of a glyphosate tolerance gene. Science 304(5674):1151–1154
Cazzonelli CI, Pogson BJ (2010) Source to sink: regulation of carotenoid biosynthesis in plants. Trends Plant Sci 15:266–274
Cazzonelli CI, Cuttriss AJ, Cossetto SB, Pye W, Crisp P, Whelan J, Finnegan EJ, Turnbull C, Pogson BJ (2009) Regulation of carotenoid composition and shoot branching in Arabidopsis by a chromatin modifying histone methyltransferase, SDG8. Plant Cell 21:39–53
Ceballos H, Sanchez T, Chavez AL, Iglesias C, Debouck D, Mafla G, Tohme J (2006) Variation in crude protein content in cassava (*Manihot esculenta* Crantz) roots. J Food Compos Anal 19: 589–593
Chai D, Lee SM, Ng JH, Yu H (2007) L-methionine sulfoximine as a novel selection agent for genetic transformation of orchids. J Biotechnol 131(4):466–472
Chakrabarti SK, Mandaokar AD, Shukla A, Pattanayak D, Naik PS, Sharma RP, Kumar PA (2000) *Bacillus thuringiensis cry1Ab* gene confers resistance to potato against *Helicoverpa armigera* (Hubner). Potato Res 43(2):143–152
Chakrabarty S, Jin M, Wu C, Chakraborty P, Xiao Y (2020) *Bacillus thuringiensis* vegetative insecticidal protein family Vip3A and mode of action against pest Lepidoptera. Pest Manag Sci 76:1612–1617
Chakraborty S, Chakraborty N, Datta A (2000) Increased nutritive value of transgenic potato by expressing a non-allergenic seed albumin gene from *Amaranthus hypochondriacus*. Proc Natl Acad Sci U S A 97:3724–3729
Chakroun M, Ferré J (2014) In vivo and in vitro binding of Vip3Aa to Spodoptera frugiperda midgut and characterization of binding sites by 125I radiolabeling. Appl Environ Microbiol 80 (20):6258–6265
Chakroun M, Banyuls N, Bel Y, Escriche B, Ferré J (2016) Bacterial vegetative insecticidal proteins (Vip) from entomopathogenic bacteria. Microbiol Mol Biol Rev 80:329–350
Chalapathi Rao ASV, Reddy AR (2008) Glutathione reductase: a putative redox regulatory system in plant cells. In: Khan NA, Singh S, Umar S (eds) Sulfur assimilation and abiotic stresses in plants. Springer, Berlin, pp 111–147
Champagne ET, Wood DF, Juliano BO, Bechtel DB (2004) The rice grain and its gross composition. In: Champagne ET (ed) Rice chemistry and technology, Cereal Chem, vol 3. Cereals & Grains Association, St Paul, MN, pp 93–96
Chan MT, Yu SM (1998) The 3′ untranslated region of a rice α-amylase gene functions as a sugar-dependent mRNA stability determinant. Proc Natl Acad Sci 95(11):6543–6547
Chan YL, Yang AH, Chen JT, Yeh KW, Chan MT (2010) Heterologous expression of taro cystatin protects transgenic tomato against Meloidogyne incognita infection by means of interfering sex determination and suppressing gall formation. Plant Cell Rep 29(3):231–238

Chandrasekaran J, Brumin M, Wolf D, Leibman D, Klap C, Pearlsman M, Sherman A (2016) Development of broad virus resistance in non-transgenic cucumber using CRISPR/Cas9 technology. Mol Plant Pathol 17:1140–1153

Chang HS, Kim NH, Park MJ, Lim SK, Kim SC, Kim JY, Kim JA, Oh HY, Lee CH, Huh K, Jeong TC, Nam DH (2003) The 5-enolpyruvylshikimate-3-phosphate synthase of glyphosate-tolerant soybean expressed in *Escherichia coli* shows no severe allergenicity. Mol Cell 15:20–26

Chang X, Lu Z, Shen Z, Peng Y, Ye G (2017) Bitrophic and tritrophic effects of transgenic cry1Ab/cry2Aj maize on the beneficial, nontarget *Harmonia axyridis* (Coleoptera: Coccinellidae). Environ Entomol 46:1171

Chanroj S, Wang G, Venema K, Zhang MW, Delwiche CF, Sze H (2012) Conserved and diversified gene families of monovalent cation/H(+) antiporters from algae to flowering plants. Front Plant Sci 3:25

Chapple C, Carpita N (1998) Plant cell walls as targets for biotechnology. Curr Opin Plant Biol 1: 179–185

Chaturvedi S (2004) Biosafety regulation: need for fine balancing. Econ Polit Wkly 14:3693–3697

Che P, Zhao ZY, Glassman K, Dolde D, Hu TX, Jones TJ, Gruis DF, Obukosia S, Wambugu F, Albertsen MC (2016) Elevated vitamin E content improves all-trans beta-carotene accumulation and stability in biofortified sorghum. Proc Natl Acad Sci U S A 113:11040–11045

Chen Y, Peumans WJ, Van-Damme EJM (2002) The Sambucus nigra type-2 ribosome inactivating protein SNA-1 exhibits in planta antiviral activity in transgenic tobacco. FEBS Lett 516:27–30

Chen LJ, Lee DS, Song ZP, Suh HS, Lu BR (2004) Gene flow from cultivated rice (Oryza sativa) to its weedy and wild relatives. Ann Bot 93(1):67–73

Chen M, Liu X, Wang Z, Song J, Qi Q, Wang PG (2005a) Modification of plant *N*-glycans processing: the future of producing therapeutic protein by transgenic plant. Med Res Rev 25: 343–360

Chen H, Tang W, Xu C, Li X, Lin Y, Zhang Q (2005b) Transgenic indica rice plants harboring a synthetic cry2A gene of *Bacillus thuringiensis* exhibit enhanced resistance against lepidopteran rice pests. Theor Appl Genet 111:1330–1337

Chen J, Brown MR, Hua G (2005c) Comparison of the localization of *Bacillus thuringiensis* Cry1A delta endotoxins and their binding proteins in larval midgut of tobacco hornworm *Manduca sexta*. Cell Tissue Res 321:123–129

Chen H, Zhang G, Zhang Q, Lin Y (2008) Effect of transgenic *Bacillus thuringiensis* rice lines on mortality and feeding behavior of rice stem borers (Lepidoptera: Crambidae). Econ Entomol 101:182–189

Chen JB, Wang SM, Jing RL, Mao XG (2009) Cloning the PvP5CS gene from common bean (*Phaseolus vulgaris*) and its expression patterns under abiotic stresses. Plant Physiol 166:12–19

Chen DX, Chen XW, Ma BT, Wang YP, Zhu LH, Li SG (2010a) Genetic transformation of rice with Pi-d2 gene enhances resistance to rice blast fungus *Magnaporthe oryzae*. Rice Sci 17:19–27

Chen HJ, Su CT, Lin CH, Huang GJ, Lin YH (2010b) Expression of sweet potato cysteine protease SPCP2 altered developmental characteristics and stress responses in transgenic *Arabidopsis* plants. Plant Physiol 167:838–847

Chen W, Qian Y, Wu X, Sun Y, Wu X, Cheng X (2014) Inhibiting replication of begomoviruses using artificial zinc finger nucleases that target viral-conserved nucleotide motif. Virus Genes 48 (3):494–501

Chen SP, Lin IW, Chen X, Huang YH, Chang SC, Lo HS, Lu HH, Yeh KW (2016) Sweet potato NAC transcription factor, Ib NAC 1, upregulates sporamin gene expression by binding the SWRE motif against mechanical wounding and herbivore attack. Plant J 86(3):234–248

Chen J, Piao Y, Liu Y, Li X, Piao Z (2018a) Genome-wide identification and expression analysis of chitinase gene family in *Brassica rapa* reveals its role in clubroot resistance. Plant Sci 270:257–267

Chen W, Liu C, Lu G, Cheng H, Shen Z, Wu K (2018b) Effects of Vip3AcAa+Cry1Ac cotton on midgut tissue in *Helicoverpa armigera* (Lepidoptera: Noctuidae). J Insect Sci 18(4):13

Chen X, Lewandowska D, Armstrong MR, Baker K, Lim TY, Bayer M, Harrower B, McLean K, Jupe F, Witek K et al (2018c) Identification and rapid mapping of a gene conferring broad-spectrum late blight resistance in the diploid potato species *Solanum verrucosum* through DNA capture technologies. Theor Appl Genet 131:1287–1297

Chenault KD, Payton ME, Melouk HA (2003) Greenhouse testing of transgenic peanut for resistance to Sclerotinia minor. Peanut Sci 30(2):116–120

Cheng X, Sardana R, Kaplan H, Altosaar I (1998) *Agrobacterium*-transformed rice plants expressing synthetic *cry1A(b)* and *cry1A(c)* genes are highly toxic to striped stem borer and yellow stem borer. Proc Natl Acad Sci U S A 95:2767–2772

Cheng X, Li F, Cai J, Chen W, Zhao N, Sun Y, Guo Y (2015) Artificial TALE as a convenient protein platform for engineering broad-spectrum resistance to begomoviruses. Viruses 7:4772–4782

Cheng P, Wang Z, Ren Y, Jin P, Ma K, Li Q, Wang B (2021) Silencing of a wheat Ortholog of Glucan synthase-like gene reduced resistance to Blumeria graminis f. sp. tritici. Front Plant Sci 12

Chida H, Nakazawa A, Akazaki H, Hirano T, Suruga K, Ogawa M, Satoh T, Kadokura K, Yamada S, Hakamata W, Isobe K (2007) Expression of the algal cytochrome c 6 gene in Arabidopsis enhances photosynthesis and growth. Plant Cell Physiol 48(7):948–957

Chinnadurai P, Stojšin D, Liu K, Frierdich GE, Glenn KC, Geng T, Schapaugh A, Huang K, Deffenbaugh AE, Liu ZL, Burzio LA (2018) Variability of CP4 EPSPS expression in genetically engineered soybean (Glycine max L. Merrill). Transgenic Res 27(6):511–524

Chinnusamy V, Zhu J, Zhu JK (2007) Cold stress regulation of gene expression in plants. Trends Plant Sci 12:444–451

Chisholm ST, Coaker G, Day B, Staskawicz BJ (2006) Host-microbe interactions: shaping the evolution of the plant immune response. Cell 124:803–814

Chitkowski RL, Turnipseed SG, Sullivan MJ, Bridges WG (2003) Field and laboratory evaluations of transgenic cottons expressing one or two *Bacillus thuringiensis* var. kurstaki Berliner proteins for management of noctuid (Lepidoptera) pests. J Econ Entomol 96:755–762

Cho SK, Kim JE, Park JA, Eom TJ, Kim WT (2006) Constitutive expression of abiotic stress inducible hot pepper CaXTH3, which encodes a xyloglucan endotransglucosylase/hydrolase homolog, improves drought and salt tolerance in transgenic *Arabidopsis* plants. FEBS Lett 580: 3136–3144

Cho E, Zhang X, Townsend MK, Selhub J, Paul L, Rosner B, Fuchs CS, Willett WC, Giovannucci EL (2015) Unmetabolized folic acid in prediagnostic plasma and the risk for colorectal cancer. J Natl Cancer Inst 12:107

Chopra R, Saini R (2014) Transformation of blackgram (*Vigna mungo* (L.) Hepper) by barley chitinase and ribosome-inactivating protein genes towards improving resistance to Corynespora leaf spot fungal disease. Appl Microbiol Biotechnol 174:2791–2800

Choudhary B, Gaur K (2010) Bt cotton in India: a country profile (ISAAA, 2010)

Choudhary SP, Yu JQ, Yamaguchi-Shinozaki K, Shinozaki K, Tran LS (2012) Benefits of brassinosteroid crosstalk. Trends Plant Sci 17(10):594–605

Christou P (2002) No credible scientific evidence is presented to support claims that transgenic DNA was introgressed into traditional maize landraces in Oaxaca, Mexico. Transgenic Res 11: 3–5

Christou P (2013) Plant genetic engineering and agricultural biotechnology 1983–2013. Trends Biotechnol 31:125–127

Cicero LL, Madesis P, Tsaftaris A, Piero AR (2015) Tobacco plants over-expressing the sweet orange tau glutathione transferases (CsGSTUs) acquire tolerance to the diphenyl ether herbicide fluorodifen and to salt and drought stresses. Phytochemistry 116:69–77

Clark WG, Fitchen JH, Beachy RN (1995) Studies of coat-protein mediated resistance to TMV using mutant CR I. The PM2 assembly defective mutant. Virology 208:485–491

Colangelo EP, Guerinot ML (2006) Put the metal to the petal: metal uptake and transport throughout plants. Curr Opin Plant Biol 9(3):322–330

Colgrave ML, Kotze AC, Huang YH, O'Grady J, Simonsen SM, Craik DJ (2008) Cyclotides: natural circular plant peptides that possess significant activity against gastrointestinal nematode parasites of sheep. Biochemistry 47:5581–5589

Collard BCY, Mackill DJ (2008) Marker-assisted selection: an approach for precision plant breeding in the twenty-first century. Philos Trans R Soc B 363:557–572

Cong L, Zheng HC, Zhang YX, Chai TY (2008) *Arabidopsis* DREB1A confers high salinity tolerance and regulates the expression of GA dioxygenases in tobacco. Plant Sci 174:156–164

Cong L, Wang C, Chen L, Liu H, Yang G, He G (2009) Expression of phytoene synthase1 and carotene desaturase crtI genes result in an increase in the total carotenoids content in transgenic elite wheat (*Triticum aestivum* L.). J Agric Food Chem 57:8652–8660

Connorton JM, Jones ER, Rodríguez-Ramiro I, Fairweather-Tait S, Uauy C, Balk J (2017) Wheat vacuolar iron transporter TaVIT2 transports Fe and Mn and is effective for biofortification. Plant Physiol 174:2434–2444

Cooper B, Lapidot M, Heick JA, Dodds JA, Beachy RN (1995) A defective movement protein of TMV in transgenic plants confers resistance to multiple viruses whereas the functional analog increases susceptibility. Virology 206:307–313

Cordeiro AM, Andrade L, Monteiro CC, Leitão G, Wigge PA, Saibo NJ (2022) Phytochrome-interacting factors: a promising tool to improve crop productivity. J Exp Bot 73(12):3881–3897

Cortina C, Culianez-Macia FA (2005) Tomato abiotic stress enhanced tolerance by trehalose biosynthesis. Plant Sci 169:75–82

Cotsaftis O, Sallaud C, Breitler JC, Meynard D, Greco R, Pereira A, Guiderdoni E (2002) Transposon-mediated generation of T-DNA-and marker-free rice plants expressing a Bt endotoxin gene. Mol Breed 10(3):165–180

Craig RP, David JE (2007) Role of receptors in Bacillus thuringiensis crystal toxin activity. Microbiol Mol Rev:255–281

Craik DJ (2009) Circling the enemy: cyclic proteins in plant defense. Trends Plant Sci 14:328–335

Cramer WA, Zhang H, Yan J, Kurisu G, Smith JL (2006) Transmembrane traffic in the cytochrome b~ 6 f complex. Annu Rev Biochem 75(1):769–790

Cramer GR, Urano K, Delrot S, Pezzotti M, Shinozaki K (2011) Effects of abiotic stress on plants: a systems biology perspective. BMC Plant Biol 11:163

Crickmore N, Zeigler DR, Feitelson J, Schnepf E, Van Rie J, Lereclus D, Baum J, Dean DH (1998) Revision of the nomenclature for *Bacillus thuringiensis* pesticidal crystal proteins. Microbiol Mol Biol Rev 62(3):807–813

Crickmore N, Zeigler DR, Schnepf E, Baum J, Bravo A, Dean DH (2013) *Bacillus thuringiensis* toxins nomenclature. www.btnomencluture.infolifesci.sussex.ac.uk/Home/Neil_Crickmore/Bt. Accessed 7 Nov 2014

Cruz J, Hidalgo-Gallego A, Lora JM, Benitez T, Pintor-Toro JA, Llobell A (1992) Isolation and characterization of three chitinases from *Trichoderma harzianum*. FEBS J 206:859–867

Cruz AB, Bianchetti RE, Alves FR, Purgatto E, Peres LE, Rossi M, Freschi L (2018) Light, ethylene and auxin signaling interaction regulates carotenoid biosynthesis during tomato fruit ripening. Front Plant Sci 9:1370

Cui X, Balcerzak M, Schernthaner J, Babic V, Datla R, Brauer EK et al (2019) An optimised CRISPR/Cas9 protocol to create targeted mutations in homoeologous genes and an efficient genotyping protocol to identify edited events in wheat. Plant Methods 15:119

Cunningham FX, Gantt E (1998) Genes and enzymes of carotenoid biosynthesis in plants. Annu Rev Plant Physiol 49:557–583

Cunningham FX Jr, Pogson B, Sun Z, McDonald KA, DellaPenna D, Gantt E (1996) Functional analysis of the beta and epsilon lycopene cyclase enzymes of Arabidopsis reveals a mechanism for control of cyclic carotenoid formation. Plant Cell 8(9):1613–1626

Cutt JR, Harpster MH, Dixon DC, Carr JP, Dunsmuir P, Klessig DF (2005) Disease response to tobacco mosaic virus in transgenic tobacco plants that constitutively express the pathogenesis-related PR1b gene. Virology 173:89–97

D'Ambrosio C, Giorio G, Marino I, Merendino A, Petrozza A, Salfi L, Stigliani AL, Cellini F (2004) Virtually complete conversion of lycopene into β-carotene in fruits of tomato plants transformed with the tomato lycopene β-cyclase (tlcy-b) cDNA. Plant Sci 166:207–214

Dafny-Yelin M, Tzfira T (2007) Delivery of multiple transgenes to plant cells. Plant Physiol 145: 1118–1128

Dai L, Wang D, Xie X, Zhang C, Wang X, Xu Y (2016) The novel gene VpPR4-1 from *Vitis pseudoreticulata* increases powdery mildew resistance in transgenic Vitis vinifera L. Front Plant Sci 7:695. https://doi.org/10.3389/fpls.2016.00695

Dale PJ, Irwin JA, Scheffler JA (1993) The experimental and commercial release of transgenic crop plants. Plant Breed 111(1):1–22

Dallal JA, Irvin JD (1978) Enzymatic inactivation of eukaryotic ribosomes by the pokeweed antiviral protein. FEBS Lett 89:257–259

Daniell H (2002) Molecular strategies for gene containment in transgenic crops. Nat Biotechnol 20: 581–586

Dary O, Hainsworth M (2008) The food fortification formulator: technical determination of fortification levels and standards for mass fortification. A2Z: The USAID Micronutrient Project/Academy for Educational Development, Washington, DC

Das G, Rao GJN (2015) Molecular marker assisted gene stacking for biotic and abiotic stress resistance genes in an elite rice cultivar. Front Plant Sci 6:698

Das B, Goswami L, Ray S, Ghosh S, Bhattacharyya S, Das S, Majumder AL (2006) Agrobacterium-mediated transformation of Brassica juncea with a cyanobacterial (Synechocystis PCC6803) delta-6 desaturase gene leads to production of gamma-linolenic acid. Plant Cell Tissue Organ Cult 86:219–231

Das G, Patra JK, Baek KH (2017) Insight into MAS: a molecular tool for development of stress resistant and quality of rice through gene stacking. Front Plant Sci 8:985

Das A, Ghosh P, Das S (2018) Expression of *Colocasia esculenta* tuber agglutinin in Indian mustard provides resistance against *Lipaphis erysimi* and the expressed protein is non-allergenic. Plant Cell Rep 37:849–863

Das MK, Sharma RS, Mishra V (2012) Induction of apoptosis by ribosome inactivating proteins. Importance of Nglycosidase activity. Appl Biochem Biotechnol 166:1552–1561

Datta SK, Bouis HE (2000) Application of biotechnology to improving the nutritional quality rice. Food Nutr Bull 21:451–456

Datta K, Vasquez A, Tu J, Torrizo L, Alam MF, Oliva N, Abrigo E, Khush GS, Datta SK (1998) Constitutive and tissue-specific differential expression of *cry1A(b)* gene in transgenic rice plants conferring resistance to rice insect pests. Theor Appl Genet 97:20–30

Datta K, Baisakh N, Oliva N, Torrizo L, Abrigo E, Tan J, Rai M, Rehana S, Al-Babili S, Beyer P, Potrykus I (2003) Bioengineered 'golden' indica rice cultivars with β-carotene metabolism in the endosperm with hygromycin and mannose selection systems. Plant Biotechnol J 1(2):81–90

Datta K, Rai M, Parkhi V, Oliva N, Tan J, Datta SK (2006) Improved 'golden' indica rice and post-transgeneration enhancement of metabolic target products of carotenoids (β-carotene) in transgenic elite cultivars (IR64 and BR29). Curr Sci 10:935–939

Datta SK, Datta K, Parkhi V, Rai M, Baisakh N, Sahoo G, Rehana S, Bandyopadhyay A, Alamgir M, Ali MS, Abrigo E (2007) Golden rice: introgression, breeding, and field evaluation. Euphytica 154:271–278

Davis MJ, Ying Z (2004) Development of papaya breeding lines with transgenic resistance to Papaya ringspot virus. Plant Dis 88:352–358

Davison J (2010) GM plants: science, politics and EC regulations. Plant Sci 178(2):94–98

de Beer A, Vivier MA (2008) Vv-AMP1 a ripening induced peptide from *Vitis vinifera* shows strong antifungal activity. BMC Plant Biol 8:75

De Cosa B, Moar W, Lee SB, Miller M, Daniell H (2001) Over expression of the Bt *cry*2Aa2 operon in chloroplasts leads to transformation of insecticidal crystals. Nat Biotechnol 19:71–74

De Greef W, Delon R, De Block M, Leemans J, Botterman J (1989) Evaluation of herbicide resistance in transgenic crops under field conditions. Nat Biotechnol 7(1):61

de la Fuente-Salcido NM, Casados-Vázquez LE, Barboza-Corona JE (2013) Bacteriocins of Bacillus thuringiensis can expand the potential of this bacterium to other areas rather than limit its use only as microbial insecticide. Can J Microbiol 59:515–522

De Maagd RA, Bravo A, Berry C, Crickmore N, Schnepf HE (2003) Structure, diversity, and evolution of protein toxins from spore-forming entomopathogenic bacteria. Annu Rev Genet 37: 409–433

De Moura FF, Moursi M, Donahue Angel M, Angeles-Agdeppa I, Atmarita A, Gironella GM, Muslimatun S, Carriquiry A (2016) Biofortified beta-carotene rice improves vitamin A intake and reduces the prevalence of inadequacy among women and young children in a simulated analysis in Bangladesh, Indonesia, and the Philippines. Am J Clin Nutr 104:769–775

De Tullio MC (2004) How does ascorbic acid prevent scurvy? A survey of the nonantioxidant functions of vitamin C. In: Asard H (ed) Vitamin C: its function and biochemistry in animals and plants. Garland Science/BIOS Scientific Publishers, London, pp 176–190

De Virgilio M, Lombardi A, Caliandro R, Fabbrini MS (2010) Ribosomeinactivating proteins: from plant defense to tumor attack. Toxins 2(11):2699–2737

de Vos CJ, Swanenburg M (2018) Health effects of feeding genetically modified (GM) crops to livestock animals: a review. Food Chem Toxicol 117:3–12

De Wit PJ, Hofman AE, Velthuis GC, Kuc JA (1985) Isolation and characterization of an elicitor of necrosis isolated from intercellular fluids of compatible interactions of *Cladosporium fulvum* (syn. *Fulvia fulva*) and tomato. Plant Physiol 77:642–647

Dean R, Van Kan JA, Pretorius ZA, Hammond-Kosack KE, Di Pietro A, Spanu PD, Rudd JJ, Dickman M, Kahmann R, Ellis J (2012) The Top 10 fungal pathogens in molecular plant pathology. Mol Plant Pathol 13:414–430

Deepak S (2012) Bt cotton in India: a review of adoption, government interventions and investment initiatives. Indian J Agric Econ 67:1–11

Deepak KV, Kimy G, Pratibha S, Rishika VJ, Dipendra KMA, Prem PS (2018) Genetically modified organisms produced enzymes: multifarious applications in food manufacturing industries. In: Bioprocessing technology in food and health. CRC Press, Boca Raton, FL, pp 73–120

del Río LA, Sandalio LM, Corpas FJ, Palma JM, Barroso JB (2006) Reactive oxygen species and reactive nitrogen species in peroxisomes. Production, scavenging, and role in cell signaling. Plant Physiol 141:330–335

Delannay X, LaVallee BJ, Proksch RK, Fuchs RL, Sims SR, Greenplate JT, Marrone PG, Dodson RB, Augustine JJ, Layton JG, Fischhoff DA (1989) Field performance of transgenic tomato plants expressing the *Bacillus thuringiensis* var kurstaki insect control protein. Bio/Technology 7:1265–1269

Dexter PB (1998) Rice fortification for developing countries. OMNI/USAID, Washington, DC

Dharmapuri S, Rosati C, Pallara P, Pallaraa P, Aquilani R, Bouvierb F, Camarab B, Giulianoa G (2002) Metabolic engineering of xanthophyll content in tomato fruits. FEBS Lett 519:30–34

Dhurua S, Gujar GT (2011) Field-evolved resistance to Bt toxin Cry1Ac in the pink bollworm, *Pectinophora gossypiella* (Saunders) (Lepidoptera: Gelechiidae), from India. Pest Manag Sci 67:898–903

Díaz-Gómez J, Twyman RM, Zhu C, Farré G, Serrano JC, Portero-Otin M, Muñoz P, Sandmann G, Capell T, Christou P (2017) Biofortification of crops with nutrients: factors affecting utilization and storage. Curr Opin Biotechnol 44:115–123

Dick RE, Quinn JP (1995) Control of glyphosate uptake and metabolism in Pseudomonas sp. 4ASW. FEMS Microbiol Lett 134(2–3):177–182

Ding SW (2010) RNA-based antiviral immunity. Nat Rev Immunol 10:632–644

Ding F, Wang M, Zhang S, Ai X (2016) Changes in SBPase activity influence photosynthetic capacity, growth, and tolerance to chilling stress in transgenic tomato plants. Sci Rep 6:32741

Diretto G, Tavazza R, Welsch R, Pizzichini D, Mourgues F, Papacchioli V, Beyer P, Giuliano G (2006) Metabolic engineering of potato tuber carotenoids through tuber-specific silencing of lycopene ε-cyclase. BMC Plant Biol 6:13

Diretto G, Al-Babili S, Tavazza R, Papacchioli V, Beyer P, Giuliano G (2007) Metabolic engineering of potato carotenoid content through tuber-specific overexpression of a bacterial mini pathway. PLoS One 2:e350

Distelfeld A, Cakmak I, Peleg Z, Ozturk L, Yazici AM, Budak H, Saranga Y, Fahima T (2007) Multiple QTL-effects of wheat GPC-B1 locus on grain protein and micronutrient concentrations. Physiol Plant 129:635–643

Dita MA, Rispail N, Prats E, Rubiales D, Singh KB (2006) Biotechnology approaches to overcome biotic and abiotic stress constraints in legumes. Euphytica 147(1):1–24

Dively GP, Rose R, Sears MK, Hellmich RL, Stanley-Horn DE, Calvin DD, Russo JM, Anderson PL (2004) Effects on monarch butterfly larvae (Lepidoptera: Danaidae) after continuous exposure to Cry1Ab-expressing corn during anthesis. Environ Entomol 33(4):1116–1125

Divya K, Jami SK, Kirti PB (2010) Constitutive expression of mustard annexin, AnnBj1 enhances abiotic stress tolerance and fiber quality in cotton under stress. Plant Mol Biol 73(3):293–308

Dixon DP, Skipsey M, Edwards R (2010) Roles for glutathione transferases in plant secondary metabolism. Phytochemistry 71:338–350

Domingo JL (2016) Safety assessment of GM plants: an updated review of the scientific literature. Food Chem Toxicol 95:12–18

Dong H, Deng Y, Mu J, Lu Q, Wang Y, Xu Y, Chu C, Chong K, Lu C, Zuo J (2007) The Arabidopsis Spontaneous Cell Death1 gene, encoding a ζ-carotene desaturase essential for carotenoid biosynthesis, is involved in chloroplast development, photoprotection and retrograde signalling. Cell Res 17(5):458–470

Donovan WP, Engleman JT, Donovan JC, Baum JA, Bunkers GJ, Chi DJ, Clinton WP, English L, Heck GR, Ilagan OM, Krasomil-Osterfeld KC (2006) Discovery and characterization of Sip1A: a novel secreted protein from Bacillus thuringiensis with activity against coleopteran larvae. Appl Microbiol Biotechnol 72(4):713–719

Dorsch JA, Candas M, Griko NB (2002) Cry1A toxins of *Bacillus thuringiensis* bind specifically to a region adjacent to the membrane-proximal extracellular domain of Bt-R1 in *Manduca sexta*: involvement of a cadherin in the entomopathogenicity of *Bacillus thuringiensis*. Insect Biochem Mol Biol 32:1025–1036

Douches DS, Westedt AL, Zarka KA, Schroeter B, Grafius EJ (1998) Potato transformation to combine natural and engineered resistance for controlling tuber moth. Hort Sci 33:1053–1056

Douglas FW, Hollis MF, Randy WD, Robert EB (1994) Yield, yield components and fibre properties of insect resistant cotton lines containing *Bacillus thuringiensis* toxin gene. Crop Sci 34:38–41

Dowd PF, Zuo WN, Gillikin JW, Johnson ET, Boston RS (2003) Enhanced resistance to *Helicoverpa zea* in tobacco expressing an activated form of maize ribosome-inactivating protein. J Agric Food Chem 51:3568–3574

Dowd PF, Rober AH, Pinkerton TS, Johnson ET, Lagrimini LM, Boston RS (2006) Relative activity of a tobacco hybrid expressing high levels of a tobacco anionic peroxidase and maize ribosome-inactivating protein against *Helicoverpa zea* and *Lasioderma serricorne*. J Agric Food Chem 54:2629–2634

Dowd PF, Johnson ET, Price NP (2012) Enhanced pest resistance of maize leaves expressing monocot crop plant-derived ribosome-inactivating protein and agglutinin. J Agric Food Chem 60:10768–10775

Dowson Day MJ, Ashurst JA, Mathias SF, Watts JW, Wilson TM, Dixon RA (1993) Plant viral leaders influence expression of a reporter gene in tobacco. Plant Mol Biol 23(1):97–109

Drakakaki G, Christou P, Stoger E (2000) Constitutive expression of soybean ferritin cDNA in transgenic wheat and rice results in increased iron levels in vegetative tissues but not in seeds. Transgenic Res 9:445–452

Drakakaki G, Marcell S, Glahn RP, Lund EK, Pariagh S, Fischer R, Christou P, Stoger E (2005) Endosperm-specific co-expression of recombinant soybean ferritin and Aspergillus phytase in maize results in significant increases in the levels of bioavailable iron. Plant Mol Biol 59:869–880

Driever SM, Simkin AJ, Alotaibi S, Fisk SJ, Madgwick PJ, Sparks CA, Jones HD, Lawson T, Parry MAJ, Raines CA (2017) Increased SBPase activity improves photosynthesis and grain yield in wheat grown in greenhouse conditions. Philos Trans R Soc B: Biol Sci 372:1730

Duan J, Cai W (2012) OsLEA3-2, an abiotic stress induced gene of rice plays a key role in salt and drought tolerance. PLoS One 7:e45117

Duan XG, Yang AF, Gao F, Zhang SL, Zhang JR (2007) Heterologous expression of vacuolar H (+)-PPase enhances the electrochemical gradient across the vacuolar membrane and improves tobacco cell salt tolerance. Protoplasma 232:87–95

Duan CG, Wang CH, Guo HS (2012) Application of RNA silencing to plant disease resistance. Silence 3:5

Duan X, Hou Q, Liu G, Pang X, Niu Z, Wang X, Zhang Y, Li B, Liang R (2018) Expression of *Pinellia pedatisecta* lectin gene in transgenic wheat enhances resistance to wheat aphids. Molecules 23:748

Dubouzet JG, Sakuma Y, Ito Y, Kasuga M, Dubouzet EG, Miura S, Seki M, Shinozaki K, Yamaguchi-Shinozaki K (2003) OsDREB genes in rice, (*Oryza sativa* L.) encode transcription activators that function in drought, high salt and cold responsive gene expression. Planta 33: 751–763

Ducreux LJ, Morris WL, Hedley PE, Shepherd T, Davies HV, Millam S, Taylor MA (2005) Metabolic engineering of high carotenoid potato tubers containing enhanced levels of beta-carotene and lutein. J Exp Bot 56(409):81–89

Duke SO (1988) Glyphosate. In: Kearney PC, Kaufman DD (eds) Herbicides, chemistry, degradation and mode of action, vol 3. Marcel Dekker, New York, pp 1–70

Duke SO, Cerdeira AL (2007) Transgenic crops for herbicide resistance. In: Transgenic crop plants 2010. Springer, Berlin, pp 133–166

Duke SO, Baerson SR, Rimando AM (2003) Herbicides: glyphosate. In: Plimmer JR, Gammon DW, Ragsdale NN (eds) Encyclopedia of agrochemicals. Wiley, New York

Duke SO, Cedergreen N, Velini ED, Belz RG (2006) Hormesis: is it an important factor in herbicide use and allelopathy. Outlooks Pest Manag 17:29–33

Dun B, Wang X, Lu W, Chen M, Wei W, Shuzhen P et al (2014) Development of highly glyphosate-tolerant tobacco by coexpression of glyphosate acetyltransferase gat and EPSPS G2-aroA genes. Crop J 2(2–3):164–169

Dunne J, Hanby AM, Poulsom R, Jones TA, Sheer D, Chin WG, Da SM, Zhao Q, Beverley PC, Owen MJ (1995) Molecular cloning and tissue expression of FAT, the human homologue of the Drosophila fat gene that is located on chromosome 4q34-q35 and encodes a putative adhesion molecule. Genomics 30:207–223

Dunwell JM (2014) Transgenic cereals: current status and future prospects. J Cereal Sci 59(3):419–434

Dutta I, Saha P, Das S (2008) Efficient *Agrobacterium*-mediated genetic transformation of oilseed mustard [*Brassica juncea* (L.) Czern.] using leaf piece explants. In Vitro Cell Dev Biol Plant 44: 401–411

Dutton A, Romies J, Bigler F (2005) Effects of Bt maize expressing cry1Ab and Bt spray on Spodoptera littoralis. Entomologia Experimentalis et Applicata 114:161–169

Duvick JP, Rood T, Rao AG, Marshak DR (1992) Purification and characterization of a novel antimicrobial peptide from maize (*Zea mays* L.) kernels. J Biol Chem 267:18814–18820

Dwivedi SL, Ceccarelli S, Blair MW, Upadhyaya HD, Are AK, Ortiz R (2016) Landrace germplasm for improving yield and abiotic stress adaptation. Trends Plant Sci 21(1):31–42

Edpuganti SL (2018) Resistance development in pink bollworm, *Pectinophora gossypiella* (Saunders) (Lepidoptera: Gelechiidae) to Bt cotton and resistance management strategies. Int J Pure Appl Biosci 6(1):1296–1302

Edwards R, Dixon DP, Walbot V (2000) Plant glutathione *S*-transferases: enzymes with multiple functions in sickness and in health. Trends Plant Sci 5:193–198

Eide D, Broderius M, Fett J, Guerinot ML (1996) A novel iron-regulated metal transporter from plants identifed by functional expression in yeast. Proc Natl Acad Sci U S A 93(11):5624–5628

Eitenmiller RR, Landen W Jr, Ye L (2016) Vitamin analysis for the health and food sciences. CRC Press, Boca Raton, FL

El-Esawi MA, Alayafi AA (2019) Overexpression of rice Rab7 gene improves drought and heat tolerance and increases grain yield in rice (*Oryza sativa* L.). Genes 10(1):56

Ellul P, Rios G, Atares A, Roig LA, Serrano R, Moreno V (2003) The expression of the *Saccharomyces cerevisiae* HAL1 gene increases salt tolerance in transgenic watermelon (*Citrullus lanatus* (Thunb.) Matsun. & Nakai.). Theor Appl Genet 107:462–469

Ely S (1993) The engineering of plants to express *Bacillus thuringiensis* d-endotoxins. Resistance to *Bacillus thuringiensis* and resistance management. *Bacillus thuringiensis* an environmental biopesticide: theory and practice. Wiley, New York

Enfissi EMA, Fraser PD, Lois LM, Boronat A, Schuch W, Bramley PM (2005) Metabolic engineering of the mevalonate and nonmevalonate isopentenyl diphosphate-forming pathways for the production of health-promoting isoprenoids in tomato. Plant Biotechnol J 3:17–27

Engel N, van den Daele K, Kolukisaoglu U, Morgenthal K, Weckwerth W, Parnik T, Keerberg O, Bauwe H (2007) Deletion of glycine decarboxylase in Arabidopsis is lethal under nonphotorespiratory conditions. Plant Physiol 144(3):1328–1335

Ervin DE, Welsh R (2006) Environmental effects of genetically modified crops: differentiated risk assessment and management. In: Regulating agricultural biotechnology: economics and policy. Springer, Boston, MA, pp 301–326

Espinosa-Ruiz A, Belles JM, Serrano R, Culianez-Macia FA (1999) *Arabidopsis thaliana* AtHAL3: a flavoprotein related to salt and osmotic tolerance and plant growth. Planta 20:529–539

Estela A, Escriche B, Ferre J (2004) Interaction of *Bacillus thuringiensis* toxins with larval midgut binding sites of *Helicoverpa armigera* (Lepidoptera: Noctuidae). Appl Environ Microbiol 70: 1378–1384

Estruch JJ, Warren GW, Mullins MA, Nye GJ, Craig JA, Koziel MG (1996) Vip3A, a novel Bacillus thuringiensis vegetative insecticidal protein with a wide spectrum of activities against lepidopteran insects. Proc Natl Acad Sci 93(11):5389–5394

Evans LT (1993) Crop evolution, adaptation, and yield. Cambridge University Press, Cambridge/New York

Evans TL (1998) Greater crop production: whence and whither? In: Waterlow JC, Armstrong DG, Fowdenand L, Riley R (eds) Feeding a world population of more than eight billion people. Oxford University Press, Cary, NC, pp 89–97

Evans IM, Boulter D (1980) Crude protein and sulphur amino acid contents of some commercial varieties of peas and beans. J Sci Food Agric 31:238–242

Evans TL, Fischer RA (1999) Yield potential: its definition, measurement, and significance. Crop Sci 39:1544–1551

Evers D, Lefevre I, Legay S, Lamoureux D, Hausman JF, Rosales RO, Marca LR, Hoffmann L, Bonierbale M, Schafleitner R (2010) Identification of drought-responsive compounds in potato through a combined transcriptomic and targeted metabolite approach. J Exp Bot 61(9): 2327–2343

Falco SC, Guida T, Locke M, Mauvais J, Sanders C, Ward RT, Webber P (1995) Transgenic canola and soybean seeds with increased lysine. Nat Biotechnol 13:577–582

Fan Q, Song A, Jiang Zhang T, Sun H, Wang Y (2016) *CmWRKY1* enhances the dehydration tolerance of chrysanthemum through the regulation of ABA-associated genes. PLoS One 11: e0150572

Fancy NN, Bahlmann AK, Loake GJ (2017) Nitric oxide function in plant abiotic stress. Plant Cell Environ 40(4):462–472

FAOSTAT (2008) Statistics on agricultural production and trade. Food and Agricultural Organization of the United Nations, Rome

Fares NH, El-Sayed AK (1998) Fine structural changes in the ileum of mice fed on endotoxin treated potatoes and transgenic potatoes. Nat Toxins 6:219–233

Fearing PL, Brown D, Vlachos D, Meghji M, Privalle L (1997) Quantitative analysis of CryIA (b) expression in Bt maize plants, tissues, and silage and stability of expression over successive generations. Mol Breed 3(3):169–176

Feng PCC, Baley GJ, Clinton WP, Bunkers GJ, Alibhai MF, Paulitz TC, Kidwell KK (2005) Glyphosate inhibits rust diseases in glyphosate-resistant wheat and soybean. Proc Natl Acad Sci U S A 102:17290–17295

Feng PCC, Clark C, Andrade GC, Balbi MC, Caldwell P (2008) The control of Asian rust by glyphosate-resistant soybeans. Pest Manag Sci 64:353–359

Fennoy SL, Baileyserres J (1993) Synonymous codon usage in *Zea mays* L nuclear genes is varied by levels of C-ending and G-ending codons. Nucleic Acids Res 21:5294–5300

Fernandez de Caleya R, Gonzalez-Pascual B, García-Olmedo F, Carbonero P (1972) Susceptibility of phytopathogenic bacteria to wheat purothionins in vitro. Appl Microbiol 23:998–1000

Fernandez LE, Aimanova KG, Gill SS, Bravo A, Soberon M (2006) A GPI-anchored alkaline phosphatase is a functional midgut receptor of Cry11Aa toxin in Aedes aegypti larvae. Biochem J 394:77–84

Ferre J, Van Rie J (2002) Biochemistry and genetics of insect resistance to Bacillus thuringiensis. Annu Rev Entomol 47:501–533

Ferreira SA, Pitz KY, Manshardt R, Zee F, Fitch M, Gonsalves D (2002) Virus coat protein transgenic papaya provides practical control of papaya ringspot virus in Hawaii. Plant Dis 86(2):101–105

Ferry N, Edwards MG, Gatehouse JA, Gatehouse AMR (2004) Plant-insect interactions: molecular approaches to insect resistance. Curr Opin Biotechnol 15:155–161

Ferry N, Edwards MG, Gatehouse J, Capell T, Christou P, Gatehouse AMR (2006) Transgenic plants for insect pest control: a forward-looking scientific perspective. Transgenic Res 15:13–19

Fillatti JJ, Kiser J, Rose R, Comai L (1987) Efficient transfer of a glyphosate tolerance gene into tomato using a binary *Agrobacterium tumefaciens* vector. Bio/Technology 5:726–730

Fischer R, Schumann D, Zimmermann S, Drossard J, Sack M, Schillberg S (1999) Expression and characterization of bispecific single-chain Fv fragments produced in transgenic plants. Eur J Biochem 262:810–816

Fischhoff DA, Bowdish KS, Perlak FJ, Marrone PG, McCormick SM, Niedermeyer JG, Dean DA, Kusano-Kretzmer K, Mayer EJ, Rochester DE, Rogers SG, Fraley RT (1987) Insect tolerant transgenic tomato plants. Bio/Technology 5:807–813

Fitzpatrick TB, Basset GJ, Borel P, Carrari F, DellaPenna D, Fraser PD, Hellmann H, Osorio S, Rothan C, Valpuesta V, Caris-Veyrat C (2012) Vitamin deficiencies in humans: can plant science help? Plant Cell 24:395–414

Flor HH (1971) Current status of gene-for-gene concept. Annu Rev Phytopathol 9:27

Food and Agriculture Organization [FAOSTAT] (2017) FAO Statistical Databases FAOSTAT. http://www.fao.org/faostat/en/#data

Foster SJ, Park TH, Pel M, Brigneti G, Sliwka J, Jagger L, van der Vossen E, Jones JDG (2009) Rpi-vnt1.1, a Tm-2(2) homolog from *Solanum venturii*, confers resistance to potato late blight. Mol Plant Microbe Interact 22:589–600

Foyer CH, Halliwell B (1976) The presence of glutathione and glutathione reductase in chloroplasts: a proposed role in ascorbic acid metabolism. Planta 133:21–25

Fraley RT, Rogers SG, Horsch RB, Sanders PR, Flick JS et al (1983) Expression of bacterial genes in plant cells. Proc Natl Acad Sci U S A 80:4803–4807

Frankard V, Ghislain M, Negrutiu I, Jacobs M (1991) High threonine producer mutant of *Nicotiana sylvestris* (Spegg. and Comes). Theor Appl Genet 82:273–282

Fraser PD, Susanne R, Cathie AS, Philippa BM, Kiano JW, Misawa N, Drake RG, Schuch W, Bramley PM (2002) Evaluation of transgenic tomato plants expressing an additional phytoene synthase in a fruit-specific manner. Proc Natl Acad Sci U S A 99:1092–1097

Fray RG, Grierson D (1993) Identification and genetic analysis of normal and mutant phytoene synthase genes of tomato by sequencing, complementation and co-suppression. Plant Mol Biol 22:589–602

Frizzi A, Huang S, Gilbertson LA, Armstrong TA, Luethy MH, Malvar TM (2008) Modifying lysine biosynthesis and catabolism in corn with a single bifunctional expression/silencing transgene cassette. Plant Biotechnol J 6(1):13–21

Fryer MJ (1992) The antioxidant effects of thylakoid vitamin E (α-tocopherol). Plant Cell Environ 15:381–392

Fujimoto H, Itoh K, Yamomoto M, Kyojuka J, Shimamoto K (1993) Insect resistant rice generated by introduction of a modified δ-endotoxin gene of *Bacillus thuringiensis*. Bio/Technology 11: 1151–1155

Fujimura M, Minami Y, Watanabe K, Tadera K (2003) Purification characterization and sequencing of a novel type of antimicrobial peptides Fa-AMP1 and Fa-AMP2 from seeds of buckwheat (*Fagopyrum esculentum* Moench.). Biosci Biotechnol Biochem 67:1636–1642

Fujisawa M, Takita E, Harada H, Sakurai N, Suzuki H, Ohyama K, Shibata D, Misawa N (2009) Pathway engineering of *Brassica napus* seeds using multiple key enzyme genes involved in ketocarotenoid formation. J Exp Bot 60:1319–1332

Gahan LJ, Gould F, Heckel DG (2001) Identification of a gene associated with Bt resistance in Heliothis virescens. Science 293(5531):857–860

Gaines TA, Zhang W, Wang D, Bukun B, Chrisholm ST, Shaner DL (2010) Gene amplification confers glyphosate resistance in Amaranthus palmeri. PNAS 107:1034

Galili G, Höfgen R (2002) Metabolic engineering of amino acids and storage proteins in plants. Metab Eng 4(1):3–11

Galili G, Amir R (2013) Fortifying plants with the essential amino acids lysine and methionine to improve nutritional quality. Plant Biotechnol J 11:211–222

Galili G, Amir R, Hoefgen R, Hesse H (2005) Improving the levels of essential amino acids and sulfur metabolites in plants. Biol Chem 386(9):817–831

Galili G, Amir R, Fernie AR (2016) The regulation of essential amino acid synthesis and accumulation in plants. Annu Rev Plant Biol 67(1):153–178

Galvez AF, Revilleza MJ, de Lumen BO, Krenz DC (2008) Enhancing the biosynthesis of endogenous methionine-rich proteins (MRP) to improve the protein quality of legumes via genetic engineering. In: Food for health in the pacific rim. Food & Nutrition Press, Inc., Trumbull, pp 540–552

Galvez LC, Banerjee J, Pinar H, Mitra A (2014) Engineered plant virus resistance. Plant Sci 228: 11–25

Gangadhar BH, Sajeesh K, Venkatesh J, Baskar V, Abhinandan K, Yu JW, Prasad R, Mishra RK (2016) Enhanced tolerance of transgenic potato plants over-expressing non-specific lipid transfer protein-1 (StnsLTP1) against multiple abiotic stresses. Front Plant Sci 7:1228

Gao AG, Hakimi SM, Mittanck CA, Wu Y, Woerner BM, Stark DM, Shah DM, Liang J, Rommens CM (2000) Fungal pathogen protection in potato by expression of a plant defensin peptide. Nat Biotechnol 18:1307–1310

Gao H, Narayanan NN, Ellison L, Bhattacharyya MK (2005) Two classes of highly similar coiled coil-nucleotide binding-leucine rich repeat genes isolated from the Rps1-k locus encode Phytophthora resistance in soybean. Mol Plant Microbe Interact 18(10):1035–1045

Gao XR, Wang GK, Su Q, Wang Y, An LJ (2007) Phytase expression in transgenic soybeans: stable transformation with a vector-less construct. Biotechnol Lett 29:1781–1787

Gapper C, Dolan L (2006) Control of plant development by reactive oxygen species. Plant Physiol 141:341–345

Garg N, Manchanda G (2009) ROS generation in plants: boon or bane? Plant Biosystems 143 (1):81–96

Garg AK, Kim JK, Owens TG, Ranwala AP, Choi YD, Kochian LV, Wu RJ (2002) Trehalose accumulation in rice plants confers high tolerance levels to different abiotic stresses. Proc Natl Acad Sci U S A 99:15898–15903

Gassmann AJ, Petzold-Maxwell JL, Keweshan RS, Dunbar MW (2011) Field-evolved resistance to Bt maize by western corn rootworm. PLoS One 6:e22629

Gatehouse JA (2008) Biotechnological prospects for engineering insect-resistant plants. Plant Physiol 146(3):881–887

Gatehouse AMR, Ferry N, Edwards MG, Bell HA (2011) Insect-resistant biotech crops and their impacts on beneficial arthropods. Philos Trans R Soc Lond B Biol Sci 366:1438–1452

Gaxiola R, Li J, Undurraga S, Dang LM, Allen GJ, Alper SL, Fink GR (2001) Drought and salt tolerant plants result from overexpression of the AVP1 H^+ pump. Proc Natl Acad Sci U S A 98: 11444–11449

Gayen D, Ali N, Ganguly M, Paul S, Datta K, Datta SK (2014) RNAi mediated silencing of lipoxygenase gene to maintain rice grain quality and viability during storage. Plant Cell Tiss Org Cult 118:229–243

Gayen D, Ali N, Sarkar SN, Datta SK, Datta K (2015) Down-regulation of lipoxygenase gene reduces degradation of carotenoids of golden rice during storage. Planta 242:353–363

Ge LF, Chao DY, Shi M, Zhu MZ, Gao JP, Lin HX (2008) Overexpression of the trehalose-6 phosphate phosphatase gene OsTPP1 confers stress tolerance in rice and results in the activation of stress responsive genes. Planta 228:191–201

Gebril S, Seger M, Muro Villanueva F, Ortega JL, Bagga S, Sengupta-Gopalan C (2015) Transgenic alfalfa (*Medicago sativa*) with increased sucrose phosphate synthase activity shows enhanced growth when grown under N_2-fixing conditions. Planta 242:1009–1024

Gerlach WL, Llewellyn D, Haseloff J (1987) Construction of a plant disease resistance gene from the satellite RNA of tobacco ringspot virus. Nature 328(6133):802–805

Ghanti SKK, Sujata KG, Kumar BMV (2011) Heterologous expression of P5CS gene in chickpea enhances salt tolerance without affecting yield. Biol Plant 55:634–640

Ghislain M, Byarugaba AA, Magembe E, Njoroge A, Rivera C, Román ML, Tovar JC, Gamboa S, Forbes GA, Kreuze JF et al (2018) Stacking three late blight resistance genes from wild species directly into African highland potato varieties confers complete field resistance to local blight races. Plant Biotechnol J 17(6):1119–1129

Ghosh S, Meli VK, Kumar A, Thakur A, Chakraborty N, Chakraborty S, Datta A (2011) The N-glycan processing enzymes α-mannosidase and β-D-1-N-acetylhexosaminidase are involved in ripening-associated softening in the non-climacteric fruits of capsicum. J Exp Bot 62:571–582

Gilbert N (2013) A hard look at GM crops. Nature 497:24–26

Giliberto L, Perrotta G, Pallara P, Weller JL, Fraser PD, Bramley PM, Fiore A, Tavazza M, Giuliano G (2005) Manipulation of the blue light photoreceptor cryptochrome 2 in tomato affects vegetative development, flowering time, and fruit antioxidant content. Plant Physiol 137: 199–208

Gill SS, Tuteja N (2010) Reactive oxygen species and antioxidant machinery in abiotic stress tolerance in crop plants. Plant Physiol Biochem 48:909–930

Gillespie S, Hodge J, Yosef S, Pandya-Lorch R (2016) Nourishing millions: stories of change in nutrition. International Food Policy Research Institute (IFPRI), Washington, DC

Gisbert C, Rus AM, Bolarin C, Lopez-Coronado JM, Arrillaga I, Montesinos C, Caro M, Serrano R, Moreno V (2000) The yeast *HAL1* gene improves salt tolerance of transgenic tomato. Plant Physiol 123:393–402

Gitlin JD (2006) Distributing nutrition. Science 314(5803):1252–1253

Giuliano G (2017) Provitamin A biofortification of crop plants: a gold rush with many miners. Curr Opin Biotechnol 44:169–180

Gleave AP, Mitra DS, Markwick NP, Morris BA, Beuning LL (1998) Enhanced expression of the shape *Bacillus thuringiensis* cry9Aa2 gene in transgenic plants by nucleotide sequence modification confers resistance to potato tuber moth. Mol Breed 4(5):459–472

Gomez I, Arenas I, Benitez I, Miranda-Rios J, Becerril B, Grande R, Almagro JC, Bravo A, Soberon M (2006) Specific epitopes of Domains II and III of *Bacillus thuringiensis* Cry1Ab toxin involved in the sequential interaction with cadherin and aminopeptidase-N receptors in *Manduca sexta*. J Biol Chem 281:34032–34039

Gomez MA, Lin ZD, Moll T, Chauhan RD, Hayden L, Renninger K, Beyene G (2019) Simultaneous CRISPR/Cas9-mediated editing of cassava eIF4E isoforms nCBP-1 and nCBP-2 reduces cassava brown streak disease symptom severity and incidence. Plant Biotechnol J 17:421–434

Gonsalves D (2006) Transgenic papaya: development, release, impact and challenges. Adv Virus Res 67:317–354

Gonzales-Salazar R, Cecere B, Ruocco M, Rao R, Corrado G (2017) A comparison between constitutive and inducible transgenic expression of the PhRIP I gene for broad-spectrum resistance against phytopathogens in potato. Biotechnol Lett 39:1049–1058

Gore J, Leonard BR, Chorch GE, Cook DR (2002) Behaviour of bollworm (Lepidoptera: Noctuidae) larvae on genetically engineered cotton. J Econ Entomol 95:763–769

Goto F, Yoshihara T, Shigemoto N, Toki S, Takaiwa F (1999) Iron fortification of rice seed by the soybean ferritin gene. Nat Biotechnol 3:282–286

Goto F, Yoshihara T, Saiki H (2000) Iron accumulation and enhanced growth in transgenic lettuce plants expressing the iron-binding protein ferritin. Theor Appl Genet 100(5):658–664

Gouiaa S, Khoudi H (2015) Co-expression of vacuolar Na^{+}/H^{+} antiporter and H^{+}-pyrophosphatase with an IRES-mediated dicistronic vector improves salinity tolerance and enhances potassium biofortification of tomato. Phytochemistry 117:537–546

Gouiaa S, Khoudi H, Leidi EO, Pardo JM, Masmoudi K (2012) Expression of wheat Na+/H+ antiporter TNHXS1 and H+-pyrophosphatase TVP1 genes in tobacco from a bicistronic transcriptional unit improves salt tolerance. Plant Mol Biol 79(1):137–155

Goulart CL, Barbosa LC, Bisch PM, von Krüger WM (2016) Catalases and PhoB/PhoR system independently contribute to oxidative stress resistance in *Vibrio cholerae* O1. Microbiology 162(11):1955–1962

Gould F (1998) Sustainability of transgenic insecticidal cultivars: integrating pest genetics and ecology. Annu Rev Entomol 43:701–726

Gourdji SM, Sibley AM, Lobell DB (2013) Global crop exposure to critical high temperatures in the reproductive period: historical trends and future projections. Environ Res Lett 8:24041

Gowda A, Rydel TJ, Wollacott AM, Brown RS, Akbar W, Clark TL, Flasinski S, Nageotte JR, Read AC, Shi X, Werner BJ (2016) A transgenic approach for controlling Lygus in cotton. Nat Commun 7(1):1–7

Graham MA, Silverstein KAT, VandenBosch KA (2008) Defensin-like genes: genomic perspectives on a diverse superfamily in plants. Crop Sci 48:S3–S11

Graham RD, Knez M, Welch RM (2012) How much nutritional iron deficiency in humans globally is due to an underlying zinc deficiency? Adv Agron 115:1–40

Grantham R, Gautier C, Gouy M (1980) Codon frequencies in 119 individual genes confirm consistent choices of degenerate bases according to genome type. Nucleic Acids Res 8:1893–1912

Gray ME (2010) Relevance of traditional integrated pest management (IPM) strategies for commercial corn producers in a transgenic agroecosystem: a bygone era? J Agric Food Chem 59: 5852–5858

Grazziotin MA, Cabral GB, Ibrahim AB, Machado RB, Aragao FJ (2020) Expression of the Arcelin 1 gene from *Phaseolus vulgaris* L. in cowpea seeds (*Vigna unguiculata* L.) confers bruchid resistance. Ann Appl Biol 176:268–274

Green JM (2012) The benefits of herbicide-resistant crops. Pest Manag Sci 68(10):1323–1331

Green JM (2014) Current state of herbicides in herbicide-resistant crops. Pest Manag Sci 70(9): 1351–1357

Green JM, Hazel CB, Forney DR, Pugh LM (2008) New multiple-herbicide crop resistance and formulation technology to augment the utility of glyphosate. Pest Manag Sci 64:332–339

Grichko VP, Glick BR (2001) Flooding tolerance of transgenic tomato plants expressing the bacterial enzyme ACC deaminase controlled by the 35S, rolD or PRB-1b promoter. Plant Physiol Biochem 39:19–25

Griffitts JS, Haslam SM, Yang T, Garczynski SF, Mulloy B, Morris H, Cremer PS, Dell A, Adang MJ, Aroian RV (2005) Glycolipids as receptors for Bacillus thuringiensis crystal toxin. Science 307(5711):922–925

Grochulski P, Masson L, Borisova S, Pusztai-Carey M, Schwartz JL, Brousseau R, Cygler M (1995) *Bacillus thuringiensis* Cry1A(a) insecticidal toxin, crystal structure and channel formation. J Mol Biol 254:447–464

Grover A, Mittal D, Negi M, Lavania D (2013) Generating high temperature tolerant transgenic plants: achievements and challenges. Plant Sci 205:38–47

Grune T, Lietz G, Palou A, Ross AC, Stahl W, Tang G, Thurnham D, Yin SA, Biesalski HK (2010) Beta-carotene is an important vitamin A source for human. J Nutr 140:2268–2285

Gu D, Liu X, Wang M, Zheng J, Hou W, Wang G, Wang J (2008) Overexpression of ZmOPR1 in *Arabidopsis* enhanced the tolerance to osmotic and salt stress during seed germination. Plant Sci 174:124–130

Gulbitti-Onarici S, Zaidi MA, Taga I, Ozcan S, Altosaar I (2009) Expression of Cry1Ac in transgenic tobacco plants under the control of a wound-inducible promoter (AoPR1) isolated from Asparagus officinalis to control Heliothis virescens and Manduca sexta. Mol Biotechnol 42(3):341–349

Gullner G, Komives T (2001) The role of glutathione and glutathione-related enzymes in plant-pathogen interactions. In: Grill D, Tausz M, Kok L (eds) Significance of glutathione in plant adaptation to the environment. Kluwer, Dordrecht, pp 207–239

Gunasekara JM, Jayasekera GA, Perera KL, Wickramasuriya AM (2017) Development of a Sri Lankan rice variety Bg 94–1 harbouring Cry2A gene of *Bacillus thuringiensis* resistant to rice leaffolder *Cnaphalocrocis medinalis*. J Nat Sci 45:2

Guo J, Litao Y, Xin L, Xiaoyan G, Lingxi J, Dabing Z (2009) Characterization of the exogenous insert and development of event-specific PCR detection methods for genetically modified Huanong No.1 Papaya. J Agric Food Chem 57:7205–7212

Guo B, Guo Y, Hong H, Jin L, Zhang L, Chang R-Z et al (2015) Co-expression of G2-EPSPS and glyphosate acetyltransferase GAT genes conferring high tolerance to glyphosate in soybean. Front Plant Sci 6(847):1–11

Gupta HS, Raman B, Agrawal PK, Mahajan V, Hossain F, Thirunavukkarasu N (2013) Accelerated development of quality protein maize hybrid through marker-assisted introgression of opaque-2 allele. Plant Breed 132(1):77–82

Gupta HS, Hossain F, Muthusamy V (2015) Biofortification of maize: an Indian perspective. Indian J Genet Plant Breed 75:1–22

Gurian-Sherman D, Union of Concerned Scientists (2009) Failure to yield-evaluating the performance of genetically engineered crops. http://www.ucsusa.org/food_and_agriculture/our-failing-food-system/genetic-engineering/failure-to-yield.html. Accessed 24 Sept 2015

Gusta M, Smyth SJ, Belcher K, Phillips PW, Castle D (2011) Economic benefits of genetically modified herbicide-tolerant canola for producers. AgBioForum 14(1):1–13

Guy C, Kaplan F, Kopka J, Selbig J, Hincha DK (2008) Metabolomics of temperature stress. Physiol Plant 132:220–235

Haake V, Zrenner R, Sonnewald U, Stitt M (1998) A moderate decrease of plastid aldolase activity inhibits photosynthesis, alters the levels of sugars and starch, and inhibits growth of potato plants. Plant J 14:147–157

Haake V, Riechmann JLCD, Pineda O, Thomashow MF, Zhang JZ (2002) Transcription factor CBF4 is a regulator of drought adaptation in *Arabidopsis*. Plant Physiol 130:639–648

Habib S, Lwin YY, Li N (2021) Down-regulation of SlGRAS10 in tomato confers abiotic stress tolerance. Genes 12(5):623

Hacham Y, Song L, Schuster G, Amir R (2007) Lysine enhances methionine content by modulating the expression of S-adenosylmethionine synthase. Plant J 51:850–861

Hacham Y, Matityahu I, Schuster G, Amir R (2008) Overexpression of mutated forms of aspartate kinase and cystathionine γ-synthase in tobacco leaves resulted in the high accumulation of methionine and threonine. Plant J 54:260–271

Hakata M, Kuroda M, Ohsumi A, Hirose T, Nakamura H, Muramatsu M, Ichikawa H, Yamakawa H (2012) Overexpression of a rice TIFY gene increases grain size through enhanced accumulation of carbohydrates in the stem. Biosci Biotechnol Biochem 76(11):2129–2134

Hall H (2006) Bt corn: is it worth the risk? The Science Creative Quarterly. http://www.scq.ubc.ca/bt-corn-is-it-worth-the-risk/

Hamadou-Charfi DB, Boukedi H, Abdelkefi-Mesrati L, Tounsi S, Jaoua S (2013) Agrotis segetum midgut putative receptor of Bacillus thuringiensis vegetative insecticidal protein Vip3Aa16 differs from that of Cry1Ac toxin. J Invertebr Pathol 114(2):139–143

Hamilton AJ, Baulcombe DC (1999) A species of small antisense RNA in posttranscriptional gene silencing in plants. Science 286(5441):950–952

Hammad IA, Abdel-Razik AB, Soliman ER, Tawfik E (2017) Transgenic potato (*Solanum tuberosum*) expressing two antifungal thionin genes confer resistance to fusarium spp. J Pharm Biol Sci 12:69–79

Hammami R, Ben Hamida J, Vergoten G, Fliss I (2009) PhytAMP: a database dedicated to antimicrobial plant peptides. Nucleic Acids Res 37:D963–D968

Hammond J, Kamo KK (1995) Effective resistance to potyvirus infection conferred by expression of antisense RNA in transgenic plants. Mol Plant Microbe interact 5:674–682

Hamshou M, Shang C, Smagghe G, Van-Damme EJM (2016) Ribosome-inactivating proteins from apple have strong aphicidal activity in artificial diet and in planta. Crop Prot 87:19–24

Han S, Craig JA, Putnam CD, Carozzi NB, Tainer JA (1999) Evolution and mechanism from structures of an ADP-ribosylating toxin and NAD complex. Nat Struct Biol 6(10):932–936

Han SM, Lee B, Won OJ, Hwang KS, Suh SJ, Kim CG, Park KW (2015) Gene flow from herbicide resistant genetically modified rice to conventional rice (Oryza sativa L.) cultivars. J Ecol Environ 38(4):397–403

Hanley-Bowdoin L, Bejarano ER, Robertson D, Mansoor S (2013) Geminiviruses: masters at redirecting and reprogramming plant processes. Nat Rev Microbiol 11:777–788

Hannon GJ (2002) RNA interference. Nature 418:244–251

Hanur VS, Reddy B, Arya VV, Rami Reddy PV (2015) Genetic transformation of tomato using Bt Cry2A gene and characterization in Indian cultivar Arka Vikas. J Agric Sci Technol 17:1805–1814

Harris FA, Furr RE, Calhoun DS (1996) Cotton insect management in transgenic Bt cotton in the Mississippi Delta, 1992–1995. Proc Beltwide Cotton Conf 2:854–858

Harris F, Dennison SR, Phoenix DA (2009) Anionic antimicrobial peptides from eukaryotic organisms. Curr Protein Pept Sci 10:585–606

Harrison EH (2005) Mechanisms of digestion and absorption of dietary vitamin A. Annu Rev Nutr 25:87–103

Harrison BD, Mayo MA, Baulcombe DC (1987) Virus resistance in transgenic plants that express cucumber mosaic virus satellite RNA. Nature 328(6133):799–802

Harrison LA, Bailey MR, Naylor MW, Ream JE, Hammond BG, Nida DL, Burnette BL, Nickson TE, Mitsky TA, Taylor ML, Fuchs RL (1996) The expressed protein in glyphosate-tolerant soybean, 5-enolypyruvylshikimate-3-phosphate synthase from Agrobacterium sp. strain CP4, is rapidly digested in vitro and is not toxic to acutely gavaged mice. J Nutr 126(3):728–740

Harrison EP, Willingham NM, Lloyd JC, Raines CA (1998) Reduced sedoheptulose-1,7-bisphosphatase levels in transgenic tobacco lead to decreased photosynthetic capacity and altered carbohydrate accumulation. Planta 204:27–36

Hartmann B, Subramaniam B, Zerner C (1999) Effect of diets containing genetically modified potatoes expressing *Galanthus nivalis* lectin on rat small intestine. Lancet 354(9187):1353–1354

Hasanuzzaman M, Fujita M (2011) Selenium pretreatment upregulates the antioxidant defense and methylglyoxal detoxification system and confers enhanced tolerance to drought stress in rapeseed seedlings. Biol Trace Elem Res 143(3):1758–1776

Hasanuzzaman M, Hossain MA, Fujita M (2011) Selenium-induced up-regulation of the antioxidant defense and methylglyoxal detoxification system reduces salinity-induced damage in rapeseed seedlings. Biol Trace Elem Res 143(3):1704–1721

Hasanuzzaman M, Nahar K, Fujita M (2014) Regulatory role of polyamines in growth, development and abiotic stress tolerance in plants. Plant adaptation to environmental change: significance of amino acids and their derivatives. Biol Trace Elem Res 10:157–1593

Hasthanasombut S, Valentine N, Supaibulwatana K et al (2010) Expression of Indica rice OsBADH1 gene under salinity stress in transgenic tobacco. Plant Biotechnol Rep 4:75–78

Hatice S, Fatih D, Nedim M, Sami D, Anne F (2017) Cry1Ac-mediated resistance to tomato leaf miner (*Tuta absoluta*) in tomato. Plant Cell Tiss Org Cult 131:65–73

Haverkort AJ, Boonekamp PM, Hutten R, Jacobsen E, Lotz LAP, Kessel GJT, Vossen JH, Visser RGF (2016) Durable late blight resistance in potato through dynamic varieties obtained by cisgenesis: scientific and societal advances in the DuRPh project. Potato Res 59:35–66

Hayashi H, Alia, Mustardy L, Deshnium P, Ida M, Murata N (1997) Transformation of *Arabidopsis thaliana* with the *codA* gene for choline oxidase; accumulation of glycine betaine and enhanced tolerance to salt and cold stress. Plant J 12:133–142

Hayashi H, Alia SA, Nonaka H et al (1998) Enhanced germination under high-salt conditions of seeds of transgenic Arabidopsis with a bacterial gene (*codA*) for choline oxidase. J Plant Res 111:357–362

Hayta S, Smedley MA, Demir SU, Blundell R, Hinchliffe A, Atkinson N et al (2019) An efficient and reproducible *Agrobacterium*-mediated transformation method for hexaploid wheat (Triticum aestivum L.). Plant Methods 15:121

He M, Yang ZY, Nie YF, Wang J, Xu P (2001) A new type of class I bacterial 5-enopyruvylshikimate-3-phosphate synthase mutants with enhanced tolerance to glyphosate. Biochim Biophys Acta (BBA)-Gen Subj 1568(1):1–6

Heap I (2014) Global perspective of herbicide-resistant weeds. Pest Manag Sci 70:1306–1315

Heap I, Duke SO (2018) Overview of glyphosate-resistant weeds worldwide. Pest Manag Sci 74 (5):1040–1049

Hefferon K (2020) Biofortification of crops using biotechnology to alleviate malnutrition. Malnutrition 9:11–12

Heineke D, Bykova N, Gardeström P, Bauwe H (2001) Metabolic response of potato plants to an antisense reduction of the P-protein of glycine decarboxylase. Planta 212(5):880–887

Heinemann JA, Massaro M, Coray DS, Agapito-Tenfen SZ, Wen JD (2014) Sustainability and innovation in staple crop production in the US Midwest. Int J Agric Sustain 12(1):71–88

Hellwald K-H, Palukaitis P (1995) Viral RNA as a potential target for two independent mechanisms of replicase-mediated resistance against cucumber mosaic virus. Cell 63:937–946

Henkes S, Sonnewald U, Badur R, Flachmann R, Stitt M (2001) A small decrease of plastid transketolase activity in antisense tobacco transformants has dramatic effects on photosynthesis and phenylpropanoid metabolism. Plant Cell 13:535–551

Hepher A, Atkinson HJ (1992) Nematode Control with Proteinase Inhibitors, European Patent Application Number, 92301890.7, Publication Number 0 502730 A1

Hernández-Campuzano B, Suárez R, Lina L, Hernández V, Villegas E, Corzo G, Iturriaga G (2009) Expression of a spider venom peptide in transgenic tobacco confers insect resistance. Toxicon 53:122–128

Hernández-Martínez P, Gomis-Cebolla J, Ferré J, Escriche B (2013a) Changes in gene expression and apoptotic response in *Spodoptera exigua* larvae exposed to sublethal concentrations of Vip3 insecticidal proteins. Sci Rep 7:1–12

Hernández-Martínez P, Hernández-Rodríguez CS, Van Rie J, Escriche B, Ferré J (2013b) Insecticidal activity of Vip3Aa, Vip3Ad, Vip3Ae, and Vip3Af from *Bacillus thuringiensis* against lepidopteran corn pests. J Invertebr Pathol 113:78–81

Herouet C, Esdaile DJ, Mallyon BA, Debruyne E, Schulz A, Currier T, Hendrickx K, van der Klis RJ, Rouan D (2005) Safety evaluation of the phosphinothricin acetyltransferase proteins encoded by the pat and bar sequences that confer tolerance to glufosinate-ammonium herbicide in transgenic plants. Regul Toxicol Pharmacol 41(2):134–149

Herrera-Estrella L, Depicker A, Montagu MV, Schell J (1983) Expression of chimeric genes transferred into plant cells using a Ti-plasmid-derived vector. Nature 303:209–213

Herrero S, Gechev T, Bakker PL, Moar WJ, de Maagd RA (2005) Bacillus thuringiensis Cry1Ca-resistant Spodoptera exigua lacks expression of one of four Aminopeptidase N genes. BMC Genomics 6(1):1–10

Hey TD, Hartley M, Walsh TA (1995) Maize ribosome-inactivating protein (b-32). Homologs in related species, effects on maize ribosomes, and modulation of activity by pro-peptide deletions. Plant Physiol 107(4):1323–1332

Hibino T, Waditee R, Araki E et al (2002) Functional characterization of choline monooxygenase, an enzyme for betaine synthesis in plants. J Biol Chem 277:41352–41360

Hilder VA, Boulter D (1999) Genetic engineering of crop plants for insect resistance–a critical review. Crop Prot 18(3):177–191

Hmida-Sayari A, Gargouri-Bouzid R, Bidani A, Jaoua L, Savouré A, Jaoua S (2005) Overexpression of Δ1-pyrroline-5-carboxylate synthetase increases proline production and confers salt tolerance in transgenic potato plants. Plant Sci 169:746–752

Ho NH, Baisakh N, Oliva N, Datta K, Frutos R, Datta SK (2006) Translational fusion hybrid Bt genes confer resistance against yellow stem borer in transgenic elite Vietnamese rice (*Oryza sativa* L.) cultivars. Crop Sci 46:781–789

Hoekenga OA (2014) Genomics of mineral nutrient biofortification: calcium, iron and zinc. In: Genomics of plant genetic resources. Springer, Dordrecht, pp 431–454

Holm PB, Kristiansen KN, Pedersen HB (2002) Transgenic approaches in commonly consumed cereals to improve iron and zinc content and bioavailability. J Nutr 132(3):514S–516S

Holmstrom KO (1998) Engineering plant adaptation to water stress. Acta Univ Agric Sueciae Agaria 84:1401–6249

Holmstrom KO, Somersalo S, Manda A, Palva TE, Welin B (2000) Improved tolerance to salinity and low temperature in transgenic tobacco producing glycine betaine. Exp Bot 51:177–185

Hong Z, Lakkineni K, Zhang Z, Verma DPS (2000) Removal of feedback inhibition of pyrroline-5-carboxylate synthetase results in increased proline accumulation and protection of plants from osmotic stress. Plant Physiol 122:1129–1136

Hong Z, Ueguchi-Tanaka M, Shimizu-Sato S, Inukai Y, Fujioka S, Shimada Y, Takatsuto S, Agetsuma M, Yoshida S, Watanabe Y, Uozu S (2002a) Loss-of-function of a rice brassinosteroid biosynthetic enzyme, C-6 oxidase, prevents the organized arrangement and polar elongation of cells in the leaves and stem. Plant J 32(4):495–508

Hong H, Datla N, Darwin WR, Patrick SC, Samuel LM, Xiao Q (2002b) High-level production of γ-linolenic acid in *Brassica juncea* using a Δ6-desaturase from *Pythium irregulare*. Plant Physiol 129:354–362

Hooper NM (1994) Families of zinc metalloproteases. FEBS Lett 354:1–6

Horvath DM, Pauly MH, Hutton SF, Vallad GE, Scott JW, Jones JB, Stall RE, Dahlbeck D, Staskawicz BJ, Tricoli D et al (2015) The pepper Bs2 gene confers effective field resistance to bacterial leaf spot and yield enhancement in Florida tomatoes. Acta Hortic 1069:47–51

Hoshikawa K, Ishihara G, Takahashi H, Nakamura I (2012) Enhanced resistance to gray mold (*Botrytis cinerea*) in transgenic potato plants expressing thionin genes isolated from Brassicaceae species. Plant Biotechnol 29:87–93

Hossain MA, Fujita M (2010) Evidence for a role of exogenous glycinebetaine and proline in antioxidant defense and methylglyoxal detoxification systems in mung bean seedlings under salt stress. Physiol Mol Biol Plants 16:19–29

Hossain DM, Shitomi Y, Moriyama K, Higuchi M, Hayakawa T, Mitsui T, Sato R, Hori H (2004) Characterization of a novel plasma membrane protein, expressed in the midgut epithelia of *Bombyx mori*, that binds to Cry1A toxins. Appl Environ Microbiol 70:4604–4612

Hossain MA, Hasanuzzaman M, Fujita M (2010) Up-regulation of antioxidant and glyoxalase systems by exogenous glycinebetaine and proline in mung bean confer tolerance to cadmium stress. Physiol Mol Biol Plants 16(3):259–272

Hossain MA, Hasanuzzaman M, Fujita M (2011) Coordinate induction of antioxidant defense and glyoxalase system by exogenous proline and glycinebetaine is correlated with salt tolerance in mung bean. Front Agric China 5:1–14

Hossain F, Muthusamy V, Pandey N, Vishwakarma AK, Baveja A, Zunjare RU, Thirunavukkarasu N, Saha S, Manjaiah KM, Prasanna BM, Gupta HS (2017) Marker-assisted introgression of opaque2 allele for rapid conversion of elite hybrids into quality protein maize. J Genet 97(1):287–298

Hotta CT, Gardner MJ, Hubbard KE, Baek SJ, Dalchau N, Suhita D, Dodd AN, Webb AA (2007) Modulation of environmental responses of plants by circadian clocks. Plant Cell Environ 30: 333–349

Howe AR, Gasser CS, Brown SM, Padgette SR, Hart J, Parker GB, Fromm ME, Armstrong CL (2002) Glyphosate as a selective agent for the production of fertile transgenic maize (*Zea mays* L.) plants. Mol Breed 10(3):153–164

Hsieh TH, Lee JT, Charng YY, Chan MT (2002) Tomato plants ectopically expressing *Arabidopsis* CBF1 show enhanced resistance to water deficit stress. Plant Physiol 130:618–626

Hua J (2009) From freezing to scorching, transcriptional responses to temperature variations in plants. Curr Opin Plant Biol 12(5):568–573

Hua G, Jurat-Fuentes JL, Adang MJ (2004) Bt-R1a extracellular cadherin repeat 12 mediates Bacillus thuringiensis Cry1Ab binding and cytotoxicity. J Biol Chem 279:28051–28056

Huang J, Hirji R, Adam L, Rozwadowski KL, Hammerlindl JK, Keller WA, Selvaraj G (2000) Genetic engineering of glycinebetaine production toward enhancing stress tolerance in plants: metabolic limitations. Plant Physiol 122:747–756

Huang J, Hu R, Rozelle S, Pray C (2005) Insect-resistant GM rice in farmers' fields: assessing productivity and health effects in China. Science 308(5722):688–690

Huang MX, Hou P, Wei Q, Xu Y, Chen F (2008) A ribosome inactivating protein (curcin 2) induced from *Jatropha curcas* can reduce viral and fungal infection in transgenic tobacco. Plant Growth Regul 54:115–123

Husain J, Tickle IJ, Wood SP (1994) Crystal structure of momordin, a type I ribosome inactivating protein from the seeds of *Momordica charantia*. FEBS Lett 342:154–158

Husaini AM, Aquil S, Bhat M, Qadri T, Kamaluddin K, Abdin MZ (2008) A high-efficiency direct somatic embryogenesis system for strawberry (Fragaria x ananassa duch.) cultivar chandler. J Crop Sci Biotechnol 11(2):107–110

Hwang JS, Lee J, Hwang B, Nam SH, Yun EY, Kim SR, Lee DG (2010) Isolation and characterization of psacotheasin, a novel knottin-type antimicrobial peptide, from *Psacothea hilaris*. J Microbiol Biotechnol 20:708–711

Iatsenko I, Boichenko I, Sommer RJ (2014) Bacillus thuringiensis DB27 produces two novel protoxins, Cry21Fa1 and Cry21Ha1, which act synergistically against nematodes. Appl Environ Microbial 80(10):3266–3275

Ibragimova SS, Kolodyazhnaya YS, Gerasimova SV, Kochetov AV (2012) Partial suppression of gene encoding proline dehydrogenase enhances plant tolerance to various abiotic stresses. Russ J Plant Physiol 59:88–96

Ibrahim MA, Griko N, Junker M, Bulla LA (2010) *Bacillus thuringiensis*: a genomics and proteomics perspective. Bioeng Bugs 1(1):31–50

Ihemere U, Narayanan NN, Sayre RT (2011) Iron biofortification and homeostasis in transgenic cassava roots expressing an algal iron assimilatory protein, FEA1. Front Plant Sci 3:171

Ilardi V, Tavazza M (2015) Biotechnological strategies and tools for Plum pox virus resistance: trans-, intra-, cis-genesis, and beyond. Front Plant Sci 6:379

Ilona P, Bouis H, Palenberg M, Moursi M, Oparinde A (2017) Vitamin A cassava in Nigeria: crop development and delivery. Afr J Food Agric Nutr Dev 17:12000–12025

Indurker S, Misra HS, Eapen S (2007) Genetic transformation of chickpea (Cicer arietinum L.) with insecticidal crystal protein gene using particle gun bombardment. Plant Cell Rep 26:755–763
Ipsaro JJ, Joshua-Tor L (2015) From guide to target: molecular insights into eukaryotic RNA-interference machinery. Nat Struct Mol Biol 22:20
Iqbal A, Khan RS, Shehryar K, Imran A, Ali F, Attia S, Shah S, Mii M (2019) Antimicrobial peptides as effective tools for enhanced disease resistance in plants. Plant Cell Tiss Org 139:1–15
Irfan M, Ghosh S, Kumar V, Chakraborty N, Chakraborty S, Datta A (2014) Insights into transcriptional regulation of β-DN-acetylhexosaminidase, an N-glycan-processing enzyme involved in ripening-associated fruit softening. J Exp Bot 65(20):5835–5848
Irfan M, Ghosh S, Meli VS, Kumar A, Kumar V, Chakraborty N, Chakraborty S, Datta A (2016) Fruit ripening regulation of α-mannosidase expression by the MADS box transcription factor ripening inhibitor and ethylene. Front Plant Sci 7:10
Irina E (2007) Experimental evidence of GMO hazards. Presentation at Scientists for a GM Free Europe. EU Parliament, Brussels
ISAAA (2018) Global status of commercialized biotech/GM crops in 2018: biotech crops continue to help meet the challenges of increased population and climate change. ISAAA Brief No. 54. ISAAA: Ithaca, NY. https://www.isaaa.org/resources/publications/briefs/54/executivesummary/pdf/B54-ExecSum-Engli sh.pdf
ISAAA (2021) Philippines approves Bt Eggplant for direct use as food, feed, or for processing. https://www.isaaa.org/blog/entry/default.asp?BlogDate=7%2F23%2F2021&fbclid=IwARIwosOFQtcEEd-NXRH9Th_bFomIETEMsbjtMdYnG7zL_10SfRETGdiiA8
ISAAA Database (2019) GM Approval Database. https://www.isaaa.org/gmapprovaldatabase/default.asp. Accessed 17 Nov 2019
Ishihara A, Asada Y, Takahashi Y, Yabe N, Komeda Y, Nishioka T, Miyagawa H, Wakasa K (2006) Metabolic changes in *Arabidopsis thaliana* expressing the feedback-resistant anthranilate synthase α subunit gene OASA1D. Phytochemistry 67(21):2349–2362
Ishimaru Y, Kim S, Tsukamoto T, Oki H, Kobayashi T, Watanabe S, Matsuhashi S, Takahashi M, Nakanishi H, Mori S, Nishizawa NK (2007) Mutational reconstructed ferric chelate reductase confers enhanced tolerance in rice to iron deficiency in calcareous soil. Proc Natl Acad Sci 104 (18):7373–7378
Ismond KP, Dolferus R, Pauw MD, Dennis ES, Good AG (2003) Enhanced low oxygen survival in *Arabidopsis* through increased metabolic flux in the fermentative pathway. Plant Physiol 132: 1292–1302
Itoh Y, Takahashi K, Takizawa H, Nikaidou N, Tanaka H, Nishihashi H, Watanabe T, Nishizawa Y (2003) Family 19 chitinase of *Streptomyces griseus* HUT6037 increases plant resistance to the fungal disease. Biosci Biotechnol Biochem 67(4):847–855
Jackson RE, Bradley JR, Van DJW (2003) Field performance of transgenic cotton expressing one or two *Bacillus thuringiensis* endotoxins against bollworm, *Helicoverpa zea* (Boddie). J Cotton Sci 7:57–64
Jacob P, Hirt H, Bendahmane A (2017) The heat-shock protein/chaperone network and multiple stress resistance. Plant Biotechnol J 15(4):405–414
Jadhav MS, Rathnasamy SA, Natarajan B, Duraialagaraja S, Varatharajalu U (2020) Study of expression of indigenous Bt cry2AX1 gene in T 3 progeny of cotton and its efficacy against *Helicoverpa armigera* (Hubner). Braz Arch Biol Technol 8:63
Jaiwa A, Darshna C, Ranjana J (2014) Genetically modified crops for developing countries. Reviewed proceedings of National Seminar. pp 324–333
James C (1997) Global status of transgenic crops in 1997. ISAAA Brief No. 5. ISAAA, Ithaca, NY, p 31. https://www.isaaa.org/purchasepublications/itemdescription.asp?ItemType=BRIEFS&Control=IB005-1997
James C (2012) Global status of commercialized biotech/GM crops. ISAAA Brief No. 44. ISAAA, Ithaca, NY

James C (2016) Global status of commercialized Biotech/GM crops. ISAAA Brief No 52. http://www.isaaa.org

James VA, Neibaur I, Altpeter F (2008) Stress inducible expression of the DREB1A transcription factor from xeric, *Hordeum spontaneum* L. in turf and forage grass (Paspalum notatum Flugge) enhances abiotic stress tolerance. Transgenic Res 17(1):93–104

James CE, Davies EV, Fothergill JL, Walshaw MJ, Beale CM, Brockhurst MA, Winstanley C (2015) Lytic activity by temperate phages of Pseudomonas aeruginosa in long-term cystic fibrosis chronic lung infections. ISME J 9(6):1391–1398

Jami SK, Clark GB, Turlapati SA, Handley C, Roux SJ, Kirti PB (2008) Ectopic expression of an annexin from Brassica juncea confers tolerance to abiotic and biotic stress treatments in transgenic tobacco. Plant Physiol Biochem 46(12):1019–1030

Jan PS, Huang HY, Chen HM (2010) Expression of a synthesized gene encoding cationic peptide cecropin B in transgenic tomato plants protects against bacterial diseases. Appl Environ Microbiol 76:769–775

Jang IC, Oh SJ, Seo JS, Choi WB, Song SI, Kim CH, Kim YS, Seo HS, Choi YD, Nahm BH, Kim JK (2003) Expression of a bifunctional fusion of the Escherichia coli genes for trehalose-6-phosphate synthase and trehalose-6-phosphate phosphatase in transgenic rice plants increases trehalose accumulation and abiotic stress tolerance without stunting growth. Plant Physiol 131: 516–524

Jason J, Archibald LK, Nwanyanwu OC, Sowell AL, Buchanan I, Larned J, Bell M, Kazembe PN, Dobbie H, Jarvis WR (2002) Vitamin A levels and immunity in humans. Clin Vaccine Immunol 9(3):616–621

Jayaraj J, Punja ZK (2007) Combined expression of chitinase and lipid transfer protein genes in transgenic carrot plants enhance resistance to foliar fungal pathogens. Plant Cell Rep 26:1539–1546

Jayaraj J, Devlin R, Punja (2008) Metabolic engineering of novel ketocarotenoid production in carrot plants. Transgenic Res 17:489–501

Jayaraman KS (2002) Poor crop management plagues Bt cotton experiment in India. Nat Biotechnol 20:10

Jenkins JN, Parrott WL, McCarty JC Jr, Callahan FE, Berberich SA, Deaton RW (1993) Growth and survival of Heliothis virescens (Lepidoptera: Noctuidae) on transgenic cotton containing a truncated form of the delta endotoxin gene from Bacillus thuringiensis. J Econ Entomol 86 (1):181–185

Jeong JS, Kim YS, Baek KH, Jung H, Ha SH, Do Choi Y, Kim M, Reuzeau C, Kim JK (2010) Root-specific expression of OsNAC10 improves drought tolerance and grain yield in rice under field drought conditions. Plant Physiol 153(1):185–197

Jhansi SR, Usha R (2013) Transgenic plants: types, benefits, public concerns and future. J Pharm Res 6:879–883

Ji X, Zhang H, Zhang Y, Wang Y, Gao C (2015) Establishing a CRISPR-Cas-like immune system conferring DNA virus resistance in plants. Nat Plants 1:15144

Jia GX, Zhu ZQ, Chang FQ, Li YX (2002) Transformation of tomato with the *BADH* gene from *Atriplex* improves salt tolerance. Plant Cell Rep 21:141–156

Jiang L, Wu J, Fan S, Li W, Dong L, Cheng Q (2015) Isolation and characterization of a novel pathogenesis-related protein gene (GmPRP) with induced expression in soybean (*Glycine max*) during infection with *Phytophthora sojae*. PLoS One 10:e0129932

Jiang M, Jiang JJ, He CM, Guan M (2016a) Broccoli plants over-expressing a cytosolic ascorbate peroxidase gene increase resistance to downy mildew and heat stress. J Plant Pathol 1:413–420

Jiang J, Jia H, Feng G, Wang Z, Li J, Gao H, Wang X (2016b) Overexpression of Medicago sativa TMT elevates the alpha-tocopherol content in Arabidopsis seeds, alfalfa leaves, and delays dark-induced leaf senescence. Plant Sci 249:93–104

Jiang K, Zhang Y, Chen Z, Wu D, Cai J, Gao X (2020) Structural and functional insights into the C-terminal fragment of insecticidal Vip3A toxin of *Bacillus thuringiensis*. Toxins 12:438

Jin L, Wang J, Guan F, Zhang J, Yu S, Liu S, Yang Y (2018) Dominant point mutation in a tetraspanin gene associated with field-evolved resistance of cotton bollworm to transgenic Bt cotton. Proc Natl Acad Sci U S A 115:11760–11765

Jinyang L, Xinxin X, Pengjun S, Bo L, Yuhong Z, Wei Z (2017) Overexpression and characterization of a novel endo-β-1,3(4)-glucanase from thermophilic fungus *Humicola insolens* Y1. Protein Exp Purif 138:63–68

Jiménez P, Tejero J, Cordoba-Diaz D, Quinto EJ, Garrosa M, Gayoso MJ, Girbés T (2015) Ebulin from dwarf elder (Sambucus ebulus L.): a mini-review. Toxins 7(3):648–658

Jo KR, Kim CJ, Kim SJ, Kim TY, Bergervoet M, Jongsma MA, Visser RGF, Jacobsen E, Vossen JH (2014) Development of late blight resistant potatoes by cisgene stacking. BMC Biotechnol 14:50

John MB (2002) 13-week dietary subchronic comparison study with Mon 863 corn in rats preceded by a 1-week baseline food consumption determination with PMI certified rodent diet #5002, December 17

Johnson AAT, Kyriacou B, Callahan DL, Carruthers L, Stangoulis J, Lombi E, Tester M (2011) Constitutive overexpression of the OsNAS gene family reveals single-gene strategies for effective iron- and zinc-biofortification of rice endosperm. PLoS One 6:e24476

Jones JDG, Dangl JL (2006) The plant immune system. Nature 444:323–329

Joshi V, Joung JG, Fei Z, Jander G (2010) Interdependence of threonine, methionine and isoleucine metabolism in plants: accumulation and transcriptional regulation under abiotic stress. Amino Acids 39:933–947

Jouzani GS, Valijanian E, Sharafi R (2017) *Bacillus thuringiensis*: a successful insecticide with new environmental features and tidings. Appl Microbiol Biotechnol 101:2691–2711

Juan JE, Nadine BC, Nalini D, Nicholas BD, Gregory WW, Michael GK (1997) Transgenic plants: an emerging approach to pest control. Nat Biotechnol 15:1–5

Jucovic M, Walters FS, Warren GW, Palekar NV, Chen JS (2008) From enzyme to zymogen: engineering Vip2, an ADP-ribosyltransferase from *Bacillus cereus*, for conditional toxicity. Protein Eng Des Sel 21:631–638

Juma C (2011) The new harvest: agricultural innovation in Africa. Oxford University Press, New York

Jurat-Fuentes JL, Adang MJ (2006) Characterization of a Cry1Ac-receptor alkaline phosphatase in susceptible and resistant *Heliothis virescens* larvae. Eur J Biochem 271:3127–3135

Kamble S, Misra HS, Mahajan SK, Eapen S (2003) A protocol for efficient biolistic transformation of mothbean Vigna aconitifolia L. Jacq. Marechal. Plant Mol Biol Report 21:457–457

Kamthan A, Kamthan M, Azam M, Chakraborty N, Chakraborty S, Datta A (2012) Expression of a fungal sterol desaturase improves tomato drought tolerance, pathogen resistance and nutritional quality. Sci Rep 2:951

Kamthan A, Chaudhuri A, Kamthan M, Datta A (2015) Small RNAs in plants: recent development and application for crop improvement. Front Plant Sci 6:208

Kamthan A, Abira C, Mohan K, Asis D (2016) Genetically modified (GM) crops: milestones and new advances in crop improvement. Theor Appl Gen 129:1639–1655

Kandoth PK, Ithal N, Recknor J, Maier T, Nettleton D, Baum TJ, Mitchum MG (2011) The soybean Rhg1 locus for resistance to the soybean cyst nematode Heterodera glycines regulates the expression of a large number of stress-and defense-related genes in degenerating feeding cells. Plant Physiol 155(4):1960–1975

Kang J, Choi H, Im M, Kim SY (2002) *Arabidopsis* basic leucine zipper proteins that mediate stress-responsive abscisic acid signaling. Plant Cell 14:343–357

Kang TJ, Loc NH, Jang MO, Yang MS (2004) Modification of the cholera toxin B subunit coding sequence to enhance expression in plants. Mol Breed 13:143–153

Kanobe MN, Rodermel SR, Bailey T, Scott MP (2013) Changes in endogenous gene transcript and protein levels in maize plants expressing the soybean ferritin transgene. Front Plant Sci 4:196

Kanzaki H, Nirasawa S, Saitoh H, Ito M, Nishihara M, Terauchi R, Nakamura I (2002) Overexpression of the wasabi defensin gene confers enhanced resistance to blast fungus (*Magnaporthe grisea*) in transgenic rice. Theor Appl Genet 105:809–814

Kar S, Basu D, Das S, Ramkrishnan NA, Mukherjee P, Nayak P, Sen SK (1996) Expression of CryIA(c) gene of Bacillus thuringiensis in transgenic chickpea plants inhibits development of pod borer (Heliothis armigera) larvae. Transgenic Res 6:177–185

Kar S, Basu D, Das S, Ramakrishnan NA, Mukherjee P, Nayak P (1997) Expression of *cry1Ac* gene of *Bacillus thuringiensis* in transgenic chickpea plants inhibits development of pod borer (*Heliothis armigera*) larvae. Transgenic Res 6:177–185

Karchi H, Shaul O, Galili G (1993) Seed specific expression of a bacterial desensitized aspartate kinase increases the production of seed threonine and methionine in transgenic tobacco. Plant J 3:721–727

Karesch H, Rolang B, Ortum MS, Ingo P (1991) Direct gene transfer to protoplast of *Arabidopsis thaliana*. Plant Cell Rep 9:571–574

Karihaloo JL, Kumar PA (2009) Bt Cotton in India. A status report, pp 1–58

Karmakar A, Bhattacharya S, Sengupta S, Ali N, Sarkar SN, Datta K, Datta SK (2019) RNAi-mediated silencing of ITPK gene reduces phytic acid content, alters transcripts of phytic acid biosynthetic genes, and modulates mineral distribution in rice seeds. Rice Sci 27(4):315–328

Kasuga M, Liu Q, Miura S, Yamaguchi-Shinozaki K, Shinozaki K (1999) Improving plant drought, salt, and freezing tolerance by gene transfer of a single stress-inducible transcription factor. Nat Biotechnol 17(3):287–291

Katta S, Talakayala A, Reddy MK, Addepally U, Garladinne M (2020) Development of transgenic cotton (Narasimha) using triple gene Cry2Ab-Cry1F-Cry1Ac construct conferring resistance to lepidopteran pest. J Biosci 45(1):1–11

Kauldhar BS, Sooch BS (2016) Tailoring nutritional and process variables for hyperproduction of catalase from a novel isolated bacterium *Geobacillus* sp. BSS-7. Microbial Cell Fact 15(7):1–16

Kaur S (2000) Molecular approaches towards development of novel *Bacillus thuringiensis* biopesticides. World J Microbiol Biotechnol 16:781–793

Kaur A, Pati PK, Pati AM, Nagpal AK (2017) In-silico analysis of cis-acting regulatory elements of pathogenesis-related proteins of *Arabidopsis thaliana* and *Oryza sativa*. PLoS One 12:9. https://doi.org/10.1371/journal.pone.0184523

Kaur R, Gupta M, Singh S, Joshi N, Sharma A (2020) Enhancing RNAi efficiency to decipher the functional response of potential genes in *Bemisia tabaci* AsiaII-1 (Gennadius) through dsRNA feeding assays. Front Physiol 11:123

Kausch AP, Hague J, Oliver M, Li Y, Daniell H, Mascia P, Watrud LS, Stewart CN (2010) Transgenic perennial biofuel feedstocks and strategies for bioconfinement. Biofuels 1:163–161

Kavi Kishor PB, Hong Z, Miao CH et al (1995) Overexpression of Δ1-pyrroline-5-carboxylate synthetase increases proline production and confers osmotolerance in transgenic plants. Plant Physiol 108:1387–1394

Keese P (2008) Risks from GMOs due to horizontal gene transfer. Environ Biosaf Res 7(3):123–149

Kerenga BK, McKenna JA, Harvey PJ, Quimbar P, Garcia-Ceron D, Lay FT, Phan TK, Veneer PK, Vasa S, Parisi K, Shafee TMA, van der Weerden NL, Hulett MD, Craik DJ, Anderson MA, Bleackley MR (2019) Salt-tolerant antifungal and antibacterial activities of the corn defensin ZmD32. Front Microbiol 10:795

Kesarwani M, Azam M, Natarajan K, Mehta A, Datta A (2000) Oxalate decarboxylase from *Collybia velutipes*: molecular cloning and its over expression to confer resistance to fungal infection in transgenic tobacco and tomato. J Biol Chem 275:7230–7238

Khan M (2015) Update on adoption of glyphosate-tolerant sugar beet in the United States. Outlooks Pest Manag 26(2):61–65

Khan GA, Bakhsh A, Riazuddin S, Husnain T (2011) Introduction of cry1Ab gene into cotton (*Gossypium hirsutum*) enhances resistance against Lepidopteran pest (*Helicoverpa armigera*). Span J Agric Res 9(1):296–302

Khanna HK, Raina SK (2002) Elite indica transgenic rice plants expressing modified Cry1Ac endotoxin of *Bacillus thuringiensis* show enhanced resistance to yellow stem borer (*Scirpophaga incertulas*). Transgenic Res 11(4):411–423

Khanom S, Jang J, Lee OR (2019) Overexpression of ginseng cytochrome P450 CYP736A12 alters plant growth and confers phenylurea herbicide tolerance in Arabidopsis. J Ginseng Res 43 (4):645–653

Kharb P, Chaughary R, Lakra N, Behl RK (2021) Transgenics for improving salt stress tolerance in legume crops chickpea and pigeon pea. Ekin J Crop Breed Genetic 7(2):106–115

Khetan SH (2001) Microbial pest control. In: Khetan SH (ed) Bacterial insecticides, Part 1. Marcel Decker, New York

Kidokoro S, Watanabe K, Ohori T, Moriwaki T, Maruyama K, Mizoi J, Myint Phyu Sin Htwe N, Fujita Y, Sekita S, Shinozaki K, Yamaguchi-Shinozaki K (2015) Soybean DREB 1/CBF-type transcription factors function in heat and drought as well as cold stress-responsive gene expression. Plant J 81(3):505–518

Kim JH, Cetiner S, Jaynes JM (1992) Enhancing the nutritional quality of crop plants: design, construction and expression of an artificial plant storage protein gene. In: Bhatnagar D, Cleveland TE (eds) Molecular approaches to improving food quality and safety. Avi Books, New York, pp 1–36

Kim SY, May GD, Park WD (1994) Nuclear protein factors binding to a class I patatin promoter region are tuber-specific and sucrose-inducible. Plant Mol Biol 26:603–615

Kim JK, Jang IC, Wu R, Zuo WN, Boston RS, Lee YH (2003) Coexpression of a modified maize ribosome-inactivating protein and a rice basic chitinase gene in transgenic rice plants confers enhanced resistance to sheath blight. Transgenic Res 12:475–484

Kim S, Kim C, Li W, Kim T, Li Y, Zaidi MA, Altosaar I (2008) Inheritance and field performance of transgenic Korean Bt rice lines resistant to rice yellow stem borer. Euphytica 164(3):829–839

Kim D, Alptekin B, Budak H (2018) CRISPR/Cas9 genome editing in wheat. Funct Integr Genomics 18(1):31–41

Kim SY, Bengtsson T, Olsson N, Hot V, Zhu LH, Ahman I (2020) Mutations in two aphid-regulated β-1,3-Glucanase Genes by CRISPR/Cas9 do not increase barley resistance to *Rhopalosiphum padi* L. Front Plant Sci 11:1043

Kis A, Tholt G, Ivanics M, Várallyay É, Jenes B, Havelda Z (2015) Polycistronic artificial miRNA-mediated resistance to Wheat dwarf virusin barley is highly efficient at low temperature. Mol Plant Pathol 17:427–437

Kis A, Hamar É, Tholt G, Bán R, Havelda Z (2019) Creating highly efficient resistance against Wheat dwarf virus in barley by employing CRISPR/Cas9 system. Plant Biotechnol J 17:1004–1006

Kisaka H, Kida T (2003) Transgenic tomato plant carrying a gene for NADP-dependent glutamate dehydrogenase (*gdh A*) from *Aspergillus nidulans*. Plant Sci 164:35–42

Kishitani S, Takanami T, Suzuki M et al (2000) Compatibility of glycine betaine in rice plants: evaluation using transgenic rice plants with a gene for peroxisomal betaine aldehyde dehydrogenase from barley. Plant Cell Environ 23:107–114

Kishor PBK, Hong Z, Miao GH, Hu CAA, Verma DPS (1995) Overexpression of Δ1-pyrroline-5-carboxylate synthetase increases proline production and confers osmotolerance in transgenic plants. Plant Physiol 108:1387–1394

Kishore GM, Jacob GS (1987) Degradation of glyphosate by Pseudomonas sp. PG2982 via a sarcosine intermediate. J Biol Chem 262(25):12164–12168

Kleter GA, Kuiper HA (2005) Assessing the environmental impact of changes in pesticide use on transgenic crops. Frontis 1:33–43

Kleter GA, Bhula R, Bodnaruk K, Carazo E, Felsot AS, Harris CA, Katayama A, Kuiper HA, Racke KD, Rubin B, Shevah Y, Stephenson GR, Tanaka K, Unsworth J, Wauchope RD, Wong SS (2007) Altered pesticide use on transgenic crops and the associated general impact from an environmental perspective. Pest Manag Sci 63:1107–1115

Klümper W, Qaim M (2014) A meta-analysis of the impacts of genetically modified crops. PLoS One 9:e111629. https://doi.org/10.1371/journal.pone.0111629

Knight P, Crickmore N, Ellar DJ (1994) The receptor for *Bacillus thuringiensis* CryIA(c) delta-endotoxin in the brush border membrane of the lepidopteran *Manduca sexta* is aminopeptidase N. Mol Microbiol 11:429–436

Knight P, Carroll PJJ, Ellar DJ (2005) Analysis of glycan structures on the 120 kDa aminopeptidase N of Manduca sexta and their interactions with Bacillus thuringiensis Cry1Ac toxin. Insect Biochem Mol Biol 34:101–112

Knight H, Thomson AJ, McWatters HG (2008) Sensitive to freezing6 integrates cellular and environmental inputs to the plant circadian clock. Plant Physiol 148:293–303

Ko SJ, Lee YH, Cha KH, An UY, Choi HS (2003) Survey and identification of virus diseases on paprika in Jeonnam province. Korean J Plant Pathol 9:342

Koßmann J, Müller-Röber B, Dyer TA, Raines CA, Sonnewald U, Willmitzer L (1992) Cloning and expression analysis of the plastidic fructose-1,6-bisphosphatase coding sequence from potato: circumstantial evidence for the import of hexoses into chloroplasts. Planta 188:7–12

Koch A, Kumar N, Weber L, Keller H, Imani J, Kogel KH (2013) Host-induced gene silencing of cytochrome P450 lanosterol C14-demethylase-encoding genes confers strong resistance to Fusarium species. Proc Natl Acad Sci U S A 110:19324–19329

Koerniati S, Sukmadjaja D, Samudra IM (2020) C synthetic gene of CryIAb-CryIAc fusion to generate resistant sugarcane to shoot or stem borer. InIOP Conf Ser Earth Environ Sci 418: 01206

Kola VS, Pichili R, Padmakumari AP, Mangrauthia SK, Balachandran SM, Madhav MS (2019) Knockdown of acetylcholinesterase (AChE) gene in rice yellow stem borer, *Scirpophaga incertulas* (Walker) through RNA interference. Agric Gene 11:100081

König A, Cockburn A, Crevel RW, Debruyne E, Grafstroem R, Hammerling U, Kimber I, Knudsen I, Kuiper HA, Peijnenburg AA, Penninks AH, Poulsen M, Schauzu M, Wal JM (2004) Assessment of the safety of foods derived from genetically modified (GM) crops. Food Chem Toxicol 42(7):1047–1088

Konstantinova T, Parvanova D, Atanassov A, Djilianov D (2002) Freezing tolerant tobacco, transformed to accumulate osmoprotectants. Plant Sci 163:157–164

Koo JC, Chun HJ, Park HC et al (2002) Over-expression of a seed specific hevein-like antimicrobial peptide from Pharbitis nil enhances resistance to a fungal pathogen in transgenic tobacco plants. Plant Mol Biol 50:441–452

Korshunova YO, Eide D, Clark WG, Guerinot ML, Pakrasi HB (1999) The IRT1 protein from Arabidopsis thaliana is a metal transporter with a broad substrate range. Plant Mol Biol 40 (1):37–44

Kota M, Daniell H, Varma S, Garczynski SF, Gould F, Moar WJ (1999) Over expression of the (Bt) Cry2Aa2 protein in chloroplasts confers resistance to plants against susceptible and Bt-resistant insects. Proc Natl Acad Sci U S A 96:1840–1845

Koul B (2020) Genetically modified (GM) crops harbouring *Bacillus thuringiensis* (BT) gene(s) to combat biotic stress caused by insect pests. In: Kumar M, Kumar V, Prasad R (eds) Phyto-microbiome in stress regulation, Environ Microbial Biotechnol. Springer, Singapore. https://doi.org/10.1007/978-981-15-2576-6_2

Koul B, Yadav R, Sanyal I, Sawant S, Sharma V, Amla DV (2012) *Cis*-acting motifs in artificially synthesized expression cassette leads to enhanced transgene expression in tomato (*Solanum lycopersicum* L.). Plant Physiol Biochem 61:131–141

Koul B, Srivastava S, Amla DV, Sanyal I (2014a) Establishment and optimization of agrobacterium-mediated transformation and regeneration of tomato (Solanum lycopersicum L.). Int J Biosci (IJB) 4(10):51–69

Koul B, Srivastava S, Sanyal I, Tripathi B, Sharma V, Amla DV (2014b) Transgenic tomato line expressing modified *Bacillus thuringiensis cry1Ab* gene showing complete resistance to two lepidopteran pests. SpringerPlus 3:84

Kourelis J, van der Hoorn RAL (2018) Defended to the nines: 25 years of resistance gene cloning identifies nine mechanisms for R protein function. Plant Cell 30:285–282

Koziel MG, Beland GL, Bowman C, Carozzi NB, Crenshaw R, Crossland L, Dawson J, Desai N, Hill M, Kadwell S, Launis K, Lewis K, Maddox D, McPherson K, Meghji MR, Merlin E, Rhodes R, Warren GW, Wright M, Evola SV (1993) Field performance of elite transgenic maize plants expressing an insecticidal protein derived from *Bacillus thuringiensis*. Bio/Technology 11:194–200

Kramkowska M, Grzelak T, Czyzewska K (2013) Benefits and risks associated with genetically modified food products. Ann Agric Environ Med 20(3):413–419

Krattinger SG, Sucher J, Selter LL, Chauhan H, Zhou B, Tang M, Upadhyaya NM, Mieulet D, Guiderdoni E, Weidenbach D, Schaffrath U (2016) The wheat durable, multipathogen resistance gene Lr34 confers partial blast resistance in rice. Plant Biotechnol J 14(5):1261–1268

Krieger IV, Revina LP, Kostina LI, Buzdin AA, Zalunin IA, Chestukhina GG, Stepanov VM (1999) Membrane proteins of *Aedes aegypti* larvae bind toxins Cry4B and Cry11A of *Bacillus thuringiensis ssp. israelensis*. Biochemistry (Moscow) 64:1163–1168

Krishna VV, Qaim M (2007) Estimating the adoption of Bt eggplant in India: who benefits from public–private partnership? Food Policy 32(5–6):523–543

Krishna VV, Qaim M (2008) Consumer attitudes toward GM food and pesticide residues in India. Appl Econ Perspect Policy 30(2):233–251

Krishna R, McDonald KA, Dandekar AM, Jackman AP, Falk B (2002) Expression of recombinant trichosanthin, a ribosome-inactivating protein, in transgenic tobacco. J Biotechnol 97:69–88

Kruger EL, Volin JC (2006) Reexamining the empirical relation between plant growth and leaf photosynthesis. Funct Plant Biol 33:421–429

Kubicek CP, Starr TL, Glass NL (2014) Plant cell wall-degrading enzymes and their secretion in plant-pathogenic fungi. Annu Rev Phytopathol 52:427–451

Kubo S, Ikeda T, Imaizumi S, Takanami Y, Mikami Y (1997) A potent plant virus inhibitor found in *Mirabilis jalapa* L. Ann Phytopathol Soc Japan 56:481–487

Kulaeva O, Kliukova M, Afonin A, Sulima A, Zhukov V, Tikhonovich I (2020) The role of plant antimicrobial peptides (AMPs) in response to biotic and abiotic environmental factors. Biol Commun 65:187–199

Kumar R (2009) Role of naturally occurring osmolytes in protein folding and stability. Arch Biochem 491:1–6

Kumar H, Kumar V (2004) Tomato expressing Cry1A(b) insecticidal protein from *Bacillus thuringiensis* protected against tomato fruit borer, *Helicoverpa armigera* (Hubner) (Lepidoptera: Noctuidae) damage in the laboratory, greenhouse and field. Crop Prot 23:135–139

Kumar A, Verma JP (2018) Does plant-microbe interaction confer stress tolerance in plants: a review? Microbiol Res 207:41–52

Kumar D, Verma HN, Tuteja N, Tewari KK (1997) Cloning and characterization of a gene encoding an antiviral protein from *Clerodendrum aculeatum* L. Plant Mol Biol 33:745–751

Kumar MA, Timm DE, Neet KE, Owen WG, Peumans WJ, Rao AG (1993) Characterization of the lectin from the bulbs of Eranthis hyemalis (winter aconite) as an inhibitor of protein synthesis. J Biol Chem 268(33):25176–25183

Kumar PA, Mandaokar A, Sreenivasu K, Chakrabarti SK, Bisaria S, Sharma SR, Kaur S, Sharma RP (1998) Insect-resistant transgenic brinjal plants. Mol Breed 4:33–37

Kumar KK, Maruthasalam S, Loganathan M, Sudhakar D, Balasubramanian P (2005) An improved *Agrobacterium* mediated transformation protocol for recalcitrant elite indica rice cultivars. Plant Mol Biol Rep 23:67–73

Kumar M, Shukla AK, Singh H, Tuli R (2009) Development of insect resistant transgenic cotton lines expressing cry1EC gene from an insect bite and wound inducible promoter. J Biotechnol 140(3–4):143–148

Kumar M, Chimote V, Singh R, Mishra GP, Naik PS, Pandey SK, Chakrabarti SK (2010) Development of Bt transgenic potatoes for effective control of potato tuber moth by using cry1Ab gene regulated by GBSS promoter. Crop Prot 29:121–127

Kumar S, Thirunavukkarasu N, Singh G, Sharma R, Kulkarni KS (2015) Biofortification for selecting and developing crop cultivars denser in iron and zinc. In: Nutrient use efficiency: from basics to advances. Springer, New Delhi, pp 237–253

Kumar V, Chattopadhyay A, Ghosh S, Irfan M, Chakraborty N, Chakraborty S, Datta A (2016a) Improving nutritional quality and fungal tolerance in soya bean and grass pea by expressing an oxalate decarboxylase. Plant Biotechnol J 14:1394–1405

Kumar S, Barone P, Smith M (2016b) Gene targeting and transgene stacking using intra genomic homologous recombination in plants. Plant Methods 12:11

Kumar K, Aggarwal C, Sapna B, Singh I, Yadava P (2018) Microbial genes in crop improvement. Crop improvement through microbial biotechnology. Elsevier, Amsterdam, pp 39–56

Kunwar S, Iriarte F, Fan Q, Evaristo da Silva E, Ritchie L, Nguyen NS, Freeman JH, Stall RE, Jones JB, Minsavage GV et al (2018) Transgenic expression of EFR and Bs2 genes for field management of bacterial wilt and bacterial spot of tomato. Phytopathology 108:1402–1411

Kumar K, Gambhir G, Dass A, Tripathi AK, Singh A, Jha AK, Yadava P, Choudhary M, Rakshit S (2020) Genetically modifed crops: current status and future prospects. Planta 251:91

Kurisu G, Zhang H, Smith JL, Cramer WA (2003) Structure of the cytochrome b6f complex of oxygenic photosynthesis: tuning the cavity. Science 302(5647):1009–10014

Kushwaha GS, Pandey N, Sinha M, Singh SB, Kaur P, Sharma S (2012) Crystal structures of a type-1 ribosome inactivating protein from *Momordica balsamina* in the bound and unbound states. Biochim Biophys Acta 1824:679–691

Kühn H, Borchert A (2002) Regulation of enzymatic lipid peroxidation: the interplay of peroxidizing and peroxide reducing enzymes. Free Radic Biol Med 33(2):154–172

Lal S, Gulyani V, Khurana P (2008) Overexpression of HVA1 gene from barley generates tolerance to salinity and water stress in transgenic mulberry (*Morus indica*). Transgenic Res 17:651–663

Lam YH, Wong YS, Wang B, Wong RNS, Yeung HM, Shaw PC (1996) Use of trichosanthin to reduce infection by turnip mosaic virus. Plant Sci 114:111–117

Lambert B, Peferoen M (1992) Insecticidal promise of *Bacillus thuringiensis*. Bioscience 42:112–122

Lapidot M, Gafny R, Ding B, Wolf S, Lucas WJ, Beachy RN (1993) A dysfunctional movement protein of tobacco mosaic virus that partially modifies the plasmodesmata and limits virus spread in transgenic plants. Plant J 4(6):959–970

Larkindale J, Vierling E (2008) Core genome responses involved in acclimation to high temperature. Plant Physiol 146:748–761

Larkindale J, Hall JD, Knight MR, Vierling E (2005) Heat stress phenotypes of Arabidopsis mutants implicate multiple signaling pathways in the acquisition of thermotolerance. Plant Physiol 138:882–897

Laurie SM, Faber M, Claasen N (2018) Incorporating orange-fleshed sweet potato into the food system as a strategy for improved nutrition: the context of South Africa. Food Res Int 104:77–85

Laustsen PG, Vang S, Kristensen T (2001) Mutational analysis of the active site of human insulin-regulated aminopeptidase. Eur J Biochem 268:98–104

Lawson T, Bryant B, Lefebvre S, Lloyd JC, Raines CA (2006) Decreased SBPase activity alters growth and development in transgenic tobacco plants. Plant Cell Environ 29:48–58

Lawo NC, Wäckers FL, Romeis J (2009) Indian Bt cotton varieties do not affect the performance of cotton aphids. PLoS One 4(3):e4804

Le Nguyen D, Heitz A, Chiche L, Castro B, Boigegrain RA, Favel A, Coletti-Previero MA (1990) Molecular recognition between serine proteases and new bioactive microproteins with a knotted structure. Biochimie 72:431–435

Leah R, Tommerup H, Svendsen I, Mundy J (1991) Biochemical and molecular characterization of three barley seed proteins with antifungal properties. J Biol Chem 266:1564–1573

Leakey AD, Ainsworth EA, Bernacchi CJ, Rogers A, Long SP, Ort DR (2009) Elevated CO2 effects on plant carbon, nitrogen, and water relations: six important lessons from FACE. J Exp Bot 60:2859–2876

LeDuc DL, Tarun AS, Montes-Bayon M, Meija J, Malit MF, Wu CP, AbdelSamie M, Chiang CY, Tagmount A, deSouza M, Neuhierl B, Böck A, Caruso J, Terry N (2004) Overexpression of selenocysteine methyltransferase in *Arabidopsis* and Indian mustard increases selenium tolerance and accumulation. Plant Physiol 135:377–383

Lee S, An G (2009) Over-expression of OsIRT1 leads to increased iron and zinc accumulations in rice. Plant Cell Environ 32(4):408–416

Lee MK, Walters FS, Hart H, Palekar N, Chen JS (2003) The mode of action of the *Bacillus thuringiensis* vegetative insecticidal protein Vip3A differs from that of Cry1Ab δ-endotoxin. Appl Environ Microbiol 69:4648–4657

Lee MK, Miles P, Chen JS (2006) Brush border membrane binding properties of Bacillus thuringiensis Vip3A toxin to Heliothis virescens and Helicoverpa zea midguts. Biochem Biophys Res Commun 339:1043–1047

Lee K, Yang K, Kang K, Kang S, Lee N, Back K (2007) Use of *Myxococcus xanthus* protoporphyrinogen oxidase as a selectable marker for transformation of rice. Pestic Biochem Physiol 88(1):31–35

Lee S, Jeon US, Lee SJ, Kim YK, Persson DP, Husted S, Schjørring JK, Kakei Y, Masuda H, Nishizawa NK, An G (2009) Iron fortification of rice seeds through activation of the nicotianamine synthase gene. Proc Natl Acad Sci U S A 106:22014–22019

Lee JH, Kim HS, Suh SC, Rhim SL (2010) Development of transgenic rice lines expressing the human lactoferrin gene. J Plant Biotechnol 37(4):556

Lee S, Persson DP, Hansen TH, Husted S, Schjoerring JK, Kim YS, Jeon US, Kim YK, Kakei Y, Masuda H, Nishizhawa NK, An G (2011) Bio-available zinc in rice seeds is increased by activation tagging of nicotianamine synthase. Plant Biotechnol J 9:865–873

Lennartz K, Plücken H, Seidler A, Westhoff P, Bechtold N, Meierhoff K (2001) HCF164 encodes a thioredoxin-like protein involved in the biogenesis of the cytochrome b 6 f complex in Arabidopsis. Plant Cell 13(11):2539–2551

Lereclus D, Lecadet MM, Ribier J, Dedonder R (1982) Molecular relationships among plasmids of Bacillus thuringiensis: conserved sequences through 11 crystalliferous strains. Mol Gen Genet MGG 186(3):391–398

Lermontova I, Grimm B (2000) Overexpression of plastidic protoporphyrinogen IX oxidase leads to resistance to the diphenyl-ether herbicide acifluorfen. Plant Physiol 122(1):75–84

Leroy T, Henry AM, Royer M, Altosaar I, Frutos R, Duris D, Philippe R (2000) Genetically modified coffee plants expressing the *Bacillus thuringiensis cry1Ac* gene for resistance to leaf miner. Plant Cell Rep 19(4):382–385

Letourneau DK, Robinson GS, Hagen JA (2003) Bt crops: predicting effects of escaped transgenes on the fitness of wild plants and their herbivores. Environ Biosaf Res 2(4):219–246

Leuber M, Orlik F, Schiffler B, Sickmann A, Benz R (2006) Vegetative insecticidal protein (Vip1Ac) of Bacillus thuringiensis HD201: evidence for oligomer and channel formation. Biochemist 45:283–288

Leyva-Guerrero E, Narayanan NN, Ihemere U, Sayre RT (2012) Iron and protein biofortification of cassava: lessons learned. Curr Opin Biotechnol 23(2):257–264

Li J, Koni PA, Ellar DJ (1996) Structure of the mosquitocidal delta-endotoxin CytB from *Bacillus thuringiensis* sp. kyushuensis and implications for membrane pore formation. J Mol Biol 257:129–152

Li Z, Jarret RL, Demski JW (1997) Engineered resistance to tomato spotted wilt virus in transgenic peanut expressing the viral nucleocapsid gene. Transgenic Res 6(4):297–305

Li G, Xu X, Xing H, Zhu H, Fan Q (2005a) Insect resistance to *Nilaparvata lugens* and *Cnaphalocrocis medinalis* in transgenic indica rice and the inheritance of GNA+sbti transgenes. Pest Manag Sci 61:390–396

Li XP, Tian AG, Luo GZ (2005b) Soybean DRE-binding transcription factors that are responsive to abiotic stresses. Theor Appl Genet 110:1355–1362

Li C, Xu N, Huang X, Wang W, Cheng J, Wu K (2007a) *Bacillus thuringiensis* Vip3 mutant proteins: insecticidal activity and trypsin sensitivity. Biocontrol Sci Technol 17:699–708

Li F, Murillo C, Wurtzel ET (2007b) Maize Y9 encodes a product essential for 15-cis-ζ-carotene isomerization. Plant Physiol 144:1181–1189
Li Y, Meissle M, Romeis J (2008) Consumption of Bt maize pollen expressing Cry1Ab or Cry3Bb1 does not harm adult green lacewings, *Chrysoperla carnea* (Neuroptera: Chrysopidae). PLoS One 3:e2909
Li P, Pei Y, Sang X, Ling Y, Yang Z, He G (2009) Transgenic indica rice expressing a Bitter melon (*Momordica charantia*) class I chitinase gene (McCHIT1) confers enhanced resistance to *Magnaporthe grisea* and *Rhizoctonia solani*. Eur J Plant Pathol 125:533–543
Li F, Wu QY, Sun YL (2010) Overexpression of chloroplastic monodehydroascorbate reductase enhanced tolerance to temperature and methyl viologen-mediated oxidative stresses. Physiol Plant 139:421–434
Li H, Jiang W, Zhang Z, Xing Y, Li F (2013) Transcriptome analysis and screening for potential target genes for RNAi-mediated pest control of the beet armyworm, *Spodoptera exigua*. PLoS One 8(6):e65931
Li M, Li Z, Li S, Guo S, Meng Q, Li G, Yang X (2014) Genetic engineering of glycinebetaine biosynthesis reduces heat-enhanced photoinhibition by enhancing antioxidative defense and alleviating lipid peroxidation in tomato. Plant Mol Biol Report 32:42–51
Li H, Guan R, Guo H, Miao X (2015a) New insights into an RNAi approach for plant defense against piercing-sucking and stem-borer insect pests. Plant Cell Environ 38(11):2277–2285
Li JB, Luan YS, Liu Z (2015b) Overexpression of SpWRKY1 promotes resistance to Phytophthora nicotianae and tolerance to salt and drought stress in transgenic tobacco. Physiol Plant 155 (3):248–266
Li N, Wang X, Ma B, Du C, Zheng L, Wang Y (2017) Expression of a Na+/H+ antiporter RtNHX1 from a recretohalophyte *Reaumuria trigyna* improved salt tolerance of transgenic *Arabidopsis thaliana*. J Plant Physiol 218:109–120
Li W, Separovic F, O'Brien-Simpson NM, Wade JD (2021) Chemically modified and conjugated antimicrobial peptides against superbugs. Chem Soc Rev 50(8):4932–4973
Lim MY, Jeong BR, Jung M, Harn CH (2016) Transgenic tomato plants expressing strawberry D-galacturonic acid reductase gene display enhanced tolerance to abiotic stresses. Plant Biotechnol Rep 10(2):105–116
Lin SC, Lo YC, Lin JY, Liaw YC (2004) Crystal structures and electron micrographs of fungal volvatoxin A2. J Mol Biol 343:477–491
Lincoln JE, Richael C, Overduin B, Smith K, Bostock R, Gilchrist DG (2002) Expression of the antiapoptotic baculovirus p35 gene in tomato blocks programmed cell death and provides broad-spectrum resistance to disease. Proc Natl Acad Sci U S A 99:15217–15221
Lindbo JA, Dougherty WG (2005) Plant pathology and RNAi: a brief history. Annu Rev Phytopathol 43:191–204
Lin H, Karki S, Coe RA et al (2016) Targeted knockdown of GDCH in rice leads to a photorespiratory-deficient phenotype useful as a building block for C4 rice. Plant Cell Physiol 57:919–932
Lindbo JA, Falk BW (2017) The impact of "coat protein-mediated virus resistance" in applied plant pathology and basic research. Phytopathology 107:624–634
Lindbo JA, Silva-Rosales L, Dougherty WG (1993) Pathogen derived resistance to potyviruses: working, but why? Semin Viral 4:369–379
Lindqvist A, Andersson S (2002) Biochemical properties of purified recombinant human β-carotene 15,15′-monooxygenase. J Biol Chem 277:23942–23948
Litz RE, Padilla G (2012) Genetic transformation of fruit trees. In: Priyadarshan PM, Schnell RJ (eds) Genomics of tree crops. Springer, Berlin, pp 117–153
Liu WK (2010) Do genetically modified plants impact arbuscular mycorrhizal fungi? Ecotoxicology 19:229–238
Liu CM, McLean PA, Sookdeo CC, Cannon FC (1991) Degradation of the herbicide glyphosate by members of the family Rhizobiaceae. Appl Environ Microbiol 57(6):1799–1804

Liu Q, Kasuga M, Sakuma Y, Abe H, Miura S, Yamaguchi-Shinozaki K, Shinozaki K (1998) Two transcription factors, DREB1 and DREB2, with an EREBP/AP2 DNA binding domain separate two cellular signal transduction pathways in drought-and low-temperature-responsive gene expression, respectively, in Arabidopsis. Plant Cell 10(8):1391–1406

Liu XD, Zhai BP, Zhang XX, Zong JM (2005) Impact of transgenic cotton on *Aphis gossypii*. Ecol Entomol 30:307–315

Liu X, Wang Z, Wang L, Wu R, Phillips J, Deng X (2009) LEA4 group genes from the resurrection plant *Boea hygrometrica* confer dehydration tolerance in transgenic tobacco. Plant Sci 176:90–98

Liu X, Brost J, Hutcheon C, Guilfoil R, Wilson AK, Leung S, Shewmaker CK, Rooke S, Nguyen T, Kiser J, De Rocher J (2012) Transformation of the oilseed crop Camelina sativa by Agrobacterium-mediated floral dip and simple large-scale screening of transformants. In Vitro Cell Develop Biol Plant 48:462–468

Liu H, Soyars CL, Li J, Fei Q, He G, Peterson BA, Meyers BC (2018) CRISPR/Cas9-mediated resistance to cauliflower mosaic virus. Plant Direct 2:e00047

Liu Y, Liu X, Wang X, Gao K, Qi W, Ren H, Hu H, Sun D, Bai J, Zheng S (2020) Heterologous expression of heat stress-responsive AtPLC9 confers heat tolerance in transgenic rice. BMC Plant Biol 20(1):1–1

Llorente B, Martinez-Garcia JF, Stange C, Rodriguez-Concepcion M (2017) Illuminating colors: regulation of carotenoid biosynthesis and accumulation by light. Curr Opin Plant Biol 37:49–55

Lloyd AM, Davis RW (1994) Functional expression of the yeast FLP/FRT site-specific recombination system in *Nicotiana tabacum*. Mol Gen Genet 242(6):653–657

Lodge JK, Kaniewski WK, Tumer NE (1993) Broad-spectrum virus resistance in transgenic plants expressing pokeweed antiviral protein. Proc Natl Acad Sci U S A 90:7089–7093

Logemann J, Jach G, Tommerup H, Mundy J, Schell J (1992) Expression of a barley ribosome-inactivating protein leads to increased fungal protection in transgenic tobacco plants. Nat Biotechnol 10:305–308

Lomonossoff GP (1995) Pathogen-derived resistance to plant viruses. Annu Rev Phytopathol 33: 323–343

Long SP, Zhu XG, Naidu SL, Ort DR (2006) Can improvement in photosynthesis increase crop yields? Plant Cell Environ 29:315–330

Lopez AB, Eck JV, Conlin BJ, Paolillo DJ, O'Neill J, Li L (2008) Effect of the cauliflower or transgene on carotenoid accumulation and chromoplast formation in transgenic potato tubers. Exp Bot 59:213–223

Lorence A, Verpoorte R (2004) Gene transfer and expression in plants. Methods Mol Biol 267:329–350

Losey JE, Rayor LS, Carter ME (1999) Transgenic pollen harms monarch larvae. Nature 399:214

Lovei GL, Andow DA, Arpaia S (2009) Transgenic insecticidal crops and natural enemies: a detailed review of laboratory studies. Environ Entomol 38:293–306

Low JW, Mwanga RO, Andrade M, Carey E, Ball AM (2017) Tackling vitamin A deficiency with biofortified sweetpotato in sub-Saharan Africa. Glob Food Secur 14:23–30

Lowe NM, Khan MJ, Broadley MR, Zia MH, McArdle HJ, Joy EJ, Ohly H, Shahzad B, Ullah U, Kabana G, Medhi R (2018) Examining the effectiveness of consuming flour made from agronomically biofortified wheat (Zincol-2016/NR-421) for improving Zn status in women in a low-resource setting in Pakistan: study protocol for a randomised, double-blind, controlled cross-over trial (BiZiFED). BMJ Open 8(4):e021364

Lu XK, Mo JM, Dong SF (2008) Effects of nitrogen deposition on forest biodiversity. Acta Ecol Sin 28:5532–5548

Lu Y, Wu K, Jiang Y, Xia B, Li P, Feng H, Wyckhuys KAG, Guo Y (2010) Mirid bug outbreaks in multiple crops correlated with wide-scale adoption of Bt cotton in China. Science 328:1151–1153

Lu Y, Wu K, Jiang Y, Guo Y, Desneux N (2012) Widespread adoption of Bt cotton and insecticide decrease promotes biocontrol services. Nature 487:362–365

Lu HP, Edwards M, Wang QZ, Zhao HJ, Fu HW, Huang JZ, Gatehouse A, Shu QY (2015) Expression of cytochrome P450 CYP81A6 in rice: tissue specificity, protein subcellular localization, and response to herbicide application. J Zhejiang Univ-SCIENCE B 16(2):113–122

Luh BS (1991) Rice-I: production, 2nd edn. Van Nostrand Reinhold, New York

Luo KM, Duan H, Zhao DG, Zheng XL, Deng W, Chen YQ, Stewart CN, McAvoy R, Jiang XN, Wu YH, He AG, Pei Y, Li Y (2007) GM-gene-deletor: fused loxP-FRT recognition sequences dramatically improve the efficiency of FLP or CRE recombinase on transgene excision from pollen and seed of tobacco plants. Plant Biotechnol J 5:263–274

Lutz KA, Knapp JE, Maliga P (2001) Expression of bar in the plastid genome confers herbicide resistance. Plant Physiol 125(4):1585–1590

Ma BL (2005) Development, yield, grain moisture and nitrogen uptake of Bt corn hybrids and their conventional near-isolines. Field Crop Res 93(2-3):199–211

Ma L, Huo R, Gao X, He D, Shao M, Wang Q (2008) Transgenic rape with hrf2 gene encoding harpinXooc resistant to *Sclerotinia sclerotinorium*. Agric Sci China 7:455–461

Ma XM, Zhang YJ, Tureckova V, Xue GP, Fernie AR, Mueller-Roeber B, Balazadeh S (2018) The NAC transcription factor SlNAP2 regulates leaf senescence and fruit yield in tomato. Plant Physiol 177:1286–1302

Maass D, Arango J, Wüst F, Beyer P, Welsch R (2009) Carotenoid crystal formation in Arabidopsis and carrot roots caused by increased phytoene synthase protein levels. PLoS One 4:e6373

Mackintosh CA, Lewis J, Radmer LE, Shin S, Heinen SJ, Smith LA, Wyckoff MN, Dill-Macky R, Evans CK, Kravchenko S, Baldridge GD (2007) Overexpression of defense response genes in transgenic wheat enhances resistance to Fusarium head blight. Plant Cell Rep 26:479–488

Macková H, Hronková M, Dobrá J, Turečková V, Novák O, Lubovská Z, Motyka V, Haisel D, Hájek T, Prášil IT, Gaudinová A (2013) Enhanced drought and heat stress tolerance of tobacco plants with ectopically enhanced cytokinin oxidase/dehydrogenase gene expression. J Exp Bot 64(10):2805–2815

Madanchi H, Shoushtari M, Kashani HH, Sardari S (2020) Antimicrobial peptides of the vaginal innate immunity and their role in the fight against sexually transmitted diseases. New Microbes New Infect 34:100627

Maeda H, Sakuragi Y, Bryant DA, DellaPenna D (2005) Tocopherols protect *Synechocystis* sp. strain PCC 6803 from lipid peroxidation. Plant Physiol 138:1422–1435

Mainieri D, Rossi M, Archinti M, Bellucci M, De Marchis F, Vavassori S, Pompa A, Arcioni S, Vitale A (2004) Zeolin. A new recombinant storage protein constructed using maize gamma-zein and bean phaseolin. Plant Physiol 136:3447–3456

Majumder S, Sarkar C, Saha P, Gotyal BS, Satpathy S, Datta K, Datta SK (2018) Bt jute expressing fused δ-endotoxin Cry1Ab/Ac for resistance to lepidopteran pests. Front Plant Sci 8:2188

Majumder S, Datta K, Datta SK (2019) Rice biofortification: high iron, zinc, and vitamin-A to fight against "hidden hunger". Agronomy 9(12):803

Malarkey T (2003) Human health concerns with GM crops. Mutat Res/Rev Mutat Res 544(2–3):217–221

Malik KA, Maqbool A (2020) Transgenic crops for biofortification. Front Sustain Food Syst 4: 571402

Malnoy M, Venisse JS, Brisset MN, Chevreau E (2003) Expression of bovine lactoferrin cDNA confers resistance to Erwinia amylovora in transgenic pear. Mol Breed 12(3):231–244

Malyshenko SI, Kondakova OA, Nazarova JV, Kaplan IB, Taliansky ME, Atabekov JG (1993) Reduction of tobacco mosaic virus accumulation in transgenic plants producing nonfunctional viral transport proteins. J Gen Virol 74:1149–1156

Mamta B, Reddy KR, Rajam MV (2016) Targeting chitinase gene of *Helicoverpa armigera* by host-induced RNA interference confers insect resistance in tobacco and tomato. Plant Mol Biol 9:281–229

Mandaokar AD, Goyal RK, Shukla A, Bisaria S, Bhalla R, Reddy VS, Chaurasia A, Sharma I, Altosaar I, Kumar PA (2000) Transgenic tomato plant resistant to fruit borer (*Helicoverpa armigera* Hubner). Crop Prot 19:307–312

Manickam S, Guranjan KN, Gopalakrishnan N (2008) Bt Cotton vis-à-vis environmental protection: an Indian perspective. Central Institute of Cotton Research, Regional Station, Coimbatore

Maqbool MA, Beshir A (2019) Zinc biofortification of maize (*Zea mays* L.): status and challenges. Plant Breed 138(1):1–28

Maqbool SB, Husnain T, Riazuddin S, Masson L, Christou P (1998) Effective control of yellow stem borer and rice leaf folder in transgenic rice indica varieties Basmati 370 and M 7 using the novel δ-endotoxin cry2A *Bacillus thuringiensis* gene. Mol Breed 4(6):501–507

Marchive C, Léon C, Kappel C, Coutos-Thévenot P, Corio-Costet MF, Delrot S, Lauvergeat V (2013) Over-expression of VvWRKY1 in grapevines induces expression of jasmonic acid pathway-related genes and confers higher tolerance to the downy mildew. PLoS One 8(1): e54185

Marco F, Bitrián M, Carrasco P, Rajam MV, Alcázar R, Tiburcio AF (2015) Genetic engineering strategies for abiotic stress tolerance in plants. In: Plant biology and biotechnology. Springer, New Delhi, pp 579–609

Marcocci ME, Amatore D, Villa S, Casciaro B, Aimola P, Franci G, Grieco P, Galdiero M, Palamara AT, Mangoni ML, Nencioni L (2020) The amphibian antimicrobial peptide temporin B inhibits *in vitro* herpes simplex virus 1 infection. Antimicrob Agents Chemother 62:e02367–ee2417

Marrs KA (1996) The functions and regulation of glutathione S-transferases in plants. Annu Rev Plant Physiol Plant Mol Biol 47:27–58

Maruyama K, Takeda M, Kidokoro S, Yamada S, Sakuma Y, Urano K, Fujita M, Yoshiwara K, Matsukura S, Morishita Y, Sasaki R, Suzuki H, Saito K, Shibata D, Shinozaki K, Yamaguchi-Shinozaki K (2009) Metabolic pathways involved in cold acclimation identified by integrated analysis of metabolites and transcripts regulated by DREB1A and DREB2A. Plant Physiol 150: 1972–1980

Marvier M, McCreedy C, Regetz J, Kareiva P (2007) A meta-analysis of effects of Bt cotton and maize on nontarget invertebrates. Science 316:1475–1477

Masuda H, Usuda K, Kobayashi T, Ishimaru Y, Kakei Y, Takahashi M, Higuchi K, Nakanishi H, Mori S, Nishizawa NK (2009) Overexpression of the barley nicotianamine synthase gene HvNAS1 increases iron and zinc concentrations in rice grains. Rice 2:155–166

Matas AJ, Gapper NE, Chung MY, Giovannoni JJ, Rose JKC (2009) Biology and genetic engineering of fruit maturation for enhanced quality and shelf-life. Curr Opin Biotechnol 20: 197–203

Matthews BF, Youssef RM (2016) Soybean cyst nematode, Heterodera glycines, infection assay using soybean roots. Bio-protocol 6(2):e1707

Mazur B, Krebbers E, Tingey S (1999) Gene discovery and product development for grain quality traits. Science 285(5426):372–375

McBride KE, Svab Z, Schaaf DJ, Hogan PS, Stalker DM, Maliga P (1995) Application of a chimeric *Bacillus* gene in chloroplasts leads to extraordinary level of an insecticidal protein in tobacco. Bio/Technology 13:362–365

McDougall P (2011) The cost and time involved in the discovery, development, and authorization of a new plant biotechnology derived trait. A Consultancy Study for Crop Life International, Midlothian

McKhann HI, Gery C, Berard A, Leveque S, Zuther E, Hincha DK, De Mita S, Brunel D, Teoule E (2008) Natural variation in CBF gene sequence, gene expression and freezing tolerance in the Versailles core collection of *Arabidopsis thaliana*. BMC Plant Biol 8:105

McMullen MD, Kresovich S, Villeda HS, Bradbury P, Li H, Sun Q, Flint-Garcia S, Thornsberry J, Acharya C, Bottoms C, Brown P (2009) Genetic properties of the maize nested association mapping population. Science 325(5941):737–740

McNall RJ, Adang MJ (2003) Identification of novel Bacillus thuringiensis Cry1Ac binding proteins in Manduca sexta midgut through proteomic analysis. Insect Biochem Mol Biol 33:999–1010

Meade T, Narva K, Storer NP, Sheets JJ, Burton SL, Woosley AT, Dow AgroSciences LLC (2017) Insect resistance management with combinations of Cry1Be and Cry1F proteins. US Patent 9: 556453

Meena PD, Awasthi RP, Chattopadhyay C, Kolte SJ, Kumar A (2010) Alternaria blight: a chronic disease in rapeseed-mustard. J Oilseed Brassica 1:1–11

Meena PD, Verma PR, Saharan GS, Borhan MH (2014) Historical perspectives of white rust caused by *Albugo candida* in oilseed Brassica. J Oilseed Brassica 5:1–41

Mehra S, Pareek A, Bandyopadhyay P, Sharma P, Burma PK, Pental D (2000) Development of transgenics in Indian oilseed mustard (*Brassica juncea*) resistant to herbicide phosphinothricin. Curr Sci 78:358–1364

Mehta RA, Cassol T, Li N, Ali N, Handa AK, Mattoo AK (2002) Engineered polyamine accumulation in tomato enhances phytonutrient content, juice quality, and vine life. Nat Biotechnol 20: 613–618

Mehta L, Dhankhar R, Gulati P, Kapoor RK, Mohanty A, Kumar S (2020a) Natural and grafted cyclotides in cancer therapy: an insight. J Pept Sci 26:e3246

Mehta BK, Muthusamy V, Zunjare RU, Baveja A, Chauhan HS, Chhabra R, Singh AK, Hossain F (2020b) Biofortification of sweet corn hybrids for provitamin-A, lysine and tryptophan using molecular breeding. J Cereal Sci 96:103093

Meli VS, Ghosh S, Prabha TN, Chakraborty N, Chakraborty S, Datta A (2010) Enhancement of fruit shelf life by suppressing N-glycan processing enzymes. Proc Natl Acad Sci U S A 107: 2413–2418

Mellacheruvu S, Tamirisa S, Reddy D, Khareedu VR (2016) Pigeonpea hybrid-proline-rich protein (CcHyPRP) confers biotic and abiotic stress tolerance in transgenic rice. Front Plant Sci 6:1–13

Melo AL, Soccol VT, Soccol CR (2016) Bacillus thuringiensis: mechanism of action, resistance, and new applications: a review. Crit Rev Biotechnol 36(2):317–326

Mendelshon M, Kough J, Vaituzis Z, Mathews K (2003) Are Bt crops safe? Nat Biotechnol 21(9): 1003–1009

Meng F, Li Y, Zang Z, Li N, Ran R, Cao Y, Li T, Zhou Q, Li W (2017) Expression of the double-stranded RNA of the soybean pod borer *Leguminivora glycinivorella* (Lepidoptera: Tortricidae) ribosomal protein P0 gene enhances the resistance of transgenic soybean plants. Pest Manag Sci 73:2447–2455

Merchant S, Bogorad L (1987) Metal ion regulated gene expression: use of a plastocyanin-less mutant of *Chlamydomonas reinhardtii* to study the Cu (II)-dependent expression of cytochrome c-552. EMBO J 6(9):2531–2535

Merdinoglu D, Wiedemann-Merdinoglu S, Coste P (2003) Genetic analysis of downy mildew resistance derived from *Muscadinia rotundifolia*. Acta Horticult 1:451–456

Mertz ET, Bates LS, Nelson OE (1964) Mutant gene that changes protein composition and increases lysine content of maize endosperm. Science 145:279–280

Mesrati AL, Rouis S, Sellami S, Jaoua S (2009) Prays oleae midgut putative receptor of *Bacillus thuringiensis* vegetative insecticidal protein Vip3LB differs from that of Cry1Ac toxin. Mol Biotechnol 43:15–19

Mesrati AL, Boukedi H, Chakroun M, Kamoun F, Azzouz H, Tounsi S (2011) Investigation of the steps involved in the difference of susceptibility of *Ephestia kuehniella* and *Spodoptera littoralis* to the *Bacillus thuringiensis* Vip3Aa16 toxin. J Invertebr Pathol 107:198–201

Mi X, Liu X, Yan H, Liang L, Zhou X, Yang J, Si H, Zhang N (2017) Expression of the *Galanthus nivalis* agglutinin (GNA) gene in transgenic potato plants confers resistance to aphids. C R Biol 340:7–12

Miao Y, Lv D, Wang P, Wang XC, Chen J, Miao C, Song CP (2006) An *Arabidopsis* glutathione peroxidase functions as both a redox transducer and a scavenger in abscisic acid and drought stress responses. Plant Cell 18:2749–2766

Midboe EG, Candas M, Bulla LA Jr (2003) Expression of a midgut-specific cadherin BT-R1 during the development of *Manduca sexta* larva. Comp Biochem Physiol B 135:125–137

Milena Valderrama A, Velásquez N, Rodríguez E, Zapata A, Abbas Zaidi M, Altosaar I, Arango R (2007) Resistance to Tecia solanivora (Lepidoptera: Gelechiidae) in three transgenic Andean varieties of potato expressing Bacillus thuringiensis Cry1Ac protein. J Econ Entomol 100 (1):172–179

Miller JK, Bradford KJ (2010) The regulatory bottleneck for biotech specialty crops. Nat Biotechnol 28(10):1012–1014

Milligan SB, Bodeau J, Yaghoobi J, Kaloshian I, Zabel P, Williamson VM (1998) The root knot resistance gene Mi from tomato is a member of the leucine zipper, nucleotide binding, leucine-rich repeat family of plant genes. Plant Cell 10:1307–1321

Minhas D, Grover A (1999) Transcript levels of genes encoding various glycolytic and fermentation enzymes change in response to abiotic stresses. Plant Sci 146:41–51

Miranda JA, Avonce N, Suárez R, Thevelein JM, Dijck PV, Iturriaga G (2007) A bifunctional TPS–TPP enzyme from yeast confers tolerance to multiple and extreme abiotic-stress conditions in transgenic *Arabidopsis*. Planta 226:1411–1421

Misztal LH, Mostowska A, Skibinska M, Bajsa J, Musial WG, Jarmolowski A (2004) Expression of modified *cry1Ac* gene of *Bacillus thuringiensis* in transgenic tobacco plants. Mol Biotechnol 26(1):17–26

Mitchell P (1999) Safety of genetically modified food questioned: interview with gene scientist, Dr. Arpad Pusztai. http://www.wsws.org/articles/1999/jun1999/gmo-j03.shtml

Mitsuhara I, Iwai T, Seo S (2008) Characteristic expression of twelve rice PR1 family genes in response to pathogen infection, wounding, and defense-related signal compounds (121/180). Mol Genet Genomics 279:415–427

Mittler R (2002) Oxidative stress, antioxidants and stress tolerance. Trends Plant Sci 7:405–410

Mittler R, Vanderauwera S, Gollery M, Van Breusegem F (2004) Reactive oxygen gene network of plants. Trends Plant Sci 9:490–498

Mlalazi B, Welsch R, Namanya P, Khanna H, Geijskes RJ, Harrison MD, Harding R, Dale JL, Bateson M (2012) Isolation and functional characterization of banana phytoene synthase genes as potential cisgenes. Planta 236:1585–1598

Mobin M, Khan NA (2007) Photosynthetic activity, pigment composition and antioxidative response of two mustard (*Brassica juncea*) cultivars differing in photosynthetic capacity subjected to cadmium stress. J Plant Physiol 164:601–610

Mody N (2017) Alterations in vitamin A/retinoic acid homeostasis in diet induced obesity and insulin resistance. Proc Nutr Soc 76:597–602

Mohamed STA (2018) Genetically engineered (modified) crops (*Bacillus thuringiensis* crops) and the world controversy on their safety. Egypt J Biol Pest Control 28:52

Molinari HBC, Marur CJ, Filho JCB (2004) Osmotic adjustment in transgenic citrus rootstock Carrizo citrange (*Citrus sinensis* Osb. X *Poncirus trifoliata* L. Raf.) overproducing proline. Plant Sci 167:1375–1381

Molvig L, Tabe LM, Eggum BO, Moore AE, Craig S, Spencer D, Higgins TJ (1997) Enhanced methionine levels and increased nutritive value of seeds of transgenic lupins (*Lupinus angustifolius* L.) expressing a sunflower seed albumin gene. Proc Natl Acad Sci U S A 94(16):8393–8398

Mondal KK, Bhattacharya RC, Koundal KR, Chatterjee SC (2006) Transgenic Indian mustard (*Brassica juncea*) expressing tomato glucanase leads to arrested growth of *Alternaria brassicae*. Plant Cell Rep 26:247–252

Montagne JA, Davis CR, Tanumihardjo SA (2009) Nutritional value of cassava for use as stable food and recent advance for improvement. Compr Rev Food Sci Food Saf 8(3):94–181

Moon YH, Song SK, Choi KW, Lee JS (1997) Expression of a cDNA encoding *Phytolacca insularis* antiviral protein confers virus resistance of transgenic potato plants. Mol Cells 7: 807–815

Moon H, Baek D, Lee B, Prasad DT, Lee SY, Cho MJ, Lim CO, Choi MS, Bahk J, Kim MO, Hong JC, Yun DJ (2002) Soybean ascorbate peroxidase suppresses Bax-induced apoptosis in yeast by inhibiting oxygen radical generation. Biochem Biophys Res Commun 290:457–462

Morán R, Garcıa R, López A, Zaldúa Z, Mena J, Garcıa M, Armas R, Somontea D, Rodríguez J, Gómez M, Pimentel E (1998) Transgenic sweet potato plants carrying the delta-endotoxin gene from *Bacillus thuringiensis* var. tenebrionis. Plant Sci 139(2):175–184

Mosa KA, Saadoun I, Kumar K, Helmy M, Dhankher OP (2016) Potential biotechnological strategies for the cleanup of heavy metals and metalloids. Front Plant Sci 7:303

Moscou MJ, Bogdanove AJ (2009) A simple cipher governs DNA recognition by TAL effectors. Science 326:1501

Muddanuru T, Polumetla AK, Maddukuri L, Mulpuri S (2019) Development and evaluation of transgenic castor (*Ricinus communis* L.) expressing the insecticidal protein *cry1Aa* of *Bacillus thuringiensis* against lepidopteran insect pests. Crop Prot 119:113–125

Mulder C, Wouterse M, Raubuch M, Roelofs W, Rutgers M (2006) Can transgenic maize affect soil microbial communities? PLoS Comput Biol 2:e128

Mundt CC (2018) Pyramiding for resistance durability: theory and practice. Phytopathology 108: 792–802

Munis MFH, Tu L, Deng F, Tan J, Xu L, Xu S, Long L, Zhang X (2010) A thaumatin-like protein gene involved in cotton fiber secondary cell wall development enhances resistance against Verticillium dahliae and other stresses in transgenic tobacco. Biochem Biophys Res Commun 393:38–44

Munné-Bosch S (2007) α-Tocopherol: a multifaceted molecule in plants. Vitam Horm 76:375–392

Munne-Bosch S, Alegre L (2003) Drought-induced changes in the redox state of α-tocopherol, ascorbate and the diterpene carnosic acid in chloroplasts of Labiatae species differing in carnosic acid contents. Plant Physiol 131:1816–1825

Munns R, Gilliham M (2015) Salinity tolerance of crops-what is the cost? New Phytol 3:668–673

Munns R, Tester M (2008) Mechanisms of salinity tolerance. Annu Rev Plant Biol 59:651–681

Murai N, Sutton DW, Murray MG, Slightom JL, Merlo DJ, Reichert NA, Sengupta-Gopalan C, Stock CA, Barker RF, Kemp JD, Hall TC (1983) Phaseolin gene from bean is expressed after transfer to sunflower via tumor-inducing plasmid vectors. Science 222:476–482

Muramoto N, Tanaka T, Shimamura T, Mitsukawa N, Hori E, Koda K, Otani M, Hirai M, Nakamura M, Imaeda K (2012) Transgenic sweet potato expressing thionin from barley gives resistance to black rot disease caused by Ceratocystis fimbriata in leaves and storage roots. Plant Cell Rep 31:987–997

Mushtaq M, Sakina A, Wani SH, Shikari AB, Tripathi P, Zaid A, Galla A, Abdelrahman M, Sharma M, Singh AK, Salgotra RK (2019) Harnessing genome editing techniques to engineer disease resistance in plants. Front Plant Sci 10:550

Musidlak O, Robert Nawrot ID, Anna Gózdzicka-Józefiak ID (2017) Which plan proteins are involved in antiviral defense? Review on in vivo and in vitro activities of selected plant proteins against viruses. Int J Mol Sci 18:2300

Muthayya S, Rah JH, Sugimoto JD, Roos FF, Kraemer K, Black RE (2013) The global hidden hunger indices and maps: an advocacy tool for action. PLoS One 8:e67860

Muthurajan R, Ramanathan V, Bansilal Shillak A, Madhuri Pralhad S, Shankarrao CN, Rahman H, Kambale R, Nallathambi J, Tamilselvan S, Madasamy P (2021) Controlled over-expression of AtDREB1A enhances tolerance against drought and salinity in rice. Agronomy 11(1):159

Muthusamy V, Hossain F, Thirunavukkarasu N, Choudhary M, Saha S, Bhat JS, Prasanna BM, Gupta HS (2013) Development of β-carotene rich maize hybrids through marker-assisted introgression of β-carotene hydroxylase allele. PLoS One 9(12):e113583

Muzhingi T, Gadaga TH, Siwela AH, Grusak MA, Russell RM, Tang GW (2011) Yellow maize with high beta-carotene is an effective source of vitamin A in healthy Zimbabwean men. Am J Clin Nutr 94:510–519

Myagmarjav D, Sukweenadhi J, Kim YJR, Genet J (2017) Molecular characterization and expression analysis of pathogenesis related protein 6 from *Panax ginseng*. Russian J Genet 53:1211–1220

Nahirñak V, Almasia NI, Fernandez PV, Hopp HE, Estevez JM, Carrari F, Vazquez-Rovere C (2012) Potato snaking-1 gene silencing affects cell division primary metabolism and cell wall composition. Plant Physiol 158:252–263

Naik G (2001) An analysis of socio-economic impact of Bt technology on Indian Cotton Farmers Centre for Management in Agriculture, Indian Institute of Management Ahemdabad

Nakajima H, Muranaka T, Ishige F, Akutsu K, Oeda K (1997) Fungal and bacterial disease resistance in transgenic plants expressing human lysozyme. Plant Cell Rep 16(10):674–679

Nakanishi H, Ogawa I, Ishimaru Y, Mori S, Nishizawa NK (2006) Iron deficiency enhances cadmium uptake and translocation mediated by the Fe2+ transporters OsIRT1 and OsIRT2 in rice. Soil Sci Plant Nutr 52(4):464–469

Nakasu EY, Edwards MG, Fitches E, Gatehouse JA, Gatehouse AM (2014) Transgenic plants expressing xACTX-Hv1a and snowdrop lectin (GNA) fusion protein show enhanced resistance to aphids. Front Plant Sci 28:673

Nakayama M, Nakajima D, Nagase T, Nomura N, Seki N, Ohara O (1998) Identification of high-molecular-weight proteins with multiple EGF-like motifs by motif-trap screening. Genomics 51: 27–34

Nakayama H, Yoshida K, Ono H, Murooka Y, Shinmyo A (2000) Ectoine, the compatible solute of *Halomonas elongata*, confers hyperosmotic tolerance in cultured tobacco cells. Plant Physiol 122:1239–1247

Nandi S, Suzuki YA, Huang J, Yalda D, Pham P, Wu L, Bartley G, Huang N, Lönnerdal B (2002) Expression of human lactoferrin in transgenic rice grains for the application in infant formula. Plant Sci 163:713–722

Nandula VK (ed) (2010) Glyphosate resistance in crops and weeds: history, development, and management. Wiley, Hoboken, pp 67–91

Nanjo T, Kobayashi M, Yoshiba Y, Kakubari Y, Yamaguchi-Shinozaki K, Shinozaki K (1999) Antisense suppression of proline degradation improves tolerance to freezing and salinity in *Arabidopsis thaliana*. FEBS Lett 461:205–210

Napoli C, Lemieux C, Jorgensen R (1990) Introduction of a chimeric chalcone synthase gene into petunia results in reversible co-suppression of homologous genes in trans. Plant Cell 2:279–289

Naqvi S, Zhu C, Farre G, Ramessar K, Bassie L, Breitenbach J, Conesa DP, Ros G, Sandmann G, Capell T, Christou P (2009) Transgenic multivitamin corn through biofortification of endosperm with three vitamins representing three distinct metabolic pathways. Proc Natl Acd Sci U S A 106:7762–7767

Narayanan N, Ihemere U, Ellery C, Sayre R (2011a) Overexpression of hydroxynitrile lyase in cassava roots elevates protein and free amino acids while reducing residual cyanogen levels. PLoS One 6:e21996

Narayanan NN, Ihemere U, Chiu W, Siritunga D, Rajamani S, Singh S, Oda S, Sayre RT (2011b) The iron assimilatory protein, FEA1, from *Chlamydomonas reinhardtii* facilitates iron-specific metal uptake in yeast and plants. Front Plant Sci 2:67

Nareshkumar EJ, MIlind MS, Sadashiv SG (2017) Estimation of yield losses caused by pest in groundnut. Agric Update 12:58–61

Návarová H, Bernsdorff Döring F, Zeier ACJ (2012) Pipecolic acid, an endogenous mediator of defense amplification and priming, is a critical regulator of inducible plant immunity. Plant Cell 24:5123–5141

Nawrot R, Barylski J, Nowicki G, Broniarczyk J, Buchwald W, Gozdzicka-Jozefiak A (2014) Plant antimicrobial peptides. Folia Microbiol 59:181–196

Nayak P, Basu D, Das S, Basu A, Ghosh D, Ramakrishnan NA, Ghosh M, Sen SK (1997) Transgenic elite indica rice plants expressing Cry1Ac δ-endotoxin of *Bacillus thuringiensis* are resistant against yellow stem borer (*Scirpophaga incertulas*). Proc Natl Acad Sci U S A 94: 2111–2116

Neeraja CN, Ravindra Babu V, Ram S, Hossain F, Hariprasanna K, Rajpurohit BS (2017) Biofortification in cereals: progress and prospects. Curr Sci 113:1050–1057

Negrão S, Schmöckel SM, Tester M (2017) Evaluating physiological responses of plants to salinity stress. Ann Bot 119(1):1–11
Nervo G, Cirillo C, Accotto GP, Vaira AM (2003) Characterization of two tomato lines highly resistant to tomato spotted wilt virus following transformation with the viral nucleoprotein gene. J Plant Pathol 85:139–144
Netherwood T, Martín-Orúe SM, O'Donnell AG, Gockling S, Graham J, Mathers JC, Gilbert HJ (2004) Assessing the survival of transgenic plant DNA in the human gastrointestinal tract. Nat Biotechnol 22(2):204–209
Newell CA, Rozman R, Hinchee MA, Lawson EC, Haley L, Sanders P, Kaniewski W, Tumer NE, Horsch RB, Fraley RT (1991) Agrobacterium-mediated transformation of *Solanum tuberosum* L. cv. 'Russet Burbank'. Plant Cell Rep 10:30–34
Ng TB, Lam SK, Fong WP (2003) A homodimeric sporamin-type trypsin inhibitor with antiproliferative. Biol Chem 834:811–815
Ni X, Tian Z, Liu J, Song B, Li J, Shi X, Xie C (2010) StPUB17, a novel potato UND/PUB/ARM repeat type gene, is associated with late blight resistance and NaCl stress. Plant Sci 178:158–169
Ni M, Ma W, Wang X, Gao M, Dai Y, Wei X, Zhang L, Peng Y, Chen S, Ding L, Tian Y (2017) Next-generation transgenic cotton: pyramiding RNAi and Bt counters insect resistance. Plant Biotechnol J 15:1204–1213
Nickson TE (2008) Planning environmental risk assessment for genetically modified crops: problem formulation for stress-tolerant crops. Plant Physiol 147(2):494–502
Nirupa N, Prasad MNV, Jami SK, Kirti PB (2007) Optimization of *Agrobacterium*-mediated overexpression of osmotin-ferritin genes in *Brassica juncea*. Transgenic Plant J 1(2):384–392
Niu QW, Lin SS, Reyes JL, Chen KC, Wu HW, Yeh SD, Chua NH (2006) Expression of artificial microRNAs in transgenic *Arabidopsis thaliana* confers virus resistance. Nat Biotechnol 24: 1420–1428
Niu X, Xiong F, Liu J et al (2014) Co-expression of ApGSMT and ApDMT promotes biosynthesis of glycine betaine in rice (*Oryza sativa* L.) and enhances salt and cold tolerance. Environ Exp Bot. https://doi.org/10.1016/j.envexpbot.2014.03.003
Niu X, Kassa A, Hu X, Robeson J, McMahon M, Richtman NM (2017) Control of western corn rootworm (*Diabrotica virgifera*) reproduction through plant-mediated RNA interference. Sci Rep 7(1):12591
Noctor G, Foyer CH (1998) Ascorbate and glutathione: keeping active oxygen under control. Annu Rev Plant Biol 49(1):249–279
Noctor G, Gomez L, Vanacker H, Foyer CH (2002) Interactions between biosynthesis, compartmentation and transport in the control of glutathione homeostasis and signalling. J Exp Bot 53: 1283–1304
Noman A, Aqeel M, Deng J, Khalid N, Sanaullah T, Shuilin H (2017) Biotechnological advancements for improving floral attributes in ornamental plants. Front Plant Sci 8:530
Nordlee JA, Taylor SL, Townsend JA (1996) Identification of a Brazil-nut allergen in transgenic soybeans. N Engl J Med 334(11):688–692
Norelli JL, Aldwinckle H, Destefano-Beltran L, Jaynes JM (1994) Transgenic Malling 26 apple expressing the attacin E gene has increased resistance to Erwinia amylovora. Euphytica 77:123–128
Nuss ET, Tanumihardjo SA (2010) Maize: a paramount staple crop in the context of global nutrition. Compr Rev Food Sci Food Saf 9:417–436
Oberschall A, Deak M, Torok K, Saa L, Vass I, Kovacs I, Feher A, Dudits D, Horvath GV (2000) A novel aldose/aldehyde reductase protects transgenic plants against lipid peroxidation under chemical and drought stresses. Plant J 24:437–446
Odell J, Caimi P, Sauer B, Russell S (1990) Site directed recombination in the genome of transgenic tobacco. Mol Gen Genet 223:369–378
OECD (1999a) Consensus document on general information concerning the genes and their enzymes that confer tolerance to glyphosate herbicide. Series on harmonization of regulatory

oversight in biotechnology Report No. 10. https://www.oecd.org/env/ehs/biotrack/46815618.pdf. Accessed 24 Sept 2015

OECD (1999b) Consensus document on general information concerning the genes and their enzymes that confer tolerance to Phosphinothricin herbicide. Series on harmonization of regulatory oversight in biotechnology. Report No. 11. https://www.oecd.org/env/ehs/biotrack/46815628.pdf. Accessed 24 Sept 2015

Oerke EC, Dehne HW (2004) Safeguarding production: losses in major crops and the role of crop protection. Crop Prot 23:275–285

Ohnishi M, Fukada-Tanaka S, Hoshino A, Takada J, Inagaki Y, Iida S (2005) Characterization of a novel Na^{+}/H^{+} antiporter gene InNHX2 and comparison of InNHX2 with InNHX1, which is responsible for blue flower coloration by increasing the vacuolar pH in the Japanese morning glory. Plant Cell Physiol 46:259–267

Ohta M, Hayashi Y, Nakashima A, Hamada A, Tanaka A, Nakamura T, Hayakawa T (2002) Introduction of a Na+/H+ antiporter gene from *Atriplex gmelini* confers salt tolerance to rice. FEBS Lett 532(3):279–282

Oliveira AP, Pampulha ME, Bennett JP (2008) A two-year field study with transgenic *Bacillus thuringiensis* maize: effects on soil microorganisms. Sci Total Environ 405:351–357

Ordiz MI, Magnenat L, Barbas CF, Beachy RN (2010) Negative regulation of the RTBV promoter by designed zinc finger proteins. Plant Mol Biol 72:621–630

Osusky M, Zhou G, Osuska L, Hancock RE, Kay WW, Misra S (2000) Transgenic plants expressing cationic peptide chimeras exhibit broad spectrum resistance to phytopathogens. Nat Biotechnol 18:1162–1166

Paal J, Henselewski H, Muth J, Meksem K, Menéndez CM, Salamini F, Ballvora A, Gebhardt C (2004) Molecular cloning of the potato Gro1-4 gene conferring resistance to pathotype Ro1 of the root cyst nematode Globodera rostochiensis, based on a candidate gene approach. Plant J 38 (2):285–297

Pacheco S, Gomez I, Arenas I, Saab-Rincon G, Rodriguez-Almazan C, Gill SS, Bravo A, Soberon M (2009) Domain II loop 3 of *Bacillus thuringiensis* Cry1Ab toxin is involved in a "ping pong" binding mechanism with *Manduca sexta* aminopeptidase-N and cadherin receptors. J Biol Chem 284:32750–32757

Paine J, Shipton C, Chaggar S, Howells R, Kennedy M, Vernon G, Wright S, Hinchliffe E, Adams J, Silverstone A, Drake R (2005) Improving the nutritional value of golden rice through increased pro-vitamin A content. Nat Biotechnol 23:482–487

Pan ZQ, Onstad DW, Nowatzki TM, Stanley BH, Meinke LJ, Flexner JL (2011) Western corn rootworm (Coleoptera: Chrysomelidae) dispersal and adaptation to single-toxin transgenic corn deployed with block or blended refuge. Environ Entomol 40:964–978

Pandey S, Fartyal D, Agarwal A, Shukla T, James D, Kaul T, Negi YK, Arora S, Reddy MK (2017) Abiotic stress tolerance in plants: myriad roles of ascorbate peroxidase. Front Plant Sci 8:581

Pang CH, Wang BS (2010) Role of ascorbate peroxidase and glutathione reductase in ascorbate–glutathione cycle and stress tolerance in plants. In: Anjum N, Chan MT, Umar S (eds) Ascorbate-glutathione pathway and stress tolerance in plants. Springer, pp 91–113

Pardey PG, Beddow JM, Kriticos DJ, Hurley TM, Park RF, Duveiller E, Sutherst RW, Burdon JJ, Hodson D (2013) Right-sizing stem-rust research. Science 340:147–148

Pardo Lopez L, Soberon M, Bravo A (2013) *Bacillus thuringiensis* insecticidal three-domain Cry toxins: mode of action, insect resistance and consequences for crop protection FEMS. Microbial Rev 37:3–22

Pardo JM, Reddy MP, Yang S, Maggio A, Huh GH, Matsumoto T, Coca MA, Paino-D'Urzo M, Koiwa H, Yun DJ, Watad AA, Bressan RA, Hasegawa PM (1998) Stress signaling through Ca^{2+}/calmodulin dependent protein phosphatase calcineurin mediates salt adaptation in plants. Proc Natl Acad Sci U S A 95:9681–9686

Pardo-Lopez L, Soberon M, Bravo A (2013) *Bacillus thuringiensis* insecticidal three-domain Cry toxins: mode of action, insect resistance and consequences for crop protection. FEMS Microbiol Rev 37:3–22

Parisi K, Shafee TMA, Quimbar P, van der Weerden NL, Bleackley MR, Anderson MA (2019) The evolution function and mechanisms of action for plant defensins. Semin Cell Dev Biol 88:107–118

Park JM, Park CJ, Lee SB, Ham BK, Shin R, Paek KH (2001) Overexpression of the tobacco Tsi1 gene encoding an EREBP/AP2-type transcription factor enhances resistance against pathogen attack and osmotic stress in tobacco. Plant Cell 13:1035–1046

Park SH, Kim C-K, Pike LM, Smith RH, Hirschi KD (2004a) Increased calcium in carrots by expression of an Arabidopsis H^+/Ca^{2+} transporter. Mol Breed 14:275–282

Park CJ, Kim KJ, Shin R, Park JM, Shin YC (2004b) Pathogenesis related protein 10 isolated from hot pepper functions as a ribonuclease in an antiviral pathway. Plant J 37:186–198

Park S, Cheng N, Pittman JK, Yoo KS, Park J, Smith RH, Hirschi KD (2005a) Increased calcium levels and prolonged shelf life in tomatoes expressing *Arabidopsis* H^+/Ca^{2+} transporters. Plant Physiol 139:1194–1206

Park S, King TS, Kim CK, Han JS, Kim S, Smith RH, Pike LM, Hirschi KD (2005b) Genetic manipulation for enhancing calcium content in potato tuber. J Agric Food Chem 53:5598–5603

Park BJ, Liu Z, Kanno A, Kameya T (2005c) Increased tolerance to salt- and water-deficit stress in transgenic lettuce (*Lactuca sativa* L.) by constitutive expression of LEA. Plant Growth Regul 45:165–171

Park EJ, Jeknic Z, Pino MT et al (2007) Glycinebetaine accumulation in chloroplasts is more effective than that in cytosol in protecting transgenic tomato plants against abiotic stress. Plant Cell Environ 30:994–1005

Park S, Elless MP, Park J, Jenkins A, Lim W, Chambers E IV, Hirschi KD (2009) Sensory analysis of calcium biofortified lettuce. Plant Biotechnol J 7:106–117

Parkash A, Ng TB, Tso WW (2002) Isolation and characterization of luffacylin, a ribosome inactivating peptide with anti-fungal activity from sponge gourd (*Luffa cylindrica*) seeds. Peptides 23:1019–1024

Parkhi V, Rai M, Tan J, Oliva N, Rehana S, Bandyopadhyay A, Torrizo L, Ghole V, Datta K, Datta SK (2005) Molecular characterization of marker-free transgenic lines of indica rice that accumulate carotenoids in seed endosperm. Mol Genet Genomics 274(4):325–336

Parmar N, Singh KH, Sharma D, Singh L, Kumar P, Nanjundan J, Khan YJ, Chauhan DK, Thakur AK (2017) Genetic engineering strategies for biotic and abiotic stress tolerance and quality enhancement in horticultural crops: a comprehensive review. 3Biotech 7(4):239

Parsa M, Kashanchi M, Zeinali A, Pourfakhraei E (2021) Overexpression of zinc finger (GpZF) promotes drought tolerance in grass pea (*Lathyrus sativus*). bioRxiv. https://doi.org/10.1101/2021.10.19.464954

Pasapula V, Shen G, Kuppu S, Paez-Valencia J, Mendoza M, Hou P, Chen J, Qiu X, Zhu L, Zhang X, Auld D, Blumwald E, Zhang H, Gaxiola R, Payton P (2011) Expression of an Arabidopsis vacuolar H^+-pyrophosphatase gene (AVP1) in cotton improves drought-and salt tolerance and increases fibre yield in the field conditions. Plant Biotechnol J 9(1):88–99

Pasquali G, Biricolti S, Locatelli F, Baldoni E, Mattana M (2008) Osmyb4 expression improves adaptive responses to drought and cold stress in transgenic apples. Plant Cell Rep 27:1677–1686

Patel KG, Mandaliya VB, Mishra GP, Dobaria JR, Radhakrishnan T (2016) Transgenic peanut overexpressing *mtlD* gene confers enhanced salinity-stress tolerance via mannitol accumulation and differential antioxidative responses. Acta Physiol Plant 38:181

Patkar RN, Chattoo BB (2006) Transgenic indica rice expressing ns-LTP-like protein shows enhanced resistance to both fungal and bacterial pathogens. Mol Breed 17:159–171

Paul JY, Khanna H, Kleidon J, Hoang P, Geijskes J, Daniells J, Zaplin E, Rosenberg Y, James A, Mlalazi B, Deo P (2017) Golden bananas in the field: elevated fruit pro-vitamin A from the expression of a single banana transgene. Plant Biotechnol J 15:520–532

Pecker I, Chamovitz D, Linden H, Sandmann G, Hirschberg J (1992) A single polypeptide catalyzing the conversion of phytoene to zeta-carotene is transcriptionally regulated during tomato fruit ripening. Proc Natl Acd Sci U S A 89:4962–4966

Peferoen M, Jansens S, Reynaerts A, Leemans J (1990) Potato plants with engineered resistance against insect attack. In: Vayda M, Park W (eds) Molecular and cellular biology of the potato. CAB, Tucson, pp 193–204

Penfield S (2008) Temperature perception and signal transduction in plants. New Phytol 179:615–628

Peng JS, Ding G, Meng S, Yi HY, Gong JM (2017) Enhanced metal tolerance correlates with heterotypic variation in SpMTL, a metallothionein-like protein from the hyperaccumulator *Sedum plumbizincicola*. Plant Cell Environ 40(8):1368–1378

Pental D, Pradhan AK, Sodhi SY, Mukhopadhyay A (1993) Variations among *Brassica juncea* cultivars for regeneration from hypocotyl explants and optimization of conditions for Agrobacterium mediated genetic transformation. Plant Cell Rep 12:462–467

Per TS, Khan NA, Reddy PS, Masood A, Hasanuzzaman M, Khan MIR, Anjum AN (2017) Approaches in modulating proline metabolism in plants for salt and drought stress tolerance: phytohormones, mineral nutrients and transgenics. Plant Physiol Biochem 115:126–140

Pérez C, Fernandez LE, Sun J, Folch JL, Gill SS, Soberón M, Bravo A (2005) *Bacillus thuringiensis* subsp. israelensis yt1Aa synergizes Cry11Aa toxin by functioning as a membrane-bound receptor. Proc Natl Acad Sci U S A 102:18303–18308

Perl A, Shaul O, Galili G (1992) Regulation of lysine synthesis in transgenic potato plants expressing a bacterial dihydrodipicolinate synthase in their chloroplasts. Plant Mol Biol 19: 815–823

Perlak FJ, Fischhoff DA (1993) Insect resistant cotton: from the laboratory to the marketplace. In: Kim L (ed) Advanced engineered pesticides. Marcel Dekker, New York, pp 199–211

Perlak FJ, Deaton RW, Armstrong TA, Fuchs RL, Sims SR, Greenplate JT, Fischhoff DA (1990) Insect-resistant cotton plants. Bio/Technology 8:939–943

Perlak FJ, Fuchs RL, Dean DA, McPherson SL, Fischhoff DA (1991) Modification of the coding sequences enhances plant expression of insect control protein genes. Proc Natl Acad Sci U S A 88:3324–3328

Perlak FJ, Stone TB, Muskopf YN, Petersen LJ, Parker GB, McPherson SA, Wyman J, Love S, Reed G, Biever D, Fischhoff DA (1993) Genetically improved potatoes: protection from damage by Colorado potato beetles. Plant Mol Biol 22:313–321

Perlak FJ, Oppenhuizen M, Gustafson K, Voth R, Sivasupramaniam S, Heering D, Carey B, Ihrig RA, Roberts JK (2001) Development and commercial use of Bollgard cotton in the USA-early promises versus today's reality. Plant J 27:489–501

Pfeiffer WH, McClafferty B (2007) Harvestplus: breeding crops for better nutrition. Crop Sci 47 (S3):S38–S115

Piao HL, Lim JH, Kim SJ, Cheong GW, Hwang I (2001) Constitutive over-expression of AtGSK1 induces NaCl stress responses in the absence of NaCl stress and results in enhanced NaCl tolerance in Arabidopsis. Planta 27:305–314

Pigott CR, Ellar DJ (2007) Role of receptors in *Bacillus thuringiensis* crystal toxin activity. Microbiol Mol Biol Rev 71:255–281

Pilion-Smits EAH, Hwang S, Lytle CM, Zhu Y, Tai JC, Bravo RC, Chen Y, Leustek T, Terry N (1999) Over expression of ATP sulfurylase in Indian mustard leads to increased selenate uptake, reduction and tolerance. Plant Physiol 119:123–132

Pixley K, Palacios-Rojas N, Babu R, Mutale R, Surles R, Simpungwe E (2013) Biofortification of maize with provitamin A carotenoids. In: Tanumihardjo S (ed) Carotenoids and human health. Springer Science + Business Media, New York, pp 271–292

Pizzo E, Di Maro A (2016) A new age for biomedical applications of Ribosome Inactivating Proteins (RIPs): from bioconjugate to nanoconstructs. J Biomed Sci 23:54

Pons E, Alquézar B, Rodríguez A, Martorell P, Genoves S, Ramon D, Rodrigo MJ, Zacar L, Pena L (2014) Metabolic engineering of β-carotene in orange fruit increases its in vivo antioxidant properties. Plant Biotechnol J 12:17–27

Poolman MG, Fell DA, Thomas S (2000) Modelling photosynthesis and its control. J Exp Bot 51:319–328

Potrykus I (2017) The GMO-crop potential for more, and more nutritious food is blocked by unjustified regulation. J Innov Knowl 2:90–96

Powell PA, Stark DM, Sanders PR, Beachy RN (1989) Protection against tobacco mosaic virus in transgenic plants that express tobacco mosaic virus antisense RNA. Proc Natl Acad Sci U S A 86:6949–6952

Powell PA, Sanders PR, Turner N, Fraley RT, Beachy RN (1990) Protection against tobacco mosaic virus infection in transgenic plants requires accumulation of coat protein rather than coat protein RNA sequences. Virology 175:124–130

Prabhu S, Dennison SR, Mura M, Lea RW, Snape TJ, Harris F (2014) Cn-AMP2 from green coconut water is an anionic anticancer peptide. J Pept Sci 20:909–915

Pradhan SK, Nayak DK, Mohanty S, Behera L, Barik SR, Pandit E, Lenka S, Anandan A (2015) Pyramiding of three bacterial blight resistance genes for broad-spectrum resistance in deepwater rice variety, Jalmagna. Rice (N Y) 8:5

Pränting M, Lööv C, Burman R, Göransson U, Andersson DI (2010) The cyclotide cycloviolacin O_2 from *Viola odorata* has potent bactericidal activity against Gram-negative bacteria. J Antimicrob Chemother 65:1964–1971

Prasad KVSK, Sharmila P, Pardha Saradhi P (2000) Enhanced tolerance of transgenic *Brassica juncea* to choline confirms successful expression of the bacterial *codA* gene. Plant Sci 159:233–242

Priem B, Gitti R, Bush CA, Gross KC (1993) Structure of ten free N-glycans in ripening tomato fruit (arabinose is a constituent of a plant N-glycan). Plant Physiol 102:445–458

Prodhan MZH, Hasan MT, Chowdhury MMI, Alam MS, Rahman ML, Azad AK (2018) Bt eggplant (*Solanum melongena* L.) in Bangladesh: fruit production and control of eggplant fruit and shoot borer (Guenée), effects on non-target arthropods and economic returns. PLoS One 13:e0205713

Purcell JP, Oppenhuizen M, Wofford T, Reed AJ, Perlak FJ (2004) The story of Bollgard cotton. In: Christou P, Klee H (eds) Handbook of plant biotechnology. Wiley, Hoboken, NJ, pp 1147–1163

Pusztai A (2002) Can science give us the tools for recognizing possible health risk of GM? Nutr Health 16:73–84

Pyott DE, Sheehan E, Molnar A (2016) Engineering of CRISPR/Cas9-mediated potyvirus resistance in transgene-free *Arabidopsis* plants. Mol Plant Pathol 17:1276–1288

Qi YB, Ye SH, Lu YT, Jin QS, Zhang XM (2009) Development of marker-free transgenic Cry1Ab rice with lepidopteran pest resistance by *Agrobacterium* mixture-mediated cotransformation. Rice Sci 16:181–186

Qian Q, Huang L, Yi R, Wang SZ, Ding Y (2014) Enhanced resistance to blast fungus in rice (Oryza sativa L.) by expressing the ribosome-inactivating protein alpha-momorcharin. Plant Sci 217–218:1–7

Qin G, Gu H, Ma L, Peng Y, Deng XW, Chen Z, Qu LJ (2007) Disruption of phytoene desaturase gene results in albino and dwarf phenotypes in Arabidopsis by impairing chlorophyll, carotenoid, and gibberellin biosynthesis. Cell Res 17(5):471–482

Qin H, Cai Y, Liu Z, Wang G, Wang J, Guo Y, Wang H (2012) Identification of QTL for zinc and iron concentration in maize kernel and cob. Euphytica 187:345–358

Qin N, Xu W, Hu L, Li Y, Wang H, Qi X, Fang Y, Hua X (2016) Drought tolerance and proteomics studies of transgenic wheat containing the maize C4 phosphoenolpyruvate carboxylase. Protoplasma 253(6):1503–1512

Qiu Y, Yu D (2009) Over-expression of the stress-induced OsWRKY45 enhances disease resistance and drought tolerance in *Arabidopsis*. Environ Exp Bot 65:35–47

Qiu L, Fan J, Zhang B, Liu L, Wang X, Lei C, Lin Y, Ma W (2017) RNA interference knockdown of aminopeptidase N genes decrease the susceptibility of *Chilo suppressalis* larvae to Cry1Ab/Cry1Ac and Cry1Ca-expressing transgenic rice. J Invertebr Pathol 145:9–12

Qiu L, Sun Y, Jiang Z, Yang P, Liu H, Zhou H, Wang X, Zhang W, Lin Y, Ma W (2019) The midgut V-ATPase subunit A gene is associated with toxicity to crystal 2Aa and crystal 1Ca-expressing transgenic rice in *Chilo suppressalis*. Insect Mol Biol 28:520–527

Qiu D, Hu W, Zhou Y, Xiao J, Hu R, Wei Q, Zhang Y, Feng J, Sun F, Sun J, Yang G (2021) TaASR1-D confers abiotic stress resistance by affecting ROS accumulation and ABA signalling in transgenic wheat. Plant Biotechnol J 19(8):1588–1601

Qu LQ, Yoshihara T, Ooyama A, Goto F, Takaiwa F (2005) Iron accumulation does not parallel the high expression level of ferritin in transgenic rice seeds. Planta 222(2):225–233

Qu J, Ye J, Fang R (2007) Artificial microRNA–mediated virus resistance in plants. J Virol 81:6690–6699

Quan R, Shang M, Zhang H (2004) Engineering of enhanced glycinebetaine synthesis improves drought tolerance in maize. Plant Biotechnol J 2:477–486

Quan R, Hu S, Zhang Z, Zhang H, Zhang Z, Huang R (2010) Overexpression of an ERF transcription factor TSRF1 improves rice drought tolerance. Plant Biotechnol J 8:476–488

Quist D, Chapela IH (2001) Transgenic DNA introgressed into traditional maize landraces in Oaxaca, Mexico. Nature 414:541–543

Rahman M, Rashid H, Shahid AA, Bashir K, Husnain T, Riazuddin S (2007) Insect resistance and risk assessment studies of advanced generations of basmati rice expressing two genes of *Bacillus thuringiensis*. Electron J Biotechnol. https://doi.org/10.2225/vol10-issue2-fulltext-3

Rai GK, Rai NP, Rathaur S, Kumar S, Major S (2013) Expression of rd29A:AtDREB1A/CBF3 in tomato alleviates drought-induced oxidative stress by regulating key enzymatic and non-enzymatic antioxidants. Plant Physiol Biochem 69:90–100

Raines CA, Paul MJ (2006) Products of leaf primary carbon metabolism modulate the developmental programme determining plant morphology. J Exp Bot 57:1857–1862

Raj SK, Rachana S, Pandey SK, Singh BP (2005) *Agrobacterium*-mediated tomato transformation and regeneration of transgenic plants expressing tomato leaf curl virus coat protein gene for resistance against TLCV infection. Curr Sci 33:1674–1679

Ralley L, Schuch W, Fraser PD, Bramley PM (2016) Genetic modification of tomato with the tobacco lycopene beta-cyclase gene produces high beta-carotene and lycopene fruit. Z Naturforsch C 71:295–301

Ramesh S, Nagadhara D, Reddy VD, Rao KV (2004) Production of transgenic indica rice resistant to yellow stem borer and sap-sucking insects, using super-binary vectors of *Agrobacterium tumefaciens*. Plant Sci 166(4):1077–1085

Rani S, Sharma V, Hada A, Bhattacharya RC, Koundal KR (2017a) Fusion gene construct preparation with lectin and protease inhibitor genes against aphids and efficient genetic transformation of *Brassica juncea* using cotyledons explants. Acta Physiol Plant 39:115

Rani S, Sharma V, Hada A, Koundal KR (2017b) Efficient genetic transformation of *Brassica juncea* with lectin using cotyledon explants. Int J Adv Biotechnol Res 7:1–12

Rao B, Rao V, Nair L, Prasad Y, Ramaragan A, Chattopadhyay C (2014) Mustard aphid infestation in India: development of forewarning models. J Environ Biol 35:683–688

Rashid MA, Hasan MK, Matin MA (2018) Socio-economic performance of Bt eggplant cultivation in Bangladesh. Bang J Agric Res 43:187–203

Ravanello MP, Ke D, Alvarez J, Huang B, Shewmaker CK (2003) Coordinate expression of multiple bacterial carotenoid genes in canola leading to altered carotenoid production. Metab Eng 5:255–263

Ray K, Bisht NC, Pental D, Burma PK (2007) Development of barnase/barstar transgenics for hybrid seed production in Indian oilseed mustard (*Brassica juncea* L. Czern & Coss) using a mutant acetolactate synthase gene conferring resistance to imidazolinone-based herbicide 'Pursuit'. Curr Sci 93(10):1390–1396

Redden R, Vardy M, Edwards D, Raman H, Batley J (2009) Genetic and morphological diversity in the Brassicas and wild relative. In: 16th Australian research assembly on Brassicas, Ballarat, Victoria. https://www.cropwildrelatives.org/fileadmin/www.cropwildrelaves.org/documents/Diversity%2520in%2520Brassica%2520CWR.pdf

Register JC III, Beachy RN (1988) Resistance to TMV in transgenic plants results from interference with an early event in infection. Virology 166:524–532

Reinhart BJ, Weinstein EG, Rhoades MW, Bartel B, Bartel DP (2002) MicroRNAs in plants. Genes Dev 16:1616–1626

Rensburg JBJ (2007) First report of field resistance by the stem borer, *Busseola fusca* (Fuller) to Bt-transgenic maize. S Afr J Plant Soil 27:147–151

Retzl B, Hellinger R, Muratspahić E, Pinto MEF, Bolzani VS, Gruber CW (2020) Discovery of a beetroot protease inhibitor to identify and classify plant derived cystine knot peptides. J Nat Prod 83:3305–3314

Rhoades JD, Loveday J (1990) Salinity in irrigated agriculture. In: Steward BA, Nielsen DR (eds) American Society of Civil Engineers, Irrigation of agricultural crops (Monograph 30). American Society of Agronomists, Washington, pp 1089–1142

Riaz N, Husnain T, Fatima T, Makhdoom R, Bashir K, Masson L, Altosaar I, Riazuddin S (2006) Development of Indica Basmati rice harboring two insecticidal genes for sustainable resistance against lepidopteran insects. South African J Bot 72(2):217–223

Riaz M, Arif MS, Ashraf MA, Mahmood R, Yasmeen T, Shakoor MB, Shahzad SM, Ali M, Saleem I, Arif M, Fahad S (2019) A comprehensive review on rice responses and tolerance to salt stress. In: Mirza H (ed) Advances in rice research for abiotic stress tolerance. Woodhead Publishing, Cambridge, pp 133–158

Riaz S, Nasir IA, Bhatti MU, Adeyinka OS, Toufiq N, Yousaf I, Tabassum B (2020) Resistance to *Chilo infuscatellus* (Lepidoptera: Pyraloidea) in transgenic lines of sugarcane expressing *Bacillus thuringiensis* derived Vip3A protein. Mol Biol Rep 47:1–10

Richard JA, Kelly I, Marion D, Pézolet M, Auger M (2002) Interaction between β-purothionin and dimyristoylphosphatidylglycerol: a 31P-NMR and infrared spectroscopic study. Biophys J 83 (4):2074–2083

Risk JM, Selter LL, Chauhan H, Krattinger SG, Kumlehn J, Hensel G, Viccars LA, Richardson TM, Buesing G, Troller A et al (2013) The wheat Lr34 gene provides resistance against multiple fungal pathogens in barley. Plant Biotechnol J11:847–854

Roberts W, Stewart TS (1979) Purification and properties of a translational inhibitor from wheat germ. Biochemistry 18:2615–2621

Robineau T, Batard Y, Nedelkina S, Cabello-Hurtado F, LeRet M, Sorokine O, Didierjean L, Werck-Reichhart D (1998) The chemically inducible plant cytochrome P450 CYP76B1 actively metabolizes phenylureas and other xenobiotics. Plant Physiol 118(3):1049–1056

Roderick H, Tripathi L, Babirye A, Wang D, Tripathi J, Urwin PE (2012) Generation of transgenic plantain (Musa spp.) with resistance to plant pathogenic nematodes. Mol Plant Pathol 13 (8):842–851

Rodrigo-Simon A, de Maagd RA, Avilla C, Bakker PL, Molthoff J, Gonzalez-Zamora JE, Ferre J (2006) Lack of detrimental effects of *Bacillus thuringiensis* Cry toxins on the insect predator *Chrysoperla carnea*: a toxicological, histopathological, and biochemical analysis. Appl Environ Microbiol 72:1595–1603

Rodríguez-Almazán C, Zavala LE, Muñoz-Garay C, Jiménez-Juárez N, Pacheco S, Masson L et al (2009) Dominant negative mutants of Bacillus thuringiensis Cry1Ab toxin function as anti-toxins: demonstration of the role of oligomerization in toxicity. PLoS One 4(5):e5545

Rodríguez M, Canales E, Borrás-Hidalgo O (2005) Molecular aspects of abiotic stress in plants. Biotechnol Appl 22:1–10

Rogers EE, Eide DJ, Guerinot ML (2000) Altered selectivity in an Arabidopsis metal transporter. Proc Natl Acad Sci U S A 97(22):12356–12360

Rojas-González JA, Soto-Súarez M, García-Díaz Á, Romero-Puertas MC, Sandalio LM, Mérida Á, Thormählen I, Geigenberger P, Serrato AJ, Sahrawy M (2015) Disruption of both chloroplastic and cytosolic FBPase genes results in a dwarf phenotype and important starch and metabolite changes in Arabidopsis thaliana. J Exp Bot 66:2673–2689

Romeis J, Dutton A, Bigler F (2004) *Bacillus thuringiensis* toxin (Cry1Ab) has no direct effect on larvae of the green lacewing *Chrysoperla carnea* (Stephens) (Neuroptera: Chrysopidae). J Insect Physiol 50(2–3):175–183

Romer S, Lubeck J, Kauder F, Steiger S, Adomat C, Sandmann G (2002) Genetic engineering of a zeaxanthin-rich potato by antisense inactivation and co suppression of carotenoid epoxidation. Metab Eng 4:263–272

Romero C, Belles JM, Vaiya JL, Serrano R, Macia FAC (1997) Expression of the yeast trechalose-6-phosphate synthetase gene in transgenic tobacco plants: pleiotropic phenotypes include drought tolerance. Planta 201:293–297

Rong J, Lu BR, Song Z, Su J, Snow AA, Zhang X, Sun S, Chen R, Wang F (2007) Dramatic reduction of crop-to-crop gene flow within a short distance from transgenic rice fields. New Phytol 173:346–353

Roosens N, Hal Bitar F, Loenders K, Angenon G, Jacobs M (2002) Overexpression of ornithine-δ-aminotransferase increases proline biosynthesis and confers osmotolerance in transgenic plants. Mol Breed 9:73–80

Rosa C, Kuo YW, Wuriyanghan H, Falk BW (2018) RNA interference mechanisms and applications in plant pathology. Annu Rev Phytopathol 56:581–610

Rosati C, Aquilani R, Dharmapuri S, Pallara P, Marusic C, Tavazza R, Bouvier F, Camara B, Giuliano G (2000) Metabolic engineering of beta-carotene and lycopene content in tomato fruit. Plant J 24:413–419

Rosenthal DM, Locke AM, Khozaei M, Raines CA, Long SP, Ort DR (2011) Over-expressing the C3 photosynthesis cycle enzyme Sedoheptulose-1-7 Bisphosphatase improves photosynthetic carbon gain and yield under fully open air CO2fumigation (FACE). BMC Plant Biol 11(1):1–2

Rosenzweig C, Elliott J, Deryng D et al (2014) Assessing agricultural risks of climate change in the 21st century in a global gridded crop model intercomparison. Proc Natl Acad Sci U S A 111: 3268–3273

Rosgen J (2007) Molecular basis of osmolyte effects on protein and metabolites. Methods Enzymol 428:459–486

Roux F, Voisin D, Badet T, Balagué C, Barlet X, Huard-Chauveau C (2014) Resistance to phytopathogens e tutti quanti: placing plant quantitative disease resistance on the map. Mol Plant Pathol 15:427–432

Roxas VP, Lodhi SA, Garrett DK, Mahan JR, Allen RD (2000) Stress tolerance in transgenic tobacco seedlings that overexpress glutathione S-transferase/glutathione peroxidase. Plant Cell Physiol 41:1229–1234

Roy-Barman S, Sautter C, Chattoo BB (2006) Expression of the lipid transfer protein Ace-AMP1 in transgenic wheat enhances antifungal activity and defense responses. Transgenic Res 15 (4):435–446

Römer S, Fraser PD, Kiano JW, Shipton CA, Misawa N, Schuch W, Bramley PM (2000) Elevation of the provitamin A content of transgenic tomato plants. Nat Biotechnol 18(6):666–669

Ruiz-Sola MÁ, Rodríguez-Concepción M (2012) Carotenoid biosynthesis in Arabidopsis: a colorful pathway. Arabidopsis Book 10:e0158

Ruiz-Sola MÁ, Coman D, Beck G, Barja MV, Colinas M, Graf A, Welsch R, Rütimann P, Bühlmann P, Bigler L, Gruissem W (2016) Arabidopsis GERANYLGERANYL DIPHOSPHATE SYNTHASE 11 is a hub isozyme required for the production of most photosynthesis-related isoprenoids. New Phytol 209(1):252–264

Ruoslahti E (1996) RGD and other recognition sequences for integrins. Annu Rev Cell Dev Biol 12:697–715

Ruta LL, Banu MA, Neagoe AD, Kissen R, Bones AM, Farcasanu IC (2018) Accumulation of Ag (I) by *Saccharomyces cerevisiae* cells expressing plant metallothioneins. Cell 7(12):266

Rühl R, Sczech R, Landes N, Pfluger P, Kluth D, Schweigert FJ (2004) Carotenoids and their metabolites are naturally occurring activators of gene expression via the pregnane X receptor. Eur J Nutr 43(6):336–343

Ryan GF (1970) Resistance of common groundsel to simazine and atrazine. Weed Sci 18(5):614–616

Sadiq M, Akram NA (2018) Field performance of transgenic drought-tolerant crop plants. In: Biochemical, physiological and molecular avenues for combating abiotic stress tolerance in plants. Academic, San Diego, pp 83–102

Saeed F, Hashmi MH, Hossain J, Ali MA, Bakhsh A (2020) Transgenic technologies for efficient insect pest management in crop plants. In: Transgenic technology based value addition in plant biotechnology. Elsevier, Amsterdam, pp 123–156

Safari E, Gororo N, Eastwood R, Lewis J, Eagles H, Ogbonnaya F (2005) Impact of Cre1, Cre8 and Cre3 genes on cereal cyst nematode resistance in wheat. Theor Appl Genet 110:567–572

Safi H, Saibi W, Alaoui MM, Hmyene A, Masmoudi K, Hanin M, Brini F (2015) A wheat lipid transfer protein (TdLTP4) promotes tolerance to abiotic and biotic stress in Arabidopsis thaliana. Plant Physiol Biochem 89:64–75

Saha B, Mishra S, Awasthi JP, Sahoo L, Panda SK (2016) Enhanced drought and salinity tolerance in transgenic mustard [*Brassica juncea* (L.) Czern & Coss.] overexpressing Arabidopsis group 4 late embryogenesis abundant gene (AtLEA4-1). Environ Exp Bot 128:99–111

Saharan GS, Verma PR, Meena PD, Kumar A (2014) White rust of crucifers: biology, ecology and management. Springer, New Delhi

Sahrawy M, Avila C, Chueca A, Cánovas FM, López-Gorgé J (2004) Increased sucrose level and altered nitrogen metabolism in Arabidopsis thaliana transgenic plants expressing antisense chloroplastic fructose-1,6-bisphosphatase. J Exp Bot 55:2495–2503

Saijo Y, Hata S, Kyozuka J, Shimamoto K, Izui K (2000) Overexpression of a single Ca^{2+} dependent protein kinase confers both cold and salt/drought tolerance on rice plants. Plant J 23:319–327

Saini RK, Nile SH, Park SW (2015) Carotenoids from fruits and vegetables: chemistry, analysis, occurrence, bioavailability and biological activities. Food Res Int 76:735–750

Saini DK, Devi P, Kaushik P (2020) Advances in genomic interventions for wheat biofortification: a review. Agronomy 10(1):62

Sakamoto A, Alia, Murata N (1998) Metabolic engineering of rice leading to biosynthesis of glycinebetaine and tolerance to salt and cold. Plant Mol Biol 38:1011–1019

Salehi Jouzani G, Seifinejad A, Saeedizadeh A, Nazarian A, Yousefloo M, Soheilivand S, Mousivand M, Jahangiri R, Yazdani M, Amiri RM, Akbari S (2008) Molecular detection of nematicidal crystalliferous Bacillus thuringiensis strains of Iran and evaluation of their toxicity on free-living and plant-parasitic nematodes. Can J Microbiol 54(10):812–822

Salehian H, Rahnama H, Dezhsetan S, Babaei S (2021) Constitutive expression of a synthetic cry1Ab gene confers resistance to potato tuber moth (Phthorimaea operculella Zeller) larva. Crop Breed Appl Biotechnol 21

Sami Z, Alemzadeh A (2016) Isolation and molecular characterization of a novel Na+/H+ antiporter gene, AlNHX2, from 'Aeluropus littoralis' and comparison of AlNHX1 and AlNHX2. Plant Omics 9(3):205

Sanahuja G, Banakar R, Twyman R, Capell T, Christou P (2011) *Bacillus thuringiensis*: a century of research, development and commercial applications. Plant Biotechnol J 9:283–300

Sanchez-Casas P, Klessig DF (1994) A salicylic acid-binding activity and a salicylic acid-inhibitable catalase activity are present in a variety of plant species. Plant Physiol 106 (4):1675–1679

Sandjaja, Jahari AB, Htet MK, Tilden RL, Soekarjo D, Utomo B, Moench-Pfanner R, Korenromp EL (2015) Vitamin A-fortified cooking oil reduces vitamin A deficiency in infants, young children and women: results from a programme evaluation in Indonesia. Public Health Nutr 18 (14):2511–2522

Sanford JC (1988) The biolistic process. Trends Biotechnol 6:299–302

Sanford JC, Johnston SA (1985) The concept of parasite derived resistance-deriving resistance genes from the parasite's own genome. J Theor Biol 113:395–405

Sangadala S, Walters FS, English LH, Adang MJ (1994) A mixture of *Manduca sexta* aminopeptidase and phosphatase enhances *Bacillus thuringiensis* insecticidal CryIA(c) toxin binding and 86Rb(+)-K+ efflux in vitro. J Biol Chem 269:10088–10092

Sanyal I, Singh AK, Kaushik M, Amla DV (2005) *Agrobacterium*-mediated transformation of chickpea (*Cicer arietinum* L.) with *Bacillus thuringiensis cry*1Ac gene for resistance against pod borer insect *Helicoverpa armigera*. Plant Sci 168:1135–1146

Sardana R, Dukiandjiev S, Giband M, Cheng X, Cowan K, Sauder C, Altosaar I (1996) Construction and rapid testing of synthetic and modified toxin gene sequences Cry1A (b & c) by expression in maize endosperm culture. Plant Cell Rep 15:677–681

Sarkar T, Radhakrishnan T, Kumar A, Mishra GP, Dobaria JR (2014a) Heterologous expression of AtDREB1A gene in transgenic peanut conferred tolerance to drought and salinity stresses. PLoS One 9(12):e110507

Sarkar P, Jana K, Sikdar SR (2014b) Overexpression of biologically safe *Rorippa indica* defensin enhances aphid tolerance in *Brassica juncea*. Planta 246:1029–1044

Sarkar T, Radhakrishnan T, Kumar A, Mishra GP, Dobaria JR (2016) Stress inducible expression of AtDREB1A transcription factor in transgenic peanut (*Arachis hypogea* L.) conferred tolerance to soil-moisture deficit stress. Front Plant Sci 7:935

Sarkar T, Thankappan R, Mishra GP, Nawade BD (2019) Advances in the development and use of DREB for improved abiotic stress tolerance in transgenic crop plants. Physiol Mol Biol Plants 25(6):1323–1334

Sarowar S, Kim YJ, Kim EN, Kim KD, Hwang BK, Islam R, Shin JS (2005) Overexpression of a pepper basic pathogenesis-related protein 1 gene in tobacco plants enhance resistance to heavy metal and pathogen stresses. Plant Cell Rep 24:216–224

Sasaki T, Yamamoto Y, Ezaki B, Katsuhara M, Ahn S, Ryan PR, Delhaize E, Matsumoto H (2004) A wheat gene encoding an aluminum-activated malate transporter. Plant J 37:645–653

Savary S, Willocquet L, Pethybridge SJ, Esker P, McRoberts N, Nelson A (2019) The global burden of pathogens and pests on major food crops. Nat Ecol Evol 3:430–439

Sawant SV, Srivastava R, Srivastava M, Singh PK, Srivastava S, Verma PC, Koul B, Shukla AK, Nautiyal CS, Amla DV (2010) A method for developing pathogen tolerant transgenic plants. Indian Patent No. 3177DEL2010

Sawardekar SV, Katageri IS, Salimath PM, Kumar PA, Kelkar VG (2017) Standardization of in-vitro genetic transformation technique in chickpea (*Cicer arietinum* L.) for pod-borer resistance. Adv Agric Res Technol J 1:2

Sayre R, Beeching JR, Cahoon EB, Egesi C, Fauquet C, Fellman J, Fregene M, Gruissem W, Mallowa S, Manary M, Maziya-Dixon B, Mbanaso A, Schachtman DP, Siritunga D, Taylor N, Vanderschuren H, Zhang P (2011) The BioCassava plus program: biofortification of cassava for sub-Saharan Africa. Annu Rev Plant Biol 62:251–272

Scandalios JG (1993) Oxygen stress and superoxide dismutases. Plant Physiol 101:7–12

Schäfer E, Bowler C (2002) Phytochrome-mediated photoperception and signal transduction in higher plants. EMBO Rep 3:1042–1048

Schaffer RJ, Ireland HS, Ross JJ, Ling TJ, David KM (2013) SEPALLATA1/2-suppressed mature apples have low ethylene, high auxin and reduced transcription of ripening-related genes. AoB Plants 5:pls047

Schmidt M, Arendt EK, Thery TLC (2019) Isolation and characterization of the antifungal activity of the cowpea defensin Cp-thionin II. Food Microbiol 82:504–514

Schmitz RJ, Amasino RM (2007) Vernalization: a model for investigating epigenetics and eukaryotic gene regulation in plants. Biochim Biophys Acta 1769:269–275

Schnepf EN, Crickmore J, Van Rie D, Lereclus J, Baum J, Feitelson DR, Zeigler DH, Dean DH (1998) *Bacillus thuringiensis* and its pesticidal crystal proteins. Microbiol Mol Biol Rev 62: 775–806

Schoonbeek H, Wang H, Stefanato FL, Craze M, Bowden S, Wallington E, Zipfel C, Ridout CJ (2015) Arabidopsis EF-Tu receptor enhances bacterial disease resistance in transgenic wheat. New Phytol 206:606–613

Schöttler MA, Tóth SZ, Boulouis A, Kahlau S (2015) Photosynthetic complex stoichiometry dynamics in higher plants: biogenesis, function, and turnover of ATP synthase and the cytochrome b 6 f complex. J Exp Bot 66(9):2373–2400

Schramm F, Larkindale J, Kiehlmann E, Ganguli A, Englich G, Vierling E, von Koskull-Doring P (2008) A cascade of transcription factor DREB2A and heat stress transcription factor HsfA3 regulates the heat stress response of Arabidopsis. Plant J 53:264–274

Scott JG, Michel K, Bartholomay LC, Siegfried BD, Hunter WB, Smagghe G, Zhu KY, Douglas AE (2013) Towards the elements of successful insect RNAi. J Insect Physiol 59:1212–1221

Sears M, Hellmich R, Stanley-Horn D, Oberhauser K, Pleasants J, Mattila H, Siegfried B, Dively G (2001) Impact of Bt corn pollen on monarch butterfly populations: a risk assessment. Proc Natl Acad Sci U S A 98(21):11937–11942

Segura A, Moreno M, Madueño F, Molina A, García-Olmedo F (1999) Snakin-1 a peptide from potato that is active against plant pathogens. Mol Plant Microbe Interact 12:16–23

Selale H, Dağlı F, Mutlu N, Doğanlar S, Frary A (2017) Cry1Ac-mediated resistance to tomato leaf miner (*Tuta absoluta*) in tomato. Plant Cell Tiss Org Cult 131:65–73

Sels J, Mathys J, De Coninck BM, Cammue BP, De Bolle MF (2008) Plant pathogenesis-related (PR) proteins: a focus on PR peptides. Plant Physiol Biochem 46:941–950

Selvapandiyan A, Reddy VS, Kumar PA, Tewari KK, Bhatnagar RK (1998) Transformation of *Nicotiana tabacum* with a native cry1Ia5 gene confers complete protection against *Heliothis armigera*. Mol Breed 4(6):473–478

Send R, Sundholm D (2007) The role of the beta-ionone ring in the photochemical reaction of rhodopsin. J Phys Chem 111:27–33

Sera T (2005) Inhibition of virus DNA replication by artificial zinc finger proteins. J Virol 79:2614–2619

Seralini GE, Cellier D, de Vendomois JS (2007) New analysis of a rat feeding study with a genetically modified maize reveals signs of hepatorenal toxicity. Arch Environ Cotam Toxicol 52(4):596–602

Séralini GE, Clair E, Mesnage R, Gress S, Defarge N, Malatesta M, Hennequin D, De Vendômois JS (2012) Long term toxicity of a Roundup herbicide and a Roundup-tolerant genetically modified maize. Food Chem Toxicol:4221–4231

Shahidi-Noghabi S, Van-Damme EJM, Smagghe G (2008) Carbohydrate binding activity of the type-2 ribosome-inactivating protein SNA-I from elderberry (*Sambucus nigra*) is a determining factor for its insecticidal activity. Phytochemistry 69:2972–2978

Shahidi-Noghabi S, Van-Damme EJM, Mahdian K, Smagghe G (2010) Entomotoxic action of Sambucus nigra agglutinin I in Acyrthosiphon pisumaphids and Spodoptera exigua caterpillars through caspase-3 like dependent apoptosis. Arch Insect Biochem Physiol 75:207–220

Shalata A, Neumann PM (2001) Exogenous ascorbic acid (vitamin C) increases resistance to salt stress and reduces lipid peroxidation. J Exp Bot 52:2207–2211

Shao E, Zhang A, Yan Y, Wang Y, Jia X, Sha L (2020) Oligomer formation and insecticidal activity of *Bacillus thuringiensis* Vip3Aa Toxin. Toxins 12:274

Sharma P, Dubey RS (2005) Drought induces oxidative stress and enhances the activities of antioxidant enzymes in growing rice seedlings. Plant Growth Regul 46(3):209–221

Sharma SS, Dietz KJ (2006) The significance of amino acids and amino acid-derived molecules in plant responses and adaptation to heavy metal stress. J Exp Bot 57:711–726

Sharma AK, Sharma MK (2009) Plants as bioreactors: recent developments and emerging opportunities. Biotechnol Adv 27(6):811–832

Sharma A, Sharma R, Imamura M, Yamakawa M, Machii H (2000) Transgenic expression of cecropin B, an antibacterial peptide from Bombyx mori, confers enhanced resistance to bacterial leaf blight in rice. FEBS Lett 484(1):7–11

Sharma HC, Sharma KK, Seetharama N, Crouch JH (2003) The utility and management of transgenic plants with *Bacillus thuringiensis* genes for protection from pests. J New Seeds 5(1):53–76

Sharma N, Park SW, Vepachedu R, Barbieri L, Ciani M, Stirpe F (2004) Isolation and characterization of an RIP (ribosome-inactivating protein)-like protein from tobacco with dual enzymatic activity. Plant Physiol 134:171–181

Sharma KK, Lavanya M, Anjaiah V (2006) *Agrobacterium*-mediated production of transgenic pigeonpea (*Cajanus cajan* L. Millsp.) expressing the synthetic *BT cry1Ab* gene. In Vitro Cell Dev Biol Plant 42:165–173

Sharma R, Sudan J, Sharma R, Kumari S, Salgotra RK, Singh R (2017) Terminal heat stress-responsive genes analysis in heat susceptible HDR77 genotype of wheat (*Triticum aestivum* L.) by using semi-quantitative RTPCR. Electron J Plant Breed 8(4):1124–1132

Shaul O, Galili G (1993) Concerted regulation of lysine and threonine synthesis in tobacco plants expressing bacterial feedback-insensitive aspartate kinase and dihydrodipicolinate synthase. Plant Mol Biol 23:759–768

Shelton AM, Hossain MJ, Paranjape V, Azad AK, Raman ML, Khan ASMM (2018) Bt eggplant project in Bangladesh: history, present status and future direction. Front Bioeng Biotechnol 6: 106

Shelton AM, Hossain MJ, Paranjape VMZH, Prodhan AK, Azad Majumder R (2019) Bt brinjal in Bangladesh: the first genetically engineered food crop in a developing country. Cold Spring Harb Perspect Biol 11:13

Shelton AM, Sarwer SH, Hossain MJ, Brookes G, Paranjape V (2020) Impact of Bt brinjal cultivation in the market value chain in five districts of Bangladesh. Front Bioeng Biotechnol 8:498

Shen JX, Wang YY, Zhang YX, Guo W, Jiao QY, Zhou AX (2018) Overexpression of the wild soybean R2R3-MYB transcription factor GsMYB15 enhances resistance to salt stress and *Helicoverpa Armigera* in transgenic Arabidopsis. Int J Mol Sci 19:1–19

Sheveleva E, Chmara W, Bohnert HJ, Jensen RG (1997) Increased salt and drought tolerance by D-ononitol production in transgenic *Nicotiana tabacum* L. Plant Physiol 115:1211–1219

Shewmaker C, Sheehy J, Daley M, Colburn S, Ke DY (1999) Seed-specific overexpression of phytoene synthase: increase in carotenoids and other metabolic effects. Plant J 20:401–412X

Shi H, Lee BH, Wu SJ, Zhu JK (2003) Overexpression of a plasma membrane Na^+/H^+ antiporter gene improves salt tolerance in *Arabidopsis thaliana*. Nat Biotechnol 21:81–85

Shi J, Gao H, Wang H, Lafitte HR, Archibald RL, Yang M, Hakimi SM, Mo H, Habben JE (2017) ARGOS 8 variants generated by CRISPR-Cas9 improve maize grain yield under field drought stress conditions. Plant Biotechnol J 15(2):207–2016

Shiferaw B, Prasanna BM, Hellin J, Bänziger M (2011) Crops that feed the world 6. Past successes and future challenges to the role played by maize in global food security. Food Secur 3(3):307–327

Shimamoto K, Terada R, Izawa T (1989) Fertile transgenic rice plants regenerated from transformed protoplasts. Nature 388:274–276

Shimamura C, Ohno R, Nakamura C, Takumi S (2006) Improvement of freezing tolerance in tobacco plants expressing a cold-responsive and chloroplast-targeting protein WCOR15 of wheat. Plant Physiol 163:213–219

Shou HX, Bordallo P, Wang K (2004) Expression of *Nicotiana* protein kinase (NPK1) enhance drought tolerance in transgenic maize. J Exp Bot 55:1013–1019

Shrivastav S (2017) BG-III cotton illegally grown in Yavatmal. The Times of India. https://timesofindia.indiatimes.com/city/nagpur/bg-iiicotton-illegally-grown-inyavatmal/articlesho

Shu Q, Ye G, Cui H, Cheng X, Xiang Y, Wu D, Gao M, Xia Y, Hu C, Sardana R, Altosaar I (2000) Transgenic rice plants with a synthetic *cry1Ab* gene from *Bacillus thuringiensis* were highly resistant to eight lepidopteran rice pest species. Mol Breed 6(4):433–439

Siegel JP (2000) Bacteria. In: Lacey LL, Kaya HK (eds) Field manual of techniques in invertebrate pathology. Kluwer Scientific Publishers, Dordrecht, Netherlands, pp 209–230

Siegfried BD, Hellmich RL (2012) Understanding successful resistance management: the European corn borer and Bt corn in the United States. GM Crops Food 3:184–193

Sieglaff DH, Ellsworth PC, Silvertooth JC, Hamilton (1999) Preliminary evaluation of the next generation of Bt cotton. http://cals.arizona.edu/pubs/crops/az1123/az1237c.pdf

Siehl DL, Castle LA, Gorton R, Chen YH, Bertain S, Cho H-J, Keenan R, Liu D, Lassner MW (2005) Evolution of a microbial acetyltransferase for modification of glyphosate: a novel tolerance strategy. Pest Manag Sci 61:235–240

Siehl DL, Castle LA, Gorton R, Keenan RJ (2007) The molecular basis of glyphosate resistance by an optimized microbial acetyltransferase. J Biol Chem 282:11446–11455

Simic D, Mladenovic DS, Zdunic Z, Jambrovic A, Ledencan T, Brkic J, Brkic I (2012) Quantitative trait Loci for biofortification traits in maize grain. J Hered 103:47–54

Simkin AJ, Gaffé J, Alcaraz JP, Carde JP, Bramley PM, Fraser PD, Kuntz M (2007) Fibrillin influence on plastid ultrastructure and pigment content in tomato fruit. Phytochemistry 68: 1545–1556

Simkin AJ, McAusland L, Headland LR, Lawson T, Raines CA (2015) Multigene manipulation of photosynthetic carbon assimilation increases CO2 fixation and biomass yield in tobacco. J Exp Bot 66:4075–4090

Simkin AJ, Lopez-Calcagno PE, Davey PA, Headland LR, Lawson T, Timm S, Bauwe H, Raines CA (2017a) Simultaneous stimulation of sedoheptulose 1,7-bisphosphatase, fructose 1,6-bisphophate aldolase and the photorespiratory glycine decarboxylase-H protein increases CO2 assimilation, vegetative biomass and seed yield in Arabidopsis. Plant Biotechnol J 15:805–816

Simkin AJ, McAusland L, Lawson T, Raines CA (2017b) Overexpression of the RieskeFeS protein increases electron transport rates and biomass yield. Plant Physiol 175:134–145

Simkin AJ, López-Calcagno PE, Raines CA (2019) Feeding the world: improving photosynthetic efficiency for sustainable crop production. J Exp Bot 70(4):1119–1140

Sindelarova M, Sindelar L (2005) Isolation of pathogenesis-related proteins from TMV infected tobacco and their influence on infectivity of TMV. Plant Prot Sci 41:52–57

Singh R, Dwivedi UN (2008) Effect of ethrel and 1-methylcyclopropene (1-MCP) on antioxidants in mango (*Mangifera indica* var. Dashehari) during fruit ripening. Food Chem 111(4):951–956

Singh Z, Jones MGK, Jones RAC (1998) Effectiveness of coat protein and defective replicase gene-mediated resistance against Australian isolates of cucumber mosaic virus. Aust J Exp Agric 38:375–383

Singh PK, Kumar M, Chaturvedi CP, Yadav D, Tuli R (2004) Development of a hybrid δ-endotoxin and its expression in tobacco and cotton for control of a polyphagous pest *Spodoptera litura*. Transgenic Res 14:1–14

Singh VV, Verma V, Pareek AK, Mathur M, Yadav R, Goyal P, Thakur AK, Singh YP, Koundal KR, Bansal KC, Mishra AK, Kumar A, Kumar S (2010) Optimization and development of regeneration and transformation protocol in Indian mustard using lectin gene from chickpea [*Cicer arietinum* (L.)]. J Plant Breed Crop Sci 1(9):306–310

Singh PK, Upadhyay SK, Krishnappa C, Sharad S, Singh R, Rai P, Singh H, Mishra M, Singh AP, Verma PC, Pillai NK, Tuli R (2011) A process preparation of a novel insecticidal chitinase toxic against whiteflies, its encoding nucleotide and application thereof. Indian Patent No.3851/DEL/2011

Singh S, Kumar NR, Maniraj R, Lakshmikanth R, Rao KY, Muralimohan N, Arulprakash T, Karthik K, Shashibhushan NB, Vinutha T, Pattanayak D, Kumar PA, Sreevathsa R (2018) Expression of Cry2Aa, a *Bacillus thuringiensis* insecticidal protein in transgenic pigeon pea confers resistance to gram pod borer, *Helicoverpa armigera*. Sci Rep 8:1–2

Singh P, George PJ, Ramtweke PW, Paul A (2020) Performance of Bollgard I with different plant spacing and in presence of refuge crops. J Pharm Phytochem 9(2):78–81

Singla-Pareek SL, Reddy MK, Sopory SK (2003) Genetic engineering of the glyoxalase pathway in tobacco leads to enhanced salinity tolerance. Proc Natl Acad Sci U S A 100:14672–14677

Singsit C, Adang MJ, Lynch RE, Anderson WF, Wang A, Cardineau G, Ozias-Akins P (1997) Expression of a *Bacillus thuringiensis cry1A*(c) gene in transgenic peanut plants and its efficacy against lesser cornstalk borer. Transgenic Res 6:169–176

Sinha P, Davis J, Saag L, Wanke C, Salgame P, Mesick J (2019) Undernutrition and tuberculosis: public health implications. J Infect Dis 219:1356–1363

Sitaram N (2006) Antimicrobial peptides with unusual amino acid compositions and unusual structures. Curr Med Chem 13:679–696

Slama I, Abdelly C, Bouchereau A, Flowers T, Savouré A (2015) Diversity, distribution and roles of osmoprotective compounds accumulated in halophytes under abiotic stress. Ann Bot 115 (3):433–447

Slater A, Scott N, Fowler M (2003) Plant biotechnology: the genetic manipulation of plants. Oxford University Press, New York, pp 179–203

Slavokhotova AA, Rogozhin EA (2020) Defense peptides from the α-hairpinin family are components of plant innate immunity. Front Plant Sci 11:465

Smart RD, Blum M, Wesseler J (2017) Trends in approval times for genetically engineered crops in the United States and the European Union. J Agric Econ 68(1):182–198

Smirnoff N (2005) Ascorbate, tocopherol and carotenoids: metabolism, pathway engineering and functions. In: Antioxidants and reactive oxygen species in plants, pp 53–86

Smyth SJ, Gusta M, Belcher K, Phillips PW, Castle D (2011a) Changes in herbicide use after adoption of HR canola in Western Canada. Weed Technol 25(3):492–500

Smyth SJ, Gusta M, Belcher K, Phillips PW, Castle D (2011b) Environmental impacts from herbicide tolerant canola production in Western Canada. Agric Syst 104(5):403–410

Sobczak M, Avrova A, Jupowicz J, Phillips MS, Ernst K, Kumar A (2005) Characterization of susceptibility and resistance responses to potato cyst nematode (Globodera spp.) infection of tomato lines in the absence and presence of the broad-spectrum nematode resistance Hero gene. Mol Plant-Microbe Interact 18(2):158–168

Soberon M, Gill SS, Bravo A (2009) Signaling versus punching hole: how do Bacillus thuringiensis toxins kill insect midgut cells? Cell Mol Life Sci 66(8):1337–1349

Soberon M, Pardo L, Monoz C, Sanchez J, Gomez I, Porta H (2010) Pore formation by toxins. In: Andeluh G, Lakey J (eds) Proteins: membrane binding and pore formation. Landes Bioscience and Springer Science, New York, pp 127–142

Somashekara R (2009) Evaluation of first and second generation Bt cotton genotypes for characterization of resistance to bollworms and tobacco caterpillar. MSc Thesis, Univ. Agric. Sci., Dharwad, Karnataka, India

Sonah H, Zhang X, Deshmukh RK, Borha MH, Fernando WG, Bélanger RR (2016) Comparative transcriptomic analysis of virulence factors in Leptosphaeria maculans during compatible and incompatible interactions with canola. Front Plant Sci 7:1784

Song WY, Wang GL, Chen LL, Kim HS, Pi LY, Holsten T, Gardner J, Wang B, Zhai WX, Zhu LH, Fauquet C, Ronald P (1995) A receptor kinase-like protein encoded by the rice disease resistance gene, Xa21. Science 270(5243):1804–1806

Sousa DA, Porto WF, Silva MZ, da Silva TR, Franco OL (2016) Influence of cysteine and tryptophan substitution on DNA-binding activity on maize α-hairpinin antimicrobial peptide. Molecules 21:1062

Souza TL, Faria JC, Aragão FJ, Del Peloso MJ, Faria LC, Wendland A, Aguiar MS, Quintela ED, Melo CL, Hungria M, Vianello RP (2018) Agronomic performance and yield stability of the RNA interference-based Bean golden mosaic virus-resistant common bean. Crop Sci 58 (2):579–591

Sridevi G, Parameswari C, Sabapathi N, Raghupathy V, Veluthambi K (2008) Combined expression of chitinase and β-1,3-glucanase genes in indica rice (*Oryza sativa* L.) enhances resistance against *Rhizoctonia solani*. Plant Sci 175:283–290

Sripriya R, Parameswari C, Veluthambi K (2017) Enhancement of sheath blight tolerance in transgenic rice by combined expression of tobacco osmotin (ap24) and rice chitinase (chi11) genes. In Vitro Cell Dev Biol Plant 53(1):12–21

Srivalli S, Khanna-Chopra R (2008) Role of glutathione in abiotic stress tolerance. In: Sulfur assimilation and abiotic stress in plants. Springer, Berlin, pp 207–225

Srivastava S, Verma HN, Srivastava A, Prasad V (2015) BDP-30, a systemic resistance inducer from Boerhaavia diffusa L., suppresses TMV infection, and displays homology with ribosome-inactivating proteins. J Biosci 40:125–135

Stanley J, Frischmuth T, Ellwood S (1990) Defective viral DNA ameliorates symptoms of geminivirus infection in transgenic plants. Proc Natl Acad Sci USA 87:6291–6295

Stark DM (1997) Risk assessment and criteria for commercial launch of transgenic plants. Proceedings of the 3rd Ifgene Workshop, Dornach, Switzerland

Stark DM, Beachy RN (1989) Protection against potyvirus infection in transgenic plants: evidence for broad spectrum resistance. Bio/Technology 7:1257–1262

Steer AC, Carapetis JR (2004) Acute hematogenous osteomyelitis in children. Pediatr Drugs 6 (6):333–346

Steeves RM, Denton ME, Barnard FC, Henry A, Lambert JM (1999) Identification of three oligosaccharide binding sites in ricin. Biochemist 38:11677–11685

Sten EV, Skov PS, Andersen SB, Torp AM, Olesen A, Bindslev-Jensen U, Lars K BJ, Bindslev-Jensen CA (2004) A comparative study of the allergenic potency of wild-type and glyphosate-tolerant gene-modified soybean cultivars. APMIS 112(1):21–28

Stephen OD, Antonio LC (2010) Transgenics for herbicide resistance. In: Transgenic crop plants. Springer, Berlin, pp 1–34

Stewart SD (2007) Bt cotton. www.utcrops.com/cotton_insects/pubs/Wi29-Btcotton

Stewart CN Jr, Adang MJ, All JN, Boerma HR, Cardineau G, Tucker D, Parrott WA (1996) Genetic transformation, recovery and characterization of fertile soybean transgenic for a synthetic *B. thuringiensis cry1Ac* gene. Plant Physiol 112:121–129

Stirpe F (2004) Ribosome-inactivating proteins. Toxicon 44(4):371–383

Stirpe F (2013) Ribosome-inactivating proteins: from toxins to useful proteins. Toxicon 67:12–16

Stockinger EJ, Skinner JS, Gardner KG, Francia E, Pecchioni N (2007) Expression levels of barley *Cbf* genes at the Frost resistance H2 locus are dependent upon alleles at Fr-H1 and Fr-H2. Plant J 51:308–321

Storer NP, Babcock JM, Schlenz M, Meade T, Thompson GD, Bing JW, Huckaba RM (2010) Discovery and characterization of field resistance to Bt maize: *Spodoptera frugiperda* (Lepidoptera: Noctuidae) in Puerto Rico. J Econ Entomol 103:1031–1038

Storozhenko S, De Brouwer V, Volckaert M, Navarrete O, Blancquaert D, Zhang GF, Lambert W, Van Der Straeten D (2007) Folate fortification of rice by metabolic engineering. Nat Biotechnol 25(11):1277–1279

Streber WR, Timmis KN, Zenk MH (1987) Analysis, cloning, and high-level expression of 2, 4-dichlorophenoxyacetate monooxygenase gene *tfdA* of *Alcaligenes eutrophus* JMP134. J Bacteriol 169(7):2950–2955

Strickland GR, Annells AJ (2005) The seasonal dynamics of arthropods in conventional, INGARD® and Bollgard II® cotton genotypes in a winter production system at Kununurra, Bentley Delivery Centre WA 6983 @ State of Western Australia 1–13

Strizhov N, Keller M, Mathur J, Koncz-Kálmán Z, Bosch D, Prudovsky E, Schell J, Sneh B, Koncz C, Ziberstein A (1996) A synthetic cry1C gene, encoding a *Bacillus thuringiensis* dendrotoxin, confers Spodoptera resistance in alfalfa and tobacco. Proc Natl Acad Sci U S A 93:15012–15017

Stuiver MH (2011) Engineering disease resistance in plants. Nature 411:865–868

Su J, Wu R (2004) Stress-inducible synthesis of proline in transgenic rice confers faster growth under stress conditions than that with constitutive synthesis. Plant Sci 166:941–948

Su J, Hirji R, Zhang L (2006) Evaluation of the stress-inducible production of choline oxidase in transgenic rice as a strategy for producing the stress-protectant glycine betaine. J Exp Bot 57: 1129–1135

Su L, Diretto G, Purgatto E, Danoun S, Zouine M, Li Z, Roustan JP, Bouzayen M, Giuliano G, Chervin C (2015) Carotenoid accumulation during tomato fruit ripening is modulated by the auxin-ethylene balance. BMC Plant Biol 15:114

Su Z, Bernardo A, Tian B, Chen H, Wang S, Ma H, Cai S, Liu D, Zhang D, Li T (2019a) A deletion mutation in TaHRC confers Fhb1 resistance to Fusarium head blight in wheat. Nat Genet 51:1099–1105

Su YL, Chen G, Chen LS, Li JZ, Wang G, He JY, Zhan TY, Li YW, Yan MT, Huang YH, Qin QW (2019b) Effects of antimicrobial peptides on serum biochemical parameters, antioxidant activity and non-specific immune responses in Epinephelus coioides. Fish Shellfish Immunol 86:1081–1087

Suarez R, Calderon C, Iturriaga G (2009) Enhanced tolerance to multiple abiotic stresses in transgenic alfalfa accumulating trehalose. Crop Sci 49:1791–1799

Subramanian A, Quim M (2010) The impact of Bt cotton on poor households in rural India. J Dev Stud 46:295–311

Sugano S, Kaminaka H, Rybka Z, Catala R, Salinas J, Matsui K, Ohme-Tkagi M, Takatsuji H (2003) Stress-responsive zinc finger gene ZPT2-3 plays a role in drought tolerance in Petunia. Planta 36:830–841

Sugino M, Hibino T, Tanaka Y, Nii N, Takabe T, Takabe T (1999) Overexpression of DnaK from a halo tolerant cyanobacterium *Aphanothece halophytica* acquires resistance to salt stress in transgenic tobacco plants. Plant Sci 146:81–88

Sui X, Zhao Y, Wang S, Duan X, Xu L, Liang R (2012) Improvement Fe content of wheat (Triticum aestivum) grain by soybean ferritin expression cassette without vector backbone sequence. J Agric Biotechnol 20:766–773

Sulpice R, Tsukaya H, Nonaka H (2003) Enhanced formation of flowers in salt-stressed Arabidopsis after genetic engineering of the synthesis of glycine betaine. Plant J 36:165–176

Sumithra K, Jutur PP, Carmel BD, Reddy AR (2006) Salinity-induced changes in two cultivars of *Vigna radiata*: responses of antioxidative and proline metabolism. Plant Growth Regul 50:11–22

Sun W, Bernard C, van de Cotte B, Van Montagu M, Verbruggen N (2001) At-HSP17.6A, encoding a small heat-shock protein in *Arabidopsis*, can enhance osmotolerance upon overexpression. Plant J 27:407–415

Sun W, Van Montagu M, Verbruggen N (2002) Small heat shock proteins and stress tolerance in plants. Biochim Biophys Acta 1577:1–9

Sun L, Yuan B, Zhang M, Wang L, Cui M, Wang Q, Leng P (2012) Fruit-specific RNAi-mediated suppression of SlNCED1 increases both lycopene and β-carotene contents in tomato fruit. J Exp Bot 63:3097–3108

Sundaresha S, Kumar AM, Rohini S, Math SA, Keshamma E, Chandrashekar SC, Udayakumar M (2010) Enhanced protection against two major fungal pathogens of groundnut, Cercospora arachidicola and Aspergillus flavus in transgenic groundnut overexpressing a tobacco b 1-3 glucanase. Eur J Plant Pathol 126:497–508

Sundaresha S, Rohini S, Appanna VK, Arthikala MK, Shanmugam NB, Shashibhushan NB, Kishore CM, Pannerselvam R, Kirti PB, Udayakumar M (2016) Co-overexpression of Brassica juncea NPR1 (BjNPR1) and Trigonella foenum-graecum defensin (Tfgd) in transgenic peanut provides comprehensive but varied protection against Aspergillus flavus and Cercospora arachidicola. Plant Cell Rep 35:1189–1203

Sunkar R, Bartels D, Kirch HH (2003) Overexpression of a stress inducible aldehyde dehydrogenase gene from *Arabidopsis thaliana* in transgenic plants improves stress tolerance. Planta 35: 452–464

Suprasanna P, Nikalje GC, Rai AN (2016) Osmolyte accumulation and implications in plant abiotic stress tolerance. In: Iqbal N, Nazar R, Khan NA (eds) Osmolytes and plants acclimation to changing environment: emerging omics technologies. Springer, New Delhi, pp 1–12

Surekha C, Beena MR, Arundhati A, Singh PK, Tuli R, Dutta-Gupta A, Kirti PB (2005) *Agrobacterium*-mediated genetic transformation of pigeonpea (*Cajanus cajan* L. Millsp.) using embryonal segments and development of transgenic plants for resistance against *Spodoptera*. Plant Sci 169:1074–1080

Sutton DW, Havstad PK, Kemp JD (1992) Synthetic cryIIIA gene from Bacillus thuringiensis improved for high expression in plants. Transgenic Res 1(5):228–236

Suzuki N, Sejima H, Tam R, Schlauch K, Mittler R (2011) Identification of the MBF1 heat-response regulon of *Arabidopsis thaliana*. Plant J 66(5):844–851

Suzuki N, Rivero RM, Shulaev V, Blumwald E, Mittler R (2014) Abiotic and biotic stress combinations. New Phytol 203:32–43

Swamy BM, Samia M, Boncodin R, Marundan S, Rebong DB, Ordonio RL, Miranda RT, Rebong AT, Alibuyog AY, Adeva CC, Reinke R (2019) Compositional analysis of genetically engineered GR2E "Golden Rice" in comparison to that of conventional rice. J Agric Food Chem 67(28):7986–7994

Swindell WR, Huebner M, Weber AP (2007) Transcriptional profiling of Arabidopsis heat shock proteins and transcription factors reveals extensive overlap between heat and non-heat stress response pathways. BMC Genomics 8:125

Syed T, Askari M, Meng Z, Li Y, Abid MA, Wei Y (2020) Current insights on vegetative insecticidal proteins (Vip) as next generation pest killers. Toxins 12:522

Tabashnik BE, Carrière Y (2017) Surge in insect resistance to transgenic crops and prospects for sustainability. Nat Biotechnol 35(10):926–935

Tabashnik BE, Carrière Y (2019) Global patterns of resistance to Bt crops highlighting pink bollworm in the United States, China, and India. J Econ Entomol 112(6):2513–2523

Tabashnik BE, Patin AL, Denney TJ, Liu YB, Carrière Y, Sims MA, Antilla L (2000) Frequency of resistance to *Bacillus thuringiensis* in field populations of pink bollworm. Proc Natl Acad Sci U S A 97:12980–12984

Tabashnik BE, Liu YB, Dennehy TJ, Sims MA, Sisterson MS, Biggs RW, Carrière Y (2002) Inheritance of resistance to Bt toxin Cry1Ac in a field-derived strain of pink bollworm (Lepidoptera: Gelechiidae). J Econ Entomol 95(5):1018–1026

Tabashnik BE, Carrière Y, Dennehy TJ, Morin S, Sisterson MS, Roush RT, Shelton AM, Zhao JZ (2003) Insect resistance to transgenic Bt crops: lessons from the laboratory and field. J Econ Entomol 96:1031–1038

Tabashnik BE, Gassman AJ, Crowder DW, Carriere Y (2008) Insect resistance to Bt crops: evidence versus theory. Nat Biotechnol 26:199–202

Tabashnik BE, Van Renzberg JJ, Carriere Y (2009) Field-evolved insect resistance to Bt crops: definition, theory, and data. J Econ Entomol 102(6):2011–2025

Tabashnik BE, Brévault T, Carrière Y (2013) Insect resistance to Bt crops: lessons from the first billion acres. Nat Biotechnol 31:510–521

Tabe L, Higgins TJ (1998) Engineering plant protein composition for improved nutrition. Trends Plant Sci 3(7):282–286

Tai TH, Dahlbeck D, Clark ET, Gajiwala P, Pasion R, Whalen MC, Stall RE, Staskawicz BJ (1999) Expression of the Bs2 pepper gene confers resistance to bacterial spot disease in tomato. Proc Natl Acad Sci U S A 96:14153–14158

Taj G, Kumar A, Bansal KC, Garg GK (2004) Introgression of osmotin gene for creation of resistance against Alternaria blight by perturbation of cell cycle machinery. Indian J Biotechnol 3:291–298

Talaat NB, Shawky BT (2016) Dual application of 24-epibrassinolide and spermine confers drought stress tolerance in maize (*Zea mays* L.) by modulating polyamine and protein metabolism. J Plant Growth Regul 35(2):518–533

Talakayala A, Katta S, Garladinne M (2020) Genetic engineering of crops for insect resistance: an overview. J Biosci 45:114

Tam JP, Wang S, Wong KH, Tan WL (2015) Antimicrobial peptides from plants. Pharmaceuticals 8:711–757

Tanaka K, Hibino T, Hayashi Y, Tanaka A, Kishitani S, Takabe T, Yokota S, Takabe T (1999) Salt tolerance of transgenic rice overexpressing yeast mitochondrial Mn-SOD in chloroplasts. Plant Sci 148:131–138

Tang Y, Liu Q, Liu Y, Zhang L, Ding W (2017) Overexpression of NtPR-Q up-regulates multiple defense-related genes in Nicotiana tabacum and enhances plant resistance to Ralstonia solanacearum. Front Plant Sci 8:1963

Tank JL, Rosi-Marshall EJ, Royer TV, Whiles MR, Griffiths NA, Frauendorf TC, Treering DJ (2010) Occurrence of maize detritus and a transgenic insecticidal protein (Cry1Ab) within the stream network of an agricultural landscape. Proc Natl Acad Sci U S A 107:17645–17650

Tarafdar AV, Kamle MA, Prakash AR, Padaria JC (2014) Transgenic plants: issues and future prospects. Biotechnology 2:1–47

Tavares LS, Santos Mde O, Viccini LF, Moreira JS, Miller RN, Franco OL (2008) Biotechnological potential of antimicrobial peptides from flowers. Peptides 29:1842–1851

Taveira GB, Mathias LS, da Motta OV, Machado OL, Rodrigues R, Carvalho AO, Teixeira-Ferreira A, Perales J, Vasconcelos IM, Gomes VM (2014) Thionin-like peptides from *Capsicum annuum* fruits with high activity against human pathogenic bacteria and yeasts. Biopolymers 102:30–39

Taylor K, Barran PE, Dorin JR (2008) Structure-activity relationships in beta defensin peptides. Biopolymers 90:1–7

Terras F, Schoofs H, Thevissen K, Osborn RW, Vanderleyden J, Cammue B, Broekaert WF (1993) Synergistic enhancement of the antifungal activity of wheat and barley thionins by radish and oilseed rape 2S albumins and by barley trypsin inhibitors. Plant Physiol 103:1311–1319

Terras FR, Eggermont K, Kovaleva V, Raikhel NV, Osborn RW, Kester A, Rees SB, Vanderleyden J, Cammue BP, Broekaert WF (1995) Small cysteine-rich antifungal proteins from radish: their role in host defense. Plant Cell 7:573–588

Tester M, Langridge (2010) Breeding technologies to increase crop production in a changing world. Science 327(5967):818–822

Thakur N, Upadhyay SK, Verma PC, Chandrashekar K, Tuli R, Singh PK (2014) Enhanced whitefly resistance in transgenic tobacco plants expressing double stranded RNA of v-ATPase A gene. PLoS One 9(3):e87235

Thakur AK, Parmar N, Singh KH, Nanjundan J (2020) Current achievements and future prospects of genetic engineering in Indian mustard (Brassica juncea L. Czern & Coss.). Planta 252(4):1–20

Thamizhmani R, Vijayachari P (2014) Association of dengue virus infection susceptibility with polymorphisms of 2′5′ oligoadenylate synthetase genes: a case-control study. Braz J Infect Dis 18:548550

Thao HT, Lan NT, Tuong HM, Thanh NV, Mau CH (2017) Expression analysis of recombinant Vigna radiata plant defensin 1 protein in transgenic tobacco plants. J Appl Biol Biotechnol 5 (1):0–7

Thipyapong P, Melkonian J, Wolfe DW, Steffens JC (2004) Suppression of polyphenol oxidase increases stress tolerance in tomato. Plant Sci 167:693–703

Thomas C (2014) Crop losses to arthropods. In: Integrated pest management reviews, pp 202–206

Thomas PE, Kaniewski WK, Lawson EC (1997) Reduced field spread of potato leafroll virus in potatoes transformed with the potato leafroll virus coat protein gene. Plant Dis 81:1447–1453

Thomashow MF (1999) Plant cold acclimation: freezing tolerance genes and regulatory mechanisms. Annu Rev Plant Physiol 50:571–599

Thompson CJ, Movva NR, Tizard R, Crameri R, Davies JE, Lauwereys M, Botterman J (1987) Characterization of the herbicide-resistance gene *bar* from *Streptomyces hygroscopicus*. EMBO J 6(9):2519–2523

Thomzik JE, Stenzel K, Stocker R, Schreier PH, Hain R, Stahl DJ (1997) Synthesis of a grapevine phytoalexin in transgenic tomatoes (*Lycopersicon esculentum* Mill.) conditions resistance against *Phytophthora infestans*. Physiol Mol Plant Pathol 51:265–278

Thorpe PE, Brown AN, Ross WC, Cumber AJ, Detre SI, Edwards DC, Davies AJ, Stirpe F (1981) Cytotoxicity acquired by conjugation of an anti-Thy1. 1 monoclonal antibody and the ribosome-inactivating protein, gelonin. Eur J Biochem 116(3):447–454

Tiger TT, Miki MC, Rolis CW, Chen LJ, Su RC, Wang CS, Jason TC (2003) Enhanced methionine and cysteine levels in transgenic rice seeds by the accumulation of sesame 2S albumin. Biosci Biotechnol Biochem 67:1699–1705

Tikhonov AN (2014) The cytochrome b6f complex at the crossroad of photosynthetic electron transport pathways. Plant Physiol Biochem 81:163–183

Tiwari S, Verma PC, Singh PK, Tuli R (2009) Plants as bioreactors for the production of vaccine antigens. Biotechnol Adv 27(4):449–467

Todaka D, Shinozaki K, Yamaguchi-Shinozaki K (2015) Recent advances in the dissection of drought-stress regulatory networks and strategies for development of drought-tolerant transgenic rice plants. Front Plant Sci 6:84

Tohidfar M, Mohammadi M, Ghareyazie B (2005) Agrobacterium-mediated transformation of cotton (*Gossypium hirsutum*) using a heterologous bean chitinase gene. Plant Cell Tiss Org Cult 83(1):83–96

Tohidfar M, Salehi Jouzani G (2008) Genetic engineering of crop plants for enhanced resistance to insects and diseases in Iran. Transgenic Plant J 2:151–156

Tohidfar M, Zare N, Jouzani GS, Efekhari SM (2013) Agrobacterium mediated transformation of alfalfa (Medicago sativa) using a synthetic cry3a gene to enhance resistance against alfalfa weevil. Plant Cell Tiss Org 113:227–235

Toledo-Ortiz G, Huq E, Rodriguez-Concepcion M (2010) Direct regulation of phytoene synthase gene expression and carotenoid biosynthesis by phytochrome-interacting factors. Proc Natl Acd Sci U S A 107:11626–11631

Tollefson J (2011) Brazil cooks up transgenic bean. Nature 478:168

Tougou M, Furutani N, Yamagishi N, Shizukawa Y, Takahata Y, Hidaka S (2006) Development of resistant transgenic soybeans with inverted repeat-coat protein genes of soybean dwarf virus. Plant Cell Rep 25(11):1213–1218

Tricoli DM, Carney KJ, Russell PF, McMaster JR, Groff DW et al (1995) Field evaluation of transgenic squash containing single or multiple virus coat protein gene constructs for resistance to cucumber mosaic virus, watermelon mosaic virus 2 and zucchini yellow mosaic virus. Nat Biotechnol 13:1458–1465

Tsatsakis AM, Nawaz MA, Tutelyan VA, Golokhvast KS, Kalantzi OI, Chung DH, Kang SJ, Coleman MD, Tyshko N, Yang SH, Chung G (2017) Impact on environment, ecosystem, diversity and health from culturing and using GMOs as feed and food. Food Chem Toxicol 107:108–121

Tselepis VH, Green V, Humphries MJ (1997) An RGD to LDV motif conversion within the disintegrin kistrin generates an integrin antagonist that retains potency but exhibits altered receptor specificity. J Biol Chem 272:21341–21348

Tu J, Zhang G, Datta K, Xu C, He Y, Zhang Q, Khush GS, Datta SK (2000) Field performance of transgenic elite commercial hybrid rice expressing *Bacillus thuringiensis* δ-endotoxin. Nat Biotechnol 18:1101–1104

Tudisco R, Lombardi P, Bovera F (2006) Genetically modified soya bean in rabbit feeding: detection of DNA fragments and evaluation of metabolic effects by enzymatic analysis. J Anim Sci 82:193–199

Tuteja N, Verma S, Sahoo RK, Raveendar S, Reddy IN (2012) Recent advances in development of marker-free transgenic plants: regulation and biosafety concern. J Biosci 37(1):167–197

Tyagi A, Singh B, Billekallu Thammegowda NK, Singh NK (2019) Shotgun metagenomics offers novel insights into taxonomic compositions, metabolic pathways and antibiotic resistance genes in fish gut microbiome. Arch Microbial 1(3):295–303

Tylleskar T, Banea M (1992) Cassava cyanogens and konzo, an upper motoneuron disease found in Africa. Lancet 339:208–211

Tylleskär T, Banea M, Bikangi N, Fresco L, Persson LA, Rosling H (1991) Epidemiological evidence from Zaire for a dietary etiology of konzo, an upper motor neuron disease. Bull World Health Organ 69(5):581

Tzfira T, Citovsky V (2006) Agrobacterium-mediated genetic transformation of plants: biology and biotechnology. Curr Opin Biotechnol 17(2):147–154

Udikeri SS (2006) Evaluation of new generation Bt cotton genotypes, sustainability of Cry protein expression, computation of ETL. Effect on aphid predators and development of IPM module for Bt Cotton under rainfed conditions. PhD Thesis, Univ. Agric. Sci., Dharwad, Karnataka, India

Uematsu K, Suzuki N, Iwamae T, Inui M, Yukawa H (2012) Increased fructose 1,6-bisphosphate aldolase in plastids enhances growth and photosynthesis of tobacco plants. J Exp Bot 63:3001–3009

Ullah A, Hussain A, Shaban M, Khan AH, Alariqi M, Gul S, Jun Z, Lin S, Li J, Jin S, Munis MF (2018) Osmotin: a plant defense tool against biotic and abiotic stresses. Plant Physiol Biochem 123:149–159

Uma B, Rani TS, Podile AR (2011) Warriors at the gate that never sleep: non-host resistance in plants. J Plant Physiol 168:2141–2152

Upadhyay SK, Singh PK (2011) Role of alkaline phosphatase in insecticidal action of Cry1Ac against Helicoverpa armigera larvae. Biotechnol Lett 33(10):2027–2036

Urnov FD, Rebar EJ, Holmes MC, Zhang HS, Gregory PD (2010) Genome editing with engineered zinc finger nucleases. Nat Rev Genet 11:636

Urwin PE, Green J, Atkinson HJ (2003) Expression of a plant cystatin confers partial resistance to Globodera, full resistance is achieved by pyramiding a cystatin with natural resistance. Mol Breed (in press)

Ushimaru T, Nakagawa T, Fujioka Y, Daicho K, Naito M, Yamauchi Y, Nonaka H, Amako K, Yamawaki K, Murata N (2006) Transgenic Arabidopsis plants expressing the rice dehydroascorbate reductase gene are resistant to salt stress. J Plant Physiol 163(11):1179–1184

USDA (2010) National Nutrient Database for Standard Reference, Release 23. Nutrient Data Laboratory, Agricultural Research Service, U.S. Department of Agriculture, Washington, DC

USDA (2015) Determinations of non-regulated status. http://www.aphis.usda.gov/biotechnology/petitions_table_pending.shtml

Usui T, Shima Y, Shimada Y, Hirano S, Burgess RW, Schwarz TL, Takeichi M, Uemura T (1999) Flamingo, a seven-pass transmembrane cadherin, regulates planar cell polarity under the control of Frizzled. Cell 98:585–595

Utkina LL, Andreev YA, Rogozhin EA, Korostyleva TV, Slavokhotova AA, Oparin PB, Vassilevski AA, Grishin EV, Egorov TA, Odintsova TI (2013) Genes encoding 4-Cys antimicrobial peptides in wheat *Triticum kiharae* Dorof. et Migush.: multimodular structural organization intraspecific variability distribution and role in defence. FEBS J 280:3594–3608

Vadlamudi RK, Ji TH, Bulla LA Jr (1993) A specific binding protein from *Manduca sexta* for the insecticidal toxin of *Bacillus thuringiensis* subsp. Berliner. J Biol Chem 268:12334–12340

Vadlamudi RK, Weber E, Ji I, Ji TH, Bulla LA Jr (1995) Cloning and expression of a receptor for an insecticidal toxin of Bacillus thuringiensis. J Biol Chem 270:5490–5494

Vaeck M, Reynaerts A, Hofte H, Jansens S, De Beuckeleer M, Dean C, Zabeau M, Van Montagu M, Leemans J (1987) Transgenic plants protected from insect attack. Nature 328: 33–37

Valaitis AP, Mazza A, Brousseau R, Masson L (1997) Interaction analyses of Bacillus thuringiensis Cry1A toxins with two aminopeptidases from gypsy moth midgut brush border membranes. Insect Biochem Mol Biol 27:529–539

Valaitis AP, Jenkins JL, Lee MK (2001) Isolation, partial characterization of Gypsy moth BTR-270 an anionic brush border membrane glycoconjugate that binds *Bacillus thuringiensis* Cry1A toxins with high affinity. Arch Insect Biochem Physiol 46:186–200

Van Baarlen P, Van Belkum A, Summerbell RC, Crous PW, Bart P (2007) Molecular mechanisms of pathogenicity: how do pathogenic microorganisms develop cross-kingdom host jumps? FEMS Micro Rev 3:239–277

Van Camp W, Capiau K, Van Montagu M, Inze D, Slooten L (1996) Enhancement of oxidative stress tolerance in transgenic tobacco plants overproducing Fe-superoxide dismutase in chloroplasts. Plant Physiol 112:1703–1714

Van Der Vossen EA, Van Der Voort JNR, Kanyuka K, Bendahmane A, Sandbrink H, Baulcombe DC (2000) Homologues of a single resistance-gene cluster in potato confer resistance to distinct pathogens: a virus and a nematode. Plant J 23(5):567–576

Van Eck J, Conlin B, Garvin DF (2007) Enhancing beta-carotene content in potato by RNAi-mediated silencing of the beta-carotene hydroxylase gene. Am J Potato Res 84:331–342

Van Parijs J, Broekaert WF, Goldstein IJ, Peumans WJ (1991) Hevein: an antifungal protein from rubber-tree (*Hevea brasiliensis*) latex. Planta 183:258–264

Vande Berg BJ, Hammer PE, Chun BL, Schouten LC, Carr B, Guo R, Peters C, Hinson TK, Beilinson V, Shekita A, Deter R (2008) Characterization and plant expression of a glyphosate-tolerant enolpyruvylshikimate phosphate synthase. Pest Manag Sci 64(4):340–345

Vanti GL, Katageri IS, Inamdar SR, Hiremathada V, Swamy BM (2018) Potent insect gut binding lectin from *Sclerotium rolfsii* impart resistance to sucking and chewing type insects in cotton. J Biotechnol 278:20–27

Vasil V, Castillo AM, Fromm ME, Vasil IK (1992) Herbicide resistant fertile transgenic wheat plants obtained by microprojectile bombardment of regenerable embryogenic callus. Bio/technology 10(6):667–674

Vendruscolo EC, Schuster I, Pileggi M, Scapim CA, Molinari HB, Marur CJ, Vieira LG (2007) Stress-induced synthesis of proline confers tolerance to water deficit in transgenic wheat. Plant Physiol 164:1367–7136

Verma AK, Chand L (2005) Agrobacterium-mediated transformation of pigeonpea (Cajanus cajan L.) with uidA and cryIA(b) genes. Physiol Mol Biol Plants 11:99–109

Verma HN, Srivastava S, Varsha-Kumar D (1996) Induction of systemic resistance in plants against viruses by a basic protein from *Clerodendrum aculeatum* leaves. Phytopathology 86:485–492

Verma C, Nanda S, Singh RK, Singh RB, Mishra S (2011) A review on impacts of genetically modified food on human health. Open Nutraceut J 4:3–11

Verma SS, Sinha RK, Jajoo A (2015) (E)-β-farnesene gene reduces *Lipaphis erysimi* colonization in transgenic *Brassica juncea* lines. Plant Signal Behav 10(7):e1042636

Vert G, Grotz N, Dedaldechamp F, Gaymard F, Guerinot ML, Briat JF, Curie C (2002) IRT1, an Arabidopsis transporter essential for iron uptake from the soil and for plant growth. Plant Cell 14 (6):1223–1233

Vestal DJ, Ranscht B (1992) Glycosyl phosphatidylinositol-anchored T-cadherin mediates calcium-dependent, homophilic cell adhesion. J Cell Biol 119:451–461

Viktorov AG (2011) Transfer of Bt corn byproducts from terrestrial to stream ecosystems. Russ J Plant Physiol 58:543–548

Villalobos MA, Bartels D, Iturriaga G (2004) Stress tolerance and glucose insensitive phenotypes in *Arabidopsis* overexpressing the CpMYB10 transcription factor gene. Plant Physiol 135:309–324

Vishnudasan D, Tripathi MN, Rao U, Khurana P (2005) Assessment of nematode resistance in wheat transgenic plants expressing potato proteinase inhibitor (PIN2) gene. Transgenic Res 14 (5):665–675

Vivanco JM, Savary BJ, Flores HE (1999) Characterization of two novel type I ribosome-inactivating proteins from the storage roots of the Andean crop *Mirabilis expansa*. Plant Physiol 119:1447–1456

Vivek BS, Krivanek AF, Palacios-Rojas N, Twumasi-Afriyie S, Diallo AO (2008) Breeding quality protein maize (QPM): protocols for developing QPM cultivars. CIMMYT, Mexico

Voinnet O (2005) Induction and suppression of RNA silencing: insights from viral infections. Nat Rev Genet 6:206–220

Vyas M, Raza A, Ali MY, Ashraf MA, Mansoor S, Shahid AA, Brown JK (2017) Knock down of whitefly gut gene expression and mortality by orally delivered gut gene-specific dsRNAs. PLoS 12(1):e0168921

Vyavhare SS (2017) BT Cotton. http://lubbock.tamu.edu/files/2017/03/Bt-Cotton_ENTO067.pdf

Wahid A, Farooq M, Hussain I, Rasheed R, Galani S (2012) Responses and management of heat stress in plants. In: Environmental adaptations and stress tolerance of plants in the era of climate change. Springer, New York, NY, pp 135–157
Waie B, Rajam MV (2003) Effect of increased polyamine biosynthesis on stress responses in transgenic tobacco by introduction of human S-adenosylmethionine gene. Plant Sci 164:727–734
Wakasa K, Hasegawa H, Nemoto H, Matsuda F, Miyazawa H, Tozawa Y, Morino K, Komatsu A, Yamada T, Terakawa T, Miyagawa H (2006) High-level tryptophan accumulation in seeds of transgenic rice and its limited effects on agronomic traits and seed metabolite profile. J Exp Bot 57(12):3069–3078
Walker BJ, South PF, Ort DR (2016) Physiological evidence for plasticity in glycolate/glycerate transport during photorespiration. Photosynth Res 129(1):93–103
Wally O, Punja ZK (2010) Genetic engineering for increasing fungal and bacterial disease resistance in crop plants. GM Crops 1:199–206
Waltz E (2009) GM crops: battlefield. Nature 461:27–32
Wan Q, Hongbo S, Zhaolong X, Jia L, Dayong Z, Yihong H (2017) Salinity tolerance mechanism of osmotin and osmotin-like proteins: a promising candidate for enhancing plant salt tolerance. Curr Genomics 18(6):553–556
Wang W, Galili G (2016) Transgenic high-lysine rice–a realistic solution to malnutrition? J Exp Bot 67(14):4009–4011
Wang WX, Vinocur B, Shoseyov O, Altman A (2000) Biotechnology of plant osmotic stress tolerance physiological and molecular considerations. In: IV international symposium on in vitro culture and horticultural breeding, pp 285–292
Wang W, Vinocur B, Altman A (2003) Plant responses to drought salinity and extreme temperatures: towards genetic engineering for stress tolerance. Planta 218:1–14
Wang W, Vinocur B, Shoseyov O, Altman A (2004a) Role of plant heat shock proteins and molecular chaperones in the abiotic stress response. Trends Plant Sci 9:244–252
Wang SY, Wu JH, Ng TB, Ye XY, Rao PF (2004b) A non-specific lipid transfer protein with antifungal and antibacterial activities from the mung bean. Peptides 25:1235–1242
Wang Y, Wisniewski M, Meilan R, Cui M, Webb R, Fuchigami L (2005a) Overexpression of cytosolic ascorbate peroxidase in tomato confers tolerance to chilling and salt stress. J Am Soc Hort Sci 130(2):167–173
Wang GR, Wu KM, Liang GM, Guo YY (2005b) Gene cloning and expression of cadherin in midgut of *Helicoverpa armigera* and its Cry1A binding region. Sci China Ser C 48:346–356
Wang Y, Deng H, Zhang X, Xiao H, Jiang Y, Song Y, Fang L, Xiao S, Zhen Y, Chen H (2009) Generation and immunogenicity of Japanese encephalitis virus envelope protein expressed in transgenic rice. Biochem Biophys Res Commun 380:292–297
Wang N, Xiao B, Xiong L (2011) Identification of a cluster of PR4-like genes involved in stress responses in rice. J Plant Physiol 168:2212–2224
Wang MB, Masuta C, Smith NA, Shimura H (2012) RNA silencing and plant viral diseases. Mol Plant Microbe Interact 25:1275–1285
Wang G, Cai G, Kong F, Deng Y, Ma N, Meng Q (2014a) Overexpression of tomato chloroplast DnaJ Protein enhances tolerance to drought stress and resistance *Pseudomonas solanacearum* in transgenic tobacco. Plant Physiol Biochem 82:95–104
Wang C, Zeng J, Li Y, Hu W, Chen L, Miao Y (2014b) Enrichment of provitamin A content in wheat (*Triticum aestivum* L.) by introduction of the bacterial carotenoid biosynthetic genes *CrtB* and *CrtI*. J Exp Bot 65:2545–2556
Wang L, Ma YK, Li NN, Zhang WB, Mao HY, Lin XF (2016) Isolation and characterization of a tonoplast Na^+/H^+ antiporter from the halophyte *Nitraria sibirica*. Biol Plant 60(1):113–122
Wang J, Wang H, Liu S, Liu L, Tay WT, Walsh TK, Wu Y (2017) CRISPR/Cas9 mediated genome editing of *Helicoverpa armigera* with mutations of an ABC transporter gene HaABCA2 confers resistance to *Bacillus thuringiensis* Cry2A toxins. Insect Biochem Mol Biol 87:147–153

Wang P, Zhao Y, Li Z et al (2018) Reciprocal regulation of the TOR kinase and ABA receptor balances plant growth and stress response. Mol Cell 69:100–112

Wang J, Ma H, Zuo Y, Yang Y, Wu Y (2020) CRISPR mediated gene knockout reveals nicotinic acetylcholine receptor (nAChR) subunit a6 as a target of spinosyns in *Helicoverpa armigera*. Pest Manag Sci 76:2925–2931

Wani W, Masoodi KZ, Zaid A, Wani SH, Shah F, Meena VS, Wani SA, Mosa KA (2018) Engineering plants for heavy metal stress tolerance. Rendiconti Lincei. Scienze Fisiche e Naturali 29(3):709–723

Ward M, Wilson LJ, Carmona CL, Turner G (1988) The oliC3 gene of Aspergillus niger: isolation, sequence and use as a selectable marker for transformation. Curr Genet 14(1):37–42

Warren GW, Koziel MG, Mullin MA, Nye GJ, Carr B, Desai NM, Kostichka K, Duck NB, Estruch JJ (1998) Auxiliary proteins for enhancing the insecticidal activity of pesticidal proteins. US patent 5:770–696

Waterhouse PM, Helliwell CA (2003) Exploring plant genomes by RNA-induced gene silencing. Nat Rev Genet 4(1):29–38

Wei GQ, Liu RS, Wang Q, Liu WY (2004) Toxicity of two type-2 ribosome-inactivating proteins (cinnamomin and ricin) to domestic silkworm larvae. Arch Insect Biochem Physiol 57:160–165. https://doi.org/10.1002/arch.20024

Wei S, Yu B, Gruber MY, Khachatourians GG, Hegedus DD, Hannoufa A (2010) Enhanced seed carotenoid levels and branching in transgenic *Brassica napus* expressing the Arabidopsis miR156b gene. J Agric Food Chem 58:9572–9578

Wei Q, Guo YJ, Cao HM, Kuai BK (2011) Cloning and characterization of an AtNHX2-like Na+/H + antiporter gene from *Ammopiptanthus mongolicus* (Leguminosae) and its ectopic expression enhanced drought and salt tolerance in *Arabidopsis thaliana*. Plant Cell, Tissue and Organ Culture (PCTOC) 105(3):309–316

Weichert H, Högy P, Mora-Ramirez I, Fuchs J, Eggert K, Koehler P, Weschke W, Fangmeier A, Weber H (2017) Grain yield and quality responses of wheat expressing a barley sucrose transporter to combined climate change factors. J Exp Bot 68(20):5511–5525

Weigel H, Manderscheid R (2012) Crop growth responses to free air CO2 enrichment and nitrogen fertilization: rotating barley, ryegrass, sugar beet and wheat. Eur J Agron 43:97–107

Welsch R, Arango J, Bar C, Salazar B, Al-Babili S, Beltran J, Chavarriaga P, Ceballos H, Tohme J, Beyer P (2010) Provitamin A accumulation in cassava (*Manihot esculenta*) roots driven by a single nucleotide polymorphism in a phytoene synthase gene. Plant Cell 22:3348–3356

Wenefrida I, Utomo HS, Blanche SB, Linscombe SD (2009) Enhancing essential amino acids and health benefit components in grain crops for improved nutritional values. Recent Pat DNA Gene Seq 3:219–225

Weng LX, Deng HH, Xu JL, Li Q, Zhang YQ, Jiang ZD, Li QW, Chen JW, Zhang LH (2011) Transgenic sugarcane plants expressing high levels of modified *cry1Ac* provide effective control against stem borers in field trials. Transgenic Res 20(4):759–772

White W, Arias-Garzon D, McMahon J, Sayre R (1998) Cyanogenesis in cassava. The role of hydroxynitrile lyase in root cyanide production. Plant Physiol 116:1219–1225

Wijaya R, Neumann GM, Condron R, Hughes AB, Polya GM (2000) Defense proteins from seed of Cassia fistula include a lipid transfer protein homologue and a protease inhibitory plant defensin. Plant Sci 159:243–255

Wilson FD, Flint HM, Deaton WR, Fischhoff DA, Perlak FJ, Armstrong TA (1992) Resistance of cotton lines containing a *Bacillus thuringiensis* toxin to pink bollworm (Lepidoptera: Gelechiidae) and other insects. J Econ Entomol 85:1516–1521

Winicov I (2000) Alfin1 transcription factor overexpression enhances plant root growth under normal and saline conditions and improves salt tolerance in alfalfa. Planta 210:416–422

Wirth J, Poletti S, Aeschlimann B, Yakandawala N, Drosse B, Osorio S, Tohge T, Fernie AR, Günther D, Gruissem W, Sautter C (2009) Rice endosperm iron biofortification by targeted and synergistic action of nicotianamine synthase and ferritin. Plant Biotechnol J 7:631–644

Wisutiamonkul A, Ampomah-Dwamena C, Allan AC, Ketsa S (2017) Carotenoid accumulation in durian (*Durio zibethinus*) fruit is affected by ethylene via modulation of carotenoid pathway gene expression. Plant Physiol Biochem 115:308–319

Wong JC, Lambert RJ, Wurtzel ET, Rocheford TR (2004) QTL and candidate genes phytoene synthase and zeta-carotene desaturase associated with the accumulation of carotenoids in maize. Theor Appl Genet 108:349–359

Wong JH, Ng TB, Cheung RCF, Ye XJ, Wang HX, Lam SK (2010) Proteins with antifungal properties and other medicinal applications from plants and mushrooms. Appl Microbiol Biotechnol 87:1–15

Wu K, Guo YY (2004) Changes in susceptibility to conventional insecticides of a Cry1Ac selected population of *Helicoverpa armigera* (Hubner) (Lepidoptera: Noctuidae). Pest Manag Sci 60: 680–684

Wu C, Fan Y, Zhang C, Olica N, Datta SK (1997) Transgenic fertile japonica rice plants expressing a modified *cryIA(b)* gene resistant to yellow stem borer. Plant Cell Rep 17:129–132

Wu KM, Geo YY, Wang WG (2000a) Field resistance evaluations of Bt transgenic cotton GK series to cotton bollworm. Acta Phytotax Sin 27:317–321

Wu SJ, Koller CN, Miller DL, Bauer LS, Dean DH (2000b) Enhanced toxicity of *Bacillus thuringiensis* Cry3A delta-endotoxin in coleopterans by mutagenesis in a receptor binding loop. FEBS Lett 473:227–232

Wu LQ, Fan Z, Guo L et al (2003) Overexpression of an Arabidopsis δ-OAT gene enhances salt and drought tolerance in transgenic rice. Chin Sci Bull 48:2594–2600

Wu L, Fan Z, Guo L, Li Y, Chen ZL, Qu LJ (2005) Over-expression of the bacterial *nhaA* gene in rice enhances salt and drought tolerance. Plant Sci 168:297–302

Wu CH, Tewari RK, Hahn EJ, Paek KY (2007) Nitric oxide elicitation induces the accumulation of secondary metabolites and antioxidant defense in adventitious roots of Echinacea purpurea. J Plant Biol 50(6):636–643

Wu H, Lv H, Li L, Liu J, Mu S, Li X, Gao J (2015) Genome-wide analysis of the AP2/ERF transcription factors family and the expression patterns of DREB genes in Moso Bamboo (*Phyllostachys edulis*). PLoS One 10(5):e0126657

Wu J, Kim SG, Kang KY, Kim JG, Park SR, Gupta R, Kim YH, Wang Y, Kim ST (2016) Overexpression of a pathogenesis-related protein 10 enhances biotic and abiotic stress tolerance in rice. Plant Pathol J 32(6):552

Wyman C, Kanaar R (2006) DNA double-strand break repair: all's well that ends well. Annu Rev Genet 40:363–383

Xiao B, Huang Y, Tang N, Xiong L (2007) Over-expression of a LEA gene in rice improves drought resistance under the field conditions. Theor Appl Genet 115:35–46

Xiaoou OD, Ronald PC (2019) Genetic engineering for disease resistance in plants: recent progress and future perspectives. Plant Physiol 180:1–13

Xie FL, Wang QL, Zhang BH (2015) Global microRNA modification in cotton (Gossypium hirsutum L). Plant Biotechnol J 13. (in press)

Xing LP, Wang HZ, Jiang ZN, Ni JL, Cao AZ, Yu L, Chen PD (2008) Transformation of wheat thaumatin-like protein gene and analysis of reactions to powdery mildew and fusarium head blight in transgenic plants. Acta Agron Sin 34:349–354

Xu D, Duan X, Wang B, Hong B, Ho TD, Wu R (1996) Expression of a late embryogenesis abundant protein gene, *HVA1*, from barley confers tolerance to water deficit and salt stress in transgenic rice. Plant Physiol 110:249–257

Xu XP, Chen CH, Fan BF, Chen ZX (2006) Physical and functional interactions between pathogen-induced Arabidopsis WRKY18, WRKY40, and WRKY60 transcription factors. Plant Cell 18:1310–1326

Xu C, Cheng J, Lin H, Lin C, Gao J, Shen Z (2018) Characterization of transgenic rice expressing fusion protein Cry1Ab/Vip3A for insect resistance. Sci Rep 8:1–8

Xue ZY, Zhi DV, Xue GP, Zhang H, Zhao YX, Xia GM (2004) Enhance salt tolerance of transgenic wheat (*Triticum aestivum* L.) expressing a vacuolar Na^+/H^+ antiporter gone with improved grain yields in saline soils in the field and a reduced level of leaf Na^+. Plant Sci 167:849–859

Yadav RC, Kamada H, Kikuchi F (1991) Genotypic and media efects on plant regeneration from cotyledonary explants of Brassica juncea (L.) Coss & Czern. Ann Biol 7(2):119–124

Yadav SK, Singla-Pareek SL, Reddy MK, Sopory SK (2005) Transgenic tobacco plants overexpressing glyoxalase enzymes resist an increase in methylglyoxal and maintain higher reduced glutathione levels under salinity stress. FEBS Lett 579:6265–6271

Yadav OP, Hossain F, Karjagi CG, Kumar B, Zaidi PH, Jat SL (2015) Genetic improvement of maize in India: retrospect and prospects. Agric Res 4:325–338

Yadav SK, Khatri K, Rathore MS, Jha B (2018) Introgression of UfCyt c6, a thylakoid lumen protein from a green seaweed *Ulva fasciata* Delile enhanced photosynthesis and growth in tobacco. Mol Biol Rep 45(6):1745–1758

Yamada T, Tozawa Y, Hasegawa H, Terakawa T, Ohkawa Y, Wakasa K (2004) Use of a feedback-insensitive α subunit of anthranilate synthase as a selectable marker for transformation of rice and potato. Mol Breed 14:363–373

Yan F, Zhang WW, Xiao H, Li SF, Cheng ZM (2007) Transgenic wheat expressing virus-derived hairpin RNA is resistant to Barley yellow dwarf virus. Yi Chuan 29:97–102

Yan H, Yan H, Li G, Gong W, Jiao H, Chen H, Ji M (2010a) Expression of human cytomegalovirus pp 150 gene in transgenic *Vicia faba* L. and immunogenicity of pp150 protein in mice. Biologicals 38:265–272

Yan J, Kandianis CB, Harjes CE, Bai L, Kim EH, Yang X, Skinner DJ, Fu Z, Mitchell S, Li Q, Fernandez MG (2010b) Rare genetic variation at *Zea mays* crtRB1 increases β-carotene in maize grain. Nat Genet 42(4):322–327

Yan S, Zhu J, Zhu W, Li Z, Shelton AM, Luo J, Cui J, Zhang Q, Liu X (2015) Pollen-mediated gene flow from transgenic cotton under greenhouse conditions is dependent on different pollinators. Sci Rep 5(1):1–9

Yan S, Qian J, Cai C, Ma Z, Li J, Yin M, Ren B, Shen J (2020) Spray method application of transdermal dsRNA delivery system for efficient gene silencing and pest control on soybean aphid *Aphis glycines*. J Pestic Sci 93:449–459

Yang RC, Xu HL, Yu WG, Lu CG, Long MS, Liu CQ, Pan NS, Chen ZL (1995) Transgenic tomato plants expressing cucumber mosaic virus coat protein (CMV-cp) and their resistance to cucumber mosaic virus (CMV). Acta Agric Jiangsu 11:42–46

Yang SX, Zhao YX, Zhang Q, He YK, Zhang H, Luo (2001) HAL1 mediate salt adaptation in *Arabidopsis thaliana*. Cell Res 11:142–148

Yang X, Wen X, Gong H (2007a) Genetic engineering of the biosynthesis of glycinebetaine enhances thermotolerance of photosystem II in tobacco plants. Planta 225:719–733

Yang X, Xiao Y, Wang X, Pei Y (2007b) Expression of a novel small antimicrobial protein from the seeds of motherwort *Leonurus japonicus* confers disease resistance in tobacco. Appl Environ Microbiol 73:939–946

Yang KZ, Xia C, Liu XL, Dou XY, Wang W, Chen LQ, Zhang XQ, Xie LF, He L, Ma X (2009) A mutation in Thermosensitive Male Sterile 1, encoding a heat shock protein with DnaJ and PDI domains, leads to thermosensitive gametophytic male sterility in Arabidopsis. Plant J 57:870–882

Yang T, Meng Y, Chen LJ, Lin HH, Xi DH (2016) The roles of alpha-momorcharin and jasmonic acid in modulating the response of *Momordica charantia* to Cucumber mosaic virus. Front Microbiol 7:179

Yanjun Z, Haixia Y, Xuejun W, Jiarui Z, Haiyan Z, Daqun L (2017) Expression analysis and functional characterization of a pathogen-induced thaumatin-like gene in wheat conferring enhanced resistance to *Puccinia triticina*. J Plant Interact 12:332–339

Ye C, Li H (2010) 20 Years of transgenic research in China for resistance to Papaya ringspot virus. Transgenic Plant J 4:58–63

Ye X, Al-Babili S, Klöti A, Zhang J, Lucca P, Beyer P, Potrykus I (2000) Engineering the provitamin A (beta-carotene) biosynthetic pathway into (carotenoid-free) rice endosperm. Science 287:303–305

Ye GY, Yao HW, Shu QY, Cheng X, Hu C, Xia YW, Gao MW, Altosaar I (2003) High levels of stable resistance in transgenic rice with a *cry1Ab* gene from *Bacillus thuringiensis* Berliner to rice leaffolder, *Cnaphalocrocis medinalis* (Guenée) under field conditions. Crop Prot 22(1): 171–178

Yemets A, Radchuk V, Bayer O, Bayer G, Pakhomov A, Baird WV, Blume YB (2008) Development of transformation vectors based upon a modified plant α-tubulin gene as the selectable marker. Cell Biol Int 32(5):566–570

Yokotani N, Ichikawa T, Kondou Y, Maeda S, Iwabuchi M, Mori M, Hirochika H, Matsui M, Oda K (2009) Overexpression of a rice gene encoding a small C2 domain protein OsSMCP1 increases tolerance to abiotic and biotic stresses in transgenic *Arabidopsis*. Plant Mol Biol 71: 391–402

Yoshida T, Sakuma Y, Todaka D, Maruyama K, Qin F, Mizoi J, Kidokoro S, Fujita Y, Shinozaki K, Yamaguchi-Shinozaki K (2008) Functional analysis of an Arabidopsis heat-shock transcription factor HsfA3 in the transcriptional cascade downstream of the DREB2A stress-regulatory system. Biochem Biophys Res Commun 368:515–521

You J, Hu H, Xiong L (2012) An ornithine 1-aminotransferase gene OsOAT confers drought and oxidative stress tolerance in rice. Plant Sci 197:59–69

Yu C-G, Mullins MA, Warren GW, Koziel MG, Estruch JJ (1997) The Bacillus thuringiensis vegetative insecticidal protein Vip3A lyses midgut epithelium cells of susceptible insects. Appl Environ Microbiol 63(2):532–536

Yu B, Lydiate DJ, Young LW, Schäfer UA, Hannoufa A (2008) Enhancing the carotenoid content of *Brassica napus* seeds by downregulating lycopene epsilon cyclase. Transgenic Res 17:573–585

Yu X, Liu T, Liang X, Tang C, Zhu J, Wang S, Li S, Deng Q, Wang L, Zheng A, Li P (2011) Rapid detection of *vip1*-type genes from *Bacillus cereus* and characterization of a novel *vip* binary toxin gene. FEMS Microbiol Lett 325:30–36

Yuan Y, Qi L, Yang J, Wu C, Liu Y, Huang L (2015) A Scutellaria baicalensis R2R3-MYB gene, SbMYB8, regulates flavonoid biosynthesis and improves drought stress tolerance in transgenic tobacco. Plant Cell Tissue Organ Cult 120(3):961–972

Yuan P, Cui S, Liu Y, Li J, Du G, Liu L (2020) Metabolic engineering for the production of fat-soluble vitamins: advances and perspectives. Appl Microbiol Biotechnol 104(3):935–951

Zaccomer B, Celller F, Boyer JC, Haenni AL, Tepfer M (1993) Transgenic plants that express genes including the 3' untranslated region of the turnip yellow mosaic virus (TYMV) genome are partially protected against TYMV infection. Gene 136:87–94

Zaidi MA, Mohammadi M, Postel S, Masson L, Altosaar I (2005) The Bt gene cry2Aa2 driven by a tissue specific ST-LS1 promoter from potato effectively controls *Heliothis virescens*. Transgenic Res 14(3):289–298

Zaidi SSEA, Briddon RW, Mansoor S (2017) Engineering dual begomovirus *Bemisia tabaci* resistance in plants. Trends Plant Sci 22:6–8

Zaitfin M, Anderson JM, Perry KL, Zhang L, Palukaitis P (1994) Specificity of replicase-mediated resistance to cucumber mosaic virus. Virology 201:200–205

Zakharian RA, Israelian IA, Agabalian AS, Tatevosian PE, Akopian SM (1979) Bacillus thuringiensis plasmid DNA. Mikrobiologiia 48:226–229

Zang X, Geng X, Wang F, Liu Z, Zhang L, Zhao Y, Tian X, Ni Z, Yao Y, Xin M, Hu Z (2017) Overexpression of wheat ferritin gene TaFER-5B enhances tolerance to heat stress and other abiotic stresses associated with the ROS scavenging. BMC Plant Biol 17(1):1–3

Zambryski P, Joos H, Genetello C, Leemans J, Van Montagu M, Schell J (1983) Ti plasmid vector for the introduction of DNA into plant cells without alteration of their normal regeneration capacity. EMBO J 2(12):2143–2150

Zamore PD (2004) Plant RNAi: how a viral silencing suppressor inactivates siRNA. Curr Biol 14: R198–R200

Zang X, Geng X, He K, Wang F, Tian X, Xin M, Yao Y, Hu Z, Ni Z, Sun Q, Peng H (2018) Overexpression of the wheat (*Triticum aestivum* L.) TaPEPKR2 gene enhances heat and dehydration tolerance in both wheat and Arabidopsis. Front Plant Sci 9:1710

Zasloff M (2002) Antimicrobial peptides of multicellular organisms. Nature 415:389–395

Zeh M, Casazza AP, Kreft O, Roessner U, Bieberich K, Willmitzer L, Hesse H (2001) Antisense inhibition of threonine synthase leads to high methionine content in transgenic potato plants. Plant Physiol 127:792–802

Zeidler D (2004) Innate immunity in *Arabidopsis thaliana*: lipopolysaccharides activate nitric oxide synthase (NOS) and induce defense genes. Proc Natl Acad Sci U S A 101:15811–15816

Zhang HX, Blumwald E (2001) Transgenic salt-tolerant tomato plants accumulate salt in foliage but not in fruit. Nat Biotechnol 19(8):765–768

Zhang Z, Huang R (2010) Enhanced tolerance to freezing in tobacco and tomato overexpressing transcription factor TERF2/LeERF2 is modulated by ethylene biosynthesis. Plant Mol Biol 73: 241–249

Zhang HX, Hodson JN, Williams JP, Blumwald E (2001) Engineering salt tolerant Brassica plants: characterization of yield and seed oil quality in transgenic plants with increased vacuolar sodium accumulation. Proc Natl Acad Sci U S A 48:12832–12836

Zhang X, Candas M, Griko NB, Taussig R, Lee AB Jr (2006) A mechanism of cell death involving an adenylyl cyclase/PKA signaling pathway is induced by the Cry1Ab toxin of *Bacillus thuringiensis*. Proc Natl Acad Sci U S A 103:9897–9902

Zhang H, Dong H, Li W, Sun Y, Chen S, Kong X (2009) Increased glycine betaine synthesis and salinity tolerance in AhCMO transgenic cotton lines. Mol Breed 23(2):289–298

Zhang Y, Shi R, Rezaul KM, Zhang F, Zou C (2010) Iron and zinc concentrations in grain and flour of winter wheat as affected by foliar application. J Agric Food Chem 58:12268–12274

Zhang S, Li N, Gao F, Yang A, Zhang J (2011a) Over-expression of TsCBF gene confers improved drought tolerance in transgenic maize. Mol Breed 26:455–465

Zhang H, Yin W, Zhao J, Jin L, Yang Y, Wu S, Tabashnik BE, Wu Y (2011b) Early warning of cotton bollworm resistance associated with intensive planting of Bt cotton in China. PLoS One 6:e22874

Zhang X, Li H, Zhang J, Zhang C, Gong P, Ziaf K, Xiao F, Ye Z (2011c) Expression of artificial microRNAs in tomato confers efficient and stable virus resistance in a cell-autonomous manner. Transgenic Res 20:569–581

Zhang YQ, Sun YX, Ye YL (2012) Zinc biofortification of wheat through fertilizer applications in different locations of China. Field Crop Res 125:1–7

Zhang L, Davies LJ, Elling AA (2015) A *Meloidogyne incognita* effector is imported into the nucleus and exhibits transcriptional activation activity in planta. Mol Plant Pathol 16(1):48–60

Zhang C, Wohlhueter R, Zhang H (2016) Genetically modified foods: a critical review of their promise and problems. Food Sci Human Wellness 5:116–123

Zhang J, Khan SA, Heckel DG, Bock R (2017a) Next-generation insect-resistant plants: RNAi-mediated crop protection. Trends Biotechnol 35(9):871–882

Zhang X, Dodds PN, Bernoux M (2017b) What do we know about NOD-like receptors in plant immunity? Annu Rev Phytopathol 55:205–229

Zhang T, Zheng Q, Yi X, An H, Zhao Y, Ma S, Zhou G (2018a) Establishing RNA virus resistance in plants by harnessing CRISPR immune system. Plant Biotechnol J 16:1415–1423

Zhang J, Hu Z, Yao Q, Guo X, Nguyen V, Li F, Chen G (2018b) A tomato MADS-box protein, SlCMB1, regulates ethylene biosynthesis and carotenoid accumulation during fruit ripening. Sci Rep 8:3413

Zhang Y, Malzahn AA, Sretenovic S, Qi Y (2019) The emerging and uncultivated potential of CRISPR technology in plant science. Nat Plants 5:778–794

Zhao TJ, Sun S, Liu Y (2006) Regulating the drought-responsive element (DRE)-mediated signaling pathway by synergic functions of transactive and trans-inactive DRE binding factors in Brassica napus. J Biol Chem 281:10752–10759

Zheng L, Cheng Z, Ai C, Jiang X, Bei X, Zheng Y, Glahn RP, Welch RM, Miller DD, Lei XG, Shou H (2010) Nicotianamine, a novel enhancer of rice iron bioavailability to humans. PLoS One 5 (4):e10190

Zheng M, Evdokimov AG, Moshiri F, Lowder C, Haas J (2020) Crystal structure of a Vip3B family insecticidal protein reveals a new fold and a unique tetrameric assembly. Protein Sci 29:824–829

Zhong Y, Ahmed S, Deng G, Fan W, Zhang P, Wang H (2019) Improved insect resistance against *Spodoptera litura* in transgenic sweet potato by overexpressing Cry1Aa toxin. Plant Cell Rep 38:1439–1448

Zhou H, Arrowsmith JW, Fromm ME, Hironaka CM, Taylor ML, Rodriguez D, Pajeau ME, Brown SM, Santino CG, Fry JE (1995) Glyphosate-tolerant CP4 and GOX genes as a selectable marker in wheat transformation. Plant Cell Rep 15(3–4):159–163

Zhou S, Chen X, Zhang X, Li Y (2008) Improved salt tolerance in tobacco plants by co-transformation of a betaine synthesis gene BADH and a vacuolar Na+/H+ antiporter gene SeNHX1. Biotechnol Lett 30(2):369–376

Zhou G, Guan R, Li Y, Chang R, Qiu L (2009) Molecular characterization of GmNHX2, a Na+/H+ antiporter gene homolog from soybean, and its heterologous expression to improve salt tolerance in Arabidopsis. Chin Sci Bull 54(19):3536–3545

Zhou Q, Yu Q, Wang Z, Pan Y, Lv W, Zhu L, Chen R, He G (2013) Knockdown of GDCH gene reveals reactive oxygen species-induced leaf senescence in rice. Plant Cell Environ 36:1476–1489

Zhou L, Wang NN, Gong SY, Lu R, Li Y, Li XB (2015) Overexpression of a cotton (*Gossypium hirsutum*) *WRKY* gene, *GhWRKY34*, in *Arabidopsis* enhances salt-tolerance of the transgenic plants. Plant Physiol Biochem 96:311–320

Zhu JK (2001) Plant salt tolerance. Trends Plant Sci 6:66–71

Zhu X, Galili G (2003) Increased lysine synthesis coupled with a knockout of its catabolism synergistically boosts lysine content and also transregulates the metabolism of other amino acids in Arabidopsis seeds. Plant Cell 15:845–853

Zhu X, Galili G (2004) Lysine metabolism is concurrently regulated by synthesis and catabolism in both reproductive and vegetative tissues. Plant Physiol 135:129–136

Zhu YX, Ou-Yang WJ, Zhang YF, Chen ZL (1996) Transgenic sweet pepper plants from *Agrobacterium*-mediated transformation. Plant Cell Rep 16:71–75

Zhu B, Su J, Chang M, Verma DP, Fan YL, Wu R (1998) Overexpression of a Δ1-pyrroline-5-carboxylate synthetase gene and analysis of tolerance to water-and salt-stress in transgenic rice. Plant Sci 139(1):41–48

Zhu XG, de Sturler E, Long SP (2007) Optimizing the distribution of resources between enzymes of carbon metabolism can dramatically increase photosynthetic rate: a numerical simulation using an evolutionary algorithm. Plant Physiol 145:513–526

Zhu C, Naqvi S, Breitenbach J, Sandman G, Christou P, Capell T (2008) Combinatorial genetic transformation generated a library of metabolic phenotypes for the carotenoid pathway in maize. Proc Natl Acad Sci U S A 105:18232–18237

Zhu XG, Long SP, Ort DR (2010) Improving photosynthetic efficiency for greater yield. Annu Rev Plant Biol 61:235–261

Zhu S, Li Y, Vossen JH, Visser RGF, Jacobsen E (2012) Functional stacking of three resistance genes against *Phytophthora infestans* in potato. Transgenic Res 21:89–99

Zhu F, Zhang P, Meng YF, Xu F, Zhang DW, Cheng J, Lin HH, Xi DH (2013) Alpha-momorcharin, a RIP produced by bitter melon, enhances defense response in tobacco plants against diverse plant viruses and shows antifungal activity in vitro. Planta 237(1):77–88

Zhu F, Yuan S, Zhang ZW, Qian K, Feng JG, Yang YZ (2016) Pokeweed antiviral protein (PAP) increases plant systemic resistance to Tobacco mosaic virus infection in *Nicotiana benthamiana*. Eur J Plant Pathol 146:541–549

Zhu F, Zhou YK, Ji ZL, Chen XR (2018) The plant ribosome-inactivating proteins play important roles in defense against pathogens and insect pest attacks. Front Plant Sci 9:146

Ziegler-Graff V, Guilford PJ, Baulcombe DC (1991) Tobacco rattle virus RNA-1 29K gene product potentiates viral movement and also affects symptom induction in tobacco. Virology 182:145–155

Zimmerli L, Hou BH, Tsai CH, Jakab G, Mauch-Mani B, Somerville S (2008) The xenobiotic beta-aminobutyric acid enhances Arabidopsis thermotolerance. Plant J 53:144–156

Zimmermann MB, Hurrell RF (2002) Improving iron, zinc and vitamin A nutrition through plant biotechnology. Curr Opin Biotechnol 13(2):142–145

Zipfel C, Felix G (2005) Plants and animals: a different taste for microbes? Curr Opin Plant Biol 8:353–360

Zou CQ, Zhang YQ, Rashid A, Ram H, Savasli E, Arisoy RZ, Ortiz-Monasterio I, Simunji S, Wang ZH, Sohu V, Hassan M (2012) Biofortification of wheat with zinc through zinc fertilization in seven countries. Plant Soil 361:119–130

Zubko E, Scutt C, Meyer P (2000) Intrachromosomal recombination between aatP regions as a tool to remove selectable marker genes from tobacco transgenes. Nat Biotechnol 18:442–445

Zulfiqar F, Akram NA, Ashraf M (2020) Osmoprotection in plants under abiotic stresses: new insights into a classical phenomenon. Planta 251(1):1–7

Zunjare RU, Hossain F, Muthusamy V, Baveja A, Chauhan HS, Bhat JS, Thirunavukkarasu N, Saha S, Gupta HS (2018) Development of biofortified maize hybrids through marker-assisted stacking of beta-carotene hydroxylase, lycopene-epsilon-cyclase and opaque2 genes. Front Plant Sci 9:178

Zuo Y, Xue Y, Lu W, Ma H, Chen M, Wu Y, Hu Z (2020) Functional validation of nicotinic acetylcholine receptor (nAChR) a6 as a target of spinosyns in *Spodoptera exigua* utilizing the CRISPR/Cas9 system. Pest Manag Sci 76:2415–2422

5 Molecular Pharming

Abstract

There are different expression platforms (transgenic plants, plant cell culture, bacteria, yeast, microalgae, animals, and animal cell culture) for the production of pharmaceuticals. The usage of plants as bioreactors for the manufacture of therapeutically and industrially important molecules like proteins, carbohydrates, and lipids in plants is termed as "molecular pharming." Plant molecular pharming has become a lucrative biotechnology industry as it is a safe, cost-effective, and eco-friendly technique with reduced risk of contamination with human and animal pathogens; allows full posttranslational modification; requires simple growth requirements with unlimited scalability in the field; and allows targeting of the recombinant proteins to different organs or subcellular compartments for accumulation without proteolysis. Over the past few years, several reports on the plant-based expression of pharmaceutically important proteins have been published. To amplify the yield and quality of biopharmaceuticals, several factors including the type of host plant, gene construct, subcellular localization, post-translational modifications, protein extraction, and downstream processing are crucial. Several plant-expressed molecules have been commercialized, and the rest are in the pipeline under clinical trials. This chapter deals with the importance, limitations, challenges, recent developments, and future prospects of molecular pharming.

Keywords

Glycosyltransferase · *Agrobacterium tumefaciens* · Transformation · Recombinant proteins · Edible vaccines

B. Koul, *Cisgenics and Transgenics*, https://doi.org/10.1007/978-981-19-2119-3_5

5.1 Introduction

The process of generating recombinant proteins in crops with sole purpose of exploiting (and eventually commercializing) the protein instead of any quality or function it imposes upon the plant is referred to as "molecular pharming" (Fischer and Schillberg 2017). The era of molecular pharming commenced in the 1980s with the production of insulin and growth hormones (human proteins) using the bacterium *Escherichia coli*, and since then, different alternative living systems such as fungi, other bacterial species, plant and animal cells in culture, and genetically modified species have been explored (Twyman, Stoger, Schillberg, Christou, and Fischer 2003). This technique varied from other genetic transformation methods like agronomic engineering where the proteins provide useful agronomic traits such as insect resistance, drought tolerance, increased yield, disease resistance, and metabolic engineering in which the expressed protein confers catalytic functions while the value-added product is a specific metabolite (Twyman, Stoger, Schillberg, Christou, and Fischer 2003).

The academic organization's enthusiasm for molecular pharming reflected many predicted and demonstrated benefits of plant species over conventional microbial based methods. The benefits involved the following: (a) the cost of forming and sustaining biopharmaceutical plants is cheaper than the cost of constructing and maintaining the facilities for fermentation processes; (b) adaptability/scalability of plants compared to yeast-based technique; and (c) the safety profile is acceptable considering the fact that plants do not favor the replication of pathogenic organisms in mammals as many crops are placed under the "generally regarded as safe" (GRAS) category (Fischer, Schillberg, Buyel, and Twyman 2013; Fischer and Buyel 2020). This has led to the excessive research articles with distinct proteins' expression in different plant-based systems, and several industries were formed with the goal of commercializing the unique technology. The company's manufacturing of non-pharmaceutical proteins achieved a decent level of success; nevertheless, the enthusiasm created by potential pioneering work was swiftly subdued by the harsh reality of commercial pragmatism in the pharmaceutical sector (Buyel 2019). In molecular pharming, the protein is typically extracted via plant cells, but that is not the case many a time. Although genetic pharming could pertain to the development of almost all types of beneficial protein, such as enzymes used industrially, reagents used in technical processes, products used in the nutrition, and materials produced using proteins, the term is typically used to refer to pharmacological substances (proteins), where protein synthesis (pharming) is also used. Molecular farming includes things like plant tissue and organ cultures (Santos, Abranches, Fischer, Sack, and Holland 2016), water plants (Everett et al. 2012), mosses (Decker and Reski 2012), phytoplankton (Rosales-Mendoza et al. 2012), and in vitro transcriptional replication. Because of the variety of these crop systems, molecular pharming encompasses a wide range of channels which could remain competitive in a wide range of markets, from technical enzymes and research reagents manufactured in bacteria and yeast to biomedical gene products in mammals, particularly Chinese ham (Schillberg, Raven, Spiegel, Rasche, and Buntru 2019).

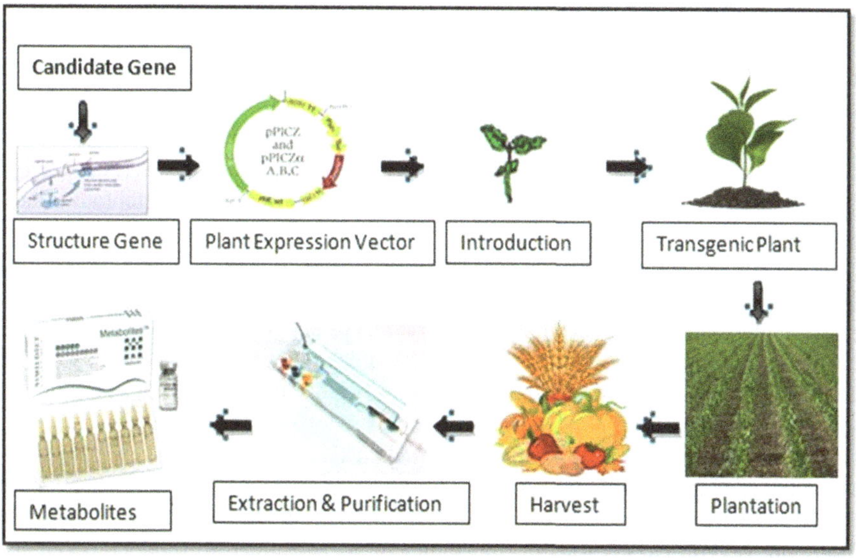

Fig. 5.1 Molecular pharming technique: a simplified illustration

The potential of succeeding in a wide range of markets highlights advantages of every plant-based system. GM crops, for instance, are less cost effective and customizable than Chinese hamster ovary (CHO) cells (Buyel, Twyman, and Fischer 2017), transient expression system enables production to expand up much quicker than that of any fermenter-based platform (Hiatt et al. 2015; Holtz et al. 2015), and plant cells allow for the manufacturing of uniquely tailored glycan structures (Buyel, Buyel, Haase, and Fischer 2015; Hiatt et al. 2015; Schillberg et al. 2017; Fischer et al. 2018). Plant-based systems are deemed fundamentally safer for medicinal products over mammalian cells since they do not enable mammalian virus replication (Hundleby, Sack, and Twyman 2018). They also respond to customer requests for "certified animal-free" products (Spiegel, Stöger, Twyman, and Buyel 2018).

With all the benefits of this pharmaceutical technique in plants, it is yet to fully implement a new generation of commercial synthetic products of protein technologies, with only a few of the products hitting the public places (Fischer and Buyel 2020; Schillberg, Raven, Spiegel, Rasche, and Buntru 2019).

The vast majority of underdeveloped countries are unable to afford the exorbitant expenses of medical treatments that arise from current procedures. As a result, we must manufacture not only novel pharmaceuticals but also lower cost copies of existing samples on the market. Molecular farming has the potential to provide cost-effective solutions to meet the growing need for biomedicines (Ahmad 2014; Alireza and Nader 2015). Figure 5.1 shows a flow diagram of the various steps involved in the molecular pharming technique.

5.2 Benefits of Molecular Pharming

In theory, recombinant proteins may be made in any living organism, and commercial techniques related to bacteria, yeasts, and mammal tissues existed even before the first molecular pharming trials. The pharmaceutical sector traditionally centered on a few conventional systems, such as *Escherichia coli, S. cerevisiae, Pichia pastoris*, yeast, and also a handful of well-studied arthropod and vertebrate (mammalian) stem cells. Plants, as relative latecomers, present a significant entrance hurdle because the regulations governing pharmaceutical manufacture have evolved around these systems (Fischer and Schillberg 2017). The bulk of medications are manufactured utilizing microorganisms in bioreactors, which are expensive to establish and sustain. Plants, on the other hand, can indeed be grown in greenhouses for a low cost of ownership. Fermentation process is not just expensive, but also tough to scale up, whereas producing multiple plants is a lot simpler. Plants, on the other hand, do not aid in the replication of mammalian viruses, which is a central issue in mammalian cultured cell procedures (Fischer et al. 2015). Plants had taken more than two decades to acquire momentum in the market, despite these advantages. There are several causes for this, but they can all be refined to three components. Initially, other systems started on early, which was followed by three decades of pressure, and their performance is evaluated making them incredibly effective, whereas plants began with poor outputs and have battled to keep up. Secondly, the companies have invested in other systems and have established a stable regulatory framework, so there is a reluctance to give up on the costly fermenters and production facilities for a somewhat emerging method and an unstable regulatory platform based on plants. Finally, there is lack of unified body campaigning for the acceptance of a particular approach which is based on plant since molecular pharming is so diverse, with several subspecies and expression methodologies competing for attention. In recent years, many of these issues have been resolved. To begin with, genomic pharming research has led to the quick creation of yield-competitive techniques. Furthermore, manufacturing inertia has also been conquered by (1) constructing guiding principles for pharmaceutical drugs produced in plant species, (2) creating rules for medications made in plant species, and (3) concentrating on goods which do not immediately strive with the existing systems. Lastly, the genomic pharming society is currently concentrating on a few numbers of technological systems, which will be described in greater depth further down.

5.3 Platforms for Molecular Pharming

The large number of genomic pharming methods highlights the fact that peptides were created in a large number of species, using entire plant species or numerous cell/tissue culture media. Each of these could work for both long-term and short-term production (in some species, including nuclear and plastid transformation). This befuddling array of options has recently converged on a few systems that happened to be the most commercially viable as described below:

5.3.1 Plants with Transgenes

The notion of molecular farming was demonstrated for the first time in the year 1989. It was then a recombinant protein that got synthesized in genetically transformed (transgenic) tobacco plant species. Transgenic plants are generated in several plants by changing explants or calluses and then recovering complete plants being selected, while in some few cases, plants can be converted directly devoid of tissue culture. The transgenic that codes for the desired recombinant protein is frequently connected to an antimicrobial or herbicide tolerance biomarker, and also the plants are reproduced under preference so as to assure that the transgene is available in each of the cells. For genetic modification, *Agrobacterium tumefaciens* or an immediate dissemination (transfer) approach is used. Transgenic plants as a platform have two major advantages, which include the potential to establish a lasting genomic reserve (resource) and the possibility for agricultural output. A seed bank, which works in a similar way as a cell pool (bank) in fermentation platforms and follows great industrial practice criteria, is critical for the development of therapeutics. Several transgenic species have already been investigated as potential mass (hosts) for genomic pharming, but two main systems have been materialized as being well established publicly. The leafy crops, including tobacco species, are the initial ones, which have the benefit of being able to yield a huge amount of biomass, enabling a wide-scale production. Due to increased water content of foliage plants, recombinant proteins should be extracted as quickly as possible upon harvesting to prevent deterioration, which necessitates the proximity of processing facilities. The latter issue has been resolved via seed crops, which are depicted by cereals. The recombinant protein is viable and could be preserved in seed crops since it is protected in a dry condition. Considering the fact that their seeds are palatable, they can be employed as an oral delivery system for vaccination and preventive therapeutics. Additionally, cereal grain molecules are secreted into protein storage sites in the digestive tract of animals, enabling oral vaccines to endure the breakdown (digestion) and interact with the immune system.

5.3.2 Suspension Cell Cultures

For the first time in 1990, recombinant protein synthesis was reported in plant cell suspension cultures. Plant cells were the first marketable discoveries (breakthroughs) in the medical companies because of their proximity to the conventional microbial based generation processes, given the fact that initial commercial uses concentrated on transgenic plants. Plants have had the same cost and confinement advantages as microbes, but they do not produce endotoxins and therefore can carry out higher eukaryotic posttranslational modifications. Because plant cells do not allow the multiplication of human infections/pathogens, they have a better safety profile than animal cells. Plant cell suspension cultures could manufacture recombinant proteins starting from the initial stage or from an established genetically transformed line by altering the cells/tissues in suspension and choosing those that have incorporated the genetically altered gene (transgene). Plant cell cultures, like other microbial based techniques, have restricted usability; nevertheless, the output (product) can be restored from the medium, enabling continual protein synthesis; however, enzymatic or mechanical procedure must be employed for the release of some bigger recombinant proteins that are held by the plant cell wall. In addition to cell culture systems, other tissue culture platforms have been established, the most prominent of which being the tobacco hairy root platform. *Agrobacterium rhizogenes* infection causes the root tissue to grow and release vast quantities of recombinant protein. Genomic hairy culture can be brought about by attacking plants with an *A. rhizogenes* species, which do also contain the genome for a recombinant protein, or the roots can be obtained via an existing genetically modified plant line. As confined broad plant systems, moss-based strategies created by German biotechnology business Greenovation GmbH as well as duckweed (*Lemna minor*) have also been established by a US-based company (US biotech company Biolex Inc.). Comparable to suspension of cell cultures, the plants are prospered/grown in a simple media and recombinant proteins are released on a regular basis. Aquatic plants, unlike most other cultivated plant cells, need light, and as such the manufacturing tanks should be lighted on a regular basis.

5.3.3 Temporal Expression Systems/Platforms

As it implies specificity which is not always reliable, this term "temporal expression" is a deluding terminology. Instead of the period of expression, the property that unifies most of these technologies is the absence of the modified gene (transgene) convergence into the plant cell genetic material. Any genetic engineering system utilized for durable/stable conversion could also accomplish temporal expression, due to the fact that genome (DNA) transmission into cells is far much productive compared to transformation (chromosomal integration). The Agrobacterium-based modifications and direct transfer systems are similar to particle bombardment, which incorporates DNA into a wide variety of cell types, but only a small proportion of those cells consolidate it, which is the reason for requiring selectable markers to favor the small percentage of optimally modified cells/tissues during rejuvenating.

This temporary expression technology is referred to as agro-infiltration, and its adaptability for industrialization has enhanced its widespread usage in the synthesis of vast amounts of recombinant proteins that must be produced quickly. So many manufacturing companies including Icon Genetics, Medicago, and Kentucky have embraced agro-infiltration-based systems in closely related species of tobacco (*Nicotiana benthamiana*), which were deployed to create antimicrobial compounds to treat influenza viral infection and other microbial disorders in few days (a matter of weeks), as opposed to the months spent using the existing system like egg-based

Table 5.1 Expression systems for the synthesis of recombinant proteins

Expression platforms	Merits	Demerits
Bacteria	Easy to modify, reduced price, high expression, scale-up simplicity, fast turnaround period, well-established regulatory procedures, and authorization	Endotoxin buildup, defective folding, and a shortage of posttranslational modifications that could affect enzyme activity
Fungi (yeast)	Easily manipulated, rapid expansion, and scalability, media and cultural environments that are simple and economical, hybrid proteins' posttranslational modifications	Because of the strong and rigid cell walls, cell rupture is difficult, proteins are hyperglycosylated, and glycosylation potential is reduced
Mammalian cells	Existing regulatory authorization and efficient folding and posttranslational modifications	High costs of culture media condition requirements, as well as limited profitability
Insects' cells	Higher expression levels, the potential to construct complex proteins including secretory, membranes, and intracellular proteins, as well as accurate folding and posttranslational alterations	Expensive culture media condition requirements, which are costly and time demanding
Plants	Rapid and low-cost growing conditions free of pathogens and bacterial toxins as well as economical posttranslational modification similar to mammalian systems	Weak glycosylation ability and adherence to regulations

system and conventional cellular expression platforms. Agro-infiltration's rates are determined by the number of cells that can be transfected by infiltrating microorganisms. Plant viruses, on the other hand, penetrate directly into cells, proliferate inside them, and spread through two ways, which include direct cell-to-cell mobility and general (systemic) transmission via the vascular bundles. Plant-infecting viruses can hence serve as vectors in the transmission and expression of the genetically transformed genes (transgenes) as they move around the plant body. Several RNA viruses of plant like mosaic viruses of tobacco, viruses infecting cowpea plant, as well as virus X in potato are capable of constructing their glycoprotein through their genetic material different from plant DNA viruses, which "headstuff" their genetic code into a preformed capsid, and this limits the percentage of exogenous DNA that can be included (Anneli et al. 2014).

Various recombinant protein expression systems as well as their potential benefits and downsides are shown in Table 5.1.

5.4 Molecular Pharming in Mammalian Organisms

The growth of biotechnology and the emergence of robotic genomic studies have cleared the path for pharming to proliferate. Furthermore, utilizing the inherent benefit of the mammalian organisms to achieve the intended objective/purpose has opened up a plethora of new possibilities. The present approach of producing biologicals is an improvement on the previous method of achieving the desired results by employing transgenic bacteria, yeast, and animal cells. Traditional processes, on the other hand, had a number of flaws, including a lack of economic feasibility, difficulties isolating and purifying products, necessity of sophisticated apparatus for the upkeep, and complication in large-scale output/production. Numerous theoretical methods, such as blood, egg white, seminal plasma, and urine, were proposed to address the market expectations for pharmaceuticals, but none of them produced sufficient results.

Several studies have revealed, for example, that blood is incapable of storing large amounts of viable antimicrobial peptides (Figueiredo and Blaszczyk 2014). Milk was shown to be the most effective model for producing and purifying recombinant proteins and nanomaterials out of all the mammalian systems studied (Houdebine 2009). The expression system that involved the use of milk from animal (e.g., cow, sheep, goats) is relatively popular, and its prevalence and versatility in our daily lives render it more acceptable. The pronuclear microinjection approach is often used to develop genetically transformed (transgenic) animals that are able to produce pharmaceutics in milk, accompanied by the use of cloning technologies to make extra progeny with the desired modified genome (Dove 2000). Unlike cell culture, animals have cellular structure that allows them to create complex proteins naturally. Furthermore, the use of animals as bioreactors reduces the costs of raw materials, capital equipment, and maintenance associated with traditional methods, lowering the unit cost per protein produced. Therapeutic proteins for disorders such as cystic fibrosis, hemophilia, osteoporosis, rheumatism, fever, HIV, and chemotherapeutic agents are produced cost-effectively in animal bioreactors. The invention of animal pharming in the "Enviropig" has led to the development of pig possessing enzyme (phytase) in its salivary glands for greater utilization of phosphorus elements in its foods, hence resolving the animal waste issue. Another successful application of this method was observed in the creation of an antithrombin molecule (ATryn), ensued from the milk of transgenic goat, which received USDA authorization in 2009. The food production security, quality of the active drug ingredient, and maintaining transgenic animals off from non-transgenic animals are among the downsides associated with animal pharming (US Department of Agriculture 2006).

5.5 Plant-Based Molecular Pharming

Plant-made pharmaceuticals is a subdiscipline of bioengineering that focuses on developing genetically engineered plants to produce medications and other compounds. Among the techniques used were the development of reliable nuclear transgenic plants, transplastomic, temporal expression via a plant virus, as well as temporal expression through *Agrobacterium*-based infiltration (Floss et al. 2007). Typically, the intended DNA containing the required genes to manufacture the medications is injected into a plant viral species, *Agrobacterium*, or directly into the plant genetic material, which is deciphered by the plant's protein-formulating machinery/system to maximize the production of pharmaceutically therapeutic agents along with other plant proteins, similar to a bioreactor. Numerous organisms have been considered as candidates, including *Lemna minor* and *Physcomitrella patens*. The plants might be raised in bioreactors, with the desired output being generated in secretory state, lowering the burden of separation processes significantly (Büttner-Mainik et al. 2011). Manufacturing therapeutics via plants is simple and efficient, with 80% fewer expenses for upkeep, security, preservation, time, and dissemination than in mammalian or microbial cell-based cultures, where scaling up synthesis and purification of proteins was historically a costly and difficult process. Plant transformation mechanisms considerably outperform typical microorganisms in terms of producing and assembling huge numbers of complex proteins including antibodies, pheromones, interleukins, serum proteins, as well as vaccines into their physiologically effective configuration through posttranslational transformation (Norris 2005). Plants do not transmit microbes that could be harmful to human health, and they do not possess proteins which are identical to the proteins made in human system. However, plants are associated to both humans and other animals in the sense that they might correctly process and structure both people's and animal's polypeptides. Plants that have been molecularly cultured possess some physiologically effective chemicals that aggregate within the plant cells and tissues and are not meant for food purposes. Although genetic modification may not impact the product on its own while improving a crop in some ways, prudence and moderation are required to preserve the ecosystem biodiversity and consumers' well-being (Norris 2005).

5.5.1 Molecular Pharming of Proteins

Glycosylation have a substantial effect on the functioning of transgenic glycoproteins, and the effects of various glycosylations on the activity of transgenic glycoproteins have been thoroughly investigated (Elliott et al. 2004; Walsh and Jefferis 2006; Jefferis 2009; Dalziel et al. 2014). Considering the significance of glycosylation in proteins, biopharmaceutical production must take into account the expression, hosts' glycosylation skills and devise technology to manage restrictions and synthesize specific glycan structures (Dicker and Strasser 2015; Reusch and Tejad 2015). Deconstruction of glycosylation routes by removing undesirable or striving glycan improvements is the initial stage in developing properly controlled and homogenized glycans on the glycoprotein peptides (Hamilton et al. 2003; Strasser et al. 2008; Yang et al. 2015). The production of missing glycosylation enzymes or processes, which yield different glycan structures which are not ordinarily present in the expression host, has been the second stage in conventional glycoengineering procedures (Hamilton et al. 2006; Castilho et al. 2010; Yang et al. 2015). Plants and primates share a similar protein-synthesizing mechanism, and plants are capable of folding and assembling complicated human recombinant proteins so easily (Ma et al. 1995; Loos and Steinkellner 2014). Plants, like mammals, can go through similar modifications, especially in terms of glycosylation, and can tolerate major changes in their N-glycan patterns without compromising the development, growth, and reproduction (Strasser et al. 2014; Steinkellner and Castilho 2015).

5.5.1.1 N-Glycan Biotechnology

Glc3Man9GlcNAc2 is transported en bloc from some kind of lipid-associated carrier to specific Asn residue within emerging polypeptides in the endoplasmic reticulum (ER), exposing the N-glycosylation convergence site AsnXSer/Thr with X representing any amino acid with the exception of proline. A sequence of carefully coordinated enzymatic processing steps convert the delivered oligosaccharide into oligomannosidic complex, or paucimannosidic N-glycans. Although the initial stages of the N-glycan distilling paths are preserved across eukaryotes, the Golgi apparatus' progression into the complicated N-glycans is chiefly dependent on the availability of particular N-glycan encoding proteins and the needed nucleotide sugars that assist as donor substrates in glycosylation reactions (Strasser 2016). As a result, changes in N-glycan structures linked to plant or mammalian derived glycoproteins are principally caused by the specific repertoire of Golgi-resident glycosyltransferases found in these species. Plants, unlike mammalian cells, carry out complex N-glycan development with exceptional uniformity and exhibit a limited range of N-glycan configurations due to a limited collection of N-glycan-processing enzymes. Plant-based glycoproteins often have their complex N-glycans changed with nonmammalian 1,2-xylose and core 1,3-fucose residues rather than core 1,6-fucose, which is commonly found on glycoproteins generated in mammalian cells. Specific hexosaminidases are thought to catalyze this process in post-Golgi compartments, resulting in paucimannosidic N-glycans, a truncated oligosaccharide with mannose residue at the end (Liebminger et al. 2011). Plants lack the 1,4-galactose and terminal sialic acid residues determined in several medicinally valuable cellular proteins in man. Additional transformations, like the creation of N-glycans (multiantennary) observed in human erythropoietin (EPO) and N-glycans with a bisecting GlcNAc found in some human serum proteins, are not performed by plants since the required N-acetylglucosaminyltransferases are missing (Shin et al. 2017).

5.5.1.2 Glycosyltransferase Expression in Plants Can Be Modified

N-glycan growth varies as the proteins travel/move across the Golgi apparatus. N-glycan processor protein converts oligomannosidic N-glycans to complicated N-glycans through a range of phases that rely on the N-glycan's availability/accessibility to Golgi processing proteins, as well as the expression and centralization of glycosylation proteins and the sugars they produce. In terms of recombinant glycoproteins designed for therapeutic application, the question of whether such potentially immunogenic N-glycan structures may be tolerated is still being debated (Bardor et al. 2003; Yao et al. 2015; Jin et al. 2008; Piron et al. 2015; Zimran et al. 2014). Nonetheless, a more humanlike N-glycosylation pattern is often desired to prevent any hazards. The expression of recombinant glycoproteins devoid of 1,2-xylose and core 1,3-fucose in plants lacking GnTI function is one technique. The transfer of 1,2-xylose and core 1,3-fucose requires the inclusion of a terminal GlcNAc residue via GnTI. Endogenous glycoproteins on Arabidopsis compound glycan 1 (cgl1) mutants and rice concentrate largely on Man5GlcNAc2, a N-glycan-processing intermediary observed in mammals and plants (Fanata et al. 2013). This oligomannosidic N-glycan has also been identified in moss-derived human galactosidase A expressed in a GnTI-deleted manufacturing burden (Shen et al. 2016) and human glucocerebrosidase expressed in an *N. crassa* production strain (Shen et al. 2016). The two recombinant proteins (glucocerebrosidase and iduronidase) were affirmed in the seed of Arabidopsis, cgl1, with substantial amounts of processed sophisticated and paucimannosidic N-glycans (He et al. 2012, 2013; Fanata et al. 2013). The unexpected results could be described by the fact that cgl1 is a dependent lethal mutant with GnTI function in the seeds (Frank et al. 2008). Other methods for removing 1,2-xylose and core 1,3-fucose have concentrated on lowering the expression of the enzymes responsible, 1,2-xylosyltransferase (XylT), core 1,3-fucosyltransferase (XylT), and core 1,3-fu (FucT). To do this, a T-DNA mutant collection was screened in Arabidopsis (Strasser et al. 2004; Schähs et al. 2007). Researchers employed homologous recombination to knock out the coding genes in the moss *Physcomitrella patens* to inactivate the XylT and FucT genes (Koprivova et al. 2004; Huether et al. 2005).

5.5.1.3 Biotechnology of O-Glycans

Proteins containing O-linked glycans are engaged in a variety of biochemical functions. This process has been found in a wide number of distinct human proteins, according to detailed O-glycosylation site-specific investigations (Steentoft et al. 2013; Hoffmann et al. 2016). Many researches on the role/benefit of O-glycans have previously employed techniques of mutation to eliminate a particular O-glycosylation region and compared the resulting mutants to O-glycosylated wild-type proteins (Novak et al. 2012; Dai et al. 2015). Changes in amino acid residues may cause conformational changes irrelevant to glycosylation, affecting how these results are evaluated. Controlled synthesis of specific O-glycan patterns will provide the way for a better understanding of O-glycan functions/processes on mammal glycoproteins, as well as improved glycoprotein treatments. Consequently, preexisting protein expression systems will need to be modified to allow for the synthesis of custom O-glycan.

5.5.1.4 O-Glycosylation in Plants

O-glycosylation, the second primary kind of protein glycosylation, is depicted by the attachment of glycan subunits/moieties to access serine/threonine debris. The first step in the construction of mucin (a form of glycan) is the transfer of N-acetylgalactosamine (GalNAc) residue from the nucleotide sugar UDP-GalNAc to the O-glycosylation site. Mucin-type O-glycans are a familiar O-linked glycan found in the secretory glycoproteins in mammalian organisms including man. The catalysis of the precise glycosylation reactions is mostly carried out in a variety of polypeptide units (GalNAc transferase family) in the Golgi body (Bennett et al. 2012). Monosaccharides as fucose, GlcNAc, galactose, and sialic acid are gradually introduced into mammals, extending the O-linked GalNAc (Tarp and Clausen 2008).

5.5.1.5 Recombinant Proteins Generated from Plants

Plants are immensely useful living systems that can be accessed to produce large percentage of polypeptides in the field of molecular agriculture. Seeds have been shown to accumulate a huge number of molecules, comprising vaccines, hormones, and other enzymes (proteins), as vehicles for pharmaceutical manufacturing throughout the last century (Fung et al. 2005; Lamphear et al. 2005; Downing et al. 2006; Nykiforuk et al. 2006; Van Droogenbroeck et al. 2007; Rademacher et al. 2008; Xie et al. 2008; He et al. 2011; Hegedus et al. 2014; Li et al. 2014, b; Ritala et al. 2014; Hensel et al. 2015; Yi et al. 2015; Montesinos et al. 2016). For this reason, several plant varieties including Arabidopsis, rice, barley, maize, tobacco, and others have all been investigated. Each species has its own set of benefits and drawbacks. Maize produces much seeds, and thus it takes up less space. Self-pollination minimizes the possibility of gene contamination in rice and barley. Oilseed legumes have a high protein content, making them ideal for collecting the necessary proteins, which have been either covalently or noncovalently bound to the surface (exterior part) of the OB, each requiring its own retrieval technological system. Table 5.2 shows the reports on molecular farming of pharmaceutically important proteins in transgenic plants.

5.5.2 Molecular Pharming of Carbohydrates

Carbohydrates are vital components of organisms and are involved in a variety of physiological and pathological processes, including cell surface identification, signal transduction, tumor metastasis, and so on. However, in comparison to nucleic acids and proteins, sugars, the third main type of biopolymers, have attracted less attention in pharmaceutical production due to a recent discovery of glycobiology's fundamentals (Cai and Li 2008; Zhang and Wang 2015). Despite this, carbohydrate drugs continued to lay the groundwork for a non-dismissible part of the therapeutics universe. The four pharmacopoeias contain roughly 131 carbohydrate medications, that highlight the prominence of carbohydrate therapeutics across both organic substances and pharmacological inert ingredients. Apart from the carbohydrate medications mentioned in the pharmacopoeia, a wide range of carbohydrate drugs have been sold and used in clinics for decades. Eighteen carbohydrate drugs that were not classified in therapeutic categories were discovered following a review of the Chinese Food and Drug Administration (CFDA) and the US Food and Drug Administration (FDA) data (CFDA 2015). Apart from *Coriolus versicolor* cellulose and lentinan, that were synthesized in Japan, almost all antivirals were sold in China

Table 5.2 Molecular farming of pharmaceutically important proteins

Crop name	Scientific name	Gene	Gene source	Useful trait	Expression (%)	References
Alfalfa	*Medicago sativa*	α-Amylase	Microorganism	Potential effect of genetically engineered organism on soil ecosystem	–	Austin et al. (1995)
Alfalfa	*Medicago sativa*	*Mannheimia haemolytica* GS60 antigen	*Mannheimia haemolytica* A1	Bovine *pneumonic pasteurellosis*	–	Lee et al. (2008)
Apple	*Malus domestica*	Syncytial virus (RSV)-F protein	Human	A runny nose and ear infections, bronchiolitis, and pneumonia	0–20 mg/g	Lau and Korban (2010)
Arabidopsis and carrot	*Arabidopsis thaliana* and *Daucus carota*	HIV-1 subtype C p24 protein	Human	HIV-1	Carrot taproot had 40% more p24_SEKDEL than Arabidopsis	Lindh et al. (2009)
Banana	*Musa paradisiaca*	Hepatitis B surface antigen	Human	Curing hepatitis B	Buoyant density of banana leaf-derived HBsAg was found to be in the range of 1.1536–1.1385 g/ml	Kumar et al. (2005)
Canola	*Brassica napus*	Hirudin	*Hirudo medicinalis*	Anticoagulant property	1%	Demain and Vaishnav (2009)
Carrot	*Daucus carota*	UreB (urease) protein	*Helicobacter pylori*	Against *helicobacter pylori*	25 μg/g in roots	Zhang and Huang (2010)
Chlamydomonas		aadA and CTB-D2	*E. coli*	*Staphylococcus aureus* infection	–	Dreesen et al. (2010)
Corn	*Zea mays*	(rGUS)	Corn	β-Glucuronidase (GUS)	0.015%	Evangelista et al. (1998)

(continued)

Table 5.2 (continued)

Crop name	Scientific name	Gene	Gene source	Useful trait	Expression (%)	References
Corn	*Zea mays*	Gastroenteritis virus vaccine	Pigs	Glycoproteins	<0.01%	Lamphear et al. (2004)
Corn	*Zea mays*	LT-B	Human	Heat-labile toxin B	–	Tacket et al. (2004)
Maize	*Zea mays*	*Avidin*	Chicken egg	Production of *avidin* in GM maize	2.3%	Hood et al. (1997)
Maize	*Zea mays*	GUS	Kernel	β-Glucuronidase (GUS)	0.015%	Evangelista et al. (1998)
Maize	*Zea mays*	Avidin	Chicken egg white	Production of avidin	2.3%	Hood et al. (1997)
Maize	*Zea mays*	Trypsin	*E. coli*	Bovine trypsin	–	Woodard et al. (2003)
Maize	*Zea mays*	Avicidin	Human	Radiolabeled anti-EpCAM murine IgG	–	Gavilondo and larrick (2000)
Maize kernels	*Zea mays*	LT-B (heat-labile toxin B)	*E. coli*	Heat-labile enterotoxin	–	Chikwamba et al. (2003)
Mustard	*Brassica* seeds	Hirudin (variant 2)	Leech	Production of hirudin	1%	Parmenter et al. (1995)
N. benthamiana	*N. benthamiana*	Hemagglutinin	Influenza virus	H5N1 influenza viruslike particle	20 μg per dose	Landry et al. (2010)
N. benthamiana	*N. benthamiana*	IgG1	Human	Tumor-derived idiotype IgG	–	Tusé et al. (2015)
N. benthamiana	*N. benthamiana*	Glycosyltransferase	Mammals	Chimeric anti-Ebola IgG cocktail	–	Chen et al. (2016)
Shiny bush	*Peperomia pellucida*	LTB	*E. coli*	Strong immune response	0.75%	Loc et al. (2010)

Potato	*N. tabacum*	Norwalk virus vaccine	Human	Level of serum antibody	<0.001%	Tacket et al. (2000)
Potato	*Solanum tuberosum*	Human lactoferrin	Human	Human lactoferrin	0.10%	Daniell et al. (2001)
Potato	*Solanum tuberosum*	Newcastle disease virus (NDV) protein	Newcastle disease virus (NDV)	Synthesis of ND-specific IgG and IgA antibodies	–	Gómez et al. (2008)
Potato	*Solanum tuberosum*	LT-B	*E. coli*	*E. coli* heat-labile enterotoxin B subunit	3.7–15.7/g	Tacket (2007)
Potato and rice	*Solanum tuberosum* and *Oryza sativa*	*Salmo salar* (SasalFN-α1) protein	Salmo salar	Pancreatic necrosis virus infection	–	Fukuzawa et al. (2010)
Potato virus X (PVX) coat protein	*Solanum tuberosum*	HIV-1	Mice	Protection against HIV	–	Marusic et al. (2001)
Rice	*Oryza sativa*	Cry j I and cry j II	Japanese cedar pollen	T-cell Proliferation and IgE-associated Th2 cytokine responses	0.5%	Takagi et al. (2005)
Rice	*Oryza sativa*	Human α1-antitrypsin	Human	Maintains appropriate levels of neutrophil elastase and other proteinases in the circulatory system	100–247 mg/l (4–10%)	McDonald et al. (2005)
Rice	*Oryza sativa*	Japanese encephalitis virus envelope protein	Japanese encephalitis virus	Increases immunogenicity	1.1–1.9 μg/mg	Wang et al. (2009)
Rice seeds	*Oryza sativa*	hIGF-1	Human	Diabetes mellitus	6.8 ± 0.5%	Xie et al. (2008)

(continued)

Table 5.2 (continued)

Crop name	Scientific name	Gene	Gene source	Useful trait	Expression (%)	References
Rice, turnip	*Oryza sativa*	hIFN-alpha	Human	Human α-interferon	–	Ceaglio et al. (2010)
Spinach	*Spinacia oleracea*	HIV-1 Tat protein tobacco mosaic virus (TMV)	MN isolate from HIV-1	Vaccine prevents development of disease	300 μg per 1 g spinach leaves	Karasev et al. (2005)
Sunflower	*Helianthus annuus*	hGH (*chimeric*)	Human	Growth hormone	0.02%	Barta et al. (1986); Leite et al. (2000)
Tobacco	*Nicotiana tabacum*	hGH (*chimeric*)	Human	Growth hormone	–	Barta et al. (1986)
Tobacco	*N. tabacum*	rhGH	Human	Trout growth factor	0.16%	Bosch et al. (1994)
Tobacco	*N. tabacum*	hEGF	Human	Human epidermal growth factor	0.001%	Higo et al. (1993); Bai et al. (2007)
Tobacco	*N. tabacum*	Epo	Human	Erythropoietin production	13.1 μg/6.3 kg	Matsumoto et al. (1995)
Tobacco	*N. tabacum*	hIFN α2b	Human	Human β-interferon	–	Edelbaum et al. (1992)
Tobacco	*N. tabacum*	hPC	Human	Human protein C	<1%	Cramer et al. (1996)
Tobacco	*N. tabacum*	MAChR	Human	Human muscarinic cholinergic receptors productions	2–2.5 pmol/mg	Mu et al. (1997)
Tobacco	*N. tabacum*	BD1	Human	Ribosome-inactivating protein	30 mg/L of culture	Francisco et al. (1997)

Tobacco	*N. tabacum*	(IL-2) and IL-4	Human	Interleukin-2 and interleukin-4	0.10 and 0.18 μg/ml	Magnuson et al. (1998)
Tobacco	*N. tabacum*	scFv	Mouse	Anticancer (lymphoma) protein	–	McCormick et al. (1999)
Tobacco	*N. tabacum*	scFv protein	Human	Human placental alkaline phosphatase	2.6–6.8%	Borisjuk et al. (1999)
Tobacco	*N. tabacum*	*Streptococcus mutans* surface protein A (SpaA)	*Streptococcus mutans*	Prevention from dental decay	0.2%	Curtiss and Cardineau (1990)
Tobacco	*N. tabacum*	Serum albumin	Human	Human serum albumin	0.02%	Sijmons et al. (1990)
Tobacco	*N. tabacum*	Human protein C	Human	Production of glucocerebrosidase	>10%	Cramer et al. (1996)
Tobacco	*N. tabacum*	Human somatotrophin	Human	Human growth hormone	7.00%	Jeffrey et al. (2000)
Tobacco	*N. tabacum*	Measles virus vaccine	Animals	Paramyxovirus surface protein hemagglutinin	–	Webster et al. (2002)
Tobacco	*N. tabacum*	IgG (hepatitis B virus)	Human	Envelope surface protein	<0.01%	Valdes et al. (2003)
Tobacco	*N. tabacum*	*Bacillus anthracis* protective antigen	Human	Anthrax vaccine	–	Watson et al. (2004)
Tobacco	*N. tabacum*	MV-H (measles virus hemagglutinin)	Influenza viruses	Detected against measles IgG	–	Webster et al. (2002)
Tobacco	*N. tabacum*	Trichosanthin-α	*Trichosanthes kirilowii*	Treat choriocarcinoma	–	Lei et al. (2006)

(continued)

Table 5.2 (continued)

Crop name	Scientific name	Gene	Gene source	Useful trait	Expression (%)	References
Tobacco	*N. tabacum*	*Plasmodium yoelii* merozoite surface protein4.5 (PyMSP4/5)	*Plasmodium yoelii*	Cures malaria	0.25%	Wang et al. (2008)
Tobacco	*N. tabacum*	Recombinant Norwalk viruslike particles (rNVs)	Norwalk virus	Cure nonbacterial gastroenteritis	0.8 mg/g leaf	Santi et al. (2008)
Tobacco	*N. tabacum*	Hepatitis B surface antigen (HBsAg)	Human	Humoral IgA and IgG response	0.01–0.1%	Kostrzak et al. (2009)
Tobacco	*N. sativa*	Potato aucuba mosaic virus (PAMV) coat protein (CP) gene	Potato	Protection to PAMV infection and a delay in symptom development	–	Appaiahgari et al. (2009)
Tobacco	*N. tabacum*	Rabies virus glycoprotein	*Vibrio cholerae*	Protein recognized by anti-cholera and anti-rabies antibody	0.4%	Roy et al. (2010)
Tobacco	*N. tabacum*	ScFv	Mouse	Anticancer (lymphoma) protein	<86%	McCormick et al. (1999)
Tobacco	*N. tabacum*	SIgA	Human	Hybrid secretory IgA-IgG	–	Weintraub et al. (2005)
Tobacco	*N. tabacum*	2G12	Human	Human anti-HIV IgG	–	Ma et al. (2015)
Tobacco and carrot cells	*N. tabacum* and *Daucus carota*	DTP subunit	*Corynebacterium diphtheriae*, *clostridium tetani*, and *Bordetella pertussis*	Diphtheria–tetanus–pertussis	–	Brodzik et al. (2009)

Tobacco chloroplast	*N. tabacum*	TetC (tetanus vaccine antigen)	Clostridium tetani	Protective antibodies produced	25%	Tregoning et al. (2004)
Tobacco chloroplasts	*N. tabacum*	Foot and mouth virus epitope VP1	*Vibrio cholerae*	Type I diabetes	0.11%	Li et al. (2006)
Tobacco potato	*N. tabacum*	Hepatitis B surface antigen	Yeast	HBsAg	–	Kong et al. (2001)
Tobacco, potato	*N. tabacum*	LT-B	*E. coli*	Heat-labile enterotoxin B	Protein accumulation in pLTB110 plants was up to 5 μg per gram and in pLTB110 plants 30 μg per gram of soluble protein	Haq et al. (1995)
Tomato	*Lycopersicon esculentum*	Rabies virus glycoprotein	Cauliflower mosaic virus	Production of edible oral vaccines	1.00%	McGarvey et al. (1995)
Tomato, potato	*Lycopersicon esculentum*	Cholera Ctox A and Ctox B subunits	*Vibrio cholerae*	Production of vaccines against cholera	0.02–0.04%	Sharma et al. (2008); Choi et al. (2005)
Tomato	*Lycopersicon esculentum*	Hepatitis B surface antigen cherry	Human	Explore the feasibility of producing HBV oral vaccine	Leaves 300 ng/g Fruit 10 ng/g	Gao et al. (2003); Thanavala et al. (2005)
Tomato	*Lycopersicon esculentum*	Pneumonic and bubonic plague		IgG concentrations were increased	4–10%	Alvarez et al. (2006)
Tomato	*Lycopersicon esculentum*	NVCP (Norwalk virus capsid protein)	Human	Glucuronyl transferase	–	Zhong et al. (2005)
Tomato	*Solanum lycopersicum*	*TBI-HBS*	Hepatitis B virus	Edible vaccine against HBV and HIV	–	Shchelkunov et al. (2004)

(continued)

Table 5.2 (continued)

Crop name	Scientific name	Gene	Gene source	Useful trait	Expression (%)	References
Tomato	*Solanum lycopersicum*	*CtrB*	*Vibrio cholerae*	Edible vaccines	Leaves 0.02% Fruits 0.04%	Jani et al. (2002)
Tomato	Pusa early dwarf (PED)	α_1-*PI*	Human	Recombinant α_1-PI	1.55%	Agarwal et al. (2008)
Faba bean	*Vicia faba*	Human cytomegalovirus (HCMV)	Human	Cures asymptomatic and persistent infections	0.005–0.015%	Yan et al. (2010)

as polysaccharides based on ancient Chinese medicinal herbs or fungus. Despite this, the FDA has not approved any of the anticancer adjuvants mentioned. Prescription for the management of plasma (blood) and cardiac illnesses accounts for 22% of all certified sugar-based therapies, positioning them second in place in terms of applications. Fondaparinux is a fully synthetic equivalent of heparin's pentasaccharide component that had been accepted in 2001 for the management of deep venous thrombosis. It is the first Xa inhibitor that specifically binds to blood antithrombin III (AT III). Heparins which have not been segregated and other low-molecular-weight heparins (LMWHs) are known as unfractionated heparin (UFH). As a result, transdermal accessibility was enhanced and the threat of contaminants was reduced (Berqqvist 2006). Voluven, a premium consumer hydroxyethyl starch plasma volume enhancer, has been licensed by Fresenius Kabi (Bad Homburg, Germany). Voluven had a lower molar substitution (0.4) and a lower molecular weight (130 kDa) than Haes (average molecular weight 200 kDa, degree of substitution 0.5; Fresenius Kabi) (Hitosugi et al. 2007). Furthermore, China manufactures a drug called swine duodenum mucopolysaccharide, which is made from the duodenum of healthy pigs and contains more than 20% glucosamine. Additionally, carbohydrates and glycomimetic medications are utilized to treat gastrointestinal ailments, viral diseases, and diabetes due to their structural or physiological qualities. As per the International Gastroenterology Society, Fibre structure, a product developed by Recip AB, is the best solution for constipation. This drug's organic wheat fiber relaxes the feces by soaking a large amount of water without causing diarrhea, which is a frequent consequence of several laxatives. The GlaxoSmithKline-registered sialic acid mimic zanamivir works by inhibiting viral neuraminidase and is extensively used in medical influenza prevention and therapy (Heneghan et al. 2014). Miglitol is a monosaccharide that mimics the structure of glucose in order to avoid glucosidase from breaking down carbohydrates (Aoki et al. 2014). Since its registration by Bayer in 1996, miglitol has been used to manage non-insulin-dependent diabetic mellitus.

5.5.2.1 Drugs Containing Carbohydrates

In Phase III trials, a variety of carbohydrate treatments are being tested. With numerous potential medicines in development, cancer is the most popular topic in contemporary carbohydrate-based pharmaceutical research and development. The unnatural tumor-related sugar-based antigens are regarded to be the most effective in eliciting antitumor immune responses. Theratope is a carbohydrate vaccine that combines a sialyl Tn oligosaccharide epitope with keyhole limpet hemocyanin (KLH) to treat metastatic colon cancer and breast cancer. Monosaccharides and their synthetic analogues are also being studied as normal metabolic enzyme inhibitors. Vevesca is an orally active amino sugar that inhibits the glycosphingolipid glucosyltransferase enzyme biosynthetic pathway. Although the FDA has approved it for the management of uncomplicated Gaucher disease (type 1) in adults, Vevesca's product used against HIV trial did work in the presymptomatic stage (Zhang and Wang 2015). It is implied that D-tagatose can also be used to minimize carbohydrate assimilation by hindering the glucose transport and blocking intestine disaccharidases (Zhang and Wang 2015). It is a potent antidiabetic and obesity-controlling medicine that produced statistically significant outcomes in an NIDDM Phase III trial led by Spherix (Jennifer 2010).

5.5.2.2 Carbohydrate Pharmaceutical Research Potential

Because most presently utilized carbohydrate medications have outstanding pharmacological activity and low toxicity, there is growing interest in developing carbohydrate prospects for innovative drug discovery. Carbohydrate medicines can be discovered and designed using a variety of methods (Zhang and Wang 2015). The most promising collection for carbohydrate drug development is naturally produced complex sugar and their disintegrated components, more specifically those made from ancient Chinese medicinal condiments and fungal species, which have a diverse spectrum of structures and bioactivities. LMWHs, alginic acid, and hyaluronic acid are only a few examples of effective carbohydrate medications emanated from aboriginal glycan configuration, which was used in hospitals for many years (Vogel et al. 2010). In addition, a plethora of nonsynthetic type of sugars (polysaccharides and oligosaccharides) are in the initial stages of the fundamental experiment, which will aid in the development of carbohydrate medications. Polysaccharide molecules from Chinese wolfberry, *Agaricus blazei* Murill, and Meretrix, for example, have entirely been well researched for their effects against cancer, as antioxidants, and for their immunomodulatory actions/effects.

5.5.2.3 Carbohydrate Manipulation

There is significant evidence that proper genomic transformation of native glycans can be used to create novel bioactive compounds, enhance the current pharmaceutics, and/or improve solubility and medicinal properties in plants. Many effective ways were developed for the improvement of complex sugars derived from several origins. Sulfation is a common polysaccharide modification process that gives glycans antiviral, antitumor, and anti-inflammation effects. In a recent study, curdlan sulfate was discovered to positively reduce the HBV infection in HepG2 and HepaRG cells, implying that it is employable as a vaccine adjuvant to prevent or treat HBV abnormalities. When compared to the unsulfated SIP, SIP-SII showed considerable suppressing efficacy for melanoma metastasis via lowering tumor adhesion and angiogenesis (Zong et al. 2013).

5.5.2.4 Drug Modification and Delivery Mediated by Carbohydrates

For pharmacological improvement and drug transport targeting, the recognition mechanisms arbitrated by saccharide ligands and their specific receptors on the respective organs, tissues, and cells have been intensively explored (Yu et al. 2009). Lactosylated microspheres, for example, have been developed to carry medicines and genes to the liver with great particularity for the asialoglycoprotein lectin conveyed on mammalian hepatocytes (Zhang and Wang 2015). Alveolar macrophage-selective drug carriers have been produced using mannose-modified liposomes that target their mannose receptors. A new lectin-directed protein-activated prodrug therapy (LEAPT) has also been developed for regioselective drug release via carbohydrate-lectin interaction-mediated localization of a glycosidase enzyme to certain cells (Davis and Robinson 2002). Because of the hydrophilic external surface and the hydrophobic interior cavity, cyclodextrins have been successfully employed to construct inclusion complexes with various hydrophobic medications to considerably improve their water solubility (Chen et al. 2014). Furthermore, both chitosan and hyaluronic acid are being studied for their novel biological capabilities as drug delivery materials (Morris et al. 2010).

5.5.2.5 Carbohydrate Vaccines

The host immune system can detect both pathogenic bacteria's surface polysaccharides and tumor-associated carbohydrate antigens, which can be exploited to make carbohydrate vaccines (Vliegenthart 2006; Oppenheimer et al. 2008). Vaccines developed from conjugated sugar molecules were found to have a positive effect in the management and prevention of several disorders in man. Many vaccines emanated from carbohydrate based on capsular sugars have been authorized for use in the prevention of bacterial and viral diseases. For instance, for the prevention of juvenile meningitis, a conjugation of pure capsular polysaccharide from *Haemophilus influenzae* type B (HiB) and tetanus toxoid has been advised and recorded in CP and EP. Using chemically generated oligosaccharide clusters as antigen epitopes, many potential antitumor and anti-HIV vaccines are currently being investigated actively. Globo-H, a synthetic hexasaccharide generated from breast cancer cells' surfaces, was covalently linked to KLH and evaluated in a phase I study with positive results. Anti-HIV activity of carbohydrate vaccines based on oligomannose clusters is also being researched (Zhang et al. 2015).

5.5.2.6 Synthesis of Oligosaccharides

Due to the scarcity of natural sources, chemical or biological synthesis is seen to be the most promising technique for obtaining adequate amounts of homogeneous and physically well-defined oligosaccharides for carbohydrate medication production. Unlike polymerase chain reaction (PCR) or solid-phase peptide synthesis (SPPS), however, the structural complexity of carbohydrates makes oligosaccharide synthesis extremely difficult without a universal automatic technique (Seeberger and Werz 2005). Currently, much work is being put into developing efficient oligosaccharide synthetic techniques, and many oligosaccharides, such as Lex, Ley, and Globo-H, have been effectively synthesized (Jeon et al. 2009). Despite their flexibility, most current chemical methods require much time and are labor intensive, with limited site- or stereoselectivity and low yield (Zhou et al. 2009). Enzymatic techniques with great efficiency and selectivity, on the other hand, are frequently constrained by the availability of the necessary enzymes. As a result, chemoenzymatic production, that integrates the plasticity of a chemical method with the efficiency and selectivity of a protein-based method, is regarded among the most successful methods for producing complex oligosaccharides. A disialyl tetrasaccharide epitope was regioselectively produced using a chemical exploitation approach arbitrated by the bacterial sialyltransferase Pd2, 6ST (Meng et al. 2014). For the effective generation of T-antigens containing fluorine and their derivatives (sialylated), a chemoenzymatic method based on bacterial 1,3-galactosyltransferase and 2,3-sialyltransferase has also been devised (Yan et al. 2013).

5.5.3 Molecular Pharming of Lipids

Lipids are a group of macromolecules (hydrophobic) involving several elements as fatty acids (FA), glycerides (GL), phospholipids (PL), sterols, sphingolipids, and prenol lipids (Fig. 5.2). Metabolic processes and storage, as well as structural components, signaling, and hormone production, are all dependent on lipids. Lipid metabolic enzymes and processes are disrupted in several ailments, including tumor, high sugar (diabetes), viral infections, and diseases related to inflammation (Milica et al. 2020). As a result of the molecular transformation, including the production of in vitro high-throughput screening tools and combinatorial chemistry, a large number of molecules with poor water solubility were selected as therapeutic prospects. Approximately 40% of the entire new therapeutic candidates are lipophilic medicines with low water solubility (Milica et al. 2020). After being taken orally, drugs face a number of obstacles on their way to the bloodstream. Orally supplied substances pass via the intestinal barrier and into systemic circulation, where they are assimilation accessible. The drug must first be metabolized in the GI environment and transformed to a genomic pattern close to the wall of intestine before it may penetrate. For lipophilic compounds with limited water solubility, this can be challenging, rendering it a restricting point in the absorption process. The oral ingestion of highly lipophilic therapeutics is affected by the existence of lipids from diet or lipid-based expression in the intestinal wall. By generating diverse colloidal forms, the drug's solubility can be increased (vesicles, micelles). Drug dissolution can be increased by the nature of lipids and the modeling of biological lipid refinement systems, leading to increased synthesis of bile salts and phospholipids (Mudie et al. 2020). Lipids have an impact on intestinal biochemistry and influx/efflux movement. Fatty excipients, according to research, can boost drug absorption in some situations by affecting P-glycoprotein (P-gp) function (Cornaire et al. 2004). Furthermore, nucleus hormonal receptors (NHR) have been demonstrated to play a vital role in lipid smuggling and metabolic rate, as well as intestinal triglyceride and drug uptake, since they regulate a group of enzymes involved in lipid/drug transport and metabolism (e.g., fatty acid-binding proteins) (Trevaskis et al. 2011). Many medications enter the general circulation system through the hepatic vein after being taken orally, while lipid-soluble substances might enter through the digestive lymph system. Many medications move through the hepatic vein on their way to the systemic bloodstream after being administered verbally, whereas highly lipophilic substances may be carried via the intestinal lymphatic system. Lipids may also help medications skip hepatic first-pass metabolism and reach the systemic circulation more quickly (Porter et al. 2007).

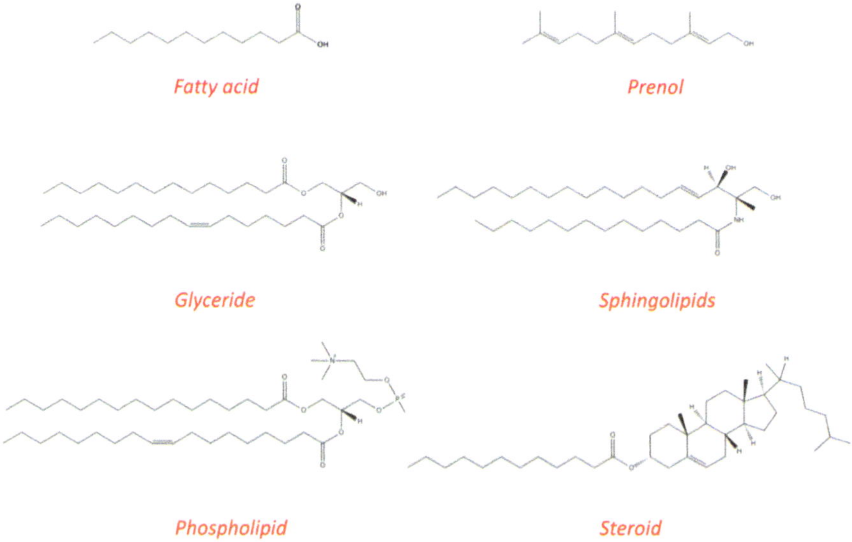

Fig. 5.2 Chemical structures of major classes of lipid

5.5.3.1 The Therapeutic Use of Lipids in Pathological Issues

Lipidomics may aid in fatty acid metabolism research as well as the understanding of the key mechanisms of different physiological disorders as well as the significance of triglycerides in sustaining homeostasis (Stephenson et al. 2017). Figure 5.3 depicts a variety of disorders induced by lipid-related metabolic problems. The development of distinctive treatment interventions for various diseases is based on the recognition of specific lipid biomarkers and a mechanism-based understanding of certain metabolic lipid-processing routes associated with various maladies. The medication is firmly attached to the fatty acid carrier in lipidic prodrugs (e.g., FA, glyceride, PL, or steroid). This bridge results in a higher lipophilicity than the drug alone, which could lead to a stronger pharmacological effect as well as other substantial advantages such as improved absorbance via cellular molecules, lengthened blood half-life, selective distribution to cells (e.g., brain, intestine), reduced liver first-pass metabolism, and overall improved clinical efficiency. According to Markovic et al. (2019), drug-lipid conjugates may also participate in biological fat transfer pathways, allowing medication delivery to specified regions.

5.5.3.2 Fatty Acids (FAs)

FAs are made up of different lengths and degrees of desaturation of hydrocarbon chains. It is used to make a variety of lipids, including phospholipids and glycerides, which have a hydrophobic FA tail, and triacylglycerides, which are generated and preserved when nutrient availability is high (Röhrig and Schulze 2016). FA biosynthesis, on the other hand, was demonstrated to show a vital role in pathological circumstances like cancer (Currie et al. 2013). Fatty acid synthase has been

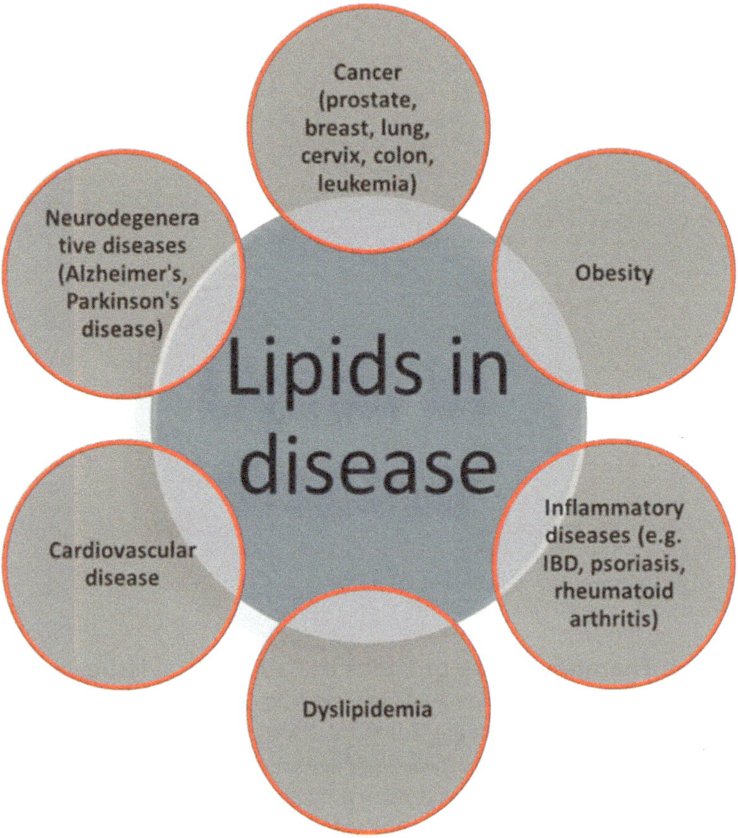

Fig. 5.3 The representation of different disorders induced by the problem of lipid metabolism

recognized as the cancer antigen OA519 in metastatic (stage 4) breast cancer and malignant neoplasm (Camassei et al. 2003), as well as in proliferating fetal tissues, implying that restoring cancer FA synthesis may result in a return to a less distinguishable embryonic state (Röhrig and Schulze 2016). FA contents may be higher in diseased tissues due to increased metabolism or adjustment to decreased lipid profile in tumor cells. In non-fasting circumstances, increased plasma levels (TG and TG-rich LP) play a role in coronary infarction, heart disease, and sometimes even fatality (Bansal et al. 2007). Regardless of HDL levels, long-term high blood TG-rich LP or residues have now been implicated in the development of coronary heart disease (Hegele et al. 2014). Nondigestible lipids, such as fructose synthetic polymers, are not digested by the intestinal wall, whereas digestible lipids, such as DG, TG, PL, FA, cholesterol, and some synthesized versions, are effective treatment composition ingredients. Long-chain TG (LCT), medium-chain TG (MCT), DG, MG, FA, PL, and others are classified according to their degree of saturation, interaction with water, and hydrocarbon chain length (long TG (LCT), medium-chain TG (MCT), and others). Mycophenolic acid (MPA), an enzyme inhibitor, was connected to the glyceride's sn-2 position in one study (Han et al. 2015). A prodrug is a type of antimicrobial (2-MPA-TG). When MPA was added to the TG deacylation–reacylation process, its lymphatic transport increased. Following systemic treatment with MPA alone and MPA-prodrug, the lymphatic transportation and systemic bioavailability of MPA in dogs were studied (Han et al. 2016).

5.5.3.3 Phospholipids

According to Kosicek and Hecimovic (2013), lipids are closely linked to the breakdown/metabolism of amyloid precursor proteins, leading to the formation of amyloid-beta peptide (A), a major part of senile plaques, which are a hallmark of Alzheimer's disease. Choline phospholipids (PLs) have been postulated as possible dementia biomarkers. With normal aging, concentrations of lysophosphatidylcholine, lyso-platelet, and choline plasmalogen-stimulating factors rise rapidly; similar but even more severe variations were observed in persons with Alzheimer's disease. Consequently, larger amounts of choline-containing phospholipids in the blood may indicate a faster effect of aging (Dorninger et al. 2018). PL, on the other hand, has been proved to be effective in treating a wide range of ailments. In ulcerative colitis, specific PL moieties have been shown to have a therapeutic benefit, and clinical studies have revealed that introducing phosphatidylcholine (PC) to the intestinal mucosa lowers inflammation (Stremmel et al. 2012). In a Phase IIA, double-blind, randomized, placebo-controlled trial incorporating 60 patients with severe active, nonsteroid-dependent ulcerative colitis, they also demonstrated that giving them 6 g of retarded-release phosphatidylcholine-rich PL for three months reduced ulcerative colitis inflammatory responses. Nutritional lecithin has been shown to protect cholic acid-fed Abcb4-deficient mice from cholestatic liver disease by significantly lowering hepatotoxicity (Lamireau et al. 2007).

5.5.3.4 Steroids

Oxidized PL is found in inflamed tissues and is hypothesized to play a role in immune response modulation (Deigner and Hermetter 2008). Although oxidized PL has been shown to have pro-inflammatory effects in the majority of studies, some PL oxidation products have been shown to exhibit anti-inflammatory properties (Feige et al. 2010). Immunological diseases such as atherosclerosis, psoriasis, multiple sclerosis, and rheumatoid arthritis may benefit from the derivatives of oxidized PL. Cholesterol is the most frequent sterol generated by mammals. It could come from the food or be produced in the body from scratch, according to Hume and Boyd (1978). It is a morphological element of the cell membrane, a constituent of numerous types of PL, and a precursor of steroid hormones, amino salts, and vitamin D, to name a few (Janowski et al. 1996). Low cholesterol levels in the blood have been associated with lung, cervical, breast, and colon cancers and leukemia, while high cholesterol levels have been connected to brain cancer (Kambach et al. 2017). Cancer of the prostate for example is a lipid-rich type of tumor (Wu, Daniels, Lee, and Monaco 2014), with the androgen receptor exhibiting a significant role in its growth. Testosterone can impact genes that code for lipogenic enzymes, resulting in increased FA and cholesterol synthesis. More PL, cholesterol, and other cell membrane components are produced as a result of enhanced lipogenesis in cancer cells. Bioactive lipids assist prostate cancer growth as well (Dang, Chen, and Hsieh 2019). High estrogen concentrations have also been linked to a higher risk of breast cancer, and cholesterol is an estrogen precursor (Murai 2015). Cholesterol is essential for the production of cell membranes in growing cancerous cells that are produced at a rapid rate in such tissues; de novo and LDL particles are major sources of obtaining via high-affinity receptor-enhanced absorption (Murai 2015; Yang et al. 2020; Kloudova et al. 2017). They have been linked to a variety of diseases, including atherosclerosis, neurodegenerative disorders, Alzheimer's, and tumorigenesis. By aggregating in the walls of aging blood arteries, oxysterols can excite monocyte and endothelium cell, as well as induce smooth muscle cells in blood vessels to grow, spread, and function as fibroblast-like cells (Gargiulo et al. 2016). Atherosclerosis is a disorder that affects millions of people worldwide and is linked to cholesterol and lipid metabolism. Atherosclerosis is caused by high cholesterol levels in the body, which leads to fat deposition in the arterial walls (Maxfield and Tabas 2005). The purpose of atherosclerosis treatment is to minimize the formation of cholesterol ester. The introduction of galactose-functionalized polyamidoamine dendrimer generation 5 to hepatocytes to deliver cholesterol ester hydrolase (CEH) is a promising therapy for reducing cholesterol esters. The increase in cholesterol ester hydrolysis caused by CEH overexpression results in the release of free cholesterol as bile acids. In atherosclerotic individuals, this method was used to prevent cholesterol esters from developing (He et al. 2017).

5.6 Conclusions

Plants have several benefits with regard to the production of recombinant proteins, but their acceptance by companies is exacerbated by the fact that mammals and other CHO cells are far more well established and described in an industrial setting, rendering plant-based procedures harder to acquire prominence. Several macromolecules, such as proteins, lipids, and carbohydrates, have been investigated in the synthesis of various therapeutics for diseases such as gastrointestinal illnesses, viral diseases, diabetes, atherosclerosis, psoriasis, multiple sclerosis, fever, rheumatoid arthritis, colon cancer, and breast cancer, among others, using molecular pharming. Thus, in order to tip the balance in favor of durable plant-derived pharmaceutical products, the greater potential of plant-based technologies may have to be completely harnessed.

References

Agarwal S, Singh R, Sanyal I, Amla DV (2008) Expression of modified gene encoding functional human alpha-1-antitrypsin protein in transgenic tomato plants. Transgenic Res 17:881–896

Ahmad K (2014) Molecular farming: strategies, expression systems and bio-safety considerations. Plant Breed 50:1–10

Alireza T, Nader RE (2015) Molecular farming in plants. In: Plants for the future, vol 21

Alvarez ML, Pinyerd HL, Crisantes JD, Rigano MM, Pinkhasov J, Walmsley AM, Mason HS, Cardineau GA (2006) Plant-made subunit vaccine against pneumonic and bubonic plague is orally immunogenic in mice. Vaccine 24:2477–2490

Anneli R, Suvi TH, Schillberg S (2014) Molecular pharming in plants and plant cell cultures: a great future ahead? Pharm Bioprocess 2(3):00–00

Aoki K, Kamiyama H, Masuda K, Kamiko K, Noguchi Y, Tajima K, Terauchi Y (2014) Effects of miglitol, vildagliptin, or their combination on serum insulin and peptide YY levels and plasma glucose, cholecystokinin, ghrelin, and obestatin levels. Endocr J 61:249–256

Appaiahgari MB, Abdin MZ, Bansal KC, Vrati S (2009) Expression of Japanese encephalitis virus envelope protein in transgenic tobacco plants. Virolo Method 162:22–29

Austin S, Bingham ET, Mathews DE, Shahan MN, Will J, Burgess RR (1995) Production and field performance of transgenic alfalfa (Medicago sativa L.) expressing alpha-amylase and manganese-dependent lignin peroxidase. Euphytica 85:381–393

Bai JY, Zeng L, Hu YL, Li YF, Lin ZP, Shang SC, Shi YS (2007) Expression and characteristic of synthetic human epidermal growth factor (hEGF) in transgenic tobacco plants. Biotechnol Lett 29(12)

Bansal S, Buring JE, Rifai N, Mora S, Sacks FM, Ridker PM (2007) Fasting compared with nonfasting triglycerides and risk of cardiovascular events in women. JAMA 298:309–316

Bardor M, Faveeuw C, Fitchette AC, Gilbert D, Galas L, Trottein F, Faye L, Leuroge P (2003) Immunoreactivity in mammals of two typical plant glycol-epitopes, core-alpha (1,3)-fucose and core xylose. Glycobiology 13:427–434

Barta A, Sommergruber K, Thompson D, Hartmuth K, Matzke M, Matzke A (1986) The expression of a nopaline synthase human growth hormone chimaeric gene in transformed tobacco and sunflower callus tissue. Plant Mol Biol 6:347–357

Bennett EP, Mandel U, Clausen H (2012) Control of mucin-type O-glycosylation: a classification of the polypeptide GalNAc-transferase gene family. Glycobiology 22:736–756

Berqqvist D (2006) Review of fondaparinux sodium injection for the prevention of venous thromboembolism in patients undergoing surgery. Vasc Health Risk Manaq 2:365–370

Borisjuk NV, Borisjuk LG, Logendra S, Petersen F, Gleba Y, Raskin I (1999) Production of recombinant proteins in plant root exudates. Nat Biotechnol 17:466–469

Bosch D, Smal J, Krebber E (1994) A trout growth hormone is expressed, correctly folded and partially glycosylated in the leaves but not the seed of transgenic plants. Transgenic Res 3:304–310

Brodzik R, Spitsin S, Pogrebnyak N, Bandurska K, Portocarrero C, Andryszak K, Koprowski H, Golovkin M (2009) Generation of plant-derived recombinant DTP subunit vaccine. Vaccine 27:3730–3734

Buyel JF (2019) Plant molecular farming–integration and exploitation of side streams to achieve sustainable biomanufacturing. Front Plant Sci 9:1893

Buyel JF, Buyel JJ, Haase C, Fischer R (2015) The impact of *Pseudomonas syringae* type III effectors on transient protein expression in tobacco. Plant Biol 17:84–492

Buyel JF, Twyman RM, Fischer R (2017) Very-large-scale production of antibodies in plants: the bilogization of manufacturing. Biotechnol Adv 35:458–465

Büttner-Mainik A, Parsons J, Jérôme H, Hartmann A, Lamer S, Schaaf A, Schlosser A, Zipfel PF, Reski R, Decker EL (2011) Production of biologically active recombinant human factor H in Physcomitrella. Plant Biotechnol J 9(3):373–383

Cai MS, Li ZJ (2008) Carbohydrate chemistry, fundamentals, reactions, synthesis. Isolations and Structures. Chemical Industry Press, Beijing, pp 315–368

Camassei FD, Cozza R, Acquaviva A, Jenkner A, Ravà L, Gareri R, Donfrancesco A, Bosman C, Vadalà P, Hadjistilianou T (2003) Expression of the lipogenic enzyme fatty acid synthase (FAS)

in retinoblastoma and its correlation with tumor aggressiveness. Investig Ophthalmol Vis Sci 44:2399–2403
Castilho A, Strasser R, Stadlmann J (2010) In planta protein sialylation through overexpression of the respective mammalian pathway. J Biol Chem 285:15923–15930
Ceaglio N, Etcheverrigaray M, Conradt HS, Grammel N, Kratje R, Oggero M (2010) Highly glycosylated human alpha interferon: an insight into a new therapeutic candidate. J Biotechnol 146:74–83
Chen J, Lu WL, Gu W, Lu SS, Chen ZP, Cai BC, Yang XX (2014) Drug-in-cyclodextrin-in-liposomes: a promising delivery system for hydrophobic drugs. Expert Opin Drug Deliv 11:565–577
Chen Y, Seepersaud R, Bensing BA, Sullam PM, Rapoport TA (2016) Mechanism of a cytosolic O-glycosyltransferase essential for the synthesis of a bacterial adhesion protein. Proc Natl Acad Sci 113(9):E1190–E1199
Chikwamba RK, Scott MP, Mejia LB, Mason HS, Wang K (2003) Localization of a bacterial protein in starch granules of transgenic maize kernels. PNAS 100:11127–11132
China Food and Drug Administration (CDF) (2015). http://www.sfda.gov.cn/WS01/CL0001/
Choi NW, Estes MK, Langridge WH (2005) Synthesis and assembly of a cholera toxin B subunit-rotavirus VP7 fusion protein in transgenic potato. Mol Biotechnol 31(3):193–202
Cornaire G, Woodley J, Hermann P, Cloarec A, Arellano C, Houin G (2004) Impact of excipients on the absorption of P-glycoprotein substrates in vitro and in vivo. Int J Pharm 278:119–131
Cramer CL, Weissenborn DL, Oishi KK, Grabau EA, Bennett S, Ponce E, Grabowski GA, Radin DN (1996) Bioproduction of human enzymes in transgenic tobacco. Ann Acad Sci 792:62–71
Currie E, Schulze A, Zechner R, Walther TC, Farese RV Jr (2013) Cellular fatty acid metabolism and cancer. Cell Metab 18:153–161
Curtiss RI, Cardineau CA (1990) World patent app WO 90/02484
Dai M, Yu C, Fang T (2015) Identification and functional characterization of glycosylation of recombinant human platelet-derived growth factor-BB in *Pichia pastoris*. PLoS One 10: e0145419
Dalziel M, Crispin M, Scanlan CN, Zitzmann N, Dwek RA (2014) Emerging principles for the therapeutic exploitation of glycosylation. Science 343(6166):1235681
Daniell H, Streatfield J, Wycoff K (2001) Medical molecular farming: production of antibodies, biopharmaceuticals and edible vaccines in plants. Trends Plant Sci 6:219–226
Dang Q, Chen YA, Hsieh JT (2019) The dysfunctional lipids in prostate cancer. Am J Clin Exp Urol 7:273–280
Davis BG, Robinson MA (2002) Drug delivery systems based on sugar-macromolecule conjugates. Curr Opin Drug Discov Dev 5:279–288
Decker EL, Reski R (2012) Glycoprotein production in moss bioreactors. Plant Cell Rep 31:453–460
Deigner HP, Hermetter A (2008) Oxidized phospholipids: emerging lipid mediators in pathophysiology. Curr Opin Lipidol 19:289–294
Demain AL, Vaishnav P (2009) Production of recombinant proteins by microbes and higher organisms. Biotechnol Adv 27:297–306
Dicker M, Strasser R (2015) Using glyco-engineering to produce therapeutic proteins. Expert Opin Biol Ther 15(10):1501–1516
Dorninger F, Moser AB, Kou J, Wiesinger C, Forss-Petter S, Gleiss A, Hinterberger M, Jungwirth S, Fischer P, Berger J (2018) Alterations in the plasma levels of specific choline phospholipids in Alzheimer's disease mimic accelerated aging. J Alzheimers Dis 62(2):841–854
Dove A (2000) Milking the genome for profit. Nat Biotechnol 18(10):1045–1048
Downing WL, Galpin JD, Clemens S (2006) Synthesis of enzymatically active human α-l-iduronidase in Arabidopsis CGL (complex glycan-deficient) seeds. Plant Biotechnol J 4:169–181
Dreesen IAJ, El Hamri GC, Fussenegger M (2010) Heat-stable oral alga-based vaccine protects mice from Staphylococcus aureus infection. J Biotechnol 145:273–280

Edelbaum O, Stein D, Holland N, Gafni Y, Livneh O, Novick D, Rubinstein M, Sela I (1992) Expression of active human interferon-beta in transgenic plants. J Interf Res 12(6):449–453
Elliott S, Chang D, Delorme E, Eris T, Lorenzini T (2004) Structural requirements for additional N-linked carbohydrate on recombinant human erythropoietin. J Biol Chem 279(16):16854–16862
Evangelista RL, Kusnadi AR, Howard JA, Nikolov ZL (1998) Process and economic evaluation of the extraction and purification of recombinant β-glucuronidase from transgenic corn. Biotechnol Prog 14:607–661
Everett KM, Dickey L, Parsons J, Loranger R, Wingate V (2012) Development of a plantmade pharmaceutical production platform. Bioprocess Int 10:16–25
Fanata WI, Lee KH, Son BH, Yoo JY, Harmoko R, Ko KS, Ramasamy NK, Kim KH, Oh DB, Jung HS, Kim JY (2013) N-glycan maturation is crucial for cytokinin-mediated development and cellulose synthesis in Oryza sativa. Plant J 73(6):966–979
Feige E, Mendel I, George J, Yacov N, Harats D (2010) Modified phospholipids as anti-inflammatory compounds. Curr Opin Lipidol 21:525–529
Figueiredo C, Blaszczyk R (2014) Genetically engineered blood pharming: generation of HLA-universal platelets derived from CD34+progenitor cells. J Stem Cells 9(3):149–161
Fischer R, Buyel JF (2020) Molecular farming: the slope of enlightenment. Biotechnol Adv 40: 107519
Fischer R, Holland T, Sack M, Schillberg S, Stoger E, Twyman RM, Buyel JF (2018) Glyco-engineering of plant-based expression systems. Adv Biochem Eng Biotechnol 175:137–166
Fischer R, Schillberg S (2017) Molecular farming. Encyclopedia of applied plant sciences, vol 2, 2nd edn. Academic Pres (Elsevier), Waltham, MA
Fischer R, Schillberg S, Buyel JF, Twyman RM (2013) Commercial aspects of pharmaceutical protein production in plants. Curr Pharm Des 19:5471–5477
Fischer S, Handrick R, Otte K (2015) The art of CHO cell engineering: a comprehensive retrospect and future perspectives. Biotechnol Adv 33(8):1878–1896
Floss DM, Falkenburg D, Conrad U (2007) Production of vaccines and therapeutic antibodies for veterinary applications in transgenic plants: an overview. Transgenic Res 16:315–332
Francisco JA (1997) Expression and characterization of bryodin 1 and a bryodin 1-based single-chain immunotoxin from tobacco cell culture. Bioconjug Chem 8:708–713
Frank CG, Sanyal S, Rush JS, Waechter CJ, Menon AK (2008) Does Rft1 flip an N-glycan lipid precursor? Nature 454(7204):E3–E4
Fukuzawa N, Tabayashi N, Okinaka Y, Furusawa R, Furuta K, Kagaya U (2010) Matsumura T production of biologically active Atlantic salmon interferon in transgenic potato and rice plants. J Biosci Bioeng 110:201–207
Fung KL, Yim YF, Tse YC (2005) Targeting and processing of membrane anchored YFP fusion proteins to protein storage vacuoles in transgenic tobacco seeds. Seed Sci Res 15(4):361–364
Gao Y, Ma Y, Li M, Cheng T, Li SW, Zhang J, Xia NS (2003) Oral immunization of animals with transgenic cherry tomatillo expressing HBsAg. World Gastroenterol 9:996–1002
Gargiulo S, Gamba P, Testa G, Leonarduzzi G, Poli G (2016) The role of oxysterols in vascular ageing. J Physiol 594:2095–2113
Gavilondo VJ, Larrick WJ (2000) Antibody engineering at the millennium. BioTechniques 29:128–145
Gómez E, Zoth SC, Carrillo E, Roux ME, Berinstein A (2008) Mucosal immunity induced by orally administered transgenic plants. Immunobiology 213:671–675
Hamilton SR, Bobrowicz P, Bobrowicz B, Davidson RC, Li H, Mitchell T, Nett JH, Rausch S, Stadheim TA, Wischnewski H, Wildt S (2003) Production of complex human glycoproteins in yeast. Science 301(5637):1244–1246
Hamilton SR, Davidson RC, Sethuraman N, Nett JH, Jiang Y, Rios S, Bobrowicz P, Stadheim TA, Li H, Choi BK, Hopkins D (2006) Humanization of yeast to produce complex terminally sialylated glycoproteins. Science 313(5792):1441–1443

Han S, Hu L, Gracia Quach T, Simpson JS, Edwards GA, Trevaskis NL, Porter CJ (2016) Lymphatic transport and lymphocyte targeting of a triglyceride mimetic prodrug is enhanced in a large animal model: studies in greyhound dogs. Mol Pharm 13:3351–3361

Han S, Hu L, Quach T, Simpson JS, Trevaskis NL, Porter CJH (2015) Constitutive triglyceride turnover into the mesenteric lymph is unable to support efficient lymphatic transport of a biomimetic triglyceride prodrug. J Pharm Sci 105:786–796

Haq TA, Mason HS, Clements JD, Arntzen CJ (1995) Oral immunization with a recombinant bacterial antigen produced in transgenic plants. Science 268:714–716

He X, Haselhorst T, von Itzstein M (2012) Production of alpha-l-iduronidase in maize for the potential treatment of a human lysosomal storage disease. Nat Commun 3:1062

He H, Lancina MG, Wang J, Korzun WJ, Yang H, Ghosh S (2017) Bolstering cholesteryl ester hydrolysis in liver: a hepatocyte-targeting gene delivery strategy for potential alleviation of atherosclerosis. Biomaterials 130:1–13

He Y, Ning T, Xie T (2011) Large-scale production of functional human serum albumin from transgenic rice seeds. Proc Natl Acad Sci U S A 108(47):19078–19083

He X, Pierce O, Haselhorst T (2013) Characterization and downstream mannose phosphorylation of human recombinant α-l-iduronidase produced in Arabidopsis complex glycan-deficient (cgl) seeds. Plant Biotechnol J 11:1034–1043

Hegedus DD, Baron M, Labbe N (2014) A strategy for targeting recombinant proteins to protein storage vacuoles by fusion to *Brassica napus* napin in napin-depleted seeds. Protein Expr Purif 95:162–168

Hegele RA, Ginsberg HN, Chapman MJ, Nordestgaard BG, Kuivenhoven JA, Averna M, Borén J, Bruckert E, Catapano AL, Descamps OS (2014) The polygenic nature of hypertriglyceridaemia: implications for definition, diagnosis, and management. Lancet Diabetes Endocrinol 2:655–666

Heneghan CJ, OnakpoyaI CMT, Spencer EA, Jones M, Jefferson T (2014) Zanamivir for influenza in adults and children: systematic review of clinical study reports and summary of regulatory comments. Br Med J 348:2547–2563

Hensel G, Floss DM, Arcalis E (2015) Transgenic production of an anti-HIV antibody in the barley endosperm. PLoS One 10(10):e0140476

Hiatt A, Pauly M, Whaley K, Qiu X, Kobinger G, Zeittin L (2015) The emergence of antibody therapies for Ebola. Hum Antibod 23:49–56

Higo K, Saito Y, Higo H (1993) Expression of a chemically synthesized gene for human epidermal growth factor under the control of cauliflower mosaic virus 35S promoter in transgenic tobacco. Biosci Biotechnol Biochem 57:1477–1481

Hitosugi T, Saito T, Suzuki S, Kubota I, Shoda E, Shimizu T, Oi Y (2007) Hydroxyethyl starch: the effect of molecular weight and degree of substitution on intravascular retention in vivo. Anesth Analg 105:724–728

Hoffmann M, Marx K, Reichl U (2016) Site-specific O-glycosylation analysis of human blood plasma proteins. Mol Cell Proteomics 15:624–641

Holtz BR, Berquist BR, Bennett LD, Kommineni VJM, Munigunti RK, White EL, Wilkerson DC, Wong K-YI, Ly LH, Marcel S (2015) Commercial-scale biotherapeutics manufacturing facility for plant-made pharmaceuticals. Plant Biotechnol J 13:1180–1190

Hood E, Witcher D, Maddock S, Meyer T, Baszczynski C, Bailey M (1997) Commercial production of Avidin from transgenic maize: characterization of transformant, production, processing, extraction and purification. Mol Breed 3:291–306

Houdebine LM (2009) Production of pharmaceutical proteins by transgenic animals. Comp Immunol Microbiol Infect Dis 32(2):107–121

Huether CM, Lienhart O, Baur A (2005) Glyco-engineering of moss lacking plant-specific sugar residues. Plant Biol (Stuttg) 7:292–299

Hume R, Boyd GS (1978) Cholesterol metabolism and steroid-hormone production. Biochem Soc Trans 6:893–898

Hundleby PAC, Sack M, Twyman RM (2018) Biosafety, risk assessment and regulation of molecular farming. In: Kermode A, Jiang L (eds) Molecular pharming: applications, challenges and emerging areas. John Wiley & Sons, Inc., Hoboken, NJ, pp 329–351

Jani D, Meena LS, Riswan-ul-Haq QM, Singh Y, Sharma AK, Tyagi AK (2002) Expression of cholera toxin B subunit in transgenic tomato plants. Transgenic Res 11:447–454

Janowski BA, Willy PJ, Devi TR, Falck JR, Mangelsdorf DJ (1996) An oxysterol signalling pathway mediated by the nuclear receptor LXR alpha. Nature 383:728–731

Jefferis R (2009) Glycosylation as a strategy to improve antibody-based therapeutics. Nat Rev Drug Discov 8:226–234

Jeffrey MS, Garcia B, Graves J, Hajdukiewicz PTJ, Hunter P, Nehra N, Paradkar V, Schlittler M, Carroll JA, Spatola L, Ward D, Ye G, Russell DA (2000) High-yield production of a human therapeutic protein in tobacco chloroplasts. Nat Biotechnol 18:333–338

Jennifer B (2010) Spherixsoars on D-tagatoseresults, starts partner hunt. Bio World Today 21:1

Jeon I, Iyer K, Danishefsky SJ (2009) A practical total synthesis of globo-H for use in anticancer vaccines. J Org Chem 74:8452–8455

Jin C, Altmann F, Strasser R, Mach L, Schahs M, Kunert R, Rademacher T, Glossl J, Steinkellner H (2008) A plant-derived human monoclonal antibody induces an anti-carbohydrate immune response in rabbits. Glycobiology 18(3):235–241

Kambach DM, Halim AS, CauerAG SQ, Tristan CA, Celiku O, Kesarwala AH, Shankavaram U, Batchelor E, Stommel JM (2017) Disabled cell density sensing leads to dysregulated cholesterol synthesis in glioblastoma. Oncotarget 8:14860–14875

Karasev AV, Foulke S, Wellens C, Rich A, Shon KJ, Zwierzynski I, Hone D, Koprowsi H, Reitz M (2005) Plant based HIV-1 vaccine candidate: Tat protein produced in spinach. Vaccine 23:1875–1880

Kloudova A, Guengerich FP, Soucek P (2017) The role of oxysterols in human cancer. Trends Endocrinol Metab 28:485–496

Kong Q, Richter L, Yang YF, Arntzen CJ, Mason HS, Thanavala Y (2001) Oral immunization with hepatitis B surface antigen expressed in transgenic plants. Proc Natl Acad Sci U S A 98:11539–11544

Koprivova A, Stemmer C, Altmann F, Hoffmann A, Kopriva S, Gorr G, Reski R, Decker EL (2004) Targeted knockouts of physcomitrella lacking plant-specific immunogenic N-glycans. Plant Biotechnol J 2(6):517–523

Kosicek M, Hecimovic S (2013) Phospholipids and Alzheimer's disease: alterations, mechanisms and potential biomarkers. Int J Mol Sci 14:1310–1322

Kostrzak A, Gonzalez MC, Guetard D, Nagaraju DB, Hobson SW, Tepfer D, Pniewski T, Sala M (2009) Oral administration of low doses of plant based HBsAg induced antigen-specific IgAs and IgGs in mice, without increasing levels of regulatory T cells. Vaccine 27:4798–4807

Kumar GB, Ganapathi TR, Revathi CJ, Srinivas L, Bapat VA (2005) Expression of hepatitis B surface antigen in transgenic banana plants. Planta 222:484–493

Lamireau T, Bouchard G, Yousef IM, Clouzeau-Girard H, Rosenbaum J, Desmoulière A, Tuchweber B (2007) Dietary lecithin protects against cholestatic liver disease in cholic acid–fed abcb4– deficient mice. Pediatr Res 61:185–190

Lamphear BJ, Jilka JM, Kesl L, Welter M, Howard JA, Streatfield SJ (2004) Acorn-based delivery system for animal vaccines: an oral transmissible gastroenteritis virus vaccine boosts lactogenic immunity in swine. Vaccine 22:2420–2424

Lamphear BJ, Barker DK, Brooks CA (2005) Expression of the sweet protein brazzein in maize for production of a new commercial sweetener. Plant Biotechnol J 3:103–114

Landry N, Ward JB, Trepanier S, Montomoll E, Dargis M, Lapini G, Vezinz PL (2010) Preclinical and clinical development of plant – made virus –like partical vaccine against avian H5N1 influenza. PLoS One 5:1–12

Lau JM, Korban SS (2010) Transgenic apple expressing an antigenic protein of the human respiratory syncytial virus. Plant Physiol 167:920–927

Lee RWH, Cornelisse M, Ziauddin A, Slack PJ, Hodgins DC, Strommer JN, Shewen PE, Lo RYC (2008) Expression of a modified Mannheimia haemolytica GS60 outer membrane lipoprotein in transgenic alfalfa for the development of an edible vaccine against bovine pneumonic pasteurellosis. J Biotechnol 135:224–231

Lei H, Qi J, Song J, Yang D, Wang Y, Zhang Y, Yang J (2006) Biosynthesis and bioactivity of trichosanthin in cultured crown gall Maximowicz. Plant Cell Rep 25:1205–1212

Leite A, Kemper EL, da Silva MJ, Luchessi AD, Siloto RMP, Bonaccorsi ED, El-Dorry HF, Arruda P (2000) Expression of correctly processed human growth hormone in seeds of transgenic tobacco plants. Mol Breed 6:47–53

Li D, Leary JO, Huang Y, Huner NPA, Jevnikar AM, Ma S (2006) Expression of cholera toxin B subunit and the B chain of human insulin as a fusion protein in transgenic tobacco plants. Plant Cell Rep 25:417–424

Li Q, Song J, Peng S, Wang JP, Qu GZ, Sederoff RR (2014) Plant biotechnology for lignocellulosic biofuel production. Plant Biotechnol J 12:74–1192. https://doi.org/10.1111/pbi.12273

Li PL, Zhang XK, Cheng YN, Li J, Xiao Y, Zhang Q, Zong A, Zhong C, Wang F (2014) Preparation and in vitro immunomodulatory effect of curdlan sulfate. Carbohyd Polym 102: 852–861

Liebminger E, Veit C, Pabst M, Batoux M, Zipfel C, Altmann F, Mach L, Strasser R (2011) β-N-acetylhexosaminidases HEXO1 and HEXO3 are responsible for the formation of paucimannosidic N-glycans in Arabidopsis thaliana. J Biol Chem 286(12):10793–10802

Lindh I, Wallin A, Kalbina I, Sävenstrand H, Engström P, Andersson S, Strid A (2009) Production of the p24 capsid protein from HIV-1 subtype C in Arabidopsis thaliana and Daucus carota using an endoplasmic reticulum directing SEKDEL sequence in protein expression constructs. Prot Express Purifi 66:46–51

Loc NH, Bach NH, Kim TG, Yang MS (2010) Tissue culture and expression of Escherichia coli heat labile enterotoxin B subunit intransgenic Peperomia pellucida. Prot Express Purif 72:82–86

Loos A, Steinkellner H (2014) Plant glyco-biotechnology on the way to synthetic biology. Front Plant Sci 5:523

Ma JK, Hiatt A, Hein M, Vine ND, Wang F, Stabila P, van Dolleweerd C, Mostov K, Lehner T (1995) Generation and assembly of secretory antibodies in plants. Science 268:716–719

Ma JK, Drossard J, Lewis D, Altmann F, Boyle J, Christou P, Cole T, Dale P, van Dolleweerd CJ, Isitt V, Katinger D (2015) Regulatory approval and a first-in-human phase I clinical trial of a monoclonal antibody produced in transgenic tobacco plants. Plant Biotechnol J 13(8):1106–1120

Magnuson NS, Linzmaier PM, Reeves R, An G, HayGlass K, Lee JM (1998) Secretion of biologically active human interleukin-2 and interleukin-4 from genetically modified tobacco cells in suspension culture. Protein Expr Purif 13:45–52

Markovic M, Ben-Shabat S, Keinan S, Aponick A, Zimmermann EM, Dahan A (2019) Lipidic prodrug approach for improved oral drug delivery and therapy. Med Res Rev 39:579–607

Marusic C, Rizza P, Lattanzi L, Mancini C, Spada M, Belardelli F, Benvenuto E, Capone I (2001) Chimeric plant virus particles as immunogens for inducing murine and human immune responses against human immunodeficiency virus type 1. J Virol 75:8434–8439

Matsumoto S, Ikura K, Ueda M, Sasaki R (1995) Characterization of a human glycoprotein (erythropoietin) produced in cultured tobacco cells. Plant Mol Biol 27:1163–1172

Maxfield FR, Tabas I (2005) Role of cholesterol and lipid organization in disease. Nature 438:612–621

McDonald KA, Hong LM, Trombly DM, Xie Q, Jackman AP (2005) Production of human alpha1-antitrypsin from transgenic rice cell culture in a membrane bioreactor. Biotechnol Prog 21:728–374

McGarvey PB, Hammond J, Dienelt MM, Hooper DC, Fu ZF, Dietzschold B, Koprowski H, Michaels FH (1995) Expression of the rabies virus glycoprotein in transgenic tomatoes. Nat Biotechnol 13:1484–1487

McCormick AA, Kumagai MH, Hanley K, Turpen TH, Hakim I, Grill LK, Tusé D, Levy S, Levy R (1999) Rapid production of specific vaccines for lymphoma by expression of the tumor-derived single-chain Fv epitopes in tobacco plants. Proc Natl Acad Sci U S A 96:703–708

Meng X, Yao W, Cheng J, Jin L, Yu H, Chen X, Wang F, Cao H (2014) Regioselective chemoenzymatic synthesis of ganglioside disialyl tetrasaccharide epitopes. J Am Chem Soc 136:5205–5208

Milica M, Shimon BS, Aaron A, Ellen MZ, Arik D (2020) Lipids and lipid-processing pathways in drug delivery and therapeutics. Int J Mol Sci 21:3248

Montesinos L, Bundo M, Izquierdo E (2016) Production of biologically active cecropin a peptide in rice seed oil bodies. PLoS One 11(1):e0146919

Morris G, Kök S, Harding S, Adams G (2010) Polysaccharide drug delivery systems based on pectin and chitosan. Biotechnol Genet Eng Rev 27:257–284

Mu JH, Chua NH, Ross EM (1997) Expression of human muscarinic cholinergic receptors in tobacco. Plant Mol Biol 34:357–362

Mudie DM, Samiei N, Marshall DJ, Amidon GE, Bergstrom CAS (2020) Selection of in vivo predictive dissolution media using drug substance and physiological properties. AAPS J:22–34

Murai T (2015) Cholesterol lowering: role in cancer prevention and treatment. Biol Chem 396:1–11

Nykiforuk CL, Boothe JG, Murray EW (2006) Transgenic expression and recovery of biologically active recombinant human insulin from Arabidopsis thaliana seeds. Plant Biotechnol 4(1):77–85

Novak J, Julian BA, Mestecky J, Renfrow MB (2012) Glycosylation of IgA1 and pathogenesis of IgA nephropathy. Semin Immunopathol 34:365–382

Norris S (2005) Molecular farming. Parliamentary information and research services, Library of Parliament. Parliament of Canada. http://www2.parl.gc.ca/Content/LOP/ResearchPublications/prb0509-e.htm

Oppenheimer SB, Alvarez M, Nnoli J (2008) Carbohydratebased experimental therapeutics for cancer, HIV/AIDS and other diseases. Acta Histochem 110:6–13

Parmenter DL, Boothe JG, van Rooijen GJ, Yeung EC, Moloney MM (1995) Production of biologically active hirudin in plant seeds using oleosin partitioning. Plant Mol Biol 29:1167–1180

Porter CJ, Trevaskis NL, Charman WN (2007) Lipids and lipid-based formulations: optimizing the oral delivery of lipophilic drugs. Nat Rev Drug Discov 6:231–248

Piron R, Santens F, De Paepe A (2015) Using GlycoDelete to produce proteins lacking plant-specific N-glycan modification in seeds. Nat Biotechnol 33:1135–1137

Rademacher T, Sack M, Arcalis E, Stadlmann J, Balzer S, Altmann F (2008) Recombinant antibody 2G12 produced in maize endosperm efficiently neutralizesHIV-1 and contains predominantly single-GlcNAc N-glycans. Plant Biotechnol J 6:189–201

Röhrig F, Schulze A (2016) The multifaceted roles of fatty acid synthesis in cancer. Nat Rev Cancer 16:732–749

Ritala A, Leelavathi S, Oksman-Caldentey KM (2014) Recombinant barley produced antibody for detection and immunoprecipitation of the major bovine milk allergen, beta-lactoglobulin. Transgenic Res 23(3):477–487

Reusch D, Tejad ML (2015) Fc glycans of therapeutic antibodies as critical quality attributes (CQAs). Glycobiology:1–30

Rosales-Mendoza S, Paz-Maldonado LM, Soria-Guerra RE (2012) Chlamydomonas reinhardtii as a viable platform for the production of recombinant proteins: current status and perspectives. Plant Cell Rep 31(3):479–494

Roy S, Tyagi A, Tiwari S, Singh A, Sawant SV, Singh PK, Tuli R (2010) Rabies glycoprotein fused with B subunit of cholera toxin expressed in tobacco plants folds into biologically active pentameric protein. Prot Expres Purifi 70:184–190

Santi L, Batchelor L, Huang Z, Hjelm B, Kilbourne J, Arntzen CJ, Chen Q, Mason HS (2008) An efficient plant viral expression system generating orally immunogenic Norwalk virus-like particles. Vaccine 26:1846–1854

Santos RB, Abranches R, Fischer R, Sack M, Holland T (2016) Putting the spotlight back on plant suspension cultures. Front Plant Sci 7:29

Seeberger PH, Werz DB (2005) Automated synthesis of oligosaccharides as a basis for drug discovery. Nat Rev Drug Discov 4:751–763

Shchelkunov SN, Saliaev RK, Ryzhova TS, Pozdniakov SG, Nesterov AE, Rekoslavskaia NI, Sumtsova VM, Pakova NV, Mishutina UO, Kopytina TV, Hammond RV (2004) Designing of a candidate edible vaccine against hepatitis B and HIV on the basis of a transgenic tomato. Vestn Ross Akad Med Nauk 11:50–55

Strasser R, Altmann F, Steinkellner H (2014) Controlled glycosylation of plant-produced recombinant proteins. Curr Opin Biotechnol 30:95–100

Schähs M, Strasser R, Stadlmann J (2007) Production of a monoclonal antibody in plants with a humanized N-glycosylation pattern. Plant Biotechnol J 5:657–663

Schillberg S, Raven N, Fischer R, Twyman RM, Schiermeyer A (2017) Contained molecular farming using plant cell and tissue cultures. Appl Bioeng: Innov Future Dir 12:5

Sharma MK, Jani D, Thungapathra M, Gautam JK, Meena LS, Singh Y, Ghosh A, Tyagi AK, Sharma AK (2008) Expression of accessory colonization factor subunit A (ACFA) of Vibrio cholerae and ACFA fused to cholera toxin B subunit in transgenic tomato (Solanum lycopersicum). J Biotechnol 135:22–27

Shen JS, Busch A, Day TS (2016) Mannose receptor-mediated delivery of mossmade α-galactosidase A efficiently corrects enzyme deficiency in Fabry mice. J Inherited Metab Dis 39:293–303

Shin Y-J, Castilho A, Dicker M, Sadio F, Vavra U, Grunwald-Gruber C, Kwon T-H, Altmann F, Steinkellner H, Strasser R (2017) Reduced paucimannosidic N-glycan formation by suppression of a specific beta-hexosaminidase from Nicotiana benthamiana. Plant Biotechnol J 15:197–206

Sijmons PC, Dekker BM, Schrammeijer B, Verwoerd TC, Van Den Elzen PJ, Hoekema A (1990) Production of correctly processed human serum albumin in transgenic plants. Bio/Technology 8 (3):217–221

Steentoft C, Bennett EP, Clausen H (2013) Glycoengineering of human cell lines using zinc finger nuclease gene targeting: SimpleCells with homogeneous GalNAc O-glycosylation allow isolation of the O-glycoproteome by one-step lectin affinity chromatography. Methods Mol Biol 1022:387–402

Steinkellner H, Castilho A (2015) N-glyco-engineering in plants: update on strategies and major achievements. Glyco-Engineering:195–212

Stephenson DJ, Hoeferlin LA, Chalfant CE (2017) Lipidomics in translational research and the clinical significance of lipid-based biomarkers. Transl Res 189:13–29

Strasser R (2016) Plant protein glycosylation. Glycobiology 26:926–939

Strasser R, Altmann F, Glössl J, Steinkellner H (2004) Unaltered complex N-glycan profiles in Nicotiana benthamiana despite drastic reduction of β1, 2-N-acetylglucosaminyltransferase I activity. Glycoconj J 21(5):275–282

Strasser R, Stadlmann J, Schähs M (2008) Generation of glyco-engineered Nicotiana benthamiana for the production of monoclonal antibodies with a homogeneous human-like N-glycan structure. Plant Biotechnol J 6:392–402

Stremmel W, Ehehalt R, Staffer S, Stoffels S, Mohr A, Karner M, Braun A (2012) Mucosal protection by phosphatidylcholine. Dig Dis (Basel Switz) 30:85–91

Spiegel H, Stöger E, Twyman RM, Buyel JF (2018) Current status and perspectives of the molecular farming landscape. In: Kermode AR, Jiang L (eds) Molecular farming: applications, challenges, and emerging areas. Chapter 1. John Wiley & Sons, Inc. https://doi.org/10.1002/9781118801512

Schillberg S, Raven N, Spiegel S, Rasche S, Buntru M (2019) Critical analysis of the commercial potential of plants for the production of recombinant proteins. Front Plant Sci 10:720

Tacket CO (2007) Plant-based vaccines against diarrheal diseases. Trans Am Clin Climatol Assoc 118:79

Tacket CO, Mason HS, Losonsky G, Estes MK, Levine MM, Arntzen CJ (2000) Human immune responses to a novel Norwalk virus vaccine delivered in transgenic potatoes. J Infect Dis 182:302–305

Tacket CO, Pasetti MF, Edelman R, Howard JA, Streatfield S (2004) Immunogenicity of recombinant LT-B delivered orally to humans in transgenic corn. Vaccine 22:4385–4389

Takagi H, Hiroi T, Yang L, Tada Y, Yuki Y, Takamura K, Ishimitsu R, Kawauchi H, Kiyono H, Takaiwa F (2005) A rice-based edible vaccine expressing multiple T cell epitopes induces oral tolerance for inhibition of Th2-mediated IgE responses. Proc Natl Acad Sci U S A 102:17525–17530

Tarp MA, Clausen H (2008) Mucin-type O-glycosylation and its potential use in drug and vaccine development. Biochim Biophys Acta 1780:546–563

Thanavala Y, Mahoney M, Pal S, Scott A, Richter L, Natarajan N, Goodwin P, Arntzen CJ, Mason HS (2005) Immunogenicity in humans of an edible vaccine for hepatitis B. Proc Natl Acad Sci U S A 102:3378–3382

Tregoning J, Maliga P, Dougan G, Nixon P (2004) New advances in the production of edible plant vaccines: chloroplast expression of a tetanus vaccine antigen, TetC. Phytochemistry 65:989–994

Trevaskis NL, Nguyen G, Scanlon MJ, Porter CJH (2011) Fatty acid binding proteins: potential chaperones of cytosolic drug transport in the enterocyte? Pharm Res 28:2176–2190

Tusé D, Ku N, Bendandi M, Becerra C, Collins R, Langford N, Sancho SI, López-Díaz de Cerio A, Pastor F, Kandzia R, Thieme F (2015) Clinical safety and immunogenicity of tumor-targeted, plant-made Id-KLH conjugate vaccines for follicular lymphoma. Biomed Res Int 2015

Twyman RM, Stoger E, Schillberg S, Christou P, Fischer R (2003) Molecular farming in plants: host systems and expression technology. Trends Biotechnol 21:570–578

US Department of Agriculture (2006) BRS fact sheet available from: USDA-APHIS

Valdes R, Gomez L, Padilla S, Brito J, Reyes B, Alvarez T, Mendoza O, Herrera O, Ferro W, Pujol M, Leal V, Linares M, Hevia Y, Garcia C, Mila L, Garcia O, Sanchez R, Acosta A, Geada D, Paez R, Vega JL, Borroto C (2003) Large scale purification of an antibody directed against hepatitis B surface antigen from transgenic tobacco plants. Biochem Biophys Res Commun 308:94–100

Vliegenthart JFG (2006) Carbohydrate based vaccines. FEBS Lett 580:2945–2950

Vogel R, Crockett RS, Oden N, Laliberte TW, Molina L (2010) Demonstration of efficacy in the treatment of dry eye disease with o.18% sodium hyaluronate ophthalmic solution (vismed, rejena). Am J Ophthalmol 149:594–601

Walsh G, Jefferis R (2006) Post translational modifications in the context of therapeutic proteins. Nat Biotechnol 24:1241–1252

Wang L, Webster DE, Campbell AE, Dry LB, Wesselingh SL, Coppel RL (2008) Immunogenicity of Plasmodium yoelii merozoite surface protein 4/5 produced intransgenic plants. Int Parasitol 38:103–110

Wang Y, Deng H, Zhang X, Xiao H, Jiang Y, Song Y, Fang L, Xiao S, Zhen Y, Chen H (2009) Generation and immunogenicity of Japanese encephalitis virus envelope protein expressed in transgenic rice. Biochem Biophys Res Commun 380:292–297

Watson J, Koya V, Leppla SH, Daniell H (2004) Expression of Bacillus anthracis protective antigen in transgenic chloroplasts of tobacco, a non-food/feedcrop. Vaccine 22:4374–4384

Webster DE, Cooney ML, Huang Z, Drew DR, Ramshaw IA, Dry IB, Strugnell RA, Martin JL, Wesselingh SL (2002) Successful boosting of a DNA measles immunization with an oral plant derived measles virus vaccine. J Virol 76:7910–7912

Weintraub JA, Hilton JF, White JM, Hoover CI, Wycoff KL, Yu L, Larrick JW, Featherstone JD (2005) Clinical trial of a plant-derived antibody on recolonization of mutans streptococci. Caries Res 39(3):241–250

Woodard SL, Mayor JM, Bailey MR, Barker DK, Love RT, Lane JR, Delaney DE, McComasWagner JM, Mallubhotla HD, Hood EE, Dangott LJ, Tichy SE, Howard JA (2003)

Maize (Zea mays) derived bovine trypsin: characterization of the first large scale, commercial protein product from transgenic plants. Biotechnol Appl Biochem 38:123–130
Wu X, Daniels G, Lee P, Monaco ME (2014) Lipid metabolism in prostate cancer. Am J Clin Exp Urol 2:111–120
Van Droogenbroeck B, Cao J, Stadlmann J (2007) Aberrant localization and underglycosylation of highly accumulating single-chain Fv-Fc antibodies in transgenic Arabidopsis seeds. Proc Natl Acad Sci U S A 104(4):1430–1435
Xie T, Qiu Q, Zhang W, Ning T, Yang W, Zheng C, Wang C, Zhu Y, Yang D (2008) A biologically active rhIGF-1 fusion accumulated in transgenic rice seeds can reduce blood glucose in diabetic mice via oral delivery. Peptides 29(11):1862–1870
Yan H, Yan H, Li G, Gong W, Jiao H, Chen H, Ji M (2010) Expression of human cytomegalovirus pp 150 gene in transgenic Vicia faba L. and immunogenicity of pp150 protein in mice. Biologicals 38:265–272
Yang W-S, Chen Y-Y, Chen P-C, Hsu H-C, Su T-C, Lin H-J, Chen M-F, Lee Y-T, Chien K-L (2020) Association between plasma N-6 polyunsaturated fatty acids levels and the risk of cardiovascular disease in a community-based cohort study. Sci Rep 9:19298
Yao J, Weng Y, Dickey A, Wang KY (2015) Plants as factories for human pharmaceuticals: applications and challenges. Int J Mol Sci 12:28549–22865
Yu WY, Zhang N, Li CJ (2009) Saccharide modified pharmaceutical nanocarriers for targeted drug and gene delivery. Curr Pharm Des 15:3826–3836
Yan J, Chen X, Wang FS, Cao HZ (2013) Chemoenzymatic synthesis of mono- and di-fluorinated Thomsen Friedenreich (T) antigens and their sialylated derivatives. Org Biomol Chem 11:842–848
Yi S, Yang J, Huang J (2015) Expression of bioactive recombinant human fibroblast growth factor 9 in oil bodies of *Arabidopsis thaliana*. Protein Expr Purif 116:127–132
Yang S, Wang M, Chen L (2015) Quantity: an isobaric tag for quantitative glycomics. Sci Rep 5: 17585
Zimran A, Duran G, Mehta A, Giraldo P, Rosenbaum H, Giona F (2014) Taliglucerase alfa 36-month clinical safety and efficacy in treatment-naive patients. J Inherit Metab Dis 1:S155
Zhang Y, Chen CC, Tan HN, Wang FS, Cao HZ (2015) Synthesis of unsymmetrical 3,6-branched Man5 oligosaccharide: a comparison between one-pot sequential glycosylation and stepwise synthesis. Carbohydr Res 401:109–114
Zhang Z, Huang R (2010) Enhanced tolerance to freezing in tobacco and tomato overexpressing transcription factor TERF2/LeERF2 is modulated by ethylene biosynthesis. Plant Mol Biol 73:241–249
Zhang Y, Wang F (2015) Carbohydrate drugs: current status and development prospect. Drug Discov Ther 9(2):79–87
Zhong R, Pena MJ, Zhou GK, Nairn CJ, Wood-Jones A, Richardson EA, Morrison WH, Darvill AG, York WS, Ye ZH (2005) The FRA8 gene, which encodes a putative glucuronyltransferase, is essential for normal secondary wall synthesis. Plant Cell 17:3390–3408
Zhou C, Xue Y, Zhang Y, Zeng Y, Ma Y (2009) Recombinant expression and characterization of Thermoanaerobacter tengcongensis thermostable α-glucosidase with regioselectivity for high-yield isomaltooligosaccharides synthesis. J Microbiol Biotechnol 19(12):1547–1556
Zong AZ, Zhao T, Zhang Y, Song X, Shi Y, Cao H, Liu C, Cheng Y, Qu X, Cao J, Wang F (2013) Anti-metastatic and antiangiogenic activities of sulfated polysaccharide of Sepiella maindroni ink. Carbohyd Polym 91:403–409

Future Prospects of GM Plants

6

Abstract

The global population is increasing by 83 million annually and is expected to cross 8.6 billion by the year 2030. To cope with this alarming situation and to feed the teeming millions, the food supplies must increase from 25 to 70% by the year 2050. In order to achieve complete food and health security, the political pledges must be put into action and quantitative targets must be set for increasing food production. The GM crop productivity achieved in the last 21 years (1996–2016) is noteworthy and indicates that the conventional crop improvement technology alone cannot meet the demands of the growing population. The continued acceptance, adoption, and cultivation of GM crops are indicators of farmer's and consumer's satisfaction towards mitigating the challenges of biotic and abiotic stresses, conservation of biodiversity, and economic, health, and social benefits. Since the commercialization (1996 to the present) of biotech crops (major: soybean, corn, cotton; minor: alfalfa, sugar beet, papaya, squash, and eggplant), the global acreage (28 countries) has reached 189.8 million hectares. The biotech crops will continue to contribute immensely towards sustainable, bountiful, and profitable agricultural yield and shall deliver significant environmental, economic, health, and social benefits to the farmers. This chapter summarizes the major contribution of GM crops towards global food security and economy, the challenges and risk associated with the release of GM crops, and the future of GM technology.

Keywords

Biotech crops · Biofortification · Risk assessment · GEAC · EPA · GM crops

B. Koul, *Cisgenics and Transgenics*, https://doi.org/10.1007/978-981-19-2119-3_6

6.1 Introduction

Genetic modification (GM) in plants is a biotechnological approach that deals with the injection of DNA obtained from a particular living organism (plant or animal) into the genome of a crop to be modified. GM plant is thus produced by transferring new genes with desirable characteristics into the plant cell, which is usually regenerated in a tissue culture medium after transformation (Venki 2016). Genetic engineering in plants commenced around 1983 by expressing a bacterial gene in a tobacco species, and tomato was the first GM food crop that was commercialized in 1994 by Calgene Company. Genes with desirable traits are inserted into the recipient host plant randomly to produce genetically modified crops (Sticklen 2015). GM crops are thus regarded as those crops whose genetic constitution is altered and modified through genetically engineered technology so as to improve the endogenic traits or to introduce novel characters that are not naturally present in the crop being transformed. The crops developed by the incorporation of foreign genetic materials into their genome by direct gene transfer or through *Agrobacterium*-mediated technique are referred to as transgenic plants (Kumar et al. 2020). The genes involved (transgenes) in this process can be obtained from different sources including plants that are sexually incompatible, viruses, fungi, bacteria, or other animal species (Jhansi and Usha 2013). The major issue with traditional plant breeding method where sexual compatibility between the plant species is necessary for the crossing is subdued with the emergence of genetic modification technique. *Agrobacterium tumefaciens* was found to be naturally capable of inserting Ti plasmid (T-DNA) into the genome of the host plant species since 1977. This has led to the recommendation of employing the T-DNA to serve as a vector for the introduction of foreign genetic materials (genes) into host plant cells (Chilton et al. 1977). The transgenes along with the existing genetic materials in the transgenic plants are transmitted to the subsequent offspring through pollination. As a result, the progenies generated are regarded transgenic because they harbor the transgenes (Kumar et al. 2018). The main objective of crop improvement is to enable the plant breeders to utilize genes from various sources for the production of crops with qualitative traits, so as to increase agricultural productivity. Transgenesis provides a more efficient way of producing crops with new desirable traits in addition to reduction in the length of time needed to create new plant cultivars. One of the basic functions of this technique is its capacity of injecting one or more genes into host plants. The production of transgenics has been positively impacting agriculture in the last few decades, estimated to have added up to 22% crop yields, leading to an increment of 68% profits gained by the farmers (Klümper and Qaim 2014). Despite the benefits offered by transgenics, these crops continue to be a matter of concern due to possibility of gene flow between transgenics and their wild relatives. This is in addition to potential negative health issues like allergic reaction and toxicity effects in humans. As a result, crops developed through this concept have been rejected publically in different parts of the globe, which has consequently ruled out their general acceptance and utilization (Kumar et al. 2020). Cisgenesis and intragenesis have been developed to serve as alternatives to the transgenesis, and in both methods, the genetic material is sourced from the gene pool of either same crop or closely related species (Rameshraddy Khan et al. 2017).

GM crops are developed in the laboratory by transfer of one or more desirable gene(s) from the donor plant/animal/bacterium, etc. species to the host plant cell nucleus (the main target). The genes used in this regard can be from plant to plant (species within the same kingdom) or from bacteria to plant (species within different kingdom). Whatever the case may be, the injected DNA (genome) must be modified for it to express in the recipient host plant accurately with high efficiency. Herbicide-tolerant genes, antigens, and Cry toxin genes can be expressed in transgenic plants. The most frequent techniques involved in the production of these crops are biolistic also known as particle gun method, *Agrobacterium*-mediated techniques, and protoplast transformation method. In the first method, the plant genome (DNA) is tied to the micro gold particles, which will later on be injected under high pressure into the targeted plant tissue/single plant cells by means of a gun. These particles shot with high acceleration will penetrate both the cell wall and cell membranes. The coated metal detached from the DNA molecule unites with the plant genome in the nucleus of the cell. This method has been used to successfully develop various GM crops such as maize and wheat. The only issue associated with this technique is the destruction of plant cell tissues during the process (Jhansi and Usha 2013).

The second method involves the use of a soil bacterium species known as *Agrobacterium tumefaciens*, which is capable of infecting plant cells using its DNA fragment. The DNA fragment is then united in the chromosome of the infected plant by means of a Ti plasmid (tumor-inducing plasmid), which pilots the plant cellular system, thereby using it to produce several copies of its own DNA pieces. The Ti plasmid replicates itself without depending on the chromosome of bacterial cell, and it has a point of transfer DNA in which a floral dip concept can be used by researchers to inoculate genes into the crops to be modified. One of the problems of this process is that it is not applicable to a number of vital food crops. The technique is useful in developing dicotyledonous species such as tobacco, tomatoes, roses, and potatoes (Kumar et al. 2018). Transgenic plants composed of alleles for bacterial enzymes have been used to remove organic pollutant, mercury, and selenium from the soils (Kumar et al. 2020).

Another technique of generating transgenic plant is protoplast transformation. This is an enzymatically enhanced method in which the cell wall of a plant is being digested while the vectors are consolidated by transfections using either chemical means or electroporation. This method has been employed in modifying the color of petunia flower to obtain a transgenic ornamental plant. It is more tedious and time consuming than the normal tissue culture method. This long culture duration often minimizes the rate of regenerating unwanted somatic mutation. As a result, protoplast transformation is not mostly employed for transgenic transformation. Nevertheless, it can be good for genome editing techniques as it can easily take up RNA and proteins. There is thus need for re-evaluation of this method for the improvement of its efficiency (Kishi-kaboshi et al. 2018).

6.2 The Current Status of Transgenic Crops

Transgenic biotechnology is considered an effective means of improving crop plants for high yield, quantity, and quality of the crops in different parts of the world (Hou et al. 2014). Currently, there has been progress in the application of transgenic biotechnologies to specifically curtail the issue of food scarcity/security for the global population (Baghbani-Arani et al. 2021). However, there are public concerns with regard to the potential risks associated with the production of genetically modified crops or transgenics including the ecological or biodiversity scourge and allergies which restrict the usage of these plants.

Despite the challenges of biosafety and ecological issues, this technique (transgenesis) has been adopted by various researchers for the quick generation of improved crop varieties and stacking of numerous desirable characters. In the past two decades, there has been increase in the area of transgenic crop cultivation all over the world. It was reported that 1.7 m hectares were used for these crops in 1996, which multiplied significantly to 191.7 m having 113-fold increase in total area (hectares) with transgenic crops in 2018 (ISAAA 2018). Different transgenic crop plants have occupied millions of hectares globally with transgenic soybean occupying the largest area with 95.9 m hectares (50%), transgenic maize 58.9 m hectares (31%), transgenic cotton 24.9 m hectares (13%), and transgenic canola 10.1 m hectares, while the rest of transgenics have occupied 1.9 m hectares (<1%) of the area (ISAAA 2018). This technology is the most quick crop cultivation technology adopted by approximately 17 million farmers across 26 countries with USD18.2 billion as the economic value all over the world (Kumar et al. 2020). The development and approval of commercializing transgenic crops were based on some improved qualitative traits such as improvement in the nutritional composition, herbicide tolerance, resistance to diseases, insect resistance, and tolerance to abiotic constraints. Of these desirable characters, herbicide tolerance covered the widest area with about 88.7 m hectares (47%) of the whole global regions covered by transgenic plants around 2017. Similarly, crops with two or more traits arranged in stack and those with insect resistance trait were reported to have covered 41% and 12% of the entire region occupied by transgenic crops (Kumar et al. 2020). To achieve the required increase in terms of yield for the transgenic crops, there is need for more than 300 million hectares of the area used for conventional crop production (Brookes and Barfoot 2016). For the required area to be obtained, less fertile lands should be included and some parts of tropical forests must be lost. However, deforestation has serious negative impacts on the environment as it leads to increase in the earth surface temperature leading to global warming (Baghbani-Arani et al. 2021).

There is significant increase in the world production of GM crops with up to 26 nations cultivating these crops in 2016, while 5.85 and 6.99 million hectares were allocated to GM products in the developed and developing nations, respectively (ISAAA 2018). Eighteen out of the 26 nations that are well known for the production of biotech products remained the pioneers in this direction by providing 50,000 h to GM crops in 2018. In 2019, the top five countries with the most biotech crops

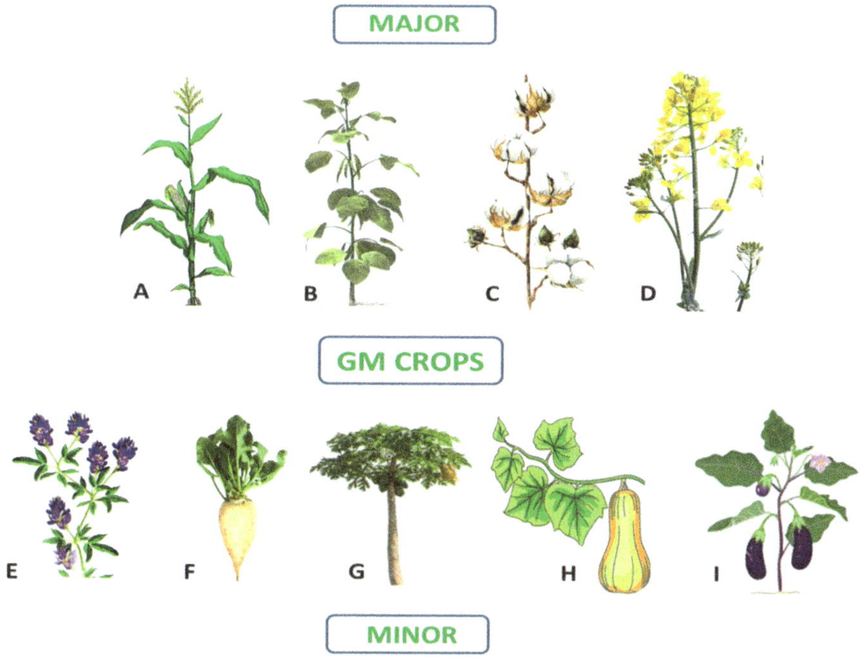

Fig. 6.1 Commercialized GM crops: (I) major crops (**a**) Maize, (**b**) Soybean, (**c**) Cotton, (**d**) Canola and (II) minor crops (**e**) SAfflower, (**f**) Sugarbeat, (**g**) Papaya, (**h**) Squash, (**i**) Brinjal

planted (the USA, Brazil, Argentina, Canada, and India) accounted for 91% of global biotech crop area. Among these, the USA produced the largest quantity of biotech products by providing 75 m hectares (39%) of the total land area used for the GM products, while the second nation was Brazil with 51.3 m hectares representing 27% of GM cultivation worldwide (Baghbani-Arani et al. 2021). The GM crops including soybean, canola, and corn species were accepted in different nations like the USA, Argentina, India, Canada, and Brazil at an average rate of 93.3%, 100%, 95%, 92.5%, and 93%, respectively, in the year 2018 (ISAAA 2018). Furthermore, about 43 GM crops including 40 species were approved globally in 2019 (Zhong et al. 2019) with decreased number of approved GM products compared to previous 2 years. The "rate of acceptance" of transgenic crops and utilization varied among the countries from 100% in Argentina to the relative adoption of such crops in Pakistan, as they are relatively restricted in Europe and there is total prohibition of utilization of the crops in Turkey (Baghbani-Arani et al. 2021). Figure 6.1 shows the commercialized major and minor GM crops based on their acreage. Table 6.1 shows the genetically modified crops approved for cultivation/food/feed, in different countries of the world. The report presented by ISAAA (2018) revealed that the total number of GM products that were approved and released to the global markets was 508. This number is composed of 30 variable plant cultivars that were transformed to improve the yield, growth, quality, quantity, and stress (biotic and abiotic) tolerance.

Table 6.1 Genetically modified crops approved in different countries

S. no.	GM crops	GM traits	Nation/ province	Authorized use
1	Soybean	Glyphosate and 2,4-DHT	Philippines	Food, feed
2	Apple	Non-borrowing phenotype AR	USA	Food, feed, cultivation
3	Maize	Glyphosate and 2,4-DHT	Argentina	Food, cultivation
4	Soybean	Glufosinate and glyphosate 2,4-DHT	Philippines, China	Food, cultivation
5	Cotton	Glyphosate and isoxaflutole HT	Brazil Taiwan	Food, feed, cultivation Food
6	Maize	Glyphosate and 2,4-DHT, glufosinate, coleopteran IP	South Korea	Food
7	Canola	Glufosinate herbicide resistance, restoration of fertility	China	Food and feed
8	Cotton	Low gossypol, antibiotic resistance	USA	Feed, food, cultivation
9	Sugarcane	Insect resistance (lepidopteran)	Brazil	Food, feed, cultivation
10	Cowpea	Insect resistance (lepidopteran)	Nigeria	Food, feed, cultivation
11	Maize	Increased ear biomass	EU	Food and feed
12	Soybean	Glufosinate and mesotrione herbicide tolerant	China	Food and feed
13	Cotton	Insect resistance (hemipteran)	Taiwan, Japan	Food
14	Maize	Glyphosate herbicide resistance, insect resistance (hemipteran)	Brazil, Argentina	Food, feed, cultivation
15	Maize	Glufosinate and glyphosate herbicide resistance, multiple insect resistance	EU	Food and feed
16	Soybean	Glyphosate herbicide resistance, drought tolerance	Brazil Argentina	Food, cultivation Food, feed, cultivation
17	Canola	Glyphosate herbicide resistance	China	Food and feed
18	Cotton	Insect resistance, visual marker, glufosinate and glyphosate herbicide resistance, antibiotic resistance	EU	Food and feed
19	Canola	Modified fatty acids, imazamox herbicide resistance	USA	Cultivation

Source: Adapted from Baghbani-Arani et al. (2021) Benefits of Transgenic Crops

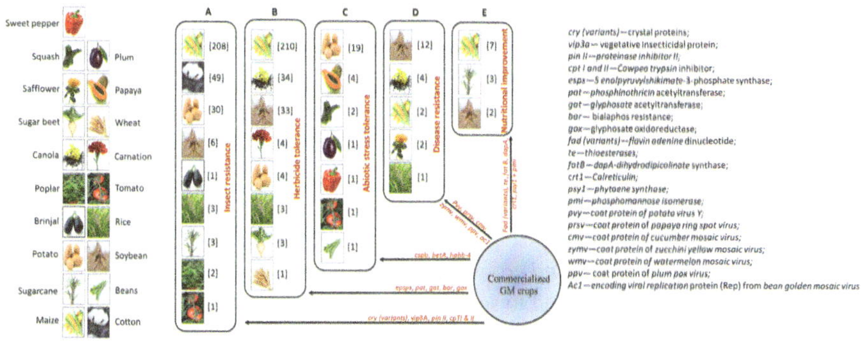

Fig. 6.2 Commercialized GM crops with various improved traits. (**a**) Insect resistance, (**b**) herbicide tolerance, (**c**) abiotic stress tolerance, (**d**) disease resistance, and (**e**) nutritional improvement. Numbers in parentheses represent numbers of commercialized transgenic events in the particular crop

Food security is one of the major challenges of the growing world population. GM biotechnology is among the reliable techniques of producing food crops of high quality and increased yield, thereby enhancing agricultural productivity globally. GM foods provide considerable benefits by upgrading the production yield and improving the nutritive composition of the crops (Jhansi and Usha 2013). Figure 6.2 shows the commercialized GM crops with various improved traits, (a) insect resistance, (b) herbicide tolerance, (c) abiotic stress tolerance, (d) disease resistance, and (e) nutritional improvement.

6.2.1 Production of Insect-Resistant Crops

Crop production has been constrained by different insect pests and diseases leading to severe losses to various economically vital crops (Reddy et al. 2009). The crops are destroyed by sap sucking or direct chewing of the leaves, stems, roots, and flowers. Insects also serve as vectors of many pathogenic organisms, which are transmitted to the plants while feeding (Rahman et al. 2012). To manage this problem, farmers depend on synthetic chemicals which are costly and have detrimental effects on the environment. The use of novel genetic modification methods like transgenesis has been of great help in developing crops that are resistant to insect pest infestation. Currently, there are ten insect-resistant transgenic crops that have been released to the public for cultivation (Kumar et al. 2020). Transgenic crops with insect resistance occupy the second largest region (23.3 m hectares) under agricultural production (ISAAA 2017). Generally, the global approved events for cultivation are 304, out of which 208 events with different IR alleles in maize have been authorized for production depending on the distribution of the insect pests. Some other crops with IR genes that have been commercialized include cotton, potato, rice, soybean, sugarcane, poplar, tomato, and brinjal with 49, 30, 3, 6, 3, 2, 1, and

1 events, respectively. *Bacillus thuringiensis*, a soil bacterium harboring *cry* gene, is highly deployed for developing transgenic plants for insect resistance trait (Table 6.2). The global status of approved commercially available Bt crops is shown in Table 6.3.

6.2.2 Herbicide-Resistant Transgenic Crops

Competition between weeds and crops for the limited resources such as water, space, nutrients, air, and light may lead to serious reduction in the yield of cultivated crops. Farmers use herbicides to eradicate weeds, but protecting the crops while destroying the weeds is the main challenge because weeds are herbaceous and selective destruction of such weeds is not possible most of the time. Development of herbicide-resistant trait in the primary crop can solve this problem by enhancing the flexible application of herbicides with robust nonselective and wide spectrum activity. The most widely used herbicides are glufosinate and glyphosate. As a result, the majority of herbicide-tolerant modified (transgenic) crops are generated to withstand the effects of chemicals (glyphosate and glufosinate). Glyphosate acts by inhibiting the main enzymes involved in the shikimate pathway of aromatic amino acid biosynthesis. Once this pathway is inhibited, glyphosate will be harmless to the living organisms including man, insects, birds, and other animals. Several glyphosate-resistant transgenic crops (cotton, tomato, tobacco, etc.) were developed by inserting aro A gene into a glyphosate synthase obtained from *Salmonella typhimurium.* Similarly, mutant gene (acetolactate synthase) from Arabidopsis was used to produce *Escherichia coli* sulfonylurea-resistant tobacco crops (Jhansi and Usha 2013). Dill et al. (2008) stated that glyphosate-tolerant soybean containing cp4epsps genetic material was commercialized as a pioneer transgenic crop tolerant to glyphosate herbicide and most of the approved glyphosate-resistant species contain this kind of gene.

6.2.3 Virus-Resistant Transgenic Crops

Transgenic technology has been used to develop crops that are resistant to virus infection. Transgenic crops resistant to mosaic viruses such as TMV-resistant tobacco and tomato crops have been developed through the insertion of coat proteins of the virus into the modified crop plant. Furthermore, some other transgenic plants that are viral resistant are rice stripe virus (RSV)-resistant rice, YMV-resistant green gram, virus-resistant potato crops, and YMV-resistant black gram (Davison 2010). Previous studies (Dunwell 2014; Cao and Li 2013; Fahim et al. 2012) show that virus resistance in cereals can be improved by expressing an artificial microRNA to develop resistance to wheat streak mosaic virus (WSMV), and resistance to maize rough dwarf disease (MRDD) was provided using a double-stranded RNA-specific endonuclease gene. Table 6.4 presents different types of genes used to produce transgenic crops that are resistant to virus infection.

Table 6.2 Insect-resistant transgenic crops using *cry* genes from *Bacillus thuringiensis*

Crop name	Family	Inserted gene	Insect targeted	Reference
Cotton (*Gossypium hirsutum* L.)	Malvaceae	*cry1Ab*	*Helicoverpa armigera*	Majeet et al. (2000)
		cry2Ab	Pink bollworm	Tabashnik et al. (2002)
		cry1A(b)	*Helicoverpa armigera*	Khan et al. (2011)
		cry1A(c)	*H. armigera*	Bakhsh et al. (2012)
		Vip3AcAa + *cry1Ac*	*H. armigera* strains	Chen et al. (2017)
		Vip3AcAa + *cry1Ac*	*H. armigera* strains	Chen et al. (2018)
		Cry1A	*Spodoptera litura*	Boonmee et al. (2019)
		Cry1Ac	*S. litura*	Siddiqui et al. (2019)
		Cry2Ab	*H. armigera* and *S. litura*	Katta et al. (2020)
		Cry1Acf + *cry2Aa*	*H. armigera*	Karthik et al. (2021)
		Cry1Fa + *cry32Aa*	*Earias fabia* and *Pectinophora gossypiella*	Razzaq et al. (2021)
		Vip3Aa	*H. armigera* and *Bemisia tabaci*	Din et al. (2021)
		cry51Aa2	*Lygus species*	Gowda et al. (2016)
		cry2AX1	*H. armigera*	Jadhav et al. (2020)
Sweet potato (*Ipomoea batatas*)	Convolvulaceae	*Cry3Aa*	Coleoptera	Peferoen et al. (1990)
		GNA	Lepidoptera	Perlak and Fischhoff (1993)
		Cowpea trypsin inhibitor	Lepidoptera	Morán et al. (1998)
		BBPs	*Phthorimaea operculella*	Murray et al. (2010)
		Cry1Aa	*S. litura*	Zhong et al. (2019)
Soybean (*Glycine max*)	Fabaceae	*cry1Ac*	Corn earworm (*Helicoverpa zea*), soybean looper (*Pseudoplusia includens*), tobacco budworm (*Heliothis virescens*), and velvet bean caterpillar (*Anticarsia gemmatalis*)	Stewart Jr et al. (1996)

(continued)

Table 6.2 (continued)

Crop name	Family	Inserted gene	Insect targeted	Reference
		cryIA(c) + Pinellia ternata agglutinin (pta) gene	Cotton bollworm	Dang and Wei (2007)
		cry8-like gene	*Holotrichia parallela*	Qin et al. (2019)
		cry8-like gene	*H. parallela*	Qin et al. (2020)
Maize (*Zea mays*)	Gramineae	*cry3Bb1*	Lepidoptera	Vaughn et al. (2005)
		cry1Ab	Lepidoptera	Koziel et al. (1993)
		cry19c	Lepidoptera	Gassmann et al. (2011)
		cry1Ab	Lepidoptera	Erasmus et al. (2019)
		cry1Ab	*Spodoptera frugiperda*	Liang et al. (2021)
Wild cabbage (*Brassica oleracea* L.)	Brassicaceae	*cry1Ab*	Diamondback moth larvae	Cao et al. (2001)
		cry1Ab	Diamondback moth larvae	Bhattacharya et al. (2002)
		cry1Ac1	*Pieris rapae*	Kim et al. (2015)
		cry11a8	*Pieris rapae*	Yi et al. (2015)
Ground nuts (*Arachis hypogea*)	Leguminosae	*cry1Ac*	*Elasmopalpus lignosellus* Zeller	Singsit et al. (1997)
		Cry1acF	*S. litura*	Keshavareddy et al. (2013)
Canola (*Brassica napus* L.)	Brassicaceae	*cry1Ac*	*Plutella xylostella* L. larvae	Himanen et al. (2009)
		cry1Ac	*S. exigua*	Liu et al. (2011)
Pigeon pea (*Cajanus cajan*)	Fabaceae	*Cry1EC*	*Helicoverpa armigera*	Surekha et al. (2005)
		cry1Ab	*H. armigera*	Verma and Chand (2005)
		cry1Ab	*Heliothis armigera*	Sharma et al. (2006)
		cry2Aa	*H. armigera*	Singh et al. (2018)
		cry1Ac + cry2Aa	*Helicoverpa armigera*	Ghosh et al. (2017)
		cry1Ac	*Helicoverpa armigera*	Ramu et al. (2012)
		cry1Ac	*Helicoverpa armigera*	Kaur et al. (2016)

Mountain/Arabian coffee (*Coffea arabica*)	Rubiaceae	*cry1Ac*	*Liriomyza* sp.	Leroy et al. (2000)
Tomato (*Lycopersicon esculentum* L.)	Solanaceae	*cry1Ac*	*Manduca sexta* L.	Fischhoff et al. (1987)
		Bt(k)	*Manduca sexta* L., *Keiferia lycopersicella*, and *Heliothis zea*	Delannay et al. (1989)
		cry1Ac	*Helicoverpa armigera* Hubner	Mandaokar et al. (2000)
		cry1Ab	*Helicoverpa armigera* Hubner	Kumar and Kumar (2004)
		Cry1Ab	*H. armigera* and *S. litura*	Koul et al. (2014)
		Cry1Ac	*H. armigera*	Koul et al. (2015)
		Cry2A	*H. armigera*	Hanur et al. (2015)
		Cry1Ac	*Tuta absoluta*	Hatice et al. (2017)
		Cry2AX1	*H. armigera*	Bamishaiye et al. (2017)
		cry1Ab	*T. absoluta*	Soliman et al. (2021)
Alfalfa (*Medicago sativa* L.)	Fabaceae	*cryIC*	*Spodoptera littoralis*	Strizhov et al. (1996)
		Synthetic cry3a gene	*Hypera postica*	Tohidfar et al. (2013)
Tobacco (*Nicotiana tabacum* L.)	Solanaceae	*cry1Aa*	*M. sexta* L.	Barton et al. (1987)
		δ-Endotoxin var. kurstaki HD1	Lepidopteran insects	Barton et al. (1987)
		cry1Ab	Tobacco hornworm (*M. sexta* L.) and budworm (*H. virescens* Fabricius)	Vaeck et al. (1987)
		cry1Ac	Tobacco hornworm (*M. sexta* L.)	Perlak et al. (1990)
		cry1Ab	Lepidopteran pests	Carozzi et al. (1992)
		cry1Ac	Tobacco budworm (*H. virescens* Fabricius)	McBride et al. (1995)
		cry1C	*Spodoptera litura* and *Spodoptera exigua*	Strizhov et al. (1996)
		cry1Ia5	*Heliothis armigera*	Selvapandiyan et al. (1998)
		cry1Aa2		Kota et al. (1999)

(continued)

Table 6.2 (continued)

Crop name	Family	Inserted gene	Insect targeted	Reference
			Heliothis virescens, Helicoverpa zea, Spodoptera exigua	
		cry2Aa2	Cotton bollworm (*H. zea* Boddie)	DeCosa et al. (2001)
		δ-Endotoxin	Polyphagous pest (*Spodoptera litura*)	Singh et al. (2004)
		cry2Aa2	*Heliothis virescens*	Zaidi et al. (2005)
		cry1Ac	*Heliothis virescens* and *Manduca sexta*	Gulbitti-Onarici et al. (2009)
		Cry2Ah1	*H. armigera*	Li et al. (2018)
Rice (*Oryza sativa* L.)	Poaceae	*cry1Ab*	Striped stem borer (*Chilo suppressalis* Walker) and leaf folder (*Cnaphalocrocis medinalis* Guenee)	Fujimoto et al. (1993)
		cry1Ac	Yellow stem borer (*S. incertulas* Walker)	Nayak et al. (1997)
		cry1Ab/ac	Striped stem borer and yellow stem borer	Cheng et al. (1998)
		cry1Ab	Yellow stem borer (*S. incertulas* Walker)	Datta et al. (1998)
		cry2A	Yellow stem borer and rice leaf folder	Maqbool et al. (1998)
		cry1B	Striped stem borer	Breitler et al. (2000)
		cry1Ab	Lepidopteran rice pests	Shu et al. (2000)
		cry1Ab *cry1Ac* Hybrid	Leaf folder (*C. medinalis* Guenee) and yellow stem borer (*S. incertulas* Walker)	Tu et al. (2000)
		cry1Ac	Yellow stem borer (*Scirpophaga incertulas*)	Khanna and Raina (2002)
		cry1Ab	Rice leaf folder, *Cnaphalocrocis medinalis*	Ye et al. (2003)
		cry1Ab/ac	Stem borer	Ramesh et al. (2004)
		cry2A	Lepidopteran rice pest	Chen et al. (2005)
		Cry2AX1	Lepidopteran	Chakraborty et al. (2016)
		Cry2A	Leaf folder	Gunasekara et al. (2017)

		Cry1Ab + vip3A	Leaf folder	Xu et al. (2018)
		Cry2AX1	Leaf folder	Manikandan et al. (2018)
		Cry2Aa + cry1Ac	*Chilo suppressalis*	Qiu et al. (2019)
		Cry1Ab/c	*H. armigera*	Fu et al. (2020)
Quaking/American aspen (*Populus tremuloides* Michx.)	Salicaceae	*cry1Aa*	Forest tent caterpillar (*Malacosoma disstria* Lasiocampidae) and gypsy moth	McCown et al. (1991))
		cry3A	*Chrysomela tremula* F. (Col.)	Cornu et al. (1996)
Sugarcane (*Saccharum officinarum* L.)	Poaceae	*cry1Ab*	Stem borer (*Diatraea saccharalis* F.)	Arencibia et al. (1997)
		cry1Ac	Stem borer in field trials	Weng et al. (2010)
		Vip3A	Lepidoptera	Riaz et al. (2020)
		Cry1Ac	Stem borer	Dessoky et al. (2020)
		Cry2Aa + cry1Ca *Cry1Ab + cry1Ac*	Shoot borer	Koerniati et al. (2020)
Brinjal (*Solanum melongena* L.)	Solanaceae	*cryIIIb*	Colorado potato beetle (*L. decemlineata* say)	Arpaia et al. (1997)
		cryIIIA	Colorado potato beetle (*L. decemlineata* say)	Jelenkovic et al. (1998)
		cry1Ab	*Leucinodes orbonalis*	Kumar et al. (1998)
Moth bean (*Vigna aconitifolia* L.)	Fabaceae	*cry1Ac*	*H. armigera*	Kamble et al. (2003)
Potato (*Solanum tuberosum* L.)	Solanaceae	*cry1Ab*	Tuber moth (*Phthorimaea operculella* Zeller)	Peferoen et al. (1990)
		cry1Ab	*Phthorimaea operculella* Zeller	Rico et al. (1998)
		cryIIIA	Colorado beetle (*Leptinotarsa decemlineata* say)	Adang et al. (1993)
		cryIIIA	*Leptinotarsa decemlineata* say	Perlak and Fischhoff (1993)
		cryIIIA	*Leptinotarsa decemlineata* say	Coombs et al. (2002)
		*cry*V Bt	Potato tuber moth (*P. operculella* Zeller)	Douches et al. (1998)

(continued)

Table 6.2 (continued)

Crop name	Family	Inserted gene	Insect targeted	Reference
		cry9Aa2	Potato tuber moth	Gleave et al. (1998)
		cry1Ab	*Helicoverpa armigera* (Hubner)	Chakrabarti et al. (2000)
		cry1Ac	*Tecia solanivora*	Valderrama et al. (2007)
		Cry3A	Colorado potato beetle	Mi et al. (2015)
		cry1Ab	*Phthorimaea operculella*	Salehian et al. (2021)
Chickpea (*Cicer arietinum* L.)	Leguminosae	*cry1Ac*	*Heliothis armigera*	Kar et al. (1997)
		cry1Ac	*H. armigera*	Sanyal et al. (2005)
		cry1Ac	*H. armigera* and *S. litura*	Indurker et al. (2007)
		cryIIAa	Pod borer	Sawardekar et al. (2017)
Jute (*Corchorus* sp.)	Malvaceae	*δ-Endotoxin cry1Ab/Ac*	Lepidoptera	Majumder et al. (2018)
Castor (*Ricinus communis*)	Euphorbiaceae	*Cry1AC*	*Achaea janata*	Muddanuru et al. (2019)
Cowpea (*Vigna unguiculata*)	Leguminosae/ Fabaceae	*Cry2Aa*	*Maruca vitrata*	Kumar et al. (2021)
		Cry1Ab	*Maruca vitrata*	Addae et al. (2020)

Table 6.3 Global status of approved and commercially available Bt crops

Crop	Transgenic event(s)	Trade name	Company	Gene (s) introduced	Targets
Cotton	LLCotton25, MON 15985	FiberMax LibertyLink Bollgard II*	Bayer CropScience	*bar, cry 1Ac, cry2Ab*	Lepidopteran pests, weeds
Cotton	DAS-21023-5, DAS-24236-5	Wide strike*	Dow AgroSciences	*pat, cry1Ac, cry1Fa*	Lepidopteran pests, weeds
Cotton	DAS-21023-5, DAS-24236-5, MONO1445–2	Wide strike* roundup/ ready*	Dow AgroSciences	*pat, cry1Ac, cry1Fa, CP4EPSPS*	Lepidopteran pests, weeds
Cotton	DAS-21023-5, DAS-24236-5, MON88913–8	Wide strike*/ roundup ready* flex	Dow AgroSciences	*pat, cry1Ac, cry1Fa, CP4EPSPS*	Lepidopteran pests, weeds
Cotton	MON531, MON1445–2	Roundup ready*, Bollgard	Monsanto	*cry1Ac, CP4EPSPS*	Lepidopteran pests, weeds
Cotton	MON88913–8, MON15985	Bollgard II*, roundup/ ready* flex	Monsanto	*CP4EPSPS, cry1Ac, cry1Ab*	Lepidopteran pests, weeds
Maize	TC1507	Herculex CB	Dow AgroSciences and Pioneer Hi Bred	*cry1Fa, pat*	Lepidopteran pests (European corn borer), weeds
Maize	TC1507	Herculex CB	Dow AgroSciences and Pioneer Hi Bred	*cry1Fa, pat*	Lepidopteran pests (European corn borer), weeds
Maize	DAS-59122-7	Herculex RW	Dow AgroSciences and Pioneer Hi Bred	*cry34Ab1/ cry35Ab1, pat*	Coleopteran pests (corn rootworms), weeds
Maize	TC1507, DAS-59122-7	Herculex XTRA	Dow AgroSciences and Pioneer Hi Bred	*cry1Fa, cry34Ab1, cry35Ab1, pat*	Lepidopteran and coleopteran pests, weeds
Maize	DAS-59122-7, TC1507, NK603	Herculex XTRA/ roundup/ ready 2	Dow AgroSciences and Pioneer Hi Bred	*pat, CP4EPSPS, cry34Ab1, cry35Ab1, cry1Fa 2*	Lepidopteran and coleopteran pests, weeds
Maize	MON89034	YieldGard VT pro	Monsanto	*cry1A105, cry2Ab2*	Lepidopteran pests
Maize	MON88017	YieldGard VT	Monsanto	*CP4EPSPS, cry3Bb1*	Coleopteran pests (corn rootworms), weeds

(continued)

Table 6.3 (continued)

Crop	Transgenic event(s)	Trade name	Company	Gene (s) introduced	Targets
Maize	MON810, MON88017	YieldGard VT triple	Monsanto	*cry1A105*, *cry2Ab2*, *cry3Bb*	Lepidopteran and coleopteran pests, weeds
Maize	MON89034, MON88017	Genuity VT triple pro	Monsanto	*cry1A105*, *cry2Ab2*, *cry3Bb*	Lepidopteran and coleopteran pests, weeds
Maize	MON89034, TC1507, MON88017, DAS-59122-7	Genuity SmartStax TM	Monsanto and Dow AgroSciences	*pat*, *CP4EPSPS*, *cry 1Fa2*, *cry1A105*, *cry2Ab*, *cry3Bb1*, *cry34Ab1*, *cry35Ab1*	Lepidopteran and coleopteran pests, weeds
Maize	BtII, GA21	Agrisure GT/CB/LL	Syngenta	*cry1Ab*, *pat*, mutant maize *EPSPS*	Lepidopteran pests (European corn borer), weeds
Maize	BtII, MIR604	Agrisure CB/LL/RW	Syngenta	*cry1Ab*, m*cry3Aa*, *pat*	Lepidopteran and coleopteran pests, weeds
Maize	GA21, BtII, MIR604	Agrisure 3000GT (GT/CB/LL/RW)	Syngenta	*pat*, *cry1Ab*, m*cry3Aa*, mutant maize *EPSPS*	Lepidopteran and coleopteran pests, weeds

6.2.4 Transgenic Crops Tolerant to Abiotic Stress

Different ecological factors collectively known as abiotic stresses including cold, drought, heat, salinity, and flooding have various ill effects on the agricultural crops, which culminate in reduced grain yield (Suzuki et al. 2014). Changes in the metabolic pathways like regulatory proteins, activation of cascade signals, altering the antioxidant defense pattern to sustain the cellular homeostasis, and production and aggregation of mixable solutes like sugars and polyamines which aid in osmotic modification/adjustment have been used by crop plants to subdue the effects of environmental stresses (Kumar et al. 2020; Gupta et al. 2015; Raza et al. 2019). Abiotic stresses also lead to adjustment in the expression of a number of genes. Currently, several crop plants like sugarcane, maize, rice, and soybean with specific number of events related to abiotic stress have been approved and commercialized (ISAAA 2019). Bacterial proteins (cold-shock protein) were applied to manage the

Table 6.4 Genes deployed for virus resistance in different crops

Gene type	Function/ description	Source (donor)	Example of crop
G5	ssDNA-binding proteins	Bacteriophage M13	Cassava
Ribonuclease	RNA degradation	Yeast	Potato, wheat, pea
Movement protein	Viral movement protein	Tomato mosaic virus, raspberry bushy dwarf virus	Tomato, raspberry
Replicase	RNA replication	Tomato yellow leaf curl virus, papaya ringspot virus, cauliflower mosaic virus, potato leafroll virus	Papaya, tomato, cassava, potato
Nuclear inclusion protein	Nuclear located protein	Papaya ringspot virus, wheat streak mosaic virus, potato virus Y	Potato, squash, wheat, melon
Coat protein	Capsid protein	Many plant viruses	Sugarcane, tomato, potato, tobacco squash, pepper, alfalfa, maize, plum, papaya beet, lettuce, melon, pineapple, pea, wheat, peanut, grape

Source: Collinge et al. (2010)

abiotic constraints like heat and water deficiency in rice and cold effect in Arabidopsis species (Castiglioni et al. 2008). They further demonstrated that pleiotropic effects do not occur due to the use of cold-shock proteins in transgenic crops. This technique has been used to develop transgenic crops that can tolerate salt and drought effects for instance, drought-tolerant maize species was produced using cspB protein in Monsanto, and it has been commercialized as MON 87460 event (Genuity® DroughtGard™) in London since 2013 (James 2013). Previous studies of ISAAA (2017) indicated that a transgenic maize characterized with insect resistance and tolerance to drought was established for illustration by farmers and significant results were recorded.

6.2.5 Transgenic Crops and Nutrition

Deficiency in vitamin was among the major public health issues that affected a significant number of preschool children and pregnant women (about 15%) in 2005 worldwide (Kumar et al. 2020). The rice crop, being a staple diet for a large number of humans globally, has been transformed genetically by introduction of three genes in which two of the genes are sourced from daffodils while one is derived from microorganisms. The production of beta-carotene which serves as the precursor for the synthesis of vitamin A has increased in the transgenic rice, which possesses a yellow-colored seed. Such golden rice cultivar can help reduce the problems associated with vitamin A deficiency in humans (Dunwell 2014).

Essential amino acids such as tryptophan (Trp) and lysine (Lys) cannot be synthesized by both animals and humans and thus should be obtained from the diet. Sindhu et al. (1997) reported that a few transgenic techniques were used to alter the amino acid contents of crop proteins by manipulating the amino acid metabolic pathways to increase the quantity of a particular essential amino acid for enhanced nutritional values. Rice and wheat crops have been generated through this technique by expressing legumin protein from lysine-rich pea plant within the endosperm. A similar achievement was reported (Bicar et al. 2008) where transgenic maize was developed through heterologous expression of lysine-rich (porcine α-lactalbumin) protein that elevated the quantity of such amino acids (lysine) by 47%. Another protein with large amount of all the essential amino acids required by man was developed through genetic manipulation of a seed storage protein derived from *Amaranthus hypochondriacus* in cereal crops.

Fats and oils are important macromolecules required for proper growth and development in humans. Genetic engineering in oil-producing crops through transgenic method was employed to improve the oilseed plants nutritionally by modifying the existing fatty acid composition of such crops so as to increase the length of time oils can be packaged without spoilage. FAO (2010) stated that oils with higher quantity of polyunsaturated fatty acids and low proportion of saturated fatty acids are regarded to be good for humans. Oils obtained from fish, sunflower, walnuts, soybean, and flaxseeds are among the beneficial oils in humans. Essential fatty acids (linoleic acid, alpha linoleic acid, and oleic acids) are also obtained from the oils. Currently, several transgenic events involving modification in the lipid composition from some oil-producing crops such as soybean, canola, and safflower have been commercialized with 12, 4, and 2 events, respectively. Usher et al. (2017) reported that high amount of omega-3 long-chain polyunsaturated fatty acids was produced through transgenic modification of *Camelina sativa* using genes obtained from microorganisms in the marine habitat. Examples of such long-chain fatty acids are eicosapentaenoic acids (EPA) and docosahexaenoic acids (DHA) similar to the fatty acids available in the fish oils.

6.2.6 Transgenic Crops with Therapeutic Functions

Proteins of medical value involved in the diagnosis and treatment of different ailments in humans can be expressed in plants using transgenic technology (Jhansi and Usha 2013). Transferring the transgenic crops from laboratory to the fields leads to the manufacture of these proteins industrially. Three types of proteins of therapeutic functions derived from transgenic crops include vaccines, antibodies, and proteins. The vaccines derived from banana and potato go against gastrointestinal infectious diseases. The antibodies produced from transgenic plants cure such diseases as cholera, rheumatoid arthritis, cancer, hepatitis B virus, HIV, malaria, herpes, and rhinovirus. Cereal grains like wheat and rice can be another potential target as anticancer antibodies that recognize cells of breast, colon, and lung cancer have been expressed in them. They are thus considered useful in the diagnosis and treatment of different diseases of humans in the future. The global GM-based events are shown in Table 6.5.

6.3 Ethical Issues and Risks Associated with Transgenic Crops

The universal acceptance and adoption of transgenic crops have brought about several nutritional, economical, health, and environmental functions to humans for the past 20 years (Zhang et al. 2016; Brookes and Barfoot 2016). Despite the tremendous benefits, several issues like environmental damages and risks associated with human consumption of these crops have been raised. Principally, five concerns are raised against the transgenics as enumerated below: (1) environmental effects of releasing GM crops into the ecosystem, (2) impact of GM seed on the universal seed markets, (3) preferences of farmer and consumer in accepting the GM products, (4) risk assessment on the biosafety of these crops, and (5) impact of universal use of GM crops on biodiversity (Sheldon 2002). These major issues have been discussed extensively (Kumar et al. 2020) in their previous studies on GM crops.

6.3.1 Transgenic Crops and Biosafety Issues

Problems related to potential health hazard in humans (allergy and toxicity) and ecological damages have been raised with respect to safety of using transgenic crops (Lovei et al. 2010). The biosafety issue of transgenic crops has perpetually been controversial. The use of Starlink maize (with Cry9c gene) in the USA as animal feed and for industrial functions and its rejection as human diet due to the potential allergic reaction, high protein stability, and possibility of its interaction with human immune system are a clear example on the safety issue of transgenic crops (Goldman 2002). This crop was later on accepted globally due to detection of *cry* genes in other food products (US EPA 2017), but direct connection between the Cry proteins and allergic reactions in consumers could not be established.

Seralini et al. (2012) reported that GM crops are potentially hazardous as they may lead to various health issues including kidney disease, increase in tumor occurrence, liver congestion and necrosis in males, as well as high death rate in female rats provided with transgenic (NK-603 Roundup Ready) maize. This study was heavily criticized due to imperfect experimental design and wrong analysis, which led to the withdrawal of the article. Several studies associated with health dangers of transgenic products have been conducted on a variety of crops like rice, soybean, maize, and wheat using poultry, rodents, cows, and frogs as experimental animals (Domingo 2016; Tsatsakis et al. 2017; De Vos and Swanenburg 2018). Most of these researches (with the exception of few ones) did not report any negative results with respect to feed containing transgenic substances on animal well-being.

Table 6.5 Global GM-based events reported (Source: ISAAA 2018)

Crop varieties	Total reported events
Brassica napus	41
Phaseolus vulgaris	1
Medicago sativa	5
Dianthus species	19
Cichorium intybus	63
Gossypium hirsutum	1
Caryophyllus	3
Agrostis stolonifera	1
Eucalyptus species	1
Linum	231
Solanum melongena	1
Usitatissimum	2
Cucumis melo	1
Zea mays	4
Carica papaya	1
Brassica rapa	48
Petunia hybrid	4
Populus species	8
Prunus domestica	2
Solanum tuberosum	2
Rosa hybrid	40
Oryza sativa	2
Tinctorius	3
Carthamus	2
Glycine max	4
Beta vulgaris	2
Cucurbita pepo	1
Saccharum species	11
Capsicum annum	1
Triticum aestivum	–
Lycopersicum esculentus	–
Nicotiana tabacum	–

Netherwood et al. (2004) have expressed another issue of antibiotic resistance of the gut microbes found in humans and other animals due to potentiality of transferring marker genes with antibiotic resistance from the transgenic food. Nevertheless, the probability of transferring this gene is very low. To overcome this concern, Tuteja et al. (2012) paid much attention on the generation of marker-free transgenic plants.

Previous studies (Yan et al. 2015; Han et al. 2015) described the ecological related issues regarding the transgenic crops. Such problems may include the transgene flow mediated by pollen grains as reported from transgenic plants to traditionally developed cultivars and to wild relatives of several crops as barleys, maize, beans, rice, and creeping grass. Another issue of importance regarding the transgenic crops and ecosystem is loss of biodiversity, which may result due to introgression of a transgene from modified crops to their wild relatives. Furthermore, transferring such genes to weedy species could pose the risk of developing novel weeds and emergence of "superweeds," very hazardous weeds with herbicide resistance trait.

6.3.2 Destruction of Resistance

Different insect-resistant (IR) and herbicide-tolerant (HT) crops have been produced through transgenic technology, and they have been into use globally to sustain crop productivity. Repeated and wide cultivation of these crops (IR and HT) may lead to the development of resistance by the targeted species (insects and weeds) because of high selection pressure. This may lead to the emergence of novel insect species and development of "superweeds" that can resist the transgenic biotechnology (Gilbert 2013). Three species of pests evolved to Bt maize: (1) *Spodoptera frugiperda* in Puerto Rico to cry1F-expressing corn; (2) *Busseola fusca* (known as stem borer of Africa) found in South Africa to cry1Ab-expressing corn; and (3) *D. virgifera* (rootworm) in America to cry3Bb-expressing corn (Rensburg 2007; Storer et al. 2010; Gassmann et al. 2011). Bagla (2010) reported on a similar field that evolved pest resistance to Bt cotton in two pest species, viz. pink bollworm (*Pectinophora gossypiella*) in India to cry1Ac-expressing cotton and *Helicoverpa zea* in the USA to cry1Ac and cry2Ab. The evolved resistance of *Spodoptera frugiperda* (fall army-worm) to Bt maize was made known in a very short time period, which led to immediate withdrawal of such crops from the field.

6.3.3 Harmful Effects on the Nontarget Organisms

Nontarget organisms that may be useful to our environment may be negatively affected by transgenic crops as reported in literature. Higher death rate of *Danaus plexippus* (monarch butterfly) larvae that were supplied with feed (milkweed leaves) containing transgenic Bt maize was reported (Losey et al. 1999). The results of this study ignited a lot of controversy among the scientists because of the nature of the experimental design (flawed) and extrapolation in the laboratory bioassays to fields. Further studies by Sears et al. (2001) and Dively et al. (2004) reported that the harmful effect of the transgenic maize on the larvae of monarch butterfly is not significant. Decrease in monarch population while adopting glyphosate-resistant modified crops was also observed in the USA and Mexico. Brower et al. (2012) described the reason for the decrease in the population of butterfly to be due to the decline in their breeding site as a result of destruction of milkweed species by the use of glyphosate. The constant use of glyphosate was also reported to be the reason for shift in the population of weeds like hemp sesbania, shatter cane, nightshade, common hemp, and nut edge (Mertens 2008).

6.4 Transgenic Crops and Patent Rights

The patentability of the genetic material/genes required to develop a crop plant with modified desirable traits is one of the major problems associated with transgenic technology. This is because the patented genes cannot be accessed by other multinational industries (Kumar et al. 2020). Majority of the GM crop events that were commercialized are generated and patented by a small number of multinational companies because of cost implication of the research and other developmental and regulatory aspects related to the invention of such genes. Business Wire (2016) reported that five important multinational companies, viz. Syngenta, DuPont, Monsanto, Groupe Limagrain, and Bayer CropScience, produced up to 70% of the commercialized GM seeds globally. The smaller business firms competing with other industries are being acquired and merged with MNCs, thereby minimizing the competition rate in the GM seed markets. This may result in overexploitation of farmers by increasing the prices of GM seeds available only in the small number of industries.

6.5 Commercialization-Related Issues

The production and release of transgenic products into markets are associated with biosafety assessment processes, which are costly and time consuming (Davison 2010). It was estimated that USD 35.01 is averagely required for safety assessment regulation, registration, and commercialization of a transgenic product (McDougall 2011). It was also estimated that the length of time required from the start of generating transgenic crop project to the final stage of lunching the product commercially was approximately 13 years. In some developed nations like those in the European Union and the USA, 5 and 7 years was the estimated average duration for the regulatory safety assessment processes concerning GM crops (Smart et al. 2017). The cost and time implications related to the production and commercialization of transgenic crops made it difficult for small business firms and public institutions to adopt this biotechnology of generating transgenic crops (Kumar et al. 2020).

6.6 Frontiers in Plant Genetic Manipulations

Different plant genetic engineering technologies including transgenesis, cisgenesis, and intragenesis have been explored to develop crop plants with a variety of quantitative and qualitative traits. In addition to the existing techniques, new biotechnological approaches have been developed and they are being polished and improved. Although these new techniques are yet to be applied to plant commercial products, they possess a potential value for the prospective GM crops (NASEM 2016). These novel approaches are enumerated as (1) genome editing, (2) synthetic DNA components and artificial chromosomes, and (3) targeted epigenetic modifications.

6.7 Advances in Genome Editing for Crop Improvement

Genome editing constitutes a variety of methods that provide means of editing, deleting, and replacing/inserting a specific DNA sequence of interest in a targeted site in the plant or animal genome. These techniques are based on the induction of cuts in double-strand break (DSB) or by means of sequence-specific nucleases (SSNs) to cleave genome at a specific site to create a single break (nick) in living organisms under study (Gaj et al. 2016; Theodora 2016). Genome editing employs sequence-specific nucleases (SSNs) to cause a sudden change in the gene of the targeted DNA sequences in plant, which can be repaired in the following ways: (1) nonhomologous end joining (NHEJ) process which results in mutation at a given site/locus and (2) homology-directed repair (HDR) process which integrates the donor molecule breakage site. Voytas and Gao (2014) reviewed the principal classes of sequence-specific nucleases applied in genome editing. They include zinc finger nucleases (ZFNs), meganucleases (MNs), transcription activator like effector nucleases (TALENs), and clustered regularly interspaced palindromic repeat (CRISPR-Cas9) nucleases. There has been rapid development in genome editing technologies with the commencement of CRISPR-Cas9 system, and a lot of breakthrough with the potential of improving this technique is expected in the future (Fig. 6.3).

Meganucleases (MNs) are less efficient in terms of edition of the target DNA sequences, which results in higher deployment of the other approaches for different purposes (Cardi 2015). Silva et al. (2011) revealed the sources of meganucleases to be naturally available in archaea, eukaryotes, and bacteria, which were the pioneer SSNs to be tested for genome editing technology. They are single proteins capable of identifying a sequence in the plant genome with at least 12 nucleotides. They cleave the intended DNA creating a double-strand break, which is usually repaired using a donor molecule through NHEJ or HDR. Previous studies by Baltes and Voytas (2014) showed that two crop plants (maize and tobacco) have been used to demonstrate genome editing through meganuclease-mediated technology.

The zinc finger medium-containing enzymes (proteins) can bind DNA and are widely distributed in nature (NASEM 2016) and sometimes serve as transcription factors. Manipulation of these domains may lead to the binding of specific DNA sequences if merged to the DNA-cutting nuclease domain with FokI protein. Urnov et al. (2010) reported that a set of ZFNs can function in tandem to trim the DNA at the required target. ZFNs are deployed to induce mutations in the plant genome through NHEJ and HDR, a similar technique with meganuclease. The first widely applied genome editing designer tool used in biology was ZFNs as they have been employed in the biotechnology of crop plants (Baltes and Voytas 2014).

Transcription activator-like effector nucleases (TALENs) constitute a new category of genome editing mechanism that determines the exact site for inserting, deleting, or substituting a specific allele so as to alter the plant genome. They have an organized domain that imitates the natural transcription activator proteins with a nuclease enzyme capable of cleaving DNA in the cells. They are considered a revolutionary tool employed in genome editing technology in many living species

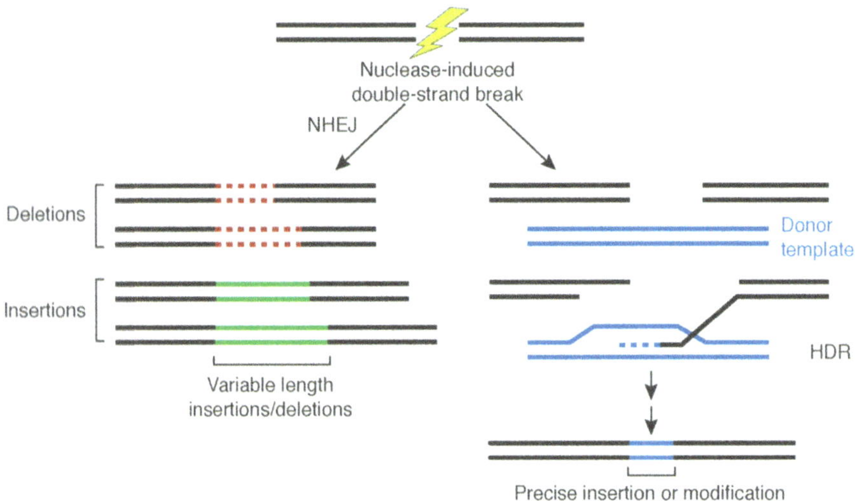

Fig. 6.3 The results of DNA cleavage through genome editing techniques in plant cell. Source: Sander and Joung (2014)

(Bhushan et al. 2016; NASEM 2016). TALEs were determined in *Xanthomonas* sp. (bacterial plant pathogen) and can be genetically engineered to bind any DNA sequence (Cardi 2015). The *Xanthomonas* sp. naturally releases TALEs into plant tissues to allow pathogenicity. Plant resistance to the pathogen will be suppressed when the TALEs bind to the promoters in the plant genes. Just like zinc finger nuclease, TALENs can be consolidated with nuclease medium FokI, which is then considered as TALENs. They are also used in pairs to influence the targeted mutations. Rice, wheat, soybean, and maize are examples of crop plants where TALENs were deployed to edit their genome.

Another important class of genome editing technology is clustered regularly interspaced short palindromic repeats (CRISPR-Cas9). This technique was highly accepted by researchers due to its reliability in food production (Huesing et al. 2016; Ishii and Araki 2016; Wolter and Putcha 2017). It has been publicized as a promising concept in crop biotechnology due to its accuracy in inserting and changing the plant genome (DNA) with required specificity, and it is very simple to implement (Aaron et al. 2018). Li et al. (2016) stated that CRISPR provides significant chances of improving crop production by minimizing both the biotic and abiotic constraints, which will result in increase in the yield potential of crops. McFadden (2017) revealed that CRISPR gene editing may differ significantly from the GMOs that were rejected in some part of the world, particularly European Union due to some ethical issues. This is because CRISPR technology could be used to present changes in the DNA intrinsic to species under study without any transgene (Lusk et al. 2018). Despite the potential of this technology in upgrading crop production, rejection of the product might cripple further production and subsequently dwindle the commercialization process (Aaron et al. 2018). Clasen et al. (2016) and Woo et al. (2015) explained that genome editing technology through CRISPR or other methods can be done in crop plants with or without external DNA completely leading to transgene-free GE crop. Specific mutations were induced using genetically ciphered TALEN pair in potato protoplasts, and after recovery, the potato mutated genes developed novel quality trait with 7 out of the 18 genetically engineered lines being free from any TALEN transgene in the genome of potato crop (Clasen et al. 2016). The in vitro Cas9-translated protein was used to direct RNA to alter the alleles in plants like tobacco, rice, lettuce, and Arabidopsis protoplasts, and all the mutated crops were recovered (Woo et al. 2015). It is evident from the exploitation of the intrinsic biological process including DNA-binding zinc finger nuclease (ZFNs), directed disintegration of DNA sequences (CRISPR/Cas9), and pathogen-oriented transcription of the host genes (TALEs), which currently allow plant genome manipulation with more precision and versatility (NASEM 2016) (Fig. 6.4).

6.8 Synthetic and Artificial Chromosomes

Advancement in understanding molecular biology tools and other biological pathways facilitates the genetic engineering of multiple genes and insertion of the whole and new biochemical processes into plants. The DNA molecules exploited in biotechnological concepts are small (<20–40 kb) and are involved in conventional cloning techniques that are very slow and tedious. Nevertheless, novel and cheap techniques of DNA synthesis and collecting them into larger molecules have been formulated (NASEM 2016). Through these methods, multigene pathways are easily constructed on a single DNA molecule as reviewed by Ellis et al. (2011). The possibility of applying these techniques in genetic engineering has been examined by Gibson et al. (2010) in which the whole genome of bacteria was synthesized. Similarly, synthetic chromosome in yeast was developed using methods as demonstrated (Annaluru et al. 2014). Up to hundreds of genes are likely to be introduced into plant chromosomes through these techniques and among the methods envisaged to achieve such development would be means of artificial minichromosomes. A chromosome inserted into the natural component of a plant chromosome inside the nucleus is referred to as artificial minichromosome. Annaluru et al. (2014) defined synthetic chromosome as the entire production of DNA that replaces the natural chromosome or replaces the whole organelle in the cell's nucleus. These genomes and chromosomes provide a means of assembling large number of alleles in a self-replicating system. Eukaryotic organisms are characterized with linear chromosomal DNA molecules containing proper sequences for segregation in the centromeres, integrity, and replication. Genes can be introduced into plants using artificial or synthetic chromosomes without any potential of disrupting the alleles on the native plant chromosomes at the point of injection.

Gaeta et al. (2012) and Birchler (2015) described the two major approaches used in artificial minichromosome investigation as "bottom-up" approach and "top-down" approach.

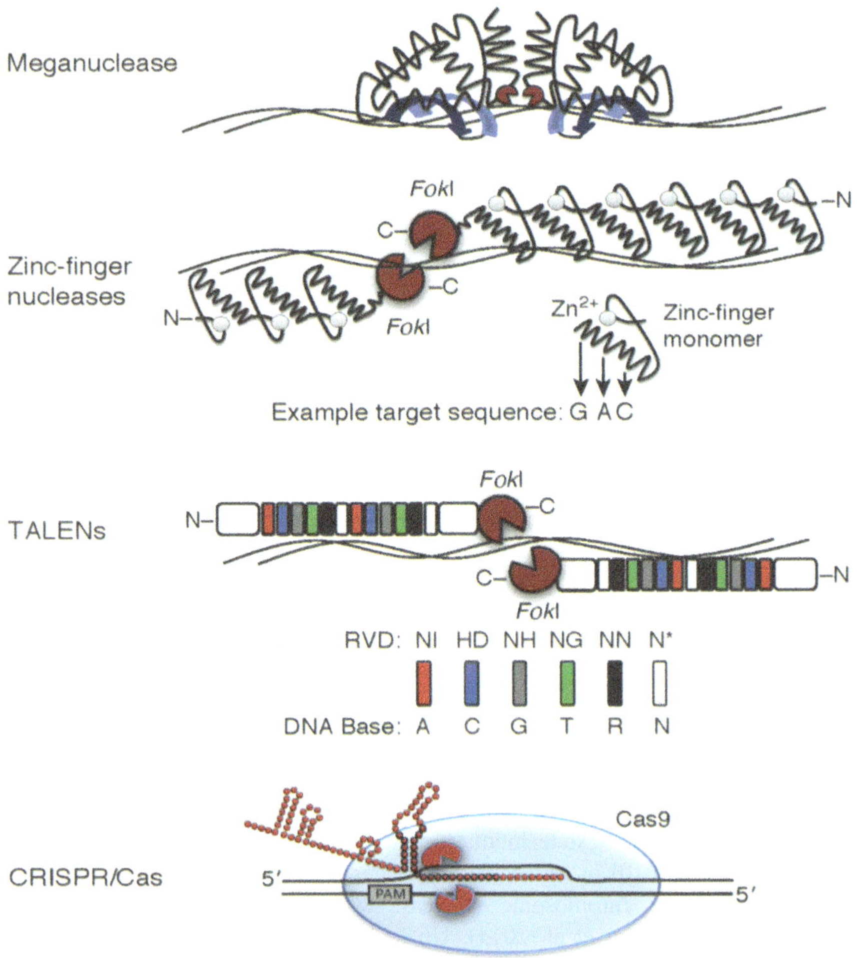

Fig. 6.4 Biotechnological approaches used in genome editing (meganucleases, zinc finger nucleases, transcription activator-like effector nucleases, and clustered regularly interspaced palindromic repeats). Source: NASEM (2016)

In the first approach, the main components of the chromosome (telomere, centromere, and origin of replication) are tacked together resulting in a de novo minichromosome. Despite the progress recorded in this approach, other extra features like epigenetic transformation of the genome nucleotides can temper with the gene expression and minichromosome replication ability in the plant cells. This issue has led to reduced routine application of this method. The second approach basically utilizes the innate chromosomes by imitating the yeast chromosome method to generate an artificial chromosome using an endogenous existing template.

There is possibility of using these approaches to clonally propagate crop plants like potato since meiosis is not needed (Birchler 2015).

6.9 Targeted Epigenetic Modification

Targeted epigenetic editing has to do with site-specific selection of an epigenetic protein to decipher the existing epigenetic structure of an endogenous genomic locus, for the purpose of transcriptional reprogramming (Rutger and Marianne 2020). Epigenetic modifications of the genome (DNA) and histone in plants and other living species are well known for their diverse roles in transcriptional regulation (Alabert and Groth 2012). These techniques play a major role in the cellular memory of transcriptional states. Insights into the biochemical processes that alter the methylation patterns of the DNA has led to an increase in knowledge on epigenetic-mediated regulation of the specific gene(s) expression. Thakore et al. (2015) mentioned that epigenetic modification technique using CRISPR-Cas9 is a revered technology that can be used directly to alter the gene expression in the cell and also to examine the essential epigenetic system of the gene regulation. The risk and safety issues associated with the alteration of epigenetic modification in the genome of a specific plant does not demonstrate significant problems. This is because the plant genomes are loaded with epigenetic modifications and the plant genomes transformed epigenetically have been in place and eaten by men since thousands of years (NASEM 2016). The safety issue is therefore the same with that of other techniques (genetically engineered or not) that helped in specific trait modifications and increase or decrease in specific gene expression.

6.10 The Future of GM Crops

GM crops demonstrate potentiality in providing solutions to the global food insecurity. However, the open field cultivation and consumption of these crops remains debatable. Increase in GM plant production to eradicate starvation is associated with hidden costs and health-related issues, and this issue remains controversial rendering the future of GM crops to be uncertain. Jhansi and Usha (2013) reported that the success of commercializing transgenic crops during 1994–2002 was due to certain vital benefits that could be obtained by commercializing the crops. Such benefits include (a) efficient weed control, (b) decreased losses due to pathogens and reduced use of pesticides, (c) reduction in postharvest damages due to improvement in the crops' shelf- life, (d) enhanced nutritional quality, and (e) efficiency in the production of hybrid seed. This will facilitate not only sustainable crop production but also keeping the environment free of insecticides and pesticides. To achieve this milestone, the seeds of superior crops must be made available to the farmers by the seed industries.

Prospectively, the GM (transgenic) crops will be deployed not only in agronomic trait improvement but also in traits related to food processing, medicine, pharmacy, and other special chemicals. There is therefore optimism in the future of transgenic crop plants as the market is expected to reach USD 37.46 billion by the end of 2027 (Coherent Market Insight 2020). The achievement of these goals depends on sustained efforts by both developed and developing nations. Farmers and general public must cooperate by responding to the current scenario. This will aid remarkably in providing abundant agricultural produce to feed the overpopulated world (Theodora 2016).

6.11 Conclusions

The use of biotechnology methods in agriculture has improved crop production by generating crops with different desirable traits such as insect resistance, herbicide tolerance, disease resistance, drought tolerance, and improvement in nutritional composition of different crop cultivars. The future perspectives of plant biotechnology are to provide vaccinations and therapeutic substances like antibiotics through genetically modified crops so as to simplify medications to the inhabitants of developing countries. Transgenic technology produces plants that put up a novel concept of synthesizing human antibodies. The development of biopharmaceuticals such as insulin to treat diabetes and that of erythropoietin to manage anemia are some of the outputs of genetic engineering technology. In addition to the current techniques, new biotechnological concepts that deal with plant genetic manipulations need to be explored for food and health security (NASEM 2016).

References

Aaron M, Shewa L, Lanier N, Heather AS, Rodolfa M, Bruce LD (2018) CRIPR versus GMOS: public acceptance and valuation. Glob Food Sec 19:71–80

Adang MJ, Brody MS, Cardineau G, Eagan N, Roush RT, Shewmaker CK, Jones A, Oakes JV, McBride KE (1993) The reconstruction and expression of a *bacillus thuringiensis cryIIIA* gene in protoplasts and potato plants. Plant Mol Biol 21:1131–1145

Addae PC, Ishiyaku MF, Tignegre JB, Ba MN, Bationo JB, Atokple IDK, Abudulai M, Dabiré-Binso CL, Traore F, Saba M, Umar ML, Adazebra GA, Onyekachi FN, Nemeth MA, Huesing JE, Beach LR, Higgins TJV, Hellmich RL, Pittendrigh BR (2020) Efficacy of a *cry1Ab* gene for control of *Maruca vitrata* (lepidoptera: Crambidae) in cowpea. J Econ Entomol 113(2):974–979

Alabert C, Groth A (2012) Chromatin replication and epigenome maintenance. Nat Rev Mol Cell Biol 13:153–167

Annaluru N, Muller H, Mitchell LA, Ramalingam S, Stracquadanio G, Richardson SM, Dymond JS, Kuang Z, Scheifele LZ, Cooper EM, Cai Y, Zeller K, Agmon N, Han JS, Hadjithomas M, Tullman J, Caravelli K, Cirelli K, Guo Z, London V, Yeluru A, Murugan S, Kandevlou K, Agier N, Fischer G, Yang K, Martin JA, Bilgel M, Bohutskyi P, Boulier KM, Capaldo BJ, Chang J, Charoen K, Choi WJ, Deng P, DiCarlo JE, Doong J, Dunn J, Feinberg JI, Fernandez C, Floria CE, Gladowski D, Hadidi P, Ishizuka I, Jabbari J, Lau CYL, Lee PA, Li S, Lin D, Linder ME, Ling J, Liu J, London M, Ma H, Mao J, McDade JE, McMillan A, Moore AM, Oh WC, Ouyang Y, Patel R, Paul M, Paulsen LC, Qiu J, Rhee A, Rubashkin MG, Soh IY, Sotuyo NE, Srinivas A, Suarez A, Wong A, Wong R, Xie WR, Xu Y, Yu AT, Koszul R, Bader JS, Boeke

JD, Chandrasegaran S (2014) Total synthesis of a functional designer eukaryotic chromosome. Science 344:55–58

Arencibia A, Vázquez RI, Prieto D, Téllez P, Carmona ER, Coego A, Hernández L, De la Riva GA, Selman-Housein G (1997) Transgenic sugarcane plants resistant to stem borer attack. Mol Breed 3(4):247–255

Arpaia S, Mennella G, Onofaro V, Perri E, Sunseri F, Rotino GL (1997) Production of transgenic eggplant (*Solanum melongena* L.) resistant to Colorado potato beetle (*Leptinotarsa decemlineata* say). Theor Appl Genet 95(3):329–334

Baghbani-Arani A, Poureisa M, Alekajbaf H, Karami R, Borz-Abad KK-D (2021) Investigating the status of transgenic crops in Iran in terms of cultivation, consumption, laws and rights in comparison with the world. Sci Rep 11:9204

Bagla P (2010) Hardy cotton-munching pests are latest blow to GM crops. Science 327:1439

Bakhsh A, Siddique S, Husnain T (2012) A molecular approach to combat spatio-temporal variation in insecticidal gene (Cry1Ac) expression in cotton. Euphytica 183(1):65–74

Baltes NJ, Voytas DF (2014) Enabling plant synthetic biology through genome engineering. Trends Biotechnol 33:120–131

Bamishaiye EI, Varanavasiappan S, Balakrishnan N, Udayasuriyan V, Sudhakar D (2017) Genetic manipulation of tomato (*Solanum lycopersicum*) cv. PKM-1 using *cry2AX1* gene for insect resistance. Int J Curr Microbiol App Sci 6(12):4309–4319

Barton KA, Whiteley HR, Yang NS (1987) *Bacillus thuringiensis* δ-endotoxin expressed in transgenic *Nicotiana tabacum* provides resistance to lepidopteran insects. Plant Physiol 85: 1103–1109

Bhattacharya N, Bhat SR, Kirti PB, Chopra VL (2002) Development of insect-resistant transgenic cabbage plants expressing a synthetic cry1A(b) gene from Bacillus thuringiensis. Curr Sci 83:146–150

Bhushan K, Pratap D, Sharma PK (2016) Transcription activator-like effector nucleases (TALENs): an efficient tool for plant genome editing. Eng Life Sci 16(4):330–337

Bicar EH, Woodman CW, Sangtong V, Peterson JM, Yang SS, Lee M, Scott MP (2008) Transgenic maize endosperm containing a milk protein has improved amino acid balance. Transgenic Res 17:59–71

Birchler JA (2015) Promises and pitfalls of synthetic chromosomes in plants. Trends Biotechnol 33: 189–194

Boonmee K, Thammasittirong SN, Thammasittirong A (2019) Molecular characterization of lepidopteran-specific toxin genes in *bacillus thuringiensis* strains from Thailand. 3. Biotech 9(4):117

Breitler JC, Marfa V, Royer M, Meynard D, Vassal JM, Vercambre B, Frutos R, Messeguer J, Gabarra R, Guiderdoni E (2000) Expression of a Bacillus thuringiensis cry1B synthetic gene protects Mediterranean rice against the striped stem borer. Plant Cell Rep 19(12):1195–1202

Brookes G, Barfoot P (2016) Environmental impacts of genetically modified (GM) crop use 1996–2015: impacts on pesticide use and carbon emissions. GM Crops Food 8:117–147

Brower LP, Taylor OR, Williams EH, Slayback DA, Zubieta RR, Ramirez MI (2012) Decline of monarch butterflies overwintering in Mexico: is the migratory phenomenon at risk? Insect Conserv Diver 5:95–100

Cao Y, Li D (2013) Impact of increased demand for animal protein products in Asian countries: implications on global food security. Anim Front 3(3):48–45

Cao J, Shelton AM, Earle ED (2001) Gene expression and insect resistance in transgenic broccoli containing a *bacillus thuringiensis cry1Ab* gene with the chemically inducible PR–1a promoter. Mol Breed 8:207–216

Cardi T (2015) Cisgenesis and genome editing: combining concepts and efforts for a smarter use of genetic resources in crop breeding. Plant Breed 135:139–147

Carozzi NB, Warren GW, Desai N, Jayne SM, Lotstein R, Rice DA, Evola S, Koziel MG (1992) Expression of a chimeric CaMV35S *bacillus thuringiensis* insecticidal protein gene in transgenic tobacco. Plant Mol Biol 20:539–548

Castiglioni P, Warner D, Bensen RJ, Anstrom DC, Harrison J, Stoecker M, Navarro S (2008) Bacterial RNA chaperones confer abiotic stress tolerance in plants and improved grain yield in maize under water-limited conditions. Plant Physiol 147(2):446–455

Chakrabarti SK, Mandaokar AD, Shukla A, Pattanayak D, Naik PS, Sharma RP, Kumar PA (2000) *Bacillus thuringiensis cry1Ab* gene confers resistance to potato against *Helicoverpa armigera* (Hubner). Potato Res 43(2):143–152

Chakraborty M, Reddy PS, Mustafa G, Rajesh G, Narasu VL, Udayasuriyan V, Rana D (2016) Transgenic rice expressing the *cry2AX1* gene confers resistance to multiple lepidopteran pests. Transgenic Res 25:665–678

Chen W, Liu C, Lu G, Cheng H, Shen Z, Wu K (2018) Effects of Vip3AcAa+Cry1Ac cotton on midgut tissue in *Helicoverpa armigera* (lepidoptera: Noctuidae). J Insect Sci 18(4):13. https://doi.org/10.1093/jisesa/iey075

Chen M, Liu X, Wang Z, Song J, Qi Q, Wang PG (2005) Modification of plant *N*-glycans processing: the future of producing therapeutic protein by transgenic plant. Med Res Rev 25: 343–360

Chen WB, Lu GQ, Cheng HM, Liu CX, Xiao YT, Xu C, Shen ZC, Soberón M, Bravo A, Wu KM (2017) Transgenic cotton co-expressing chimeric Vip3AcAa and Cry1Ac confers effective protection against Cry1Ac-resistant cotton bollworm. Transgenic Res 6:763–774

Cheng X, Sardana R, Kaplan H, Altosaar I (1998) *Agrobacterium*-transformed rice plants expressing synthetic *cryIA(b)* and *cryIA(c)* genes are highly toxic to striped stem borer and yellow stem borer. Proc Natl Acad Sci U S A 95:2767–2772

Chilton MD, Drummond MH, Merlo DJ, Sciaky D, Montoya AL, Gordon MP et al (1977) Stable incorporation of plasmid DNA into higher plant cells: the molecular basis of crown gall tumorigenesis. Cell 11:263–271

Clasen BM, Stoddard TJ, Luo S, Demorest ZL, Li J, Cedrone F, Tibebu R, Davison S, Ray EE, Daulhac A, Coffman A, Yabandith A, Retterath A, Haun W, Baltes NJ, Mathis L, Voytas DF, Zhang F (2016) Improving cold storage and processing traits in potato through targeted gene knockout. Plant Biotechnol J 14:169–176

Coherent Market Insight (2020) Genetically Modified Crops Market to Reach USD 37.46 Billion at a CAGR of 8.7%, globally, by end of 2027. https://www.coherentmarketinsights.com/insight/request-sample/3495

Collinge DB, Hans JL, Ole J, Lund S, Michael Lyngkjær F (2010) Engineering pathogen resistance in crop plants: current trends and future prospects. Annu Rev Phytopathol 48:269–291

Coombs J, Douches D, Li W, Grafius E, Pett W (2002) Combining engineered (Bt-*cry3A*) and natural resistance mechanisms in potato for control of Colorado potato beetle. J Amer Soc Hort Sci 127:62–68

Cornu D, Leple JC, Bonade M, Ross A, Augustin S, Delplanque A, Jouanin L, Pilate G (1996) Expression of proteinase inhibitor and a *bacillus thuringiensis* δ-endotoxin in transgenic poplars. In: Proceedings IUFRO meeting on somatic cell genetics and molecular genetics of trees. Kluwer, Dordrecht, pp 131–136

Dang W, Wei ZM (2007) An optimized agrobacterium-mediated transformation for soybean for expression of binary insect resistance genes. Plant Sci 173(4):381–389

Datta K, Vasquez A, Tu J, Torrizo L, Alam MF, Oliva N, Abrigo E, Khush GS, Datta SK (1998) Constitutive and tissue-specific differential expression of *cryIA(b)* gene in transgenic rice plants conferring resistance to rice insect pests. Theor Appl Genet 97:20–30

Davison J (2010) GM plants: science, politics and EC regulations. Plant Sci 178:94–98

De Vos CJ, Swanenburg M (2018) Health effects of feeding genetically modified (GM) crops to livestock animals: a review. Food Chem Toxicol 117:3–12

DeCosa B, Moar W, Lee SB, Miller M, Daniell H (2001) Overexpression of the Bt Cry2Aa2 operon in chloroplasts leads to formation of insecticidal crystals. Nat Biotechnol 19:71–74

Delannay X, LaVallee BJ, Proksch RK, Fuchs RL, Sims SR, Greenplate JT, Marrone PG, Dodson RB, Augustine JJ, Layton JG, Fischhoff DA (1989) Field performance of transgenic tomato

plants expressing the *bacillus thuringiensis* var kurstaki insect control protein. Biotech 7:1265–1269
Dessoky ES, Ismail RM, Elarabi NI, Abdelhadi AA, Abdallah NA (2020) Improvement of sugarcane for borer resistance using Agrobacterium mediated transformation of cry1Ac gene. GM Crops & Food 12(1):47–56
Dill GM, CaJacob CA, Padgette SR (2008) Glyphosate-resistant crops: adoption, use and future considerations. Pest Manag Sci 64:326–331
Din SU, Azam S, Rao AQ, Shad M, Ahmed M, Gul A, Latif A, Ali MA, Husnain T, Shahid AA (2021) Development of broad-spectrum and sustainable resistance in cotton against major insects through the combination of Bt and plant lectin genes. Plant Cell Rep 40(4):707–721
Dively GP, Rose R, Sears MK, Hellmich RL, Stanley-Horn DE, Calvin DD, Russo JM, Anderson PL (2004) Effects on monarch butterfly larvae (Lepidoptera: Danaidae) after continuous exposure to Cry1Ab-expression corn during anthesis. Environ Entomol 33:1116–1125
Domingo JL (2016) Safety assessment of GM plants: an updated review of the scientific literature. Food Chem Toxicol 95:12–18
Douches DS, Westedt AL, Zarka K, Schroeter B, Grafius EJ (1998) Potato transformation to combine natural and engineered resistance for controlling tuber moth. Hort Sci 33:1053–1056
Dunwell JM (2014) Transgenic cereals: current status and future prospects. J Cereal Sci 59:419–434
Ellis T, Adie T, Baldwin GS (2011) DNA assembly for synthetic biology: from parts to pathways and beyond. Integr Biol (Camb) 3(2):109–118
Erasmus R, Rialet P, Hannalene DP, Angelika H, Miluse T, Annemie E, Johnnie VDB (2019) Introgression of a *cry1Ab* transgene into open pollinated maize and its effect on cry protein concentration and target pest survival. PLoS One 14(12):e0226476. https://doi.org/10.1371/journal.pone.0226476
Fahim M, Millar AA, Wood CC, Larkin PJ (2012) Resistance to Wheat streak mosaic virus generated by expression of an artificial polycistronic microRNA in wheat. Plant Biotechnol J 10(2):150–163
FAO (2010) Fats and fatty acids in human nutrition. Report of an expert consultation. Rome: Food and Agricultural Organisation of the United Nations. 10–14 November 2008, Geneva. FAO Food and Nutrition Paper 91. https://agris.fao.org/agris-search/search.do?recordID=XF2016049106
Fischhoff DA, Bowdish KS, Perlak FJ, Marrone PG, McCormick SM, Niedermeyer JG, Dean DA, Kusano-Kretzmer K, Mayer EJ, Rochester DE, Rogers SG, Fraley RT (1987) Insect tolerant transgenic tomato plants Bio/Tech 5:807–813
Fu J, Liu B (2020) Exogenous Cry1Ab/c protein recruits different endogenous proteins for its function in plant growth and development. Front Bioeng Biotechnol 8:685
Fujimoto H, Itoh K, Yamomoto M, Kyojuka J, Shimamoto K (1993) Insect resistant rice generated by introduction of a modified δ-endotoxin gene of *bacillus thuringiensis*. Biotech 11:1151–1155
Gaeta RT, Masonbrink RE, Krishnaswamy L, Zhao C, Birchler JA (2012) Synthetic chromosome platforms in plants. Annu Rev Plant Biol 63:307–330
Gaj T, Sirk SJ, Shui SL, Liu J (2016) Genome-editing technologies: principles and applications. Cold Spring Harbor Persp Biol 8(12):a023754
Gassmann AJ, Petzold-Maxwell JL, Keweshan RS, Dunbar MW (2011) Field-evolved resistance to Bt maize by western corn rootworm. PLoS One 6:e22629. https://doi.org/10.1371/journal.pone.0022629
Ghosh G, Ganguly S, Purohit A, Chaudhuri RK, Das S, Chakraborti D (2017) Transgenic pigeonpea events expressing Cry1Ac and Cry2Aa exhibit resistance to Helicoverpa armigera. Plant Cell Rep 36(7):1037–1051
Gibson DG, Glass JI, Lartigue C, Noskov VN, Chuang RY, Algire MA, Benders GA, Montague MG, Ma L, Moodie MM, Merryman C, Vashee S, Krishnakumar R, Assad-Garcia N, Andrews-Pfannkoch C, Denisova EA, Young L, Qi Z-Q, Segall-Shapiro TH, Calvey CH, Parmar PP,

Hutchinson CA, Smith HO, Venter JC (2010) Creation of a bacterial cell controlled by a chemically synthesized genome. Science 329:52–56

Gilbert N (2013) A hard look at GM crops. Nature 497:24–26

Gleave AP, Mitra DS, Markwick NP, Morris BAM, Beuning LL (1998) Enhanced expression of the *bacillus thuringiensis cry9Aa2* gene in transgenic plants by nucleotide sequence modification confers resistance to potato tuber moth. Mol Breed 4:459–472

Goldman E (2002) Netwatch. Science 298(5598):1521–1522

Gowda A, Rydel TJ, Wollacott AM, Brown RS, Akbar W, Clark TL, Flasinski S, Nageotte JR, Read AC, Shi X, Werner BJ (2016) A transgenic approach for controlling Lygus in cotton. Nat Commun 7(1):1–7

Gulbitti-Onarici S, Zaidi MA, Taga I, Ozcan S, Altosaar I (2009) Expression of *cry1Ac* in transgenic tobacco plants under the control of a wound-inducible promoter (AoPR1) isolated from *Asparagus officinalis* to control *Heliothis virescens* and *Manduca sexta*. Mol Biotechnol 42(3): 341–349

Gunasekara JM, Jayasekera GA, Perera KL, Wickramasuriya AM (2017) Development of a Sri Lankan rice variety Bg 94-1 harbouring Cry2A gene of Bacillus thuringiensis resistant to rice leaffolder [Cnaphalocrocis medinalis (Guenée)]. J Nat Sci Found Sri Lanka 45(2)

Gupta B, Tripathi AK, Joshi R, Pareek A, Singla-Pareek SL (2015) Designing climate smart future crops employing signal transduction components. In: Pandey GK (ed) Elucidation of abiotic stress signaling in plants: functional genomics perspectives. Springer, New York, pp 393–414

Han SM, Lee B, Won OJ, Hwang KS, Suh SJ, Kim C, Park KW (2015) Gene flow from herbicide resistant genetically modified rice to conventional rice (*Oryza sativa* L.) cultivars. J Ecol Environ 38:397–403

Hanur VS, Reddy B, Arya VV, Rami Reddy PV (2015) Genetic transformation of tomato using Bt *cry2A* gene and characterization in Indian cultivar Arka Vikas. J Agr Sci Tech 17:1805–1814

Hatice S, Fatih D, Nedim M, Sami D, Anne F (2017) Cry1Ac-mediated resistance to tomato leaf miner (Tuta absoluta) in tomato. Plant Cell Tissue Organ Cult 131:65–73

Himanen SJ, Nerg AM, Nissinen A, Stewart CN Jr, Poppy GM, Holopainen JK (2009) Elevated atmospheric ozone increases concentration of insecticidal *bacillus thuringiensis* (Bt) *Cry1Ac* protein in Bt *Brassica napus* and reduces feeding of a Bt target herbivore on the non-transgenic parent. Environ Pollut 157(1):181–185

Hou H, Atlihan N, Lu ZX (2014) New biotechnology enhances the application of cisgenesis in plant breeding. Front Plant Sci 5:389

Huesing JE, Andres D, Braverman MP, Burns A, Felsot AS, Harrigan GG (2016) Global adoption of genetically modified (GM) crops: challenges for the public sector. J Agric Food Chem 64: 394–402. https://doi.org/10.1021/acs.jafc.5b05116

Indurker S, Misra HS, Eapen S (2007) Genetic transformation of chickpea (Cicer arietinum L.) with insecticidal crystal protein gene using particle gun bombardment. Plant Cell Rep 26:755–763

ISAAA (2017) Biotech country facts and trends: China. http://isaaa.org/resources/publications/biotech_country_facts_and_trends/download/FactsandTrends-China.pdf

ISAAA (2018) Global status of commercialized biotech/GM crops in 2018: executive brief. ISAAA, Ithaca, NY

ISAAA (2019) GM approval database. http://www.isaaa.org/gmapprovaldatabase/default.asp

Ishii T, Araki M (2016) A future scenario of the global regulatory landscape regarding genome-edited crops. GM Crops Food 8:44–56

Jadhav MS, Rathnasamy SA, Natarajan B, Duraialagaraja S, Varatharajalu U (2020) Study of expression of indigenous Bt cry2AX1 gene in T 3 progeny of cotton and its efficacy against *Helicoverpa armigera* (Hubner). Braz Arch Biol Technol 8:63

James C (2013) Global Status of Commercialized Biotech/GM Crops: 2013. ISAAA Brief No.46. ISAAA: Ithaca, NY. https://www.isaaa.org/resources/publications/briefs/46/

Jelenkovic G, Billings S, Chen Q, Lashomb J, Hamilton G, Ghidiu G (1998) Transformation of eggplant with synthetic *cryIIIA* gene produces a high level of resistance to the Colorado potato beetle. J Am Soc Hortic Sci 123:19–25

Jhansi SR, Usha R (2013) Transgenic plants: types, benefits, public concerns and future. J Pharm Res 6:879–883

Kamble S, Misra HS, Mahajan SK, Eapen S (2003) A protocol for efficient biolistic transformation of mothbean *Vigna aconitifolia* L. Jacq Marechal Plant Mol Biol Rep 21:457–457

Kar S, Basu D, Das S, RamKrishnan NA, Mukherjee P, Nayak P (1997) Expression of *cry1Ac* gene of *bacillus thuringiensis* in transgenic chickpea plants inhibits development of pod borer (*Heliothis armigera*) larvae. Transgenic Res 6:177–185

Karthik K, Negi J, Rathinam M, Saini N, Sreevathsa R (2021) Exploitation of novel Bt ICPs for the management of *Helicoverpa armigera* (Hübner) in cotton (*Gossypium hirsutum* L.): a transgenic approach. Front Microbiol 12:661212

Katta S, Talakayala A, Reddy MK, Addepally U, Garladinne M (2020) Development of transgenic cotton (Narasimha) using triple gene Cry2Ab-Cry1F-Cry1Ac construct conferring resistance to lepidopteran pest. J Biosci 45(1):1–11

Kaur A, Sharma M, Sharma C, Kaur H, Kaur N, Sharma S, Arora R, Singh I, Sandhu JS (2016) Pod borer resistant transgenic pigeon pea (Cajanus cajan L.) expressing cry1Ac transgene generated through simplified Agrobacterium transformation of pricked embryo axes. Plant Cell Tissue Org Cult 127(3):717–727

Keshavareddy GS, Rohini SV, Ramu S, Sundaresha ARV, Kumar P, Ananda K, Udayakumar M (2013) Transgenics in groundnut (*Arachis hypogaea* L.) expressing *cry1AcF* gene for resistance to *Spodoptera litura* (F.). Physiol Mol Biol Plants 19(3):343–352

Khan GA, Bakhsh A, Riazuddin S, Husnain T (2011) Introduction of cry1Ab gene into cotton (Gossypium hirsutum) enhances resistance against Lepidopteran pest (Helicoverpa armigera). Span J Agric Res 9(1):296–302

Khanna HK, Raina SK (2002) Elite indica transgenic rice plants expressing modified *cry1Ac* endotoxin of *bacillus thuringiensis* show enhanced resistance to yellow stem borer (*Scirpophaga incertulas*). Transgenic Res 11(4):411–423

Kim YH, Issa MS, Cooper AM, Zhu KY (2015) RNA interference: applications and advances in insect toxicology and insect pest management. Pestic Biochem Physiol 120:109–117

Kishi-Kaboshi M, Aida R, Sasaki K (2018) Genome engineering in ornamental plants: current status and future prospects. Plant Physiol Biochem 131:47–52

Klümper W, Qaim M (2014) A meta-analysis of the impacts of genetically modified crops. PLoS One 9:e111629

Koerniati S, Sukmadjaja D, Samudra IM (2020) C synthetic gene of CryIAb-CryIAc fusion to generate resistant sugarcane to shoot or stem borer. IOP Conf Ser Earth Environ Sci 418:012069

Kota MDH, Varma S, Gareznski F, Moar WJ (1999) Overexpression of *bacillus thuringiensis* (Bt)cry2Aa2 protein in chloroplasts confers resistance to plants against susceptible and Bt-resistant insects. Proc Natl Acad Sci U S A 96:1840–1845

Koul B, Srivastava S, Sanyal I, Tripathi B, Sharma V, Amla DV (2014) Transgenic tomato line expressing modified *bacillus thuringiensis cry1Ab* gene showing complete resistance to two lepidopteran pests. Springer Plus 3(1):1–13

Koul B, Yadav R, Sanyal I, Amla DV (2015) Comparative performance of modified full-length and truncated *bacillus thuringiensis-cry1Ac* genes in transgenic tomato. Springer Plus 4(1):1–14

Koziel MG, Beland GL, Bowman C, Carozzi NB, Crenshaw R, Crossland L, Dawson J, Desai N, Hill M, Kadwell S, Launis K (1993) Field performance of elite transgenic maize plants expressing an insecticidal protein derived from Bacillus thuringiensis. Bio/Technology 11 (2):194–200

Kumar H, Kumar V (2004) Tomato expressing Cry1A(b) insecticidal protein from *bacillus thuringiensis* protected against tomato fruit borer, *Helicoverpa armigera* (Hubner) (lepidoptera: Noctuidae) damage in the laboratory, greenhouse and field. Crop Prot 23:135–139

Kumar PA, Mandaokar A, Sreenivasu K, Chakrabarti SK, Bisaria S, Sharma SR, Kaur S, Sharma RP (1998) Insect-resistant transgenic brinjal plants. Mol Breed 4(1):33–37

Kumar K, Aggarwal C, Sapna B, Singh I, Yadava P (2018) Microbial genes in crop improvement. Crop improvement through microbial biotechnology. Elsevier, Amsterdam, pp 39–56

Kumar K, Gambhir G, Dass A, Tripathi AK, Singh A, Jha AK, Yadava P, Choudhary M, Rakshit S (2020) Genetically modified crops: current status and future prospects. Planta 251(4):1–27

Kumar A, Jaiwal R, Sreevathsa R, Chaudhary D, Jaiwal PK (2021) Transgenic cowpea plants expressing *bacillus thuringiensis* Cry2Aa insecticidal protein imparts resistance to *Maruca vitrata* legume pod borer. Plant Cell Rep 40(3):583–594

Leroy T, Henry AM, Royer M, Altosaar I, Frutos R, Duris D, Philippe R (2000) Genetically modified coffee plants expressing the *bacillus thuringiensis cry1Ac* gene for resistance to leaf miner. Plant Cell Rep 19:382–389

Li M, Li X, Zhou Z, Wu P, Fang M, Pan X, Lin Q, Luo Q, Wu G, Li H (2016) Reassessment of the four yield-related genes Gn1a, DEP1, GS3, and IPA1, in rice using a CRISPR/Cas9 system. Front Plant Sci 7:377

Li S, Wang Z, Zhou Y, Li C, Wang G, Wang H, Zhang J, Liang G, Lang Z (2018) Expression of cry2Ah1 and two domain II mutants in transgenic tobacco confers high resistance to susceptible and Cry1Ac-resistant cotton bollworm. Sci Rep 8(1):1–11

Liang J, Zhang D, Li D, Zhao SY, Wang CY, Xiao YT, Xu D, Yang YZ, Li GP, Wang LL, Gao Y, Yang XQ, Yuan H, Liu J, Zhang XJ, Wu KM (2021) Expression profiles of Cry1Ab protein and its insecticidal efficacy against the invasive fall armyworm for Chinese domestic GM maize DBN9936. J Integr Agric 20(3):792–803

Liu H, Guo X, Naeem MS, Liu D, Xu L, Zhang W, Tang G, Zhou W (2011) Transgenic Brassica napus L. lines carrying a two gene construct demonstrate enhanced resistance against Plutella xylostella and Sclerotinia sclerotiorum. Plant Cell Tissue Organ Cult 106(1):143–151

Losey JE, Rayor LS, Carter ME (1999) Transgenic pollen harms monarch larvae. Nature 399:214

Lovei GL, Bøhn T, Hilbeck A (2010) Biodiversity, ecosystem services and genetically modified organisms third world network, Malaysia

Lusk JL, McFadden BR, Wilson N (2018) Do consumers care how a genetically engineered food was created or who created it? Food Policy 78:81–90

Majeet A, Husnain T, Riazuddin S (2000) Transformation of virus-resistant genotype of Gossypium hirsutum L. with pesticidal gene. Plant Biotechnol 17(2):105–110

Majumder S, Sarkar C, Saha P, Gotyal BS, Satpathy S, Datta K, Datta SK (2018) Bt jute expressing fused δ-endotoxin Cry1Ab/ac for resistance to lepidopteran pests. Front Plant Sci 8:2188

Mandaokar AD, Goyal RK, Shukla A, Bisaria S, Bhalla R, Reddy VS, Chaurasia A, Sharma I, Altosaar I, Kumar PA (2000) Transgenic tomato plant resistant to fruit borer (*Helicoverpa armigera* Hubner). Crop Prot 19:307–312

Manikandan R, Sathish S, Balakrishnan N, Balasubramani V, Sudhakar D, Udayasuriyan V (2018) Agrobacterium mediated transformation of indica rice with synthetic cry2AX1 gene for resistance against rice leaf folder. J Pure Appl Microbiol 8(4):3135–3142

Maqbool SB, Husnain T, Riazuddin S, Masson L, Christou P (1998) Effective control of yellow stem borer and rice leaf folder in transgenic rice indica varieties basmati 370 and M7 using the novel δ-endotoxin cry2A Bacillus thuringiensis gene. Mol Breed 4(6):501–507

Mcbride KE, Svab Z, Schaaf DJ, Hogan PS, Stalker DM, Maliga P (1995) Amplification of a chimeric *bacillus thuringiensis* gene in chloroplasts leads to an extraordinary level of an insecticidal protein in tobacco. Bio/Technol 13:362–365

McCown BH, McCabe DE, Russell DR, Robison DJ, Barton KA, Raffa KF (1991) Stable transformation of Populus and incorporation of pest resistance by electric discharge particle acceleration. Plant Cell Rep 9:590–594

McDougall P (2011) The cost and time involved in the discovery, development and authorisation of a new plant biotechnology derived trait. In: A Consultancy Study for Crop Life International, Pathhead, Midlothin, UK. https://croplife.org/wp-content/uploads/2014/04/Getting-a-Biotech-Crop-to-Market-PhillipsMcDougall-Study.pdf

McFadden BR (2017) The unknowns and possible implications of mandatory labeling. Trends Biotechnol 35(1):1–3

Mertens M (2008) Assessment of environmental impacts of genetically modified plants. BfN-Skripten 217. Federal Agency for Nature Conservation, New York. https://www.bfn.de/fileadmin/MDB/documents/service/skript217.pdf

Mi X, Ji X, Yang J, Liang L, Si H, Wu J et al (2015) Transgenic potato plants expressing *cry3A* gene confer resistance to Colorado potato beetle. C R Biol 338(7):443–450

Morán R, García R, López A, Zaldúa Z, Mena J, García M, Armas R, Somonte D, Rodríguez J, Gómez M, Pimentel E (1998) Transgenic sweet potato plants carrying the delta-endotoxin gene from *bacillus thuringiensis* var. tenebrionis. Plant Sci 139:175–184

Muddanuru T, Polumetla AK, Maddukuri L, Mulpuri S (2019) Development and evaluation of transgenic castor (*Ricinus communis* L.) expressing the insecticidal protein Cry1Aa of *bacillus thuringiensis* against lepidopteran insect pests. Crop Prot 119:113–125

Murray C, Markwick NP, Kaji R, Poulton J, Martin H, Christeller JT (2010) Expression of various biotin-binding proteins in transgenic tobacco confers resistance to potato tuber moth, Phthorimaea operculella (Zeller)(fam. Gelechiidae). Transgenic Res 19(6):1041–1051

NASEM (National Academies of Sciences and Engineering Medicine) (2016) Genetically engineered crops: experiences and prospects. National Academies Press, pp 353–404

Nayak P, Basu D, Das S, Basu A, Ghosh D, Ramakrishnan NA, Ghosh M, Sen SK (1997) Transgenic elite indica rice plants expressing Cry1Ac δ-endotoxin of *bacillus thuringiensis* are resistant against yellow stem borer (*Scirpophaga incertulas*). Proc Natl Acad Sci U S A 94: 2111–2116

Netherwood T, Martín-Orúe SM, O'Donnell AG, Gockling S, Graham J, Mathers JC, Gilbert HJ (2004) Assessing the survival of transgenic plant DNA in the human gastrointestinal tract. Nat Biotechnol 22:204–209

Peferoen M, Jansens S, Reynaerts A, Leemans J (1990) Potato plants with engineered resistance against insect attack. In: Vayda M, Park W (eds) Molecular and cellular biology of the potato. CAB, Tucson, pp 193–204

Perlak FJ, Fischhoff DA (1993) Insect resistant cotton: from the laboratory to the marketplace. In: Kim L (ed) Advanced engineered pesticides. Marcel Dekker, New York, pp 199–211

Perlak FJ, Deaton RW, Armstrong TA, Fuchs RL, Sims SR, Greenplate JT, Fischhoff DA (1990) Insect-resistant cotton plants. Bio/Tech 8:939–943

Qin D, Liu XY, Miceli C, Zhang Q, Wang PW (2019) Soybean plants expressing the Bacillus thuringiensis cry8-like gene show resistance to Holotrichia parallela. BMC Biotechnol 19 (1):1–2

Qin D, Liu X-Y, Miceli C, Zhang Q, Wang P-w (2020) Soybean plants expressing the *bacillus thuringiensis* cry8-like gene show resistance to *Holotrichia parallela*. BMC Biotechnol 19:66

Qiu L, Sun Y, Jiang Z, Yang P, Liu H, Zhou H, Wang X, Zhang W, Lin Y, Ma W (2019) The midgut V-ATPase subunit A gene is associated with toxicity to crystal 2Aa and crystal 1Ca-expressing transgenic rice in Chilo suppressalis. Insect Mol Biol 28(4):520–527

Rahman M, Hussain K, Khan MA, Bakhsh A, Rao AQ (2012) An insight of cotton leaf curl virus: a devastating plant pathogenic begomovirus. Pure Appl Bio 1:52–58

Ramesh S, Nagadhara D, Reddy VD, Rao KV (2004) Production of transgenic indica rice resistant to yellow stem borer and sap-sucking insects, using super-binary vectors of *agrobacterium tumefaciens*. Plant Sci 166:1077–1085

Rameshraddy Khan JA, Mahesh S, Shankar AG (2017) Cisgenesis: an approach for crop improvement. Int J Pure Appl Biosci 5(3):245–250

Ramu SV, Rohini S, Keshavareddy G, Gowri Neelima M, Shanmugam NB, Kumar AR, Sarangi SK, Ananda Kumar P, Udayakumar M (2012) Expression of a synthetic cry1AcF gene in transgenic Pigeon pea confers resistance to Helicoverpa armigera. J Appl Entomol 136(9):675–687

Raza A, Razzaq A, Mehmood SS, Zou X, Zhang X, Lv Y, Xu J (2019) Impact of climate change on crops adaptation and strategies to tackle its outcome: a review. Plan Theory 8:34

Razzaq A, Ali A, Zafar MM, Nawaz A, Xiaoying D, Pengtao L, Qun G, Ashraf M, Ren M, Gong W, Youlu Y (2021) Pyramiding of cry toxins and methanol producing genes to increase insect resistance in cotton. GM Crops & Food 12(1):382–395

Reddy DVR, Sudarshanaw MR, Fuchs M, Rao NC, Thottappilly G (2009) Genetically engineered virus-resistant plants in developing countries: current status and future prospects. Adv Virus Res 75:185–220

Rensburg JBJ (2007) First report of field resistance by the stem borer, Busseola fusca (Fuller) to Bt-transgenic maize. S Afr J Plant Soil 27:147–151

Riaz S, Nasir IA, Bhatti MU, Adeyinka OS, Toufiq N, Yousaf I, Tabassum B (2020) Resistance to *Chilo infuscatellus* (lepidoptera: Pyraloidea) in transgenic lines of sugarcane expressing *bacillus thuringiensis* derived Vip3A protein. Mol Biol Rep 47(4):2649–2658

Rico E, Ballester V, Mensua JL (1998) Survival of two strains of *Phthorimaea operculella* (lepidoptera: Gelechiidae) reared on transgenic potatoes expressing a *bacillus thuringiensis* crystal protein. Agronomie 18:151–155

Rutger AFG, Marianne GR (2020) Advances of epigenetic editing. Curr Opin Chem Biol 57:75–81

Salehian H, Rahnama H, Dezhsetan S, Babaei S (2021) Constitutive expression of a synthetic *cry1Ab* gene confers resistance to potato tuber moth (*Phthorimaea operculella* Zeller) larva. Crop Breed Appl Biotechnol 21

Sander JD, Joung JK (2014) TALENs: a widely applicable technology for targeted genome editing. Nat Rev Mol Cell Biol 14:49–55

Sanyal I, Singh AK, Kaushik M, Amla DV (2005) *Agrobacterium*-mediated transformation of chickpea (*Cicer arietinum* L.) with *bacillus thuringiensis cry1Ac* gene for resistance against pod borer insect *Helicoverpa armigera*. Plant Sci 168:1135–1146

Sawardekar SV, Katageri IS, Salimath PM, Kumar PA, Kelkar VG (2017) Standardization of in-vitro genetic transformation technique in chickpea (*Cicer arietinum* L.) for pod-borer resistance. Adv Agric Res Technol J 1 2

Sears MK, Hellmich RL, Stanley-Horn DE, Oberhauser KS, Pleasants JM, Mattila HR, Siegfried BD, Dively GP (2001) Impact of Bt corn pollen on monarch butterfly populations: a risk assessment. Proc Natl Acad Sci U S A 98:11937–11942

Selvapandiyan A, Reddy VS, Kumar PA, Tewari KK, Bhatnagar RK (1998) Transformation of *Nicotiana tabacum* with a native *cry1Ia5* gene confers complete protection against *Heliothis armigera*. Mol Breed 4(6):473–478

Seralini GE, Clair E, Mesnage R, Gress S, Defarge N, Malatesta M, Hennequin D, de Vendômois JS (2012) Long term toxicity of a roundup herbicide and a roundup-tolerant genetically modified maize. Food Chem Toxicol 50:4221–4231

Sharma KK, Lavanya K, Anjalah A (2006) *Agrobacterium tumefaciens*-mediated production of transgenic pigeonpea (*Cajanus cajan* [L.] mill sp.) expressing the synthetic BT *cry1Ab* gene. In Vitro Cell Dev Biol Plant 42:165–173

Sheldon IM (2002) Regulation of biotechnology: will we ever 'freely'trade GMOs? Eur Rev Agric Econ 29(1):155–176

Shu Q, Ye G, Cui H, Cheng X, Xiang Y, Wu D, Gao M, Xia Y, Hu C, Sardana R, Altosaar I (2000) Transgenic rice plants with a synthetic *cry1Ab* gene from *bacillus thuringiensis* were highly resistant to eight lepidopteran rice pest species. Mol Breed 6(4):433–439

Siddiqui HA, Asif M, Asad S, Naqvi RZ, Ajaz S, Umer N, Anjum N, Rauf I, Sarwar M, Arshad M, Amin I (2019) Development and evaluation of double gene transgenic cotton lines expressing Cry toxins for protection against chewing insect pests. Sci Rep 9(1):1–7

Silva G, Poirot L, Galetto R, Smith J, Montoya G, Duchateau P, Pâques F (2011) Meganucleases and other tools for targeted genome engineering: perspectives and challenges for gene therapy. Curr Gen Ther 11(1):11–27

Sindhu AS, Zheng ZW, Murai N (1997) The pea seed storage protein legumin was synthesized, processed and accumulated stably in transgenic rice endosperm. Plant Sci 130:189–196

Singh PK, Kumar M, Chaturvedi CP, Yadav D, Tuli R (2004) Development of a hybrid δ-endotoxin and its expression in tobacco and cotton for control of a polyphagous pest *Spodoptera litura*. Trans Res 14:1–14

Singh S, Kumar NR, Maniraj R, Lakshmikanth R, Rao KY, Muralimohan N, Arulprakash T, Karthik K, Shashibhushan NB, Vinutha T, Pattanayak D (2018) Expression of Cry2Aa, a Bacillus thuringiensis insecticidal protein in transgenic pigeon pea confers resistance to gram pod borer Helicoverpa armigera. Sci Rep 8(1):1–12

Singsit C, Adang MJ, Lynch RE, Anderson WF, Wang A, Cardineau G, Ozias-Akins P (1997) Expression of *bacillus thuringiensis cry1A*(c) gene in transgenic peanut plants and its efficacy against lesser cornstalk borer. Trans Res 6:169–176

Smart RD, Blum M, Wesseler J (2017) Trends in approval times for genetically engineered crops in the United States and the European Union. J Agric Econ 68:182–198

Soliman HI, Abo-El-Hasan FM, El-Seedy AS, Mabrouk YM (2021) Agrobacterium-mediated transformation of tomato (Lycopersicon esculentum mill.) using a synthetic cry1ab gene for enhanced resistance against Tuta absoluta (Meyrick). J Microbiol Biotechnol Food Sci:67–74

Stewart CN Jr, Adang MJ, All JN, Boerma HR, Cardineau G, Tucker D, Parrott WA (1996) Genetic transformation, recovery and characterization of fertile soybean transgenic for a synthetic *B. thuringiensis cry1Ac* gene. Plant Physiol 112:121–129

Sticklen M (2015) Advances in crop synthetic biology: a platform that bridges the expertise of crop molecular biologists, engineers, software developers and mathematical modelers. Adv Crop Sci Tech 3(171):2

Storer NP, Babcock JM, Schlenz M, Meade T, Thompson GD, Bing JW, Huckaba RM (2010) Discovery and characterization of field resistance to Bt maize: *Spodoptera frugiperda* (lepidoptera: Noctuidae) in Puerto Rico. J Econ Entomol 103:1031–1038

Strizhov N, Keller M, Mathur J, Koncz-Kálmán Z, Bosch D, Prudovsky E, Schell J, Sneh B, Koncz C, Zilberstein A (1996) A synthetic *cryIC* gene, encoding a *bacillus thuringiensis* delta-endotoxin, confers Spodoptera resistance in alfalfa and tobacco. Proc Natl Acad Sci U S A 93(26):15012–15017

Surekha C, Beena MR, Arundhati A, Singh PK, Tuli R, Dutta-Gupta A, Kirti PB (2005) *Agrobacterium*-mediated genetic transformation of pigeonpea (*Cajanus cajan* L. Millsp.) using embryonal segments and development of transgenic plants for resistance against *Spodoptera*. Plant Sci 169:1074–1080

Suzuki N, Rivero RM, Shulaev V, Blumwald E, Mittler R (2014) Abiotic and biotic stress combinations. New Phytol 203:32–43

Tabashnik BE, Dennehy TJ, Sims MA, Larkin K, Head GP, Moar WJ, Carrière Y (2002) Control of resistant pink bollworm by transgenic cotton with *bacillus thuringiensis* toxin *Cry2Ab*. Appl Environ Microbiol 68:3790–3794

Thakore PI, Black JB, Hilton IB, Gersbach CA (2015) Editing the epigenome: technologies for programmable transcription and epigenetic modulation. Nat Methods 13:127–137

Theodora CA (2016) Cisgenesis and genome editing: combining concepts and efforts for a smarter use of genetic resources in crop breeding. Plant Breed 135:139–147

Tohidfar M, Zare N, Jouzani GS, Efekhari SM (2013) Agrobacterium mediated transformation of alfalfa (Medicago sativa) using a synthetic cry3a gene to enhance resistance against alfalfa weevil. Plant Cell Tiss Org 113:227–235

Tsatsakis AM, Nawaz MA, Kouretas D, Balias G, Savolainen K, Tutelyan VA, Golokhvast KS, Lee JD, Yang SH, Chung G (2017) Environmental impacts of genetically modified plants: a review. Food Chem Toxicol 156:818–833

Tu J, Zhang G, Datta K, Xu C, He Y, Zhang Q, Khush GS, Datta SK (2000) Field performance of transgenic elite commercial hybrid rice expressing *bacillus thuringiensis* δ–endotoxin. Nat Biotech 18:1101–1104

Tuteja N, Verma S, Sahoo R, Raveendar S, Reddy I (2012) Recent advances in development of marker free transgenic plants: regulation and biosafety concern. J Biosci 37:167–197

Urnov FD, Rebar EJ, Holmes MC, Zhang HS, Gregory PD (2010) Genome editing with engineered zinc finger nucleases. Nat Rev Genet 11(9):636–646
US EPA (2017) Starlink™ corn regulatory information. Pesticides. US EPA. https://www3.epa.gov/pesticides/chem_search/reg_actions/pip/starlink_corn.htm#proposal
Usher S, Han L, Haslam RP, Michaelson LV, Sturtevant D, Aziz M, Chapman KD, Sayanova O, Napier JA (2017) Tailoring seed oil composition in the real world: optimising omega-3 long chain polyunsaturated fatty acid accumulation in transgenic Camelina sativa. Sci Rep 7:6570
Vaeck M, Reynaerts A, Hofte H, Jansens S, De Beuckeleer M, Dean C, Zabeau M, Van Montagu M, Leemans J (1987) Transgenic plants protected from insect attack. Nature 328: 33–37
Valderrama AM, Velásquez N, Rodríguez E, Zapata A, Abbas Zaidi M, Altosaar I, Arango R (2007) Resistance to *Tecia solanivora* (lepidoptera: Gelechiidae) in three transgenic Andean varieties of potato expressing *bacillus thuringiensis* Cry1Ac protein. J Econ Entomol 100(1): 172–179
Vaughn T, Cavato T, Brar G, Coombe T, DeGooyer T, Ford S, Groth M, Howe A, Johnson S, Kolacz K, Pilcher C (2005) A method of controlling corn rootworm feeding using a Bacillus thuringiensis protein expressed in transgenic maize. Crop Sci 45(3):931–938
Venki R (2016) GM plants question and answers. The Royal Society pp. 1–40
Verma AK, Chand L (2005) *Agrobacterium*-mediated transformation of pigeonpea (*Cajanus cajan* L.) with *uid*A and *cryIA(b)* genes. Physiol Mol Biol Plant 11:99–109
Weng LX, Deng HH, Xu JL, Li Q, Zhang YQ, Jiang ZD, Li QW, Chen JW, Zhang LH (2010) Transgenic sugarcane plants expressing high levels of modified *cry1Ac* provide effective control against stem borers in field trials. Transgenic Res 20(4):759–772
Wolter F, Puchta H (2017) Knocking out consumer concerns and regulator's rules: efficient use of CRISPR/Cas ribonucleoprotein complexes for genome editing in cereals. Genome Biol 18:43
Woo JW, Kim J, Kwon SI, Corvalán C, Cho SW, Kim H, Kim SG, Kim ST, Choe S, Kim JS (2015) DNA-free genome editing in plants with preassembled CRISPR-Cas9 ribonucleoproteins. Nat Biotechnol 33:1159–1161
Xu C, Cheng J, Lin H, Lin C, Gao J, Shen Z (2018) Characterization of transgenic rice expressing fusion protein Cry1Ab/Vip3A for insect resistance. Sci Rep 8(1):1–8
Yan S, Zhu J, Zhu W, Li Z, Shelton AM, Luo J, Cui J, Zhang Q, Liu X (2015) Pollen-mediated gene flow from transgenic cotton under greenhouse conditions is dependent on different pollinators. Sci Rep 5:15917
Ye GY, Yao HW, Shu QY, Cheng X, Hu C, Xia YW, Gao MW, Altosaar I (2003) High levels of stable resistance in transgenic rice with a *cry1Ab* gene from *bacillus thuringiensis* Berliner to rice leaffolder, *Cnaphalocrocis medinalis* (Guenée) under field conditions. Crop Prot 22:171–178
Yi D, Cui S, Yang L, Fang Z, Liu Y, Zhuang M, Zhang Y (2015) Influences of Cry1Ac broccoli on larval survival and oviposition of diamondback moth. J Insect Sci 15(1):3
Zaidi MA, Mohammadi M, Postel S, Masson L, Altosaar I (2005) The Bt gene *cry2Aa2* driven by a tissue specific ST–LS1 promoter from potato effectively controls *Heliothis virescens*. Trans Res 14(3):289–298
Zhang C, Wohlhueter R, Zhang H (2016) Genetically modified foods: a critical review of their promise and problems. Food Sci Hum Wellness 5:116–123
Zhong Y, Ahmed S, Deng G, Fan W, Zhang P, Wang H (2019) Improved insect resistance against *Spodoptera litura* in transgenic sweetpotato by overexpressing Cry1Aa toxin. Plant Cell Rep 38(11):1439–1448

The manufacturer's authorised representative in the EU is Springer Nature Customer Service Centre GmbH, Europaplatz 3, 69115 Heidelberg, Germany. If you have any concerns regarding our products, please contact ProductSafety@springernature.com

Printed and bound by CPI Group (UK) Ltd, Croydon, CR0 4YY
30/09/2026
02225239-0018